Diffusions, Markov Processes, and Martingales

Volume 2: Itô Calculus

2nd Edition

Diffusions, Markov Processes, and Martingales

Volume 2: ITÔ CALCULUS

2nd Edition

L. C. G. ROGERS

School of Mathematical Sciences, University of Bath

and

DAVID WILLIAMS

Department of Mathematics, University of Wales, Swansea

CAMBRIDGE
UNIVERSITY PRESS

CAMBRIDGE
UNIVERSITY PRESS

University Printing House, Cambridge CB2 8BS, United Kingdom

Cambridge University Press is part of the University of Cambridge.

It furthers the University's mission by disseminating knowledge in the pursuit of education, learning and research at the highest international levels of excellence.

www.cambridge.org
Information on this title: www.cambridge.org/9780521775939

First published 1979 by John Wiley & Sons Ltd, Chichester
Second edition 1994 published by John Wiley & Sons Ltd
Reissued by Cambridge University Press 2000
9th printing 2013

A catalogue record for this publication is available from the British Library

ISBN 978-0-521-77593-9 Paperback

Preface

(*a*) *Welcome back on board!* You will have noticed that for this second leg of your journey, there are two pilots rather than one. D.W. is sure that you will be as delighted as he is that control is being shared with L.C.G.R.—amongst so many other things, just the man for a Wiley excursion!

We apologize for the considerable delay in departure. Anyone who knows what has been happening to British universities will need no further explanation, and will share our sadness.

(*b*) The book is meant to help the research student reach the stage where he or she can begin both to think up and tackle new problems and to read the up-to-date literature across a wide spectrum; and to persuade him or her that it is worth the effort.

We can say that we ourselves find the subject sufficiently good fun to have enjoyed the task of writing. (We even had some amusement from typing the manuscript ourselves with the very basic non-mathematical word-processor VIEW on the BBC micro. Occasionally, we got into trouble when trying to use global editing to substitute the most commonly occurring phrases for shorthand versions of our own devising. But, in the main, we were very satịto's formulaied!)

(*c*) Chapter IV, *Introduction to Itô calculus*, is particularly concerned with developing the theory of the stochastic integral (of a previsible process) with respect to a continuous semimartingale, and with giving a large number of applications. Chung and (Ruth) Williams [1] would make a splendid companion volume for this chapter.

Chapter V, *Stochastic differential equations and diffusions*, presents first the theory of SDEs: existence and uniqueness for strong and weak solutions, martingale problems, etc. It has an extended treatment of 1-dimensional diffusions, and a huge attempt to introduce the very fashionable subject of stochastic differential geometry. Strongly recommended 'parallel' reading for this chapter: McKean's sparkling book [1] and the authoritative Ikeda and Watanabe [1].

Chapter VI, *The general theory*, presents *la théorie générale*: dual previsible projections, the Meyer decomposition theorem, the general integral, etc., with a chunky piece on excursions. The literature on the general theory is dominated by the masterly account by Dellacherie and Meyer ([1]) who created so much of it. Dellacherie's own very fine survey article [3], Jacod [2], Metivier and Pellaumail [1] should also be consulted.

In everything, the Russian literature, as represented by such important volumes as Gikhman and Skorokhod [1] and Liptser and Shiryayev [1], has its own characteristic style and special value.

(*d*) The book has a large bibliography, but this represents a small and rather haphazard selection of what we should have included. We apologize for the enormous number of very important papers which are omitted.

Numerous important topics are omitted too, or given treatment far too brief for their true significance. (Reviewers who find the previous sentence handy are free to use it without acknowledgement.) So, here are some guidelines on what you might move on to when your reading of our book is done.

(*e*) (i) *Large deviations.* The recent appearance of books by Stroock [4] and the grandmaster himself, Varadhan [1], would have made any efforts from us look silly. This is the only reason for our omission of this topic and for the (otherwise scandalous) omission from the bibliography of the historic papers by Donsker and Varadhan and by Ventcel and Freidlin.

(ii) *Malliavin calculus.* See § V.36.

(iii) *Large deviations and Malliavin calculus.* See Bismut [4], and also Elworthy and Truman [1, 2] for important work which provided motivation. Keep a look out for forthcoming work by Léandre.

(iv) *Markov processes.* The value of the classics mentioned in Volume 1—Blumenthal and Getoor [1], Getoor [1], and Meyer [3]—remains as great as ever. Sharpe [1] is sure to be a definitive account, as (of course) is that provided by later volumes of Dellacherie and Meyer [1]. (Volume 4 of the latter has arrived just as we are posting off the final proofs. Splendid to look forward to reading it!). The volumes in the 'Seminars on stochastic processes' (Çinlar, Chung and Getoor [1]) are important state-of-the-art reports.

Ethier and Kurtz [1] is a valuable source for much theory, for the establishment of weak-convergence results, etc. Liggett's account [1] of one of the most important application areas, interacting particle systems, is magisterial.

For a profound study of the relationship between Markov processes and semimartingales, see Çinlar, Jacod, Protter and Sharpe [1].

For applications to potential theory and complex analysis, see Doob [3], Durrett [1], and Port and Stone [3]. Two papers by Lyons [1, 2] are very much recommended.

(v) *Quantum theory.* So much has been achieved in interrelating quantum theory and probability that one hardly knows where to begin, but an excellent lead-in is provided by de Witte-Morette and Elworthy [1].

It is essential to realize that some of the finest work on probability is being done by people who are first and foremost mathematical physicists or functional analysts. See Simon [1, 2], Davies and Simon [1], Aizenmann and Simon [1], and the literature you can trace through them.

Local time and self-intersection local time have come to play a big part in the construction of quantum fields. See Geman and Horowitz [1], Rosen [1], Geman, Horowitz and Rosen [1], Le Gall [2, 3], Yor [4, 5] and then Dynkin [5, 6, 7] to begin your study in this area.

Whatever the philosophical problems, Nelson's *stochastic mechanics* is certainly prompting very interesting mathematics. See Nelson [1, 2] and Carlen [1, 2].

A fascinating theory of *non-commutative stochastic integrals* and of non-commutative SDEs has been created by Hudson and Parthasarathy. Meyer [1, 2] is a splendid attempt to make probabilists informed and involved.

(vi) *Measure-valued diffusions, random media, etc.*. Durrett [2] and Dawson and Gärtner [1] can be your 'open sesame' to what is sure to be one of the richest of Aladdin's caves.

(vii) *The Séminaires.* It is impossible to overstate our indebtedness to the famous *Séminaires de Probabilités*, originated by Meyer and developed by him (with help from Dellacherie and Weil) into an absolutely indispensable handbook, and now maintained as such in Azéma and Yor's expert hands. *Séminaire* XX: *Springer Lecture Notes in Mathematics Volume* 1204, contains an index to the series so far.

(*f*) *Further acknowledgements.* The work on this book has been done at the Universities of Wales (Swansea), Warwick and Cambridge, all of which deserve our thanks.

Most was done at Swansea where both of us spent very happy times. Special thanks to Aubrey Truman, Peter Townsend and Betty Williams.

We thank our colleagues at Cambridge for their warm welcome; and are pleased to acknowledge the help and advice we have received from many, especially Frank Adams, Keith Carne, David Kendall and James Norris.

Our best thanks to Sheila Williams, amanuensis extraordinary, who is just about to rediscover after a long period that there are such things as a dining-room table and a sideboard in the Williams household.

And, of course, our thanks to Charlotte Farmer, Robert Hambrook and the other staff of Wiley for making sure that it has become a reality; and to copy editors, and to wonderfully accurate typesetters.

Cambridge, October 1986

Chris Rogers
David Williams

Added, April 2000
Our thanks too to the staff of C.U.P., especially David Tranah, and also to the wonderfully accurate typesetters for their superb 'invisible mending'.

Contents

CHAPTER V. STOCHASTIC DIFFERENTIAL EQUATIONS AND DIFFUSIONS

Some Frequently Used Notation

$\mathscr{P}$,	the previsible σ-algebra; § IV.6.
b $\mathscr{P}$,	the space of bounded previsible processes; § IV.6
b $\mathscr{E}$,	the space of bounded elementary processes; § IV.6.
FV_0,	the space of (adapted) finite-variation processes null at zero; § IV.7.
IV_0,	the space of (adapted) integrable-variation processes null at zero; § IV.7.
IV $\mathscr{M}_0$,	the space of integrable-variation martingales null at zero; § IV.8.
lb $\mathscr{P}$,	the space of locally bounded previsible processes; § IV.10.
$\mathscr{M}_{0,\text{loc}}$,	the space of local martingales null at zero; § IV.11.
FV $\mathscr{M}_{0,\text{loc}}$,	the space of finite-variation local martingales null at zero; § IV.11.
UI $\mathscr{M}_0$,	the space of uniformly-integrable martingales null at zero; § IV.11.
$\mathscr{S}$,	the space of semimartingales; § IV.15.
$\mathscr{M}_0^2$,	the space of L^2-bounded martingales null at zero; § IV.24.
c $\mathscr{M}_0^2$,	the space of continuous L^2-bounded martingales null at zero; § IV.24.
d $\mathscr{M}_0^2$,	the space of purely discontinuous L^2-bounded martingales null at zero; § IV.24.
m $\mathscr{G}$,	($\mathscr{G}$ a σ-algebra), the space of $\mathscr{G}$-measurable functions.
b$\mathscr{G}$,	the space of bounded $\mathscr{G}$-measurable functions.
$[M]$, $[X]$,	quadratic-variation processes; §§ IV.26, 30, VI.36–38.
$\langle M \rangle$,	§ VI.34.
${}^{o}X$,	the optional projection of X; § VI.7
A^p,	the dual previsible projection of A; §§ VI.1, 21, 23.
$\equiv$,	(which) is defined to equal.
$\uparrow\uparrow$,	$S_n \uparrow\uparrow t$ means: $S_n \leqslant S_{n+1} < t$ and $S_n \to t$.

$\mathbb{N} \equiv \{1, 2, 3, \ldots\}$, $\mathbb{Z}^+ \equiv \{0, 1, 2, \ldots\}$, $\mathbb{R}^+ \equiv [0, \infty)$, $\mathbb{R}^{++} \equiv (0, \infty)$.

CHAPTER IV

Introduction to Itô Calculus

Here, we give the gist of the 'martingale and stochastic integral' method, and illustrate its use via a large number of fully-worked examples. We do not apologize for sometimes advertising the method by showing how it can obtain results which are well known and elementary. Thus, for example, we take the trouble to prove some standard results about the humble Markov chain with finite state-space. But we have also tried to bring into this chapter applications which are less elementary, and which hint at the excitement of the subject today.

TERMINOLOGY AND CONVENTIONS

R-processes and L-processes

We now use the term *R-process* on $[0, \infty)$ to signify a process all of whose paths are right-continuous on $[0, \infty)$ with limits from the left on $(0, \infty)$. Thus an R-process is what was called in Volume 1 a Skorokhod process, and what is called elsewhere a càdlàg process, or a corlol process, or whatever. An R-function or R-path on $[0, \infty)$ is defined via the obvious analogous definition.

The *L-processes* on $(0, \infty)$, all of whose paths are left-continuous with limits from the right, will now begin to feature largely in the theory.

Usual conditions, etc.

Everywhere in this chapter, we work with a set-up $(\Omega, \mathcal{F}, \{\mathcal{F}_t\}, \mathbf{P})$ *satisfying the 'usual conditions'. See (II.67).*

All martingales (and 'finite-variation processes', and 'semimartingales') will be taken to be R-processes. Because we are assuming that the usual conditions hold, this is in order. See (II.67).

We shall also always assume that a process $\{X_t : t \geq 0\}$ *is jointly measurable; that is, the map* $(t, \omega) \mapsto X_t(\omega)$ *is measurable with respect to* $\mathcal{B}(\mathbb{R}^+) \times \mathcal{F}$.

Recall that the process X is said to be adapted if X_t *is* $\mathcal{F}_t$*-measurable for every* $t \geq 0$.

Important convention about time 0

Our stochastic integrals will be defined over intervals $(0, t]$ open at 0. Thus, the value of the integral at time 0 will be 0. This differs from the convention in Dellacherie and Meyer [1]. As explained there, time 0 plays *le rôle du diable.* We consign it to Hell.

In accordance with this convention, the parameter set for our previsible processes will be the open interval $(0, \infty)$.

1. SOME MOTIVATING REMARKS

1. Itô integrals. One of our main tasks is to define the Itô integral

$$\int H\,dX,$$

where H and X are stochastic processes of appropriate classes.

We shall regard this integral as a new stochastic process, often written $H \bullet X$, and shall use the alternative notations:

$$(H \bullet X)(t, \omega) = \left(\int_{(0,t]} H_s\,dX_s \right)(\omega). \tag{1.1}$$

We shall often use differential notation, in which we can rewrite equation (1.1) as a 'stochastic differential equation':

$$d(H \bullet X) = H\,dX. \tag{1.2}$$

The theory is now essentially complete in the sense that it is known exactly what conditions need to be imposed on our integrand H and integrator X:

> *The essential requirement on the integrand H is that it be 'previsible'.*
> *The integrator X must be a 'semimartingale'.*

The most important example of a previsible process is provided by an *adapted L-process.* Indeed, the adapted L-processes 'generate' the previsible processes, as will be explained later. Let H be an adapted L-process. Then H_s is known to the observer at time s. The reason that H is 'previsible' is (roughly speaking) that, for a stopping time $T > 0$, H_T is known immediately before time T because

$$H_T = \lim_{s \uparrow\uparrow T} H_s.$$

The simplest adapted L-process is the process

$$H = 1(S, T],\tag{1.3}$$

where S and T are stopping times with $S < T$. Thus,

$$H(t, \omega) = \begin{cases} 1 & \text{if } S(\omega) < t \leqslant T(\omega), \\ 0 & \text{otherwise.} \end{cases}$$

Let M be a martingale. Then, for H as in equation (1.3), the obvious definition of $H \bullet M$ is:

$$(H \bullet M)_t = M_{T \wedge t} - M_{S \wedge t},$$

and we can easily show that *$H \bullet M$ is a martingale.*

From this simple case develops *the most fundamental property of Itô integrals*:

(*1.4*) ITÔ INTEGRALS PRESERVE LOCAL MARTINGALES. It will do no harm to give a precise statement of this now. The reader new to the subject will not know what Theorem 1.5 means, but it will help him or her to know that it is one of the main landmarks in our route through the subject.

(*1.5*) **FUNDAMENTAL THEOREM.** *If H is a locally bounded previsible process and M is a local martingale, then $H \bullet M$ exists and is a local martingale.*

We could very easily explain now what a locally bounded previsible process is. We could also easily explain what a local martingale is; indeed, let us do it:

(*1.6*) DEFINITION (local martingale): *A process M is a* local martingale *if M_0 is $\mathcal{F}_0$ measurable and there exists an increasing sequence of stopping times (T_n) with $T_n \uparrow \infty$ such that each 'stopped' process*

$$\{M_{T_n \wedge t} - M_0 : t \geqslant 0\}$$

is a martingale.

What we cannot explain in a short space is what $H \bullet M$ means in the generality of Theorem 1.5. But the discrete-time setting explains why Theorem 1.5 is true.

(*1.7*) *A discrete-time analogue.* Let $(\Omega, \mathcal{F}, \{\mathcal{F}_n\}, \mathbf{P})$ be a discrete-time set-up, and let M be an associated martingale. Let H be a bounded process previsible in the sense that

$$Z_{n-1} = H_n \in \mathrm{b}\mathcal{F}_{n-1} \quad (n \in \mathbb{N}).$$

Define:

$$(H \bullet M)_n = \sum_{k=1}^{n} H_k (M_k - M_{k-1}) = \sum_{k=1}^{n} Z_{k-1} (M_k - M_{k-1})$$

$$(H \bullet M)_0 = 0.$$

Then $H \bullet M$ is a martingale.

Proof. To show that a process N is a martingale, we need only show that

$$\mathbf{E}[N_n - N_{n-1} | \mathscr{F}_{n-1}] = 0, \quad n \in \mathbb{N}.$$

But, since $Z_{n-1} \in \mathrm{b}\mathscr{F}_{n-1}$,

$$\begin{aligned}\mathbf{E}[(H \bullet M)_n - (H \bullet M)_{n-1} | \mathscr{F}_{n-1}] &= \mathbf{E}[Z_{n-1}(M_n - M_{n-1}) | \mathscr{F}_{n-1}] \\ &= Z_{n-1}\mathbf{E}[M_n - M_{n-1} | \mathscr{F}_{n-1}] = 0. \quad \square\end{aligned}$$

As was mentioned earlier, *the general 'integrator' X will be a semimartingale.* This means that X *may be written in the form*

(1.8) $$X = X_0 + M + A,$$

where X_0 is $\mathscr{F}_0$ measurable, M is a local martingale null at 0, *and A is an adapted process with paths of finite variation, also null at* 0.

In this chapter, we present the full theory for two special cases of great importance:

(i) *the case in which $X = A$, a process with paths of finite variation;*
(ii) *the case in which the paths of X are continuous.*

This will allow us to develop many of the main applications. The general theory is given in Chapter VI.

If A is a process with paths of finite variation, then (for a bounded measurable process H) we can define $H \bullet A$ as the Stieltjes integral for each ω:

$$(H \bullet A)(t, \omega) = \int_{(0,t]} H(s, \omega)\, dA(s, \omega).$$

Though no new concept of integration is involved here, the theory is extremely useful because of what Theorem 1.5 says in this context:

(*1.9*) THEOREM. *Let H be a locally bounded previsible process, and let M be a local martingale with paths of finite variation. Then $H \bullet M$, as defined by the Stieltjes integral, is a local martingale.*

If M is a (path-) continuous local martingale, then the paths of M generally will not have finite variation. Indeed, the only paths of finite variation will be constant! Thus the integral $H \bullet M$ (where H is a locally bounded previsible process) is a true extension of the Stieltjes integral. The very existence of the integral is inextricably tied up with its calculus, that is, with the *integration-by-parts* formula and the *pièce de résistance* of the theory, *Itô's formula.*

2. Integration by parts. The most important integral associated with a local martingale M is the integral $\int_{(0,t]} M_{s-} dM_s$. The adapted L-process $M_- = \{M_{s-} : s > 0\}$ is previsible and also locally bounded, so the integral exists. More generally, if X and Y are semimartingales, then the Itô integral $\int X_{s-} dY_s$ may be defined. Now the integral $\int X_{s-}\,dY_s$ is analogous to a sum of the form

(2.1) $$\sum x_{k-1}(y_k - y_{k-1}).$$

The summation-by-parts formula for such sums:

(2.2) $$x_n y_n - x_0 y_0 = \sum_{k=1}^{n} x_{k-1}(y_k - y_{k-1}) + \sum_{k=1}^{n} y_{k-1}(x_k - x_{k-1}) + \sum_{k=1}^{n} (x_k - x_{k-1})(y_k - y_{k-1})$$

suggests the fundamental *integration-by-parts formula for semimartingales*:

(2.3) $$X_t Y_t - X_0 Y_0 = \int_{(0,t]} X_{s-}\,dY_s + \int_{(0,t]} Y_{s-}\,dX_s + \int_{(0,t]} dX_s\,dY_s.$$

But what sense are we to make of the last term in (2.3)?

It is easy to believe (and to prove!) that if X and Y have paths of finite variation, then the correct interpretation is as follows:

(2.4) $$\int_{(0,t]} dX_s\,dY_s = \sum_{0<s\leqslant t} (X_s - X_{s-})(Y_s - Y_{s-}).$$

What happens if X and Y are path-continuous martingales? To gain insight into this situation, and into more general situations, we again look at a discrete analogue.

(*2.5*) *A discrete-time analogue.* Let $M = \{M_n \colon n \geqslant 0\}$ and $N = \{N_n \colon n \geqslant 0\}$ be martingales on a set-up $(\Omega, \mathscr{F}, \{\mathscr{F}_n\}, \mathbf{P})$ such that for all n, $\mathbf{E}M_n^2 < \infty$, $\mathbf{E}N_n^2 < \infty$. Take $x_k = M_k(\omega)$, $y_k = N_k(\omega)$ in (2.2) to see that

$$M_n N_n - M_0 N_0 = \sum_{k=1}^{n} M_{k-1}(N_k - N_{k-1}) + \sum_{k=1}^{n} N_{k-1}(M_k - M_{k-1}) + \sum_{k=1}^{n} \Delta M_k \Delta N_k,$$

where $\Delta M_k = M_k - M_{k-1}$ and $\Delta N_k = N_k - N_{k-1}$. Now (compare (1.6)):

$$U_n = \sum_{k=1}^{n} M_{k-1}(N_k - N_{k-1})$$

defines a martingale U because

$$\mathbf{E}[U_n - U_{n-1} \mid \mathscr{F}_{n-1}] = M_{n-1}\mathbf{E}[N_n - N_{n-1} \mid \mathscr{F}_{n-1}] = 0.$$

Hence, if we put

$$[M, N]_n = \sum_{k=1}^{n} \Delta M_k \Delta N_k,$$

then

$$M_n N_n - M_0 N_0 - [M, N]_n \textit{ is a martingale.}$$

In particular, on taking $N=M$, we see that

$$V_n = M_n^2 - M_0^2 - \sum_{k=1}^{n} (M_k - M_{k-1})^2 \text{ is a martingale}$$

because V is the 'stochastic integral':

$$V_n = \sum_{k=1}^{n} 2M_{k-1}(M_k - M_{k-1}). \qquad \square$$

With the above discrete-time analogue very much in mind, we shall now develop the interpretation of (2.3) for the case when $X=Y=B$, a Brownian motion on $\mathbb{R}$. This is the only case we shall consider in this section. Since we shall be providing all details for the general continuous local martingale later, a sketched argument will suffice here. Let us write for $t \geqslant 0$,

$$t_k^n \equiv t \wedge (k2^{-n}).$$

We have:

$$B(t)^2 - B(0)^2 = \sum_{k \geqslant 1} 2B(t_{k-1}^n)(B(t_k^n) - B(t_{k-1}^n)) + [B]_t^n,$$

where

(2.6) $$[B]_t^n \equiv \sum_{k \geqslant 1} (B(t_k^n) - B(t_{k-1}^n))^2.$$

For each t, there are only finitely many non-zero terms in (2.6), and these are independent random variables. Moreover, since $B_{s+h} - B_s$ has the normal distribution of mean zero and variance h,

$$B_{s+h} - B_s \sim N(0, h),$$

we have:

$$\mathbf{E}[(B_{s+h} - B_s)^2] = h, \quad \text{var}\,[(B_{s+h} - B_s)^2] = h^2 \text{var}\,(\xi^2),$$

where $\xi \sim N(0, 1)$. Hence, for fixed t, $[B]_t^n$ has mean t and variance $O(2^{-n})$, so that:

(2.7) $$[B]_t^n \to t \text{ with probability } 1.$$

More is true. By an obvious modification of our argument for the discrete-time analogue, we can show that for each n,

(2.8) $$V_t^n \equiv B(t)^2 - B(0)^2 - [B]_t^n = \sum_{k=1}^{n} 2B(t_{k-1}^n)(B(t_k^n) - B(t_{k-1}^n))$$

is a martingale. But

$$V_t^n - V_t^{n+1} = [B]_t^{n+1} - [B]_t^n,$$

so that $[B]^{n+1} - [B]^n$ is a martingale. Doob's L^2 inequality (II.52.6) now allows us to strengthen (2.7) to the following result, quoted at I.11.1:

(*2.9*) (*Lévy's quadratic variation result*). *With probability* 1, $[B]_t^n \to t$ *uniformly on compact intervals.*

See Exercise 2.17. It now follows from (2.8) that, with probability 1,

(2.10) $$\lim_n \sum_{k=1}^{n} 2B(t_{k-1}^n)(B(t_k^n) - B(t_{k-1}^n)) = (B(t)^2 - t) - (B(0)^2 - 0),$$

uniformly on compact intervals.

The correct interpretation of (2.3) when $X = Y = B$ is now clear:

(*2.11*) *The Itô integral* $\int_{(0,t]} 2B_{s-}\,dB_s$ *is the martingale:*

(2.12) $$\int_{(0,t]} 2B_{s-}\,dB_s = (B(t)^2 - t) - (B(0)^2 - 0), \text{ and}$$

(2.13) $$\int_{(0,t]} (dB_s)^2 = [B]_t = t.$$

We sometimes write (2.13) in differential notation:

(2.14) $$(dB_t)^2 = dt.$$

Of course, since B is a continuous process, the B_{s-} in (2.12) is equal to B_s. However, it was worth leaving the '$s-$' in (2.12) to emphasize that the integral must be interpreted as the limit at (2.10). This situation is quite different from that in Riemann integration. Thus it is tautological that:

(2.15) $$\lim \sum (B(t_{k-1}^n) + B(t_k^n))(B(t_k^n) - B(t_{k-1}^n)) = B(t)^2 - B(0)^2,$$

which corresponds to the Stratonovich integral (written in this book with a ∂):

$$\int_{(0,t]} 2B_s\,\partial B_s = B(t)^2 - B(0)^2.$$

The Stratonovich integral is useful, particularly in applications to stochastic differential geometry. But the Itô integral is the one we need at present because (as you are aware by now!) the Itô integral preserves local martingales. Until further notice, we shall deal only with the Itô integral.

The reason for the discrepancy between (2.10) and (2.15) is of course that the Brownian motion path has non-zero 'quadratic variation' $[B]$. As was explained at I.11, this implies:

(*2.16*) LEMMA. *Almost all Brownian motion paths have infinite variation on every time interval.*

Proof. Suppose that b is a continuous function of bounded variation on $[0, t]$. Then:

$$\sum_{k=1}^{n} (b(t_k^n) - b(t_{k-1}^n))^2 \leqslant S^n(t) V^n(t)$$

where

$$S^n(t) \equiv \sup_k |b(t_k^n) - b(t_{k-1}^n)|,$$

$$V^n(t) \equiv \sum_k |b(t_k^n) - b(t_{k-1}^n)|.$$

Now $V^n(t)$ is bounded by the total variation of b on $[0, t]$, and since b is uniformly continuous on $[0, t]$, $S^n(t) \to 0$ as $n \to \infty$. Hence:

$$\lim \sum (b(t_k^n) - b(t_{k-1}^n))^2 = 0,$$

for all t. This contradicts (2.9). □

(*2.17*) **Exercise.** Prove (2.9). *Hint.* Let $U_n(t) \equiv [B]_t^{n+1} - [B]_t^n$. Then, by II.43.3,

$$\mathbf{P}[\sup_{s \leqslant t} |U_n(s)| \geqslant \varepsilon] \leqslant \varepsilon^{-2} \mathbf{E}[\sup_{s \leqslant t} |U_n(s)|^2] \leqslant 4\varepsilon^{-2} \mathbf{E}[U_n(t)^2].$$

Show that $\mathbf{E}[U_n(t)^2] = O(2^{-n})$. □

3. Itô's formula for Brownian motion. The celebrated *Itô's formula for Brownian motion* says that for $f \in C^2$,

(3.1) $$f(B_t) - f(B_0) = \int_{(0,t]} f'(B_s)\,dB_s + \int_{(0,t]} \tfrac{1}{2} f''(B_s)\,ds.$$

This will be derived from (2.11), to which it reduces when $f(x) = x^2$. The first integral in (3.1), an Itô integral, is a suitable limit:

(3.2) $$\lim \sum f'(B(t_{i-1}^n))(B(t_i^n) - B(t_{i-1}^n)).$$

The second integral in (3.1) is a Riemann integral for each ω.

We often write (3.1) in differential form:

(3.3) $$df(B_t) = f'(B_t)\,dB_t + \tfrac{1}{2} f''(B_t)\,dt,$$

and regard (3.3) as a Taylor expansion with the rules:

(3.4) $$(dB_t)^2 = dt, \quad (dB_t)^3 = 0.$$

A useful extension of (3.3) is the following: for $f \in C^{1,2}$:

(3.5) $$df(t, B_t) = \frac{\partial f}{\partial t} dt + \frac{\partial f}{\partial B} dB_t + \tfrac{1}{2} \frac{\partial^2 f}{\partial B^2} dt,$$

so that we need to supplement (3.4) with the rules:

$$(dt)^2 = 0, \quad dtdB = 0. \tag{3.6}$$

For calculations and estimates, we need martingales, and Itô's formula can provide us with them. We shall see many illustrations of this.

We saw in McKean's proof (§ I.16) of the iterated logarithm law and in calculation of the distribution of hitting times in § I.9 how important it is that, for $\theta \in \mathbb{R}$, the '*Brownian exponential*'

$$M_t^\theta = \exp(\theta B_t - \tfrac{1}{2}\theta^2 t)$$

is a martingale. If we apply (3.5) with $f(t, x) = \exp(\theta x - \frac{1}{2}\theta^2 t)$, we obtain:

$$dM_t^\theta = M_t^\theta(-\tfrac{1}{2}\theta^2 dt + \theta dB_t + \tfrac{1}{2}\theta^2 dt) = \theta M_t^\theta dB_t,$$

so that M_t^θ, as a stochastic integral relative to B, is certainly a local martingale. How to prove by Itô calculus that it is a true martingale, we shall see later.

Exponential martingales have a very prominent rôle in the theory, and it is important to be able to solve equations such as

$$dM_t^\theta = \theta M_t^\theta dB_t, \quad M_0^\theta = 1.$$

We show how to solve such equations in § 19 for the finite variation case, and in § 37 for the continuous case. See Doléans [2] for the general case.

4. A rough plan of the chapter. The remainder of the chapter is divided into five parts as follows.

Part 2 introduces some essential concepts and methods. Previsible processes make their first appearance as the integrands of our theory. (We shall get to know them better, and see them feature in other starring rôles, in Chapter VI.) Fundamental Lemma 5.3 establishes the most basic result on preservation of the local martingale property by Itô integration. Extensions of this lemma, either by continuity arguments or by 'localization', are what lead to the Fundamental Theorem 1.5. The substantial portion of Part 2 devoted to localization is absolutely essential. We have tried to make it run as smoothly as possible, so don't shirk your duty!

Part 3 develops the elementary stochastic calculus of finite-variation processes to a point where we can do calculations relating to a Markov chain with finite state–space. This should help attune your intuition to a new way of thinking about results you know well, and thereby sharpen it for more challenging things. (You could skip Part 3 on a first reading, but we would not recommend this.) There is a lot more to the calculus of finite-variation processes, as we shall see when we come to study dual previsible projections in Chapter VI.

Part 4 presents the so-called L^2 theory of stochastic integrals. This superbly elegant theory due to Kunita and (S.) Watanabe extends the Fundamental Lemma 5.3 by exploiting various Hilbert-space isometries.

Part 5 completes the development of the stochastic integral with respect to a continuous semimartingale, and proves Itô's formula for this context.

Part 6, 'Applications of Itô's formula', contains Lévy's Theorem, the Cameron–Martin–Girsanov Theorem, Tanaka's formula, Trotter's Theorem, and many others of the results which make the subject come to life.

2. SOME FUNDAMENTAL IDEAS: PREVISIBLE PROCESSES, LOCALIZATION, etc

Previsible processes

5. Basic integrands $Z(S, T]$. An Itô integral will be a *process*

$$H \bullet X = \int H_s \, dX_s,$$

where H is an '*integrand*', and X is an '*integrator*'. All integrators will be special kinds of R-processes. We shall see that:

The only sensible integrands are previsible processes.
The only sensible integrators are semimartingales.

As is usual in integration theory, we build things up in stages, starting with integrands of a particularly simple type, namely, step-functions (linear combinations of indicator functions). Of course, for our stochastic step-functions, both the intervals of constancy and the values taken on those intervals must be allowed to depend on ω in an intelligent way.

Let $(\Omega, \mathcal{F}, \{\mathcal{F}_t\}, \mathbf{P})$ be a filtered probability space satisfying the usual conditions. Let S and T be stopping times, $S \leqslant T$, and let $Z \in \mathrm{b}\mathcal{F}_S$. (Recall that $\mathcal{F}_S$ is the pre-S σ-algebra, and $\mathrm{b}\mathcal{F}_S$ is the space of bounded $\mathcal{F}_S$-measurable functions on Ω.) The most basic type of integrand is the process

$$H = Z(S, T]. \tag{5.1}$$

This signifies that, for $t < \infty$,

$$H(t, \omega) = \begin{cases} Z(\omega) & \text{if } S(\omega) < t \leqslant T(\omega), \\ 0 & \text{otherwise.} \end{cases}$$

Note that $T(\omega)$ may be infinite. Then for any integrator X we make the obvious definition:

$$(H \bullet X)_t \equiv \int_{(0,t]} H_s \, dX_s \equiv \int_0^t H_s \, dX_s \equiv Z(X_{T \wedge t} - X_{S \wedge t}). \tag{5.2}$$

In full, (5.2) means:

$$(H \bullet X)(t, \omega) = Z(\omega)(X_{T(\omega) \wedge t}(\omega) - X_{S(\omega) \wedge t}(\omega)).$$

The reason for insisting that S and T are stopping times and that $Z \in \mathrm{b}\mathcal{F}_S$ will now be made clear.

(5.3) FUNDAMENTAL LEMMA. *Let H be as at* (5.1). *Let M be a uniformly integrable martingale. Then $H{\bullet}M$ is a uniformly integrable martingale.*

Moral: 'Itô integrals preserve etc.'

Proof. Because of Lemma II.77.6, we need only prove that if R is a stopping time, then

$$\mathbf{E}|(H{\bullet}M)_R| < \infty, \quad \mathbf{E}(H{\bullet}M)_R = 0.$$

Now if $\|Z\| \equiv \sup_\omega |Z(\omega)|$, then

$$\mathbf{E}|(H{\bullet}M)_R| \leqslant \|Z\|\ \mathbf{E}\{|M_{T\wedge R}| + |M_{S\wedge R}|\}.$$

We know that $\mathbf{E}[M_\infty | \mathcal{F}_{T\wedge R}] = M_{T\wedge R}$ (by the Optional Sampling Theorem II.77.5), and that conditional expectation is a contraction mapping on L^1. Hence

$$\mathbf{E}|(H{\bullet}M)_R| \leqslant 2\|Z\|\ E|M_\infty| < \infty.$$

Our next task is to prove that $\mathbf{E}(H{\bullet}M)_R = 0$; in other words, that for $Z \in \mathrm{b}\mathcal{F}_S$,

(5.4) $$\mathbf{E}Z(M_{T\wedge R} - M_{S\wedge R}) = 0.$$

It is enough to prove this for $Z = I_A$, $A \in \mathcal{F}_S$, because we can then obtain (5.4) for all $Z \in b\mathcal{F}_S$ by the monotone-class theorem. But if $Z = I_A$, $A \in \mathcal{F}_S$, then (5.4) merely states that

(5.5) $$\mathbf{E}(M_{T_A\wedge R} - M_{S_A\wedge R}) = 0$$

where S_A is the stopping time:

(5.6) $$S_A \equiv \begin{cases} S & \text{on } A, \\ \infty & \text{on } \Omega\setminus A. \end{cases}$$

Result (5.5) is now obvious, again from the Optional Stopping Theorem. □

To a large extent, stochastic-integral theory consists of extensions of Lemma 5.3. Two techniques of generalization are used: *monotone-class methods* and *localization.* But it took the most remarkable insight on the part of Meyer, Kunita, Watanabe, Doléans, . . . , to see how these methods may be exploited to what we shall see in §16 to be their fullest possible extent.

6. Previsible processes on $(0, \infty)$, $\mathcal{P}$, $\mathrm{b}\mathcal{P}$, $\mathrm{b}\mathcal{E}$. The natural time-parameter set for previsible processes is the interval $(0, \infty)$. We consider a process with time-parameter set $(0, \infty)$ as a map:

$$H\colon (0, \infty) \times \Omega \to \mathbb{R}.$$

(*6*) DEFINITION (previsible σ-algebra $\mathscr{P}$, previsible process). *The* previsible σ-algebra $\mathscr{P}$ *on* $(0, \infty)\times\Omega$ *is defined to be the smallest* σ-*algebra on* $(0, \infty)\times\Omega$ *such that every adapted L-process is* $\mathscr{P}$-*measurable. A process with time-parameter set* $(0, \infty)$ *is called* previsible *if it is a* $\mathscr{P}$-*measurable map from* $(0, \infty)\times\Omega$ *to* $\mathbb{R}$ *(or to a more general state-space).*

Let us now show that our basic integrand $Z(S, T]$ of § 5 is previsible.

(*6.1*) LEMMA. *Suppose that* S *and* T *are stopping times with* $S\leqslant T\leqslant\infty$. *Let* $Z\in b\mathscr{F}_S$, *and let* $H=Z(S, T]$. *Then* H *is previsible.*

Proof. The process H is certainly an L-process, so we need only show that H is adapted. Now

$$H=\lim Z[S_n, T_n), \quad S_n=S+n^{-1}, \quad T_n=T+n^{-1}.$$

So it is enough to prove that $Z[U, V)$ is adapted, where U and V are stopping times with $U\leqslant V$, and $Z\in b\mathscr{F}_U$. But then, for any Borel subset B of $\mathbb{R}\backslash\{0\}$,

$$\{Z[U, V)_t\in B\}=[\{Z\in B\}\cap\{U\leqslant t\}]\cap\{V>t\}.$$

By definition of $\mathscr{F}_U$, the set in square brackets [.] is in $\mathscr{F}_t$, and since V is a stopping time, $\{V>t\}=\Omega\backslash\{V\leqslant t\}$ is also in $\mathscr{F}_t$. □

(*6.2*) *Definition of* $b\mathscr{E}$. Let $b\mathscr{E}$ (the $\mathscr{E}$ stands for 'elementary') be the vector space spanned by our basic integrands. It is easy (if rather messy) to show that the typical element of $b\mathscr{E}$ may be written in the form

$$\sum_{i=1}^{n} Z_i(S_i, T_i] \tag{6.3}$$

where (S_i) and (T_i) are sequences of stopping times such that $S_1\leqslant T_1\leqslant S_2\leqslant T_2, \ldots$, and $Z_i\in b\mathscr{F}(S_i)$. (See Exercise 6.6 below). What is needed to complete preparations for applications of the Monotone Class Theorem II.3.1 is the fact that

$$\sigma(b\mathscr{E})=\mathscr{P}. \tag{6.4}$$

In other words, $\mathscr{P}$ *is the smallest* σ-*algebra on* $(0, \infty)\times\Omega$ *with respect to which each process in* $b\mathscr{E}$ *is measurable.*

Proof of (6.4). We need only show that every bounded adapted L-process is $\sigma(b\mathscr{E})$-measurable. But if H is any bounded left-continuous process, then

$$H=\lim_{k\uparrow\infty}\lim_{n\uparrow\infty}\sum_{i=2}^{nk} H_{(i-1)/n}\left(\frac{i-1}{n}, \frac{i}{n}\right],$$

and if H is also adapted, then $H_{(i-1)/n}\in b\mathscr{F}_{(i-1)/n}$. □

The Monotone-Class Theorem together with (6.6) now yields the following result.

(6.5) LEMMA *Let $\mathscr{H}$ be a vector space of bounded processes with time-parameter set $(0, \infty)$. Each element of $\mathscr{H}$ is to be regarded as a function on $(0,\infty) \times \Omega$.*

Suppose that $\mathscr{H}$ satisfies the following three conditions:

(i) *Constant functions are in $\mathscr{H}$.*
(ii) *If (H_n) is a sequence of elements of $\mathscr{H}$ which converges uniformly on $(0, \infty)\times\Omega$ to a function H, then H is in $\mathscr{H}$.*
(iii) *If (H_n) is a uniformly bounded sequence of nonnegative elements of $\mathscr{H}$ and $H_n \uparrow H$, then $H \in \mathscr{H}$.*

Then if $\mathscr{H}$ contains every element of $\mathrm{b}\mathscr{E}$, $\mathscr{H}$ contains every bounded previsible process on $(0, \infty)$.

(*6.6*) **Exercise.** Prove that the set of elements of the form (6.3) is a vector space, and, indeed, an algebra.

Hints. Suppose that S and T are stopping times with $S \leqslant T$, and that U and V are stopping times with $U \leqslant V$. Suppose that $Z \in \mathrm{b}\mathscr{F}_S$, and $Y \in \mathrm{b}\mathscr{F}_U$. You need only show that

(*a*) $Y(U, V]+Z(S, T]$ may be written in the form (6.3),
(*b*) $Y(U, V]Z(S, T]$ may be written in the form (6.3).

Part (*b*) is easier because

$$Y(U, V]Z(S, T]=YZ(U \vee S, (U \vee S) \vee (T \wedge V)]$$

and Y and Z are both in $\mathrm{b}\mathscr{F}_{U \vee S}$. It should now be clear how to attempt part (*a*). □

The next result follows from Fundamental Lemma (5.3) by linearity.

(*6.7*) LEMMA. *Let $H \in \mathrm{b}\mathscr{E}$ (with H written in the form (6.3)). Let M be a uniformly integrable martingale. Define*

$$(H\bullet M)_t \equiv \sum_{i=1}^{n} Z_i(M_{T_i \wedge t} - M_{S_i \wedge t}).$$

Then $H\bullet M$ is a uniformly integrable martingale.

To conclude this section, we leave you to do the following exercise which will be used when we make a very substantial study of previsible processes in Chapter VI.

(*6.8*) **Exercise**—*important*! Prove that $\mathscr{P}$ is generated by all sets of the form

$$(u_\Lambda, \infty) \equiv \{(t, \omega): t > u, \omega \in \Lambda\}$$

where $u \geqslant 0$ and $\Lambda \in \mathscr{F}_u$. Note that this is exactly right intuitively.

Hint. We know from the proof of Lemma 6.4 that $\mathscr{P}$ is generated by processes of the form $Z(s, \infty)$, where $Z \in \mathrm{b}\mathscr{F}_s$.

(*6.9*) *Corollary of Exercise*:

$$\mathscr{P} = \{(T, \infty): T \text{ a stopping time}\}.$$

Finite-variation and integrable-variation processes

7. FV_0 and IV_0 processes. Processes of finite variation (FV processes) and processes of integrable variation (IV processes) are important building blocks for the theory.

(*7.1*) *Convention.* It is standard terminology that *FV processes and IV processes* are adapted. The corresponding concepts in which the adapted requirement is removed are also important in the theory; they are *raw FV processes* and *raw IV processes.* Thus '*raw*' *means* '*drop the "adapted" requirement from the definition.*' (As Meyer has pointed out, the situation is not different from that concerning 'signed measure', 'skew field', etc.)

(*7.2*) DEFINITION (FV_0 process, variation V_X). *An FV_0* process *is an adapted R-process $\{X_t: t \geqslant 0\}$ such that each path $t \mapsto X(t, \omega)$ is of finite variation, and $X_0 = 0$. Thus, for each t and each ω, the* variation $V_X(t, \omega)$ *of $s \mapsto X(s, \omega)$ over* $(0, t]$:

$$V_X(t, \omega) = \int_{(0, t]} |dX(s, \omega)| = \sup \sum_{i=1}^{n} |X(s_i, \omega) - X(s_{i-1}, \omega)|$$

is finite. The supremum is taken over all partitions: $0 = s_0 < s_1 < \ldots < s_n = t$ of $[0, t]$.

We shall write FV_0 for the space of FV_0 processes.

(*7.3*) '*Stochastic integrals*' *relative to* FV_0 *processes.* If H is a bounded $\mathscr{B}(0, \infty) \times \mathscr{F}$-measurable process, and X is a raw FV_0 process, define the Stieltjes integral:

$$(H \cdot X)(t, \omega) \equiv \int_{(0, t]} H(s, \omega)\, dX(s, \omega).$$

(*7.4*) DEFINITION (IV_0 process). *An* IV_0 process *is an* FV_0 *process X such that*

$$\|X\|_V = \mathbf{E} V_X(\infty, \omega) < \infty.$$

Of course,

$$V_X(\infty, \omega) = \uparrow \lim_{t \uparrow \infty} V_X(t, \omega).$$

We write IV_0 *for the space of* IV_0 *processes.*

8. Preservation of the martingale property. Here is the promised first monotone-class extension of Lemma 6.7.

(*8.1*) THEOREM. *If* $H \in \mathrm{b}\mathscr{P}$ *and* $M \in \mathrm{IV}\mathscr{M}_0$, *then* $H \bullet M \in \mathrm{IV}\mathscr{M}_0$. ($\mathrm{IV}\mathscr{M}_0$ denotes the space of *martingales* in IV_0.)

Note. Since

$$M^* \equiv \sup_t |M_t| \leqslant V_M(\infty),$$

we have $\mathbf{E}M^* < \infty$, so our IV_0 martingale M is certainly uniformly integrable.

Proof. Let M be an IV_0 martingale. Let $\mathscr{H}$ be the class of bounded previsible processes H such that $H \bullet M$ is a martingale. (Since $\sup_t |(H \bullet M)_t| \leqslant \|H\| \, V_M(\infty)$, where $\|H\|$ is the supremum norm of H, it is clear that $H \bullet M$ is a uniformly integrable process.) Lemma 6.7 tells us that $\mathscr{H}$ contains $\mathrm{b}\mathscr{E}$.

Obviously $\mathscr{H}$ contains constants, so property (i) of Lemma 6.5 holds. We can establish properties (ii) and (iii) simultaneously. For suppose that (H_n) is a uniformly bounded sequence of elements of $\mathscr{H}$ and that the limit

$$H(t, \omega) = \lim H_n(t, \omega)$$

exists for all t and all ω. The dominated-convergence theorem shows that, for every t,

$$(H_n \bullet M)_t \to (H \bullet M)_t \text{ in } L^1(\Omega, \mathscr{F}, \mathbf{P}).$$

Since $H_n \bullet M$ is a martingale, we have for $s < t$,

$$\mathbf{E}[(H_n \bullet M)_t \,|\, \mathscr{F}_s] = (H_n \bullet M)_s \quad \mathbf{P}\text{-a.s.} \tag{8.2}$$

Because $(H_n \bullet M)_t \to (H \bullet M)_t$ in L^1 and $\mathbf{E}[\,.\,|\,\mathscr{F}_s]$ is a contraction on L^1, the left-hand side of (8.2) converges in L^1 to $\mathbf{E}[(H \bullet M)_t \,|\, \mathscr{F}_s]$. The right-hand side of (8.2) converges in L^1 to $(H \bullet M)_s$. Hence

$$\mathbf{E}[(H \bullet M)_t \,|\, \mathscr{F}_s] = (H \bullet M)_s \quad \mathbf{P}\text{-a.s.},$$

and $(H \bullet M)$ is a martingale. (Note that $H \bullet M$ is automatically right continuous by the way it is defined as a Stieltjes integral.) □

Localization

9. $\boldsymbol{H(0, T], X^T}$. The boundedness and integrability assumptions in Theorem 8.1 are too stringent for practical purposes. To obtain a more useful result we must relax the hypotheses of that theorem to their 'localized' versions. Of course, the conclusion of the theorem will then only hold 'locally'.

Let us think about ways of expressing the idea of 'localizing' a 'global' equation

$$d(H \bullet X) = H\,dX \text{ on } (0, \infty) \tag{9.1}$$

to a 'local' equation

$$d(H \bullet X) = H\,dX \text{ on } (0, T], \tag{9.2}$$

where T is a stopping time.

(*9.3*) DEFINITION ($H(0, T]$). *Let T be a stopping time, and H any process $\{H_t: t>0\}$. With (9.2) in mind, it is natural to introduce the process $H(0, T]$, where*

$$(H(0, T])(t, \omega) \equiv \begin{cases} H(t, \omega) & \text{if } 0<t\leqslant T(\omega), \\ 0 & \text{if } t>T(\omega). \end{cases}$$

We note that if H is previsible, then so is $H(0, T]$.

(*9.4*) DEFINITION (X^T). *Let T be a stopping time, and X any process $\{X_t: t\geqslant 0\}$. With (9.2) in mind, it is natural to introduce a process X^T such that (in a formal sense)*

$$dX^T(t, \omega) \equiv \begin{cases} dX(t, \omega) & \text{if } 0\leqslant t\leqslant T(\omega), \\ 0 & \text{if } t>T(\omega). \end{cases}$$

The obvious way to do this is to define (with precision):

$$X^T(t, \omega) \equiv \begin{cases} X(t, \omega) & \text{if } 0\leqslant t\leqslant T(\omega), \\ X(T(\omega), \omega) & \text{if } t>T(\omega). \end{cases}$$

The localization of (9.1) to (9.2) now reads as follows:

(9.5) $$(H\bullet X)^T = H(0, T]\bullet X^T.$$

It has now become standard to view integrators as processes X rather than as 'measures' dX which they induce. We have therefore been forced to present two different kinds of localization:

one for integrands H based on the map $H\mapsto H(0, T]$,
one for integrators X based on the map $X\mapsto X^T$.

Because we are interested in the 'measure' induced by X on the interval $(0, \infty)$ open at 0, we can concentrate on integrators which are null at zero.

10. Localization of integrands, $\mathrm{lb}\mathscr{P}$. Let $\mathscr{H}$ be a family of previsible processes with the property that if $H\in\mathscr{H}$, then $H(0, T]$ is in $\mathscr{H}$ for every stopping time T.

(*10.1*) DEFINITION ($\mathrm{l}\mathscr{H}$). *A process $H=\{H_t: t>0\}$ is in $\mathrm{l}\mathscr{H}$ if there exists a sequence (T_n) of stopping times with $T_n\uparrow\infty$ such that, for all n, $H(0, T_n]\in\mathscr{H}$.*
Note that $\mathscr{H}\subseteq\mathrm{l}\mathscr{H}$, and that $\mathrm{l}H$ is stable under maps $H\mapsto H(0, T]$.

By far the most important example is that in which $\mathscr{H}=\mathrm{b}\mathscr{P}$. Then $\mathrm{l}\mathscr{H}=\mathrm{lb}\mathscr{P}$, *the space of locally bounded previsible processes.*

(*10.2*) LEMMA. *Let H be an adapted L-process for which* $\limsup_{t\downarrow 0}|H_t|<\infty$. *Then $H\in\mathrm{lb}\mathscr{P}$.*

Proof. Let $T_n\equiv\inf\{t: |H_t|>n\}$. □

(*10.3*) *Terminology* ('reducing sequence', etc.). If T is a stopping time such that $H(0, T] \in \mathscr{H}$, we say that T *reduces* H (into $\mathscr{H}$). By a *reducing sequence* for H in $l\mathscr{H}$, we mean a sequence (T_n) of stopping times such that $T_n \uparrow \infty$ and $H(0, T_n] \in \mathscr{H}$ for all n.

(*10.4*) **Important Exercise.** Prove that $l(l\mathscr{H}) = l\mathscr{H}$. *Hint.* If $H \in l(l\mathscr{H})$, then we can find $U_n \uparrow \infty$, $V_{n,k} \uparrow U_n$ such that $H(0, V_{n,k}] \in \mathscr{H}$. Choose $k(n)$ such that $\mathbf{P}(V_{n,k(n)} > U_n - 2^{-n}) > 1 - 2^{-n}$. Now set

$$T_n = \inf\{V_{r,k(r)}: r \geqslant n\}.$$

11. Localization of integrators, $\mathscr{M}_{0,\text{loc}}$, FV$\mathscr{M}_{0,\text{loc}}$, etc.. Let $\mathscr{X}_0$ be a family of adapted R-processes null at zero, such that if $X \in \mathscr{X}_0$, and T is any stopping time, then $X^T \in \mathscr{X}_0$. We say that $\mathscr{X}_0$ is *stable under stopping*.

(*11.1*) DEFINITION ($\mathscr{X}_{0,\text{loc}}$). *A process* $X = \{X_t : t \geqslant 0\}$ *is in* $\mathscr{X}_{0,\text{loc}}$ *if there exists a sequence* (T_n) *of stopping times with* $T_n \uparrow \infty$ *such that, for all* n, $X^{T_n} \in \mathscr{X}_0$.

Note that $X_0 = 0$ for X in $\mathscr{X}_{0,\text{loc}}$. Clearly, $\mathscr{X}_0 \subseteq \mathscr{X}_{0,\text{loc}}$.

(*11.2*) *Terminology* ('reducing sequence', etc.). If T is a stopping time such that $X^T \in \mathscr{X}_0$, we say that T *reduces* X (into $\mathscr{X}_0$). By a *reducing sequence* for $X \in \mathscr{X}_{0,\text{loc}}$, we mean a sequence (T_n) with $T_n \uparrow \infty$ such that each T_n reduces X.

(11.3) *Points to notice.* If T reduces X, and S is any stopping time, then $T \wedge S$ reduces X because $X^{T \wedge S} = (X^T)^S$ and $\mathscr{X}_0$ is stable under stopping. A similar remark applies to localization of integrands because $(H(0, T])(0, S] = H(0, T \wedge S]$.

Suppose that $H \in l\mathscr{H}$ and $X \in \mathscr{X}_{0,\text{loc}}$. Let (T_n) be a reducing sequence for H (into $\mathscr{H}$), and let (S_n) be a reducing sequence for X (into $\mathscr{X}_0$). Then the sequence $(T_n \wedge S_n)$ is a reducing sequence both for H and for X.

(11.4) *Notations.* For our set-up $(\Omega, \mathscr{F}, \{\mathscr{F}_t\}, \mathbf{P})$, let:

$\mathscr{M}_0 \equiv$ space of all martingales null at 0,
UI$\mathscr{M}_0 \equiv$ space of all uniformly integrable martingales null at 0,
FV$\mathscr{M}_0 \equiv$ space of all finite-variation martingales null at 0,
IV$\mathscr{M}_0 \equiv$ space of all integrable-variation martingales null at 0.

(11.5) *Further notation.* Each of the four spaces in (11.4) is stable under stopping. (See Lemma 12.4 below.) We may therefore apply to each the extension $\mathscr{X}_0 \mapsto \mathscr{X}_{0,\text{loc}}$ and obtain the spaces:

$$\mathscr{M}_{0,\text{loc}},\ \text{UI}\mathscr{M}_{0,\text{loc}},\ \text{FV}\mathscr{M}_{0,\text{loc}},\ \text{IV}\mathscr{M}_{0,\text{loc}}.$$

There is an obvious terminology. Thus:

$\mathscr{M}_{0,\text{loc}}$ = space of *local martingales null at* 0,
IV$\mathscr{M}_{0,\text{loc}}$ = space of *local martingales null at* 0 which are *locally of integrable variation.*

We emphasize that $IV\mathcal{M}_{0,\,loc} = (IV\mathcal{M}_0)_{loc}$, so that both the 'martingale null at 0' property and the integrable-variation property have to be interpreted locally. A similar comment applies to $UI\mathcal{M}_{0,\,loc}$. Of course, the FV property is already 'local', so that:

$FV\mathcal{M}_{0,\,loc}$ = space of *local martingales null at* 0 *with paths of finite variation.*

12. Nil desperandum! Things are not as bad as they seem!

(*12.1*) THEOREM.
(i) $\mathcal{M}_{0,\,loc} = UI\mathcal{M}_{0,\,loc}$.
(ii) $FV\mathcal{M}_{0,\,loc} = IV\mathcal{M}_{0,\,loc}$.
So this leaves us with just two 'local' spaces (so far . . . !).

It is easy to recognize elements of $\mathcal{M}_{0,\,loc}$ and of $FV\mathcal{M}_{0,\,loc}$. It is technically important that if $M \in \mathcal{M}_{0,\,loc}$ (respectively, $FV\mathcal{M}_{0,\,loc}$) then we can find a reducing sequence (T_n) which reduces M 'further' into $UI\mathcal{M}$ (respectively, $IV\mathcal{M}$).

We held over from the last section the proofs that the four spaces at (11.4) are stable under stopping. That result is closely related to Theorem 12.1, and it is convenient to prove these results together.

The few short proofs which follow are nice illustrations of techniques which you cannot do without. Please do not skip them in an attempt to rush on to more 'interesting' things.

Let us begin by reminding ourselves of the rôle of uniformly integrability by recalling Doob's martingale-convergence and optional-sampling theorems into the following result.

(*12.2*) THEOREM (Doob). (i) (Martingale Convergence Theorem II.69). *Let M be any R-martingale bounded in L^1. Then $M_\infty \equiv \lim M_t$ exists with probability* 1. *We have L^1 convergence of M_t to M_∞ if and only if the martingale M is uniformly integrable.*

(ii) (Optional Sampling Theorem II.77.5). *Let M be a uniformly integrable martingale, and let S and T be stopping times with $S \leqslant T$ $(\leqslant \infty)$. Then*

$$\mathbf{E}[M_T|\mathcal{F}_S] = M_S \quad (\text{a.s. } \mathbf{P}).$$

Recall the following corollary of the Optional Sampling Theorem.

COROLLARY (see § II.44). *Let M be a uniformly integrable martingale. Then the family of random variables $\{M_T : T \text{ a stopping time}\}$ is uniformly integrable.*

The next result is an immediate consequence of Theorem 12.2.

(*12.3*) LEMMA. *Let K be a positive real. Let M be any martingale. Then the stopped process*

$$M^K \equiv \{M_{t\wedge K}: t \geqslant 0\}$$

is a uniformly integrable martingale. If W and U are stopping times with $W \leqslant U \leqslant K$, then

$$\mathbf{E}[M_U | \mathcal{F}_W] = M_W.$$

We can now prove the following lemma:

(*12.4*) LEMMA. *The spaces $\mathcal{M}_0$, $\mathrm{UI}\mathcal{M}_0$, $\mathrm{FV}\mathcal{M}_0$, $\mathrm{IV}\mathcal{M}_0$ at* (11.4) *are all stable under stopping.*

Proof. Let $M \in \mathcal{M}_0$. We must prove that if T is any stopping time, and $s \leqslant t$, then

$$\mathbf{E}[M_{T\wedge t} | \mathcal{F}_s] = M_{T\wedge s}.$$

It is enough to prove the cleaner result: *if U and V are stopping times such that for some real number K in* $(0, \infty)$, *$U \leqslant K$ and $V \leqslant K$, then*:

(12.5)
$$\mathbf{E}[M_U | \mathcal{F}_V] = M_{U\wedge V}.$$

But the martingale $M^U \equiv \{M_{U\wedge t}: t \geqslant 0\}$ is uniformly integrable, so by Theorem 12.2(ii),

$$\mathbf{E}[M_U | \mathcal{F}_V] = \mathbf{E}[M^U_\infty | \mathcal{F}_V] = (M^U)_V = M_{U\wedge V}.$$

That the remaining three spaces are stable under stopping is now trivial. □

Proof of Part (*i*) *of Theorem 12.1.* This is implicit in Lemma 12.3: for if (T_n) is a reducing sequence for M into $\mathcal{M}_0$, then $(T_n \wedge n)$ reduces M into $\mathrm{UI}\mathcal{M}_0$. □

Proof of Part (*ii*) *of Theorem 12.1.* Suppose that $M \in \mathrm{FV}\mathcal{M}_{0,\mathrm{loc}}$. Because of part (i) of the theorem, we can (and do) immediately reduce the problem to the case where M is a uniformly integrable martingale null at zero.

So assume that M is a uniformly integrable martingale null at zero with paths of finite variation. Let $V_M(t)$ denote the variation of M on $(0, t]$. For $n \in \mathbb{N}$, define the stopping time

$$T_n \equiv \inf\{t: V_M(t) > n\}.$$

(T_n is the first-entry time into an open set of an adapted R-process.) Then

$$V_M(T_n) \leqslant n + |M(T_n) - M(T_n-)| \leqslant n + |M(T_n)| + |M(T_n-)|.$$

But $|M(T_n-)| \leqslant n$. (Recall that $M(0) = 0$.) Also,

$$\mathbf{E}|M(T_n)| \leqslant \mathbf{E}|M(\infty)|$$

because of the optional sampling theorem. Hence, M^{T_n} has integrable variation. It is, of course, obvious that $T_n \uparrow \infty$. □

(*12.6*) *Discussion of* (*12.5*). Equation (12.5) states that the projections $\mathbf{E}[\cdot|\mathscr{F}_U]$ and $\mathbf{E}[\cdot|\mathscr{F}_V]$ commute, the product being $\mathbf{E}[\cdot|\mathscr{F}_{U\wedge V}]$. See Yor [1] and the papers referred to therein for some fascinating work on commutators.

13. Extending stochastic integrals by localization. On several occasions, we shall meet the following situation.

(*13.1*) *Set-up.* We have a collection $\mathscr{H}$ of 'integrands', and a collection $\mathscr{X}_0$ of 'integrators', a collection $\mathscr{I}_0$ of 'integrals', and a map

$$(H, X) \mapsto H \cdot X \tag{13.2}$$

of $\mathscr{H} \times \mathscr{X}_0$ into $\mathscr{I}_0$. The elements of $\mathscr{H}$ are previsible processes (on $(0, \infty)$), and $\mathscr{H}$ is stable under the map $H \mapsto H(0, T]$, for every stopping time T. The elements of $\mathscr{X}_0$ and $\mathscr{I}_0$ are adapted R-processes null at 0, and $\mathscr{X}_0$ and $\mathscr{I}_0$ are stable under stopping. Finally, for $H \in \mathscr{H}$, $X \in \mathscr{X}_0$, and any stopping time T, we have

$$(H \cdot X)^T = H(0, T] \cdot X^T. \tag{13.3}$$

See equation (9.5).

(*13.4*) **Example.** The example we have so far studied is that in which

$$\mathscr{H} = \mathrm{b}\mathscr{P}, \quad \mathscr{X}_0 = \mathrm{IV}\mathscr{M}_0, \quad \mathscr{I}_0 = \mathrm{IV}\mathscr{M}_0,$$

and $(H, X) \mapsto H \cdot X$ is the stochastic Stieltjes integral. Of course, (13.3) is a standard property of Stieltjes integrals. □

Now suppose that for a general set-up (13.1),

$$H \in \mathrm{l}\mathscr{H}, \quad X \in \mathscr{X}_{0,\,\mathrm{loc}}.$$

We know from our discussion at (11.3) that we can find a sequence (T_n) of stopping times which is simultaneously reducing for H (into $\mathscr{H}$) and for X (into $\mathscr{X}_0$). Fix the sequence (T_n) for the moment.

For each n, $H(0, T_n] \cdot X^{T_n}$ is defined, and, from (13.3) with $T = T_{n-1}$, we have (for $n \geq 2$) the consistency condition:

$$(H(0, T_n] \cdot X^{T_n})^{T_{n-1}} = H(0, T_{n-1}] \cdot X^{T_{n-1}}.$$

Thus it is clear that we can define a process $H \cdot X$ with parameter set $[0, \infty)$ by

$$(H \cdot X)^{T_n} = H(0, T_n] \cdot X^{T_n}, \quad \text{for all } n.$$

Let (S_n) be another sequence of stopping times which is reducing both for H and for X. Then

$$(H(0, T_n] \cdot X^{T_n})^{S_n} = H(0, T_n \wedge S_n] \cdot X^{T_n \wedge S_n} = (H(0, S_n] \cdot X^{S_n})^{T_n}.$$

It is therefore clear that we would obtain the same process $H \cdot X$ from (S_n) as from (T_n).

The essential point is therefore the following.

(*13.5*) LEMMA. *The map at (13.2) extends uniquely to a map of* $\mathrm{l}\mathcal{H}\times\mathcal{X}_{0,\,\mathrm{loc}}$ *into* $\mathcal{I}_{0,\,\mathrm{loc}}$ *such that (13.3) holds for this extension.*

Of course, for Example 13.4, the extension is still given by the Stieltjes integral. But, taking into account that

$$\mathrm{FV}\,\mathcal{M}_{0,\,\mathrm{loc}}=\mathrm{IV}\,\mathcal{M}_{0,\,\mathrm{loc}},$$

we now have a really useful result, quoted as Theorem 1.8.

(*13.6*) THEOREM. *If* $H\in\mathrm{lb}\,\mathcal{P}$ *and* $M\in\mathrm{FV}\,\mathcal{M}_{0,\,\mathrm{loc}}$, *and* $H\bullet M$ *is the Stieltjes integral, then* $H\bullet M\in\mathrm{FV}\,\mathcal{M}_{0,\,\mathrm{loc}}$.

It goes without saying that if, for example, the map at (13.2) is bilinear, then the extension provided by (13.5) will again be bilinear.

(*13.7*) *Discussion of 'modulo evanescence'.* It will frequently happen that the previsible integrand H is defined only up to 'indistinguishability', or 'modulo evanescence'. (Processes Y and Z are said to be *indistinguishable* if $\mathbf{P}(Z_t=Y_t\ \forall t)=1$, and a process V is said to be *evanescent* if $\mathbf{P}(V_t=0\ \forall t)=1$.) This causes no difficulty, because every evanescent process is previsible (as a consequence of our assumption that all processes are jointly measurable), and because, as we shall see, the stochastic integral of an evanescent process is evanescent.

(*13.8*) LEMMA. *Every evanescent process is previsible.*

Proof. Fix $\Lambda\in\mathcal{F}$, with $\mathbf{P}(\Lambda)=0$. It is well-known that if $f: S_1\to S_2$ is any map, and if $\mathcal{C}$ is a collection of subsets of S_2, then $f^{-1}(\sigma(\mathcal{C}))=\sigma(f^{-1}(\mathcal{C}))$. We apply this to the case where $S_1=\mathbb{R}^{++}\times\Lambda$, $S_2=\mathbb{R}^{++}\times\Omega$, f is the natural injection, and $\mathcal{C}=\{(s,t]\times F: 0<s<t,\ F\in\mathcal{F}\}$. Then $\sigma(\mathcal{C})=\mathcal{B}(\mathbb{R}^{++})\times\mathcal{F}$, and each set in $f^{-1}(\mathcal{C})$, being of the form $(s,t]\times(F\cap\Lambda)$, is previsible, since it is left-continuous and adapted. (Recall that every null set is in each $\mathcal{F}_t$!) Hence any measurable process X such that $X(t,\omega)=0$ for all t, for all $\omega\notin\Lambda$, is previsible. □

14. Local martingales, $\mathcal{M}_{\mathrm{loc}}$, and the Fatou lemma. The appropriate definition of a local martingale is the following.

(*14.1*) DEFINITION (local martingale, $\mathcal{M}_{\mathrm{loc}}$). *A process* $M=\{M_t: t\geqslant 0\}$ *is called a* local martingale, *and we write* $M\in\mathcal{M}_{\mathrm{loc}}$, *if*

(i) M_0 *is* $\mathcal{F}_0$*-measurable,*

(ii) $\{M_t-M_0: t\geqslant 0\}\in\mathcal{M}_{0,\,\mathrm{loc}}$.

Note that *no integrability condition is imposed on* M_0. *Because of this, it need not be possible to 'reduce' a local martingale to be a martingale.*

If $M \in \mathscr{M}_{\text{loc}}$ and $H \in \text{lb}\,\mathscr{P}$, we shall always use the definition:

$$(14.2) \qquad (H \bullet M)_t \equiv \int_{(0,t]} H_s\, d(M_s - M_0),$$

so that for integrators, we can always concentrate on local martingales null at 0. (Of course, so far we can only make sense of (14.2) if M has paths of finite variation.)

It is obvious that every martingale is a local martingale. We shall recall at (14.5), and study further in § VI.33, a famous example of an L^2 bounded continuous local martingale which is not a martingale.

As explained earlier, we do need practicable ways of proving that certain local martingales are true martingales. The following result is often effective in this tightening-up process, particularly when used in conjunction with the exponential formula of Doléans. We shall see this combination in action in §§ 21 and 37.

(*14.3*) LEMMA (to be referred to as 'the Fatou lemma'). *Let M be a nonnegative local martingale such that* $\mathbf{E}M_0 < \infty$. *Then M is a supermartingale.*

(*14.4*) *Preliminary remarks.* As stated at II.41, the monotone convergence theorem extends immediately to conditional expectations: if $\eta_n \geqslant 0$, and $\eta_n \uparrow \eta$, then

$$\mathbf{E}[\eta_n \mid \mathscr{G}] \uparrow \mathbf{E}[\eta \mid \mathscr{G}].$$

Fatou's lemma therefore applies to conditional expectations: if $\xi_n \geqslant 0$ then

$$\mathbf{E}[\liminf \xi_n \mid \mathscr{G}] \leqslant \liminf \mathbf{E}[\xi_n \mid \mathscr{G}].$$

(Put $\eta_n \equiv \inf_{r \geqslant n} \xi_r$.)

Proof of Lemma 14.3. Let (T_n) be a reducing sequence for $\{M_t - M_0\}$. Then, for $t > s$,

$$\begin{aligned}\mathbf{E}[M_{t\wedge T_n} \mid \mathscr{F}_s] &= \mathbf{E}[M_0 \mid \mathscr{F}_s] + \mathbf{E}[M_{t\wedge T_n} - M_0 \mid \mathscr{F}_s]\\ &= M_0 + M_{s\wedge T_n} - M_0 = M_{s\wedge T_n}.\end{aligned}$$

Apply Fatou's lemma:

$$\mathbf{E}[M_t \mid \mathscr{F}_s] \leqslant M_s. \qquad \square$$

(*14.5*) *Discussion.* In § III.31, we studied at some length a very important example of a nonnegative supermartingale which is a local martingale but not a martingale. Take B to be Brownian motion in $\mathbb{R}^3$ with $|B_0| = 1$, and let $M_t \equiv 1/|B_t|$.

The sequence (T_n), where $T_n \equiv \inf\{t: |M_t| > n\}$ is reducing for M, and for $t > s$,

$$\mathbf{E}[M_{t \wedge T_n} | \mathscr{F}_s] = M_{s \wedge T_n},$$

but

$$\mathbf{E}[M_t | \mathscr{F}_s] < M_s.$$

The explanation is that given in §III.31.7: the variables $\{M_{t \wedge T_n} : n \in \mathbb{N}\}$ are not uniformly integrable. In terminology which we shall be studying later, M is not of class (D) on $[0, t]$. This example will be studied again in Chapter VI.

Semimartingales as integrators

15. Semimartingales, $\mathscr{S}$. We now define the most important type of process in stochastic integral theory.

(*15.1*) DEFINITION (semimartingale, $\mathscr{S}$). *As was mentioned at (1.7), a process X is a* semimartingale (*relative to our* $(\Omega, \mathscr{F}, \{\mathscr{F}_t\}, \mathbf{P})$ *set-up) if X is an adapted R-process which may be written in the form*:

(15.2) $$X = X_0 + M + A, \quad M \in \mathscr{M}_{0,\text{loc}}, \quad A \in \text{FV}_0.$$

Thus, M is a local martingale null at 0, and A is a process null at 0 with paths of finite variation. We emphasize that the decomposition (15.2) need not be unique: there are martingales of finite variation! We write $\mathscr{S}$ for the space of semimartingales, and $\mathscr{S}_0$ for the space of semimartingales null at 0.

The concept of semimartingales was originally arrived at in an *ad hoc* fashion, as the class of integrands in Itô integrals was generalized further and further—by Kunita and Watanabe, and then by Meyer and his school.

If X is our semimartingale at (15.2), then with our convention about time 0, we have, for a locally bounded previsible H,

(15.3) $$(H \bullet X)_t = (H \bullet M)_t + (H \bullet A)_t.$$

The last integral at (15.3) is the Stieltjes integral. Thus (modulo questions about uniqueness which you will have spotted) the problem of defining $H \bullet X$ amounts to the problem of defining $H \bullet M$. The Fundamental Theorem that Itô integrals preserve local martingales now yields:

(15.4) *Itô integrals preserve semimartingales.*

In symbols:

(15.5) *if* $H \in \text{lb}\,\mathscr{P}$ *and* $X \in \mathscr{S}$, *then* $H \bullet X \in \mathscr{S}_0$.

Here is an obvious 'coordinate' definition.

(*15.6*) *Semimartingales with values in* $\mathbb{R}^n$. A process $X=(X^i: i=1, \dots, n)$ with values in $\mathbb{R}^n$ is called a semimartingale if each coordinate process X^i is a semimartingale.

The class $\mathscr{S}$ of semimartingales embraces very many of the most important processes for applications. *A diffusion process* on $\mathbb{R}^n$ defined via a stochastic differential equation relative to Brownian motion is, *ipso facto*, a semimartingale. Every *Lévy process* on $\mathbb{R}^n$ (= process with stationary independent increments) is a semimartingale. See § VI.2.

Another key fact is that $\mathscr{S}$ *is stable under a wide range of natural operations*. Of course, (15.4) expresses one especially important case. *Itô's formula* implies another:

(*15.7*) *If* X *is a semimartingale with values in* $\mathbb{R}^n$, *and* $f \in C^\infty(\mathbb{R}^n)$, *then* $f(X) \in \mathscr{S}$.

Because of this, it is natural to make the following definition, this time coordinate-free.

(*15.8*) *Semimartingales with values in manifolds*. Let E be a C^∞ manifold. A process X with values in E is a semimartingale (relative to our $(\Omega, \mathscr{F}, \{\mathscr{F}_t\}, \mathbf{P})$ set-up) if and only if, for each $f \in C^\infty(E)$, $f(X) \in \mathscr{S}$.

We have just seen that one can define a semimartingale on E in a natural way. It is not at all obvious how to define a *martingale* on E! We return to this in § V.33.

16. Integrators. Recall from § 6 that the space $\mathrm{b}\mathscr{E}$ consists of bounded previsible processes of the form

$$H = \sum_{i=1}^{n} Z_i (S_i, T_i] \tag{16.1}$$

where $n \in \mathbb{N}$, $\{S_i: i=1, \dots, n\}$ and $\{T_i: i=1, \dots, n\}$ are sequences of stopping times such that $0 \leqslant S_1 \leqslant T_1 \leqslant \dots \leqslant S_n \leqslant T_n$, and $Z_i \in \mathrm{b}\mathscr{F}(S_i)$ for each i. For $H \in \mathrm{b}\mathscr{E}$, define $\|H\| \equiv \sup_{t,\omega} |H(t, \omega)|$. Let X be an adapted R-process. For H as at (16.1), define:

$$(H * X)_t = \sum_{i=1}^{n} Z_i (X_{T_i \wedge t} - X_{S_i \wedge t}). \tag{16.2}$$

(*16.3*) DEFINITION (integrator). *An adapted R-process X is called an* integrator *if the hypothesis*:

(*a*) (H_n) *is a sequence in* $\mathrm{b}\mathscr{E}$ *with* $\|H_n\| \to 0$
implies the consequence,
(*b*) $\forall t$, $\forall \varepsilon > 0$, $\mathbf{P}[|(H_n * X)_t| > \varepsilon] \to 0$ as $n \to \infty$.

The hypothesis (*a*) involves a very strict form of convergence, and the consequence (*b*) involves only the weak 'convergence in probability' of $(H_n * X)_t$ to 0 for each t. If X is not an integrator, no sensible extension of the map $H \mapsto H * X$

from b$\mathscr{E}$ to b$\mathscr{P}$ is possible. However, if X is an integrator, then one can develop a theory of the stochastic integral map $H \mapsto H*X$ on b$\mathscr{P}$ as part of the theory of vector-valued measures. See Bichteler [1].

As previously mentioned, we have the following remarkable tribute to the insight of Meyer, Kunita, Watanabe, Doléans, . . . , who had worked in *ad hoc* fashion.

(*16.4*) THEOREM (Bichteler, Dellacherie, Kussmaul, Metivier and Pellaumail):

X is an integrator if and only if X is a semimartingale.

Thus $H*X$ and $H\bullet X$ are in exact agreement. Theorem 16.4 is a remarkably beautiful result, and it clinches the absolute correctness of the semimartingale concept. However, we do not prove it here. The *ad hoc* methods are in any case essential to the theory, even if one has Theorem 16.4; and so we follow the classical development. For proof of Theorem 16.4, see Part 4 of Chapter VIII of Dellacherie and Meyer [1].

One crucial stability property of the space $\mathscr{S}$ of semimartingales, already known in the classical theory, finds an elegant explanation in Theorem 16.4.

Let $\tilde{\mathbf{P}}$ *be a probability measure on* $(\Omega, \mathscr{F})$ *equivalent to* $\mathbf{P}$. Thus, each of $\mathbf{P}$ and $\tilde{\mathbf{P}}$ is absolutely continuous relative to the other. Or again, the null sets of $\mathbf{P}$ and $\tilde{\mathbf{P}}$ are the same. Note that the set-up $(\Omega, \mathscr{F}, \{\mathscr{F}_t\}, \tilde{\mathbf{P}})$ satisfies the usual conditions.

It is obvious from the definition of integrator that:

(*16.5*) *X is a $\tilde{\mathbf{P}}$ integrator if and only if X is a $\mathbf{P}$ integrator.*

Of course, a $\mathbf{P}$ integrator means as integrator relative to the set-up $(\Omega, \mathscr{F}, \{\mathscr{F}_t\}, \mathbf{P})$.

In view of Theorem 16.4, we deduce the result:

(*16.6*) *X is a $\tilde{\mathbf{P}}$ semimartingale if and only if X is a $\mathbf{P}$ semimartingale.*

This is not at all obvious from the definition of a semimartingale. However, it matters very much for the theory that if X is a $\tilde{\mathbf{P}}$ semimartingale, then we can explicitly exhibit X as a $\mathbf{P}$ semimartingale by obtaining a decomposition of the form (15.2). In § 38, we do this for continuous semimartingales.

Likelihood ratios

17. Martingale property under change of measure. Likelihood ratios are important both in the probability theory and the statistics of random processes. We give here a result which we shall apply in several contexts. It is obviously a step towards a direct proof of (16.6).

Let $(\Omega, \mathscr{F}^\circ, \{\mathscr{F}_t^\circ\})$ be a filtered space. Let $\mathbf{P}$ and $\tilde{\mathbf{P}}$ be two equivalent probability measures on $(\Omega, \mathscr{F}^\circ)$. The set-up $(\Omega, \mathscr{F}^\circ, \{\mathscr{F}_t^\circ\})$ has the same augmentation $(\Omega, \mathscr{F}, \{\mathscr{F}_t\})$ under $\tilde{\mathbf{P}}$ as under $\mathbf{P}$. Some obvious terminology and notation will be used. Thus a $\tilde{\mathbf{P}}$ martingale means a martingale relative to $(\Omega, \mathscr{F}, \{\mathscr{F}_t\}, \tilde{\mathbf{P}})$, and $\tilde{\mathbf{E}}$ denotes expectation relative to $\tilde{\mathbf{P}}$.

(*17.1*) THEOREM

(i) *There exists a* $\mathbf{P}$ *martingale* ρ *such that* ρ *is an R-process and for all* t, ρ_t *is a version of the likelihood ratio* $d\tilde{\mathbf{P}}/d\mathbf{P}$ *on* $(\Omega, \mathscr{F}_t)$.

(ii) $\mathbf{P}[\forall t, \rho_t > 0, \rho_{t-} > 0] = 1$.

(iii) M *is a* $\tilde{\mathbf{P}}$ *martingale if and only if* ρM *is a* $\mathbf{P}$ *martingale.*

(iv) M *is a* $\tilde{\mathbf{P}}$ *local martingale if and only if* ρM *is a* $\mathbf{P}$ *local martingale.*

Proof. (i) Let $\rho_\infty \equiv d\tilde{\mathbf{P}}/d\mathbf{P}$ on $\mathscr{F}$, and let ρ be an R-process modification of the $\mathbf{P}$ martingale:

$$\rho_t \equiv \mathbf{E}[\rho_\infty | \mathscr{F}_t].$$

For $\Lambda \in \mathscr{F}_t$,

$$\mathbf{E}[\rho_t; \Lambda] = \mathbf{E}[\rho_\infty; \Lambda] = \tilde{\mathbf{P}}(\Lambda),$$

so that ρ_t is a version of $d\tilde{\mathbf{P}}/d\mathbf{P}$ on $(\Omega, \mathscr{F}_t)$.

(ii) Let

$$\zeta \equiv \inf\{t \colon \rho_t = 0 \text{ or } \rho_{t-} = 0\}.$$

By Theorem II.78.1, $\mathbf{P}[\rho_t = 0, \forall t \geqslant \zeta] = 1$. But for each fixed t, since $\mathbf{P}$ is absolutely continuous relative to $\tilde{\mathbf{P}}$, we have

$$\mathbf{P}[\rho_t = 0] = 0.$$

Hence, $\mathbf{P}[\zeta = \infty] = 1$.

(iii) Let M be a $\tilde{\mathbf{P}}$ martingale. Then for $s \leqslant t$ and $\Lambda \in \mathscr{F}_s$, we have:

$$\tilde{\mathbf{E}}[M_t; \Lambda] = \tilde{\mathbf{E}}[M_s; \Lambda],$$

so that

$$\mathbf{E}[\rho_t M_t; \Lambda] = \mathbf{E}[\rho_s M_s; \Lambda].$$

Hence ρM is a $\mathbf{P}$ martingale, and the 'only if' part is proved. The 'if' part is really the same result with $(\mathbf{P}, \tilde{\mathbf{P}}, \rho^{-1})$ replacing $(\tilde{\mathbf{P}}, \mathbf{P}, \rho)$.

(iv) Let M be a $\tilde{\mathbf{P}}$ local martingale. By replacing M by $\{M_t - M_0 : t \geqslant 0\}$, we can, and do, assume that $M_0 = 0$. Let (T_n) be a reducing sequence for M for $\tilde{\mathbf{P}}$, so that each process M^{T_n} is a $\tilde{\mathbf{P}}$ martingale. Under $\mathbf{P}$, (T_n) is again a sequence of stopping times with $T_n \uparrow \infty$ a.s. By part (iii), $M^{T_n} \rho$ is a $\mathbf{P}$ martingale, so that

$$(\rho M)^{T_n} = (M^{T_n} \rho)^{T_n}$$

is a $\mathbf{P}$ martingale. The result follows. □

3. THE ELEMENTARY THEORY OF FINITE-VARIATION PROCESSES

18. Itô's formula for FV functions. Here we work with classical (deterministic) functions on $[0, \infty)$.

Let x (mapping t to $x(t)$) be an FV function on $[0, \infty)$. It is again part of our definition of an FV function that it be an R-function. Recall that we can write x as

$$x = x_0 + x^+ - x^-,$$

where x^+ and x^- are increasing (= non-decreasing) R-functions. Indeed, if $V_x(t)$ is the variation of x over $(0, t]$, we can take

$$x_t^+ = \tfrac{1}{2}[x_t - x_0 + V_x(t)], \quad x_t^- = -\tfrac{1}{2}[x_t - x_0 - V_x(t)].$$

We write

$$\Delta x_s = x_s - x_{s-}, \quad s > 0.$$

The function x induces a signed Stieltjes measure μ_x on $(0, \infty)$ with

$$\mu_x(u, v] = x_v - x_u.$$

The measure μ_x has an atom Δx_s at any point $s > 0$, where $\Delta x_s \neq 0$. We can decompose x into the sum of its continuous and atomic parts as follows:

$$x = x_0 + x^c + x^a, \quad x_t^a = \sum_{0 < s \leqslant t} \Delta x_s.$$

Integration by parts. Let x and y be two FV functions on $[0, \infty)$. We can define the product signed measure $\mu_x \times \mu_y$ on $(0, \infty) \times (0, \infty)$. From a picture showing the square $(0, t] \times (0, t]$ together with its main diagonal,

$$(0, t] \times (0, t] = \{(u, v)\colon 0 < v \leqslant t; 0 < u < v\} \cup \{(u, v)\colon 0 < u \leqslant t; 0 < v < u\}$$
$$\cup \{(u, v)\colon 0 < u = v \leqslant t\}.$$

Taking $\mu_x \times \mu_y$ measures in the above equation, we see that

$$(x_t - x_0)(y_t - y_0) = \int_{(0,t]} (x_{v-} - x_0) dy_v + \int_{(0,t]} (y_{u-} - y_0) dx_u + [x, y]_t$$

where:

(18.1)
$$[x, y]_t \equiv \sum_{0 < s \leqslant t} \Delta x_s \Delta y_s.$$

On rearranging, we obtain the *integration by parts formula*:

(18.2)
$$x_t y_t - x_0 y_0 = \int_{(0,t]} x_{s-}\, dy_s + \int_{(0,t]} y_{s-}\, dx_s + [x, y]_t$$

which we write in differential notation as

(18.3)
$$d(xy) = x_- dy + y_- dx + d[x, y].$$

(*18.4*) THEOREM (Itô's formula for FV functions). *Let $f \in C^1(\mathbb{R})$, and let x be an FV function on $[0, \infty)$. Then*

$$f(x_t) - f(x_0) = \int_{(0,t]} f'(x_{s-})\,dx_s + \sum_{0<s\leqslant t} \{f(x_s) - f(x_{s-}) - f'(x_{s-})\Delta x_s\}. \tag{18.5}$$

If we write F_t for $f(x_t)$, then we may write (18.5) in differential notations as

$$dF_t = f'(x_{t-})dx_t + \{\Delta F_t - f'(x_{t-})\Delta x_t\}. \tag{18.6}$$

Remarks. (*a*) On an interval $[0, t_0]$, x is bounded, and f' is bounded (by K, say) on the range of x. Hence,

$$\sum_{0<s\leqslant t_0} |f(x_s) - f(x_{s-})| \leqslant K \sum_{0<s\leqslant t_0} |x_s - x_{s-}| < \infty$$

and

$$\sum_{0<s\leqslant t_0} |f'(x_{s-})\Delta x_s| \leqslant K \sum_{0<s\leqslant t_0} |\Delta x_s| < \infty.$$

(*b*) Because of the inequalities just given, it is possible—for this FV situation—to reformulate (18.5) in neater ways. However, (18.5) is the correct form for generalizations, and we stick to that.

Proof of Theorem 18.4. Fix the FV function x. Let $\mathscr{A}$ be the family of functions in $C^1(\mathbb{R})$ such that (18.5) holds. Then $\mathscr{A}$ is clearly a vector space. But it is also true that $\mathscr{A}$ is an *algebra*. For let $f \in \mathscr{A}$, $g \in \mathscr{A}$, and put $h = fg$. Write $F_t = f(x_t)$, $G_t = g(x_t)$ and $H_t = h(x_t)$. Then

$$dF = f'(x_-)dx + \{\Delta F - f'(x_-)\Delta x\},$$
$$dG = g'(x_-)dx + \{\Delta G - g'(x_-)\Delta x\}.$$

By the integration by parts formula,

$$dH = F_-\,dG + G_-\,dF + [F, G], \tag{18.7}$$

and consideration of the atoms in (18.7) now tells us that

$$\Delta H = F_-\,\Delta G + G_-\,\Delta F + \Delta F\,\Delta G.$$

So, from (18.7) we obtain

$$\begin{aligned} dH &= \{f(x_-)g'(x_-) + g(x_-)f'(x_-)\}dx \\ &\quad + \{F_-\Delta G + G_-\Delta F + \Delta F \Delta G - h'(x_-)\Delta x\} \\ &= h'(x_-)dx + \{\Delta H - h'(x_-)\Delta x\}, \end{aligned}$$

so that $h \in \mathscr{A}$. Since it is trivial that $\mathscr{A}$ contains the function $f(x) = x$, it follows that $\mathscr{A}$ contains all polynomials.

Now let f be any element of $C^1(\mathbb{R})$. It is enough to take $t_0 \in (0, \infty)$ and prove (18.5) for all $t \leqslant t_0$. Now for some N, $x(t) \in [-N, N]$ for all $t \leqslant t_0$. Moreover, we can find polynomials p_n such that

$$p_n \to f \text{ uniformly on } [-N, N],$$
$$p_n' \to f' \text{ uniformly on } [-N, N].$$

Since (18.5) is true for each p_n, it is now trivial that, for $t \leqslant t_0$, (18.5) is true for f. □

Here is the n-dimensional generalization of Theorem 18.4.

(*18.8*) THEOREM (Itô's formula, continued). *Suppose that:*

$$x = (x^1, x^2, \ldots, x^n)$$

is an $\mathbb{R}^n$*-valued function, each component of which is an FV function on* $[0, \infty)$. *Let* f *be a function on* $\mathbb{R}^n$ *with continuous first-order partial derivatives* $D_i f$. *Then:*

$$f(x_t) - f(x_0) = \int_{(0,t]} D_i f(x_{s-})\, dx_s^i + \sum_{0<s\leqslant t} \{f(x_s) - f(x_{s-}) - D_i f(x_{s-}) \Delta x_s^i\}.$$

Of course, Einstein's summation convention is being used here, so we sum over the repeated index i in an expression such as $D_i f(x_{s-}) dx_s^i$.

We skip the proof, which is an extension of the method used to prove Theorem 18.4.

19. The Doléans exponential $\mathscr{E}(x.)$. 'Exponential martingales' already featured briefly in §3. We shall see that they play an important part in the theory—as likelihood ratios, amongst other things. Here, we study the Doléans exponential $\mathscr{E}(x.)$ of an FV function x.

Let x be an FV function on $[0, \infty)$. The function $y = \mathscr{E}(x)$ is defined to be the unique solution to the equation

(19.1) $$dy_t = y_{t-}\, dx_t, \quad y_0 = 1.$$

In integral form, equation (19.1) becomes

(19.2) $$y_t = 1 + \int_{(0,t]} y_{s-}\, dx_s.$$

Let y be any solution of (19.1). Since the solution is clearly $\exp(x_t)$ if x is continuous, it is natural in the general case to consider

$$z_t = y_t \exp(-x_t^c).$$

Since $d[y, \exp(-x^c)] = 0$, we have

$$\begin{aligned} dz_t &= \exp(-x_t^c)\{-y_t dx_t^c + dy_t\} \\ &= \exp(-x_t^c)\{-y_t dx_t^c + y_{t-}\, dx_t^c + y_{t-}\, dx_t^a\}. \end{aligned}$$

Since $y_t = y_{t-}$ except on a countable set, which has zero measure with respect to $\exp(-x_t^c)dx_t^c$, we have

$$dz_t = \exp(-x_t^c) y_{t-}\, dx_t^a = z_{t-}\, dx_t^a.$$

We see that z is purely discontinuous with

$$z_t - z_{t-} = z_{t-}\Delta x_t, \quad \text{so that } z_t = z_{t-}(1 + \Delta x_t).$$

It follows (see Exercise 19.6 below) that

$$z_t = \prod_{0 < s \leq t} (1 + \Delta x_s). \tag{19.3}$$

We have proved that the unique solution y of (19.1) is given by the explicit formula

$$y_t = \mathscr{E}(x_t) = \exp(x_t^c) \prod_{0 < s \leq t} (1 + \Delta x_s). \tag{19.4}$$

□

Yor's addition formula. Let x and y be two FV functions on $[0, \infty)$. Let $f = \mathscr{E}(x)$ and $g = \mathscr{E}(y)$. Then:

$$df_t = f_{t-}\, dx_t, \quad dg_t = g_{t-}\, dy_t, \quad d[f, g]_t = f_{t-} g_{t-}\, d[x, y]_t.$$

By the differentiation by parts formula (18.3),

$$\begin{aligned} d(f_t g_t) &= f_{t-}\, dg_t + g_{t-}\, df_t + d[f, g]_t \\ &= f_{t-} g_{t-}(dy_t + dx_t + d[x, y]_t). \end{aligned}$$

Hence:

$$\mathscr{E}(x)\mathscr{E}(y) = \mathscr{E}(x + y + [x, y]). \tag{19.5}$$

□

(*19.6*) **Exercise.** Let x be purely discontinuous ($x = x^a$). Prove that the unique solution of

$$dz = z_-\, dx, \quad z_0 = 1 \tag{19.7}$$

is given by (19.3). (*Hint.* Let w be another solution. Apply Itô's formula to show that $d(w/z) = 0$. Explain carefully why this always holds, with the convention that $(0/0) = 1$.)

Applications to Markov chains with finite state-space

20. Martingale problems. Let I be a finite set. Let Q be a conservative Q-matrix on I, so that

$$q_{ij} \geq 0 \ (i \neq j); \quad q_i \equiv -q_{ii} = \sum_{j \neq i} q_{ij}.$$

When more convenient, we shall write $Q(i,j)$ for q_{ij}. Then $\{P(t)\}$, where $P(t)=\exp(tQ)$, defines a Markov transition function on I which is obviously Feller–Dynkin (see §III.6.5) if we take the discrete topology on I. Theorem III.7.17 therefore tells us that we can find a Markov chain X on I with transition function $\{P(t)\}$, with all paths R-functions. Let us spell this out in detail.

(*20.1*) THEOREM. *Let* Ω *be the space of all R-functions from* $[0,\infty)$ *to* I. *For* $\omega\in\Omega$, *put* $X_t(\omega)\equiv\omega(t)$. *Define:*

$$\mathscr{F}_t^\circ\equiv\sigma\{X_s: s\leqslant t\},\quad \mathscr{F}^\circ\equiv\sigma\{X_s: 0\leqslant s<\infty\}.$$

Then there exists a probability measure $\mathbf{P}^i$ *on* $(\Omega, \mathscr{F}^\circ)$ *such that:*

(20.2 (i)) $\mathbf{P}^i[X_0=i]=1$;

(20.2 (ii)) *for* $n\in\mathbb{N}$, $i_1,i_2,\ldots,i_n\in I$, *and* $0<t_1<t_2<\ldots<t_n$,

$$\mathbf{P}^i[X(t_k)=i_k,\ 1\leqslant k\leqslant n]=\prod_{k=1}^{n}p(t_k-t_{k-1},i_{k-1},i_k),$$

where $i_0=i$, $t_0=0$.

Of course, there is at most one measure $\mathbf{P}^i$ satisfying conditions (20.2).

In our study of *Dynkin's formula* (see §III.10), we saw the importance of the fact that, every function f on I

(20.3)
$$M_t^f\equiv f(X_t)-f(X_0)-\int_0^t (Qf)(X_s)\,ds$$

is a martingale with respect to $(\Omega, \mathscr{F}^\circ, \{\mathscr{F}_t^\circ\}, \mathbf{P}^i)$. We are now going to show that this martingale property (together with the fact that $\mathbf{P}^i[X_0=i]=1$) characterizes the Markovian measure $\mathbf{P}^i$.

(*20.4*) *The martingale problem for* Q *starting from* i. A probability measure $\mathbf{P}$ on $(\Omega, \mathscr{F}^\circ)$ is said to *solve the martingale problem for* Q *starting from* i if:

(20.4 (i)) $\mathbf{P}[X_0=i]=1$,

(20.4 (ii)) for every function f on I, M^f, as defined at (20.3), is a martingale relative to $(\Omega, \mathscr{F}^\circ, \{\mathscr{F}_t^\circ\}, \mathbf{P})$.

We have just reminded ourselves that

(20.5) $\mathbf{P}^i$ *solves the martingale problem for* Q *starting from* i.

We shall now prove:

(*20.6*) THEOREM. *The measure* $\mathbf{P}^i$ *is the unique solution to the martingale problem for* Q *starting from* i.

We shall see later how important this type of result is.

(*20.7*) *Note on the usual conditions.* Most of the theory of martingales and stochastic integrals is developed under the assumption that the usual conditions hold. Now, if **P** solves the martingale problem for Q starting from i, then the set-up $(\Omega, \mathcal{F}^\circ, \{\mathcal{F}_t^\circ\}, \mathbf{P})$ will not satisfy the usual conditions.

But look at things as follows. Fix s and t, $s<t$. Let v lie in (s, t). If **E** denotes expectation with respect to the particular **P** which we are studying, we have

$$\mathbf{E}[M_t^f \mid \mathcal{F}_v^\circ] = M_v^f, \quad \mathbf{P}\text{-a.s.}$$

Let $v \downarrow\downarrow s$. Using the backward-martingale convergence theorem and the fact that M^f is *right-continuous by definition*, we have

$$\mathbf{E}[M_t^f \mid \mathcal{F}_{s+}^\circ] = M_s^f, \quad \mathbf{P}\text{-a.s.},$$

so that each M^f is a martingale relative to the usual **P**-augmentation of $(\Omega, \mathcal{F}^\circ, \{\mathcal{F}_t^\circ\}, \mathbf{P})$. This simply recalls Lemma II.67.10. □

Proof of Theorem 20.6. Let **P** solve the martingale problem for Q starting from i. To prove that $\mathbf{P} = \mathbf{P}^i$, we need only show that, for $0<t<u$, and for every function f on I,

$$\mathbf{E}[f(X_u) \mid \mathcal{F}_t^\circ] = (P_{u-t}f)(X_t), \quad \mathbf{P}\text{-a.s.} \tag{20.8}$$

For then, taking $f = I_k$,

$$\mathbf{P}[X_u = k \mid \mathcal{F}_t^\circ] = p(u-t, X_t, k), \tag{20.9}$$

so that

$$\mathbf{P}[X_t = j;\ X_u = k \mid \mathcal{F}_t^\circ] = I_j(X_t)p(u-t, X_t, k) = I_j(X_t)p(u-t, j, k).$$

Next, changing (u, k, t) to (t, j, s) in (20.9), we obtain

$$\mathbf{P}[X_t = j;\ X_u = k \mid \mathcal{F}_s^\circ] = p(t-s, X_s, j)p(u-t, j, k)$$

and so on. We can build up (20.2 (ii)).

To prove (20.8), we need only show that

(20.10) $N = \{N_t : 0 \leqslant t \leqslant u\}$ *is a martingale relative to* **P**,

where

$$N_t = (P_{u-t}f)\ (X_t). \tag{20.11}$$

Then (20.8) simply says

$$\mathbf{E}[N_u \mid \mathcal{F}_t^\circ] = N_t.$$

For $0 \leqslant t \leqslant u$ and $j \in I$, let

$$g(t, j) = (P_{u-t}f)\,(j).$$

Then $N_t = g(t, X_t)$. But since $P_s = \exp(sQ)$,

$$\frac{\partial g}{\partial t} + Qg = 0,$$

where, of course,

$$(Qg)(t, j) \equiv \sum_k Q(j, k) g(t, k).$$

The next lemma shows that N is a *local* martingale relative to $\mathbf{P}$. But $N = \{N_t : 0 \leqslant t \leqslant u\}$ is uniformly bounded by $\|f\|_\infty$ (see (20.11)). Hence N is a *true* martingale, and the proof of Theorem 20.6 is complete. □

(*20.12*) LEMMA. *Let* $\mathbf{P}$ *solve the martingale problem for* Q *starting from* i. *Let* g *be a function on* $[0, \infty) \times I$:

$$(t, j) \mapsto g(t, j)$$

such that $(\partial/\partial t)\, g(t, j)$ *is continuous for every* j. *Then*:

$$g(t, X_t) - g(0, X_0) - \int_0^t \left(\frac{\partial g}{\partial s} + Qg \right)(s, X_s)\, ds$$

is a local martingale relative to $(\Omega, \mathscr{F}^\circ, \{\mathscr{F}_t^\circ\}, \mathbf{P})$.

Proof. Since

$$g(t, X_t) = \sum_j g(t, j) I_j(X_t),$$

linearity ensures that it is enough to prove the lemma when g has the form

$$g(t, j) = h(t) f(j),$$

where h is a function with continuous derivative. We know that

$$d(f(X_t)) = dM^f + (Qf)(X_t) dt.$$

Now apply the integration by parts formula to g to find that

$$dg(t, X_t) = \{h'(t) f(X_t) + h(t)(Qf)(X_t)\} dt + h(t) dM_t^f;$$

and since $h(t) dM_t^f$ is the differential of a local martingale, the result follows. □

21. Probabilistic interpretation of Q. We continue to work with our chain X. Since Theorem 20.6 shows that all properties of X are derivable from martingale properties, it must be possible to perform the basic calculations on Markov chains by Itô calculus. Let us see how to do this.

Let f be a function on I, and suppose $\lambda > 0$. By taking $g(t, j) = e^{-\lambda t} f(j)$ in

Lemma 20.12, we see that

(21.1) $$M^{f,\lambda} \equiv e^{-\lambda t} f(X_t) - f(X_0) - \int_0^t e^{-\lambda s}(Q-\lambda)f(X_s)\,ds$$

is a bounded martingale.

In this section, we mean by 'martingale' a martingale relative to $\mathbf{P}^i$ for every i (or $\mathbf{P}^\mu$ for every initial law μ). The fact that $M^{f,\lambda}$ is bounded is obvious, and since it a local martingale because of Lemma 20.12, it is a true martingale. Of course, the result (21.1) is just a special case of (III.10.10).

If $f = I_b$ for some state b, we write $M^{b,\lambda}$ for $M^{f,\lambda}$. Now let a be some fixed state. Let $\tau \equiv \inf\{t: X_t \neq X_0\}$. Then, for $b \neq a$,

$$0 = \mathbf{E}^a M_\tau^{b,\lambda} = \mathbf{E}^a[e^{-\lambda\tau}; X_\tau = b] - \mathbf{E}^a \int_0^\tau e^{-\lambda s} q_{ab}\,ds.$$

Thus

(21.2) $$\mathbf{E}^a[e^{-\lambda\tau}; X_\tau = b] = q_{ab}(1 - \mathbf{E}^a[e^{-\lambda\tau}])/\lambda.$$

On summing (21.2) over $b \neq a$, we obtain

$$\mathbf{E}^a[e^{-\lambda\tau}] = q_a(1 - \mathbf{E}^a[e^{-\lambda\tau}])/\lambda,$$

so that $\mathbf{E}^a[e^{-\lambda\tau}] = q_a/(\lambda + q_a)$. We have proved that

(21.3) *under $\mathbf{P}^a$, τ has the exponential distribution of rate q_a.*

It is now immediate from (21.2) that we have in addition:

(21.4) *under $\mathbf{P}^a$, X_τ is independent of τ, and*

$$\mathbf{P}^a[X_\tau = b] = q_{ab}/q_a \quad (b \neq a).$$

As usual, M^f denotes $M^{f,0}$, and M^b denotes $M^{b,0}$.

Let $H_t^a = I_a(X_{t-})$ for $t > 0$. Then H^a is previsible on $(0, \infty)$, and so for $a \neq b$,

(21.5) $(H^a \bullet M^b)_t = J_{ab}(t) - q_{ab} L_a(t)$ *is a local martingale,*

where

(21.6) $$J_{ab}(t) = \#\,\{s: 0 < s \leq t,\ X_{s-} = a,\ X_s = b\}$$

(21.7) $$L_a(t) = \text{meas}\,\{s: 0 < s \leq t,\ X_s = a\}.$$

Thus $J_{ab}(t)$ is the number of jumps from a to b during $(0, t]$, and $L_a(t)$ is the time spent in a before time t. Let

(21.8) $$U_{ab}(t) = (H^a \bullet M^b)_t = J_{ab}(t) - q_{ab} L_a(t).$$

We know that U_{ab} is a local martingale. We would like to show that it is a *true* martingale. For $t \leqslant t_0$, we have $L_a(t) \leqslant t_0$, $J_{ab}(t) \leqslant J_{ab}(t_0)$. Hence,

$$\sup_{t \leqslant t_0} |U_{ab}(t)| \leqslant J_{ab}(t_0) + q_{ab} t_0,$$

and it is enough to show that

$$\mathbf{E} J_{ab}(t_0) < \infty. \tag{21.9}$$

We can prove a much stronger result than (21.9) as follows. Suppose $\alpha > 0$. Introduce the Doléans exponential

$$Y(t) = \mathscr{E}(\alpha U_{ab})(t) = \exp(-\alpha q_{ab} L_a(t))(1+\alpha)^{J(t)} \tag{21.10}$$

where $J(t) = J_{ab}(t)$ for convenience. Then

$$dY(t) = Y(t-)\alpha dU_{ab}(t).$$

The process $Y(t-)$ is an adapted L-process, and so is locally bounded previsible (see 10.2)). Since U_{ab} is a local martingale, Y is a local martingale. But Y is positive, so (see (14.3)) *Y is a positive supermartingale.* Hence

$$1 = \mathbf{E} Y(0) \geqslant \mathbf{E} Y(t) \geqslant \exp(-\alpha q_{ab} t) \mathbf{E}(1+\alpha)^{J(t)}$$

and so

$$\mathbf{E}(1+\alpha)^{J(t)} \leqslant \exp(\alpha t q_{ab}), \quad J = J_{ab}, \tag{21.11}$$

and (21.9) follows. □

Discussion of (21.11). Suppose that our chain spends an infinite amount of time in state a ($L_a(\infty) = \infty$). Then define

$$\gamma_t = \inf\{u : L_a(u) > t\}.$$

Then $\{J_{ab}(\gamma_t)\}$ is a Poisson process of rate q_{ab}. This is part of Itô excursion theory—see Chapter VI. Since $\gamma_t \geqslant t$, this makes (21.11) obvious. □

Let us collect together what we have proved.

(*21.12*) LEMMA. *Suppose $a \neq b$. Let $J_{ab}(t)$ be the number of jumps from a to b during time interval* $(0, t]$. *Let $L_a(t)$ be the time spent by X in a before time t. Then (under each* $\mathbf{P}^{\mu}$)

$$U_{ab}(t) = J_{ab}(t) - q_{ab} L_a(t) \text{ is a martingale.}$$

As stated at (III.57.15), this perfectly captures the rôle of q_{ab} as a *jump intensity* from a to b.

The jumping of Markov processes is the subject of the theory of *Lévy kernels.* For some important applications described in the next few sections, we need the following more complete description of the role of Q as the Lévy kernel of X.

(*21.13*) LEMMA. *Let g be a function*

$$g: (0, \infty) \times I \times I \times \Omega \to \mathbb{R}$$

with the following two properties:

(i) *for all i, j, the map $(t, \omega) \mapsto g(t, i, j, \omega)$ is locally bounded previsible;*
(ii) *for all t, ω and i, $g(t, i, i, \omega) = 0$.*

Then:

$$N_t^g(\omega) \equiv \sum_{0<s\leq t} g(s, X_{s-}, X_s, \omega) - \int_{(0,t]} \sum_b Q(X_{s-}, b) g(s, X_{s-}, b, \omega)\, ds$$

defines a local martingale N^g.

Proof. If we write our local martingale U_{ab} (with $a \neq b$) in the alternative form:

$$U_{ab}(t) = \sum_{0<s\leq t} I_a(X_{s-}) I_b(X_s) - \int_{(0,t]} I_a(X_{s-}) Q(X_{s-}, b)\, ds,$$

then it is clear that

$$N_t^g(\omega) = \sum_{a\neq b} \int_{(0,t]} g(s, a, b, \omega)\, dU_{ab}(t, \omega);$$

and the result follows. □

Example. If f is a function on I, and we have

$$g(t, i, j, \omega) = f(j) - f(i) \quad \text{for all } t, \omega,$$

then

$$N_t^g = M_t^f = f(X_t) - f(X_0) - \int_{(0,t]} (Qf)(X_s)\, ds$$

$$= \sum_i \sum_j [f(j) - f(i)] U_{ij}(t).$$

(*21.14*) *Remark.* It is possible to relax somewhat the assumption that g be locally bounded. □

The next result is one of the remarkable series of martingale-representation results which are important throughout the subject. There is a general principle that uniqueness of the solution to the martingale problem implies martingale representation. In fact, it is extremality of the solution rather than uniqueness which is the key thing. The general principle involved here owes its origins to Dellacherie, and was developed by Yor and, particularly, Jacod. We return to it in § V.25.

(*21.15*) THEOREM. *Let μ be an 'initial' law on I. Let N be any local martingale relative to* $(\Omega, \mathscr{F}^\mu, \{\mathscr{F}^\mu_t\}, \mathbf{P}^\mu)$ *such that $N_0=0$. Then:*

$$N = N^g$$

for some function g on $(0, \infty) \times I \times I \times \Omega$ *such that:*

(i) *for all i, j, the map* $(t, \omega) \mapsto g(t, i, j, \omega)$ *is previsible;*

(ii) *for all t, ω and i,* $g(t, i, i, \omega) = 0$.

As is implicit in Remark 21.14, we cannot say that g is necessarily locally bounded.

22. Likelihood ratios and some key distributions. There are many results about finite chains which are intuitively obvious, but which are awkward to prove rigorously by more classical methods. Several of these results are presented time and time again in the applied literature. Let us see how Itô calculus can be used to make intuition precise.

To lead into things, let us construct a discrete-time problem.

(*22.1*) *A discrete-time example.* A particle moves in discrete time on a finite set I. For $n=0, 1, \ldots, X_n$ denotes the particle's position at time n. The starting position is a given state i_0. We have two hypotheses about the particle's motion:

Null Hypothesis: X is a Markov chain with one-step transition matrix P (supposed known);

Alternative Hypothesis: X is a Markov chain with one-step transition matrix $\tilde{P}$ (supposed known);

and for convenience, we assume that for all i, j, $p_{ij}>0$ and $\tilde{p}_{ij}>0$. We write

$$r(i, j) \equiv \tilde{p}_{ij}/p_{ij}.$$

Let ρ be the likelihood-ratio process:

$$\rho_n \equiv \prod_{k=1}^{n} r(X_{k-1}, X_k). \qquad (22.2)$$

Thus ρ_n is the probability of obtaining the observed sequence $X_0, X_1, \ldots, X_n$ under the Alternative Hypothesis divided by the corresponding probability under the Null Hypothesis. We are interested in describing the evolution of ρ via a 'stochastic differential equation'. It is clear that the appropriate SDE is the following:

$$\Delta\rho_n \equiv \rho_n - \rho_{n-1} = \rho_{n-1}[r(X_{n-1}, X_n) - 1]. \qquad (22.3)$$

Equation (22.3) is the Doléans equation:

$$\Delta\rho_n = \rho_{n-1}\Delta N_n,$$

where N is the martingale (under the Null Hypothesis):

$$N_n = \sum_{k=1}^{n} \{r(X_{k-1}, X_k) - 1\}.$$

Equation (22.2) gives the martingale ρ as the Doléans exponential

$$\rho = \mathscr{E}(N). \qquad \square$$

The problem which we are now going to consider for the continuous-parameter case is more general in kind.

Let X be an R-process with values in a finite set I.

Null Hypothesis: X is a time-homogeneous Markov chain with known Q-matrix Q and known initial law μ. The law $\mathbf{P}$ of X under the Null Hypothesis is therefore characterized by the fact that $\mathbf{P} \circ X_0^{-1} = \mu$, and for each function f on I,

$$M_t^f \equiv f(X_t) - f(X_0) - \int_{(0,t]} (Qf)(X_s)\, ds$$

is a $\mathbf{P}$ local martingale.

For our *Alternative Hypothesis*, we take some arbitrary (but fixed) law $\tilde{\mathbf{P}}$ equivalent to $\mathbf{P}$ with the same initial law: $\tilde{\mathbf{P}} \circ X_0^{-1} = \mu$. (*We make no Markovian assumption about* $\tilde{\mathbf{P}}$.)

We let ρ be the likelihood-ratio martingale for this situation, as described in Theorem 17.1. We wish to regard ρ as an exponential martingale of Doléans type, so we let N be the local martingale defined by

$$dN_t = (\rho_{t-})^{-1} d\rho_t, \quad \rho_t = \mathscr{E}(N).$$

(Recall from Theorem 17.1(ii) that ρ_{t-} cannot be zero, and that, by Lemma 10.2, $\{\rho_{t-} : t > 0\} \in \mathrm{lb}\mathscr{P}$.) We know from Theorem 21.15 that $N = N^g$ for some g, so that

$$N_t^g(\omega) = \sum_{0 < s \leq t} g(s, X_{s-}, X_s, \omega)$$

$$- \int_{(0,t]} \sum_b Q(X_{s-}, b) g(s, X_{s-}, b, \omega)\, ds.$$

(*22.4*) THEOREM (Girsanov formula for chains). *For each function f on I,*

$$\tilde{M}_t^f \equiv f(X_t) - f(X_0) - \int_{(0,t]} (\tilde{Q}f)(s, X_s)\, ds$$

is a $\tilde{\mathbf{P}}$ local martingale, where

$$(\tilde{Q}f)(t, i, \omega) = \sum_j Q(t, i, j, \omega) \{f(j) - f(i)\},$$

where, for $i \neq j$,

(22.5) $$\tilde{Q}(t, i, j, \omega) = Q(i, j)\{1 + g(t, i, j, \omega)\}.$$

(*22.6*) *Exercise.* Try out your Itô calculus! Prove this theorem by verifying that

(22.7) $$d(\rho_t \tilde{M}_t^f) = \rho_{t-} dM_t^f + \tilde{M}_{t-}^f d\rho_t + \rho_{t-} dN_t^h,$$

where
$$h(t, i, j, \omega) = g(t, i, j, \omega)\{f(j) - f(i)\}.$$

Thus $\rho \tilde{M}^f$ is a **P** local martingale, and, by Theorem 17.1, $\tilde{M}^f$ is a $\tilde{\mathbf{P}}$ local martingale. The proof of (22.7) is given at (22.12) below.

(*22.8*) *Example.* Suppose that $\tilde{Q}$ is a Q-matrix such that $\tilde{Q}(i, j) > 0$ if and only if $Q(i, j) > 0$, and that $\tilde{\mathbf{P}}$ is the law of the Markov chain on I with initial law μ and Q-matrix $\tilde{Q}$.

[*Technical note.* The law $\tilde{\mathbf{P}}$ will not be equivalent to **P** on the whole σ-algebra $\mathscr{F}_\infty^\circ$, but it will be equivalent to **P** on $\mathscr{F}_u^\circ$ for every u. Thus you should pretend that we are now working with the finite time-parameter set $[0, u]$ for some u; and then everything is fine. We shall consider this kind of thing at much greater length in §38.]

We see from (22.5) that we need to take

$$g(t, i, j, \omega) = g(i, j) = \begin{cases} \tilde{Q}(i, j)/Q(i, j) - 1 & \text{if } i \neq j, \\ 0 & \text{if } i = j. \end{cases}$$

Thus

(22.9) $$\rho_t = \mathscr{E}(N) = \tilde{\zeta}_t / \zeta_t,$$

where, with $q(a) = \sum_{b \neq a} Q(a, b)$, and $\tilde{q}(a) = \sum_{b \neq a} \tilde{Q}(a, b)$,

(22.10) $$\zeta_t = \exp\left\{ -\int_{(0, t]} q(X_s)\, ds \right\} \prod_{\substack{0 < s \leq t \\ X_{s-} \neq X_s}} Q(X_{s-}, X_s)$$

and $\tilde{\zeta}_t$ is defined analogously. Of course, ζ_t is the 'likelihood' of the observed outcome under **P**. (*Exercise.* Relative to what measure?!)

Proof of (22.9). Let U_{ab} continue to denote the martingale

$$U_{ab}(t) = J_{ab}(t) - q_{ab} L_a(t) \quad (a \neq b).$$

Then, as we know from (21.10), or directly from the Doléans formula,

$$\mathscr{E}(\alpha U_{ab})(t) = \exp(-\alpha q_{ab} L_a(t))(1 + \alpha)^{J_{ab}(t)}.$$

Now

$$N = \sum_{a \neq b}\sum g_{ab} U_{ab}(t).$$

For $a\neq b$, $c\neq d$, and $(a,b)\neq(c,d)$, the processes U_{ab} and U_{cd} have no common jumps. Yor's exponential formula (19.5) now yields:

$$\mathscr{E}(N)_t=\prod_{a\neq b}\prod\mathscr{E}(g_{ab}U_{ab})_t=\zeta_t/\zeta_t. \qquad \square$$

(*22.11*) *Example* (Generalized Feynman–Kac formula for chains). This is just a reformulation of the above. Let X be a Markov chain on our finite set I with Q-matrix Q. Fix $t>0$. We want the joint law of the variables:

$$\{L_k(t)\colon k\in I;\ J_{ab}(t)\colon a,b\in I,\ a\neq b;\ X(t)\},$$

or, equivalently, the joint moment-generating function:

$$\hat{P}_t(i,j)\equiv\mathbf{E}^i\left\{\exp\left(-\sum\gamma_k L_k(t)\right)\prod_{a\neq b}\prod\theta_{ab}^{J_{ab}(t)};\ X(t)=j\right\},$$

where $\gamma_k>0$, $0<\theta_{ab}<1$. Then:

$$\hat{P}_t=\exp(t\hat{Q}),$$

where

$$\hat{q}_{ab}=\theta_{ab}q_{ab}\quad(a\neq b)$$

$$\hat{q}_a=q_a+\gamma_a.$$

Intuitively, we obtain a chain $\hat{X}$ by killing X at rate γ_a while X is in a, and killing X with probability $(1-\theta_{ab})$ each time it jumps from a to b.

Proof. Instead of trying to derive this result from those of Example 22.8 (you can do this as an exercise), we indicate a direct argument which will give you more practice in Itô's formula. You should therefore check through the following.

We are going to simplify things by fixing a pair (a,b) of distinct states, taking $\gamma_k=0$ if $k\neq a$, and $\theta_{cd}=1$ unless $(c,d)=(a,b)$. One can show by Yor's formula (as in Example 22.8) that this case may be extended to the general one.

For simplicity, write

$$q=q_{ab},\quad L=L_a,\quad I=I_a,\quad J=J_{ab},\quad U=U_{ab},\quad \theta=\theta_{ab},\quad \alpha=\theta-1.$$

Write

$$\exp[-\gamma L(t)]\theta^{J(t)}f(X_t)=W_tN_tF_t,$$

where

$$N_t=\mathscr{E}(\alpha U)_t=\exp[-\alpha qL(t)]\theta^{J(t)},$$

$$W_t=\exp[-(\gamma-\alpha q)L(t)],\quad F_t=f(X_t).$$

We have

$$dN_t=\alpha N_{t-}\,dU,$$

$$dW_t=-(\gamma-\alpha q)W_tI(X_t)\,dt,$$

$$dF_t\doteq(Qf)(X_t)\,dt,$$

the $\doteq$ signifying that the two sides differ by the derivative of a local martingale. Thus

$$d[N, F] = \alpha N_-(f(b) - f(a))dJ$$
$$\doteq \alpha N_-(f(b) - f(a))qI(X_t)\,dt,$$

and

$$d(NF) \doteq N_-(Qf)(X_t)\,dt + \alpha q N_-(f(b) - f(a))I(X_t)\,dt.$$

Hence, recalling that W is continuous, etc.,

$$d(WNF) \doteq WN(\hat{Q}f)(X_t)\,dt,$$

where

$$\hat{Q}(i, j) = Q(i, j) + \delta_a(i)\alpha Q(a, b)\delta_b(j) - \gamma\delta_a(i)\delta_a(j).$$

It is now easy to argue that

$$W_t N_t F_t - W_0 N_0 F_0 - \int_0^t W_s N_s(\hat{Q}f)(X_s)\,ds$$

is a martingale, and taking expectations, we have

$$\hat{\mathbf{P}}_t f - f_0 = \int_0^t \hat{\mathbf{P}}_s \hat{Q} f\,ds.$$

You can prove directly that $\{\hat{\mathbf{P}}_t\}$ is a semigroup—just use the Markov property and the additive-functional nature of L and J. Then the argument is finished. □

(*22.12*) *Solution of Exercise 22.6.* We have

$$[\rho, \tilde{M}^f]_t = \sum_{0<s\leq t} \Delta\rho_s \Delta\tilde{M}^f_s = \sum_{0<s\leq t} \rho_{s-}\Delta N_s \Delta\tilde{M}^f_s$$
$$= \sum_{0<s\leq t} \rho_{s-} g(s, X_{s-}, X_s, \omega)\{f(X_s) - f(X_{s-})\}.$$

You can see that the definition of $\tilde{Q}$ was therefore chosen so that

$$\int_{(0,t]} \rho_{s-}(d\tilde{M}^f_s - dM^f_s) + [\rho, \tilde{M}^f]_t = \int_{(0,t]} \rho_{s-}\,dN^h_s.$$

Finally, therefore,

$$\rho_t \tilde{M}^f_t - \rho_0 \tilde{M}^f_0 = \int_{(0,t]} \{\rho_{s-}\,d\tilde{M}^f_s + \tilde{M}^f_{s-}\,d\rho_s + d[\rho, \tilde{M}^f]_s\}$$
$$= \int_{(0,t]} \{\rho_{s-}\,dM^f_s + \rho_{s-}\,dN^h_s + \tilde{M}^f_{s-}\,d\rho_s\}.$$

4. STOCHASTIC INTEGRALS: THE L^2 THEORY

23. Orientation. We promised to present in this chapter a self-contained theory of stochastic integrals with respect to continuous semimartingales. We do so.

However—and this should not surprise you too greatly after our earlier remarks—the theory of the stochastic integral relative to a local martingale M which is not of finite variation rests on two things. The first is the existence of a *quadratic-variation* process $[M]$, which is a suitable limit:

$$[M](t) = \lim_n \sum_i \{M(t_i^n) - M(t_{i-1}^n)\}^2, \tag{23.1}$$

where $t_i^n = t \wedge (i2^{-n})$. The second is the *integration-by-parts formula*:

$$M^2 - [M] = 2\int M_- dM. \tag{23.2}$$

Indeed, we must go through four steps.

Step 1. Establish the existence of $[M]$.
Step 2. Construct the stochastic integral under suitable boundedness conditions. We know about $H{\bullet}M$ for bounded elementary H from §6. The approximation procedures which allow us to extend the integral to a wider class of H rely on the existence of $[M]$.
Step 3. Apply localization procedures.
Step 4. Establish (23.2) and thence derive Itô's formula.

Now, Steps 1 and 4 are greatly simplified by the assumption that M is continuous; which is why we can present that case of the theory at this early stage. However, for Steps 2 and 3, this 'continuity' assumption is a complete irrelevance, and therefore certainly does not result in any simplification.

We believe that there are considerable advantages to be derived from the strategy which we now adopt. First, we give a clear statement of a fundamental theorem on the existence of $[M]$. Then, assuming this result, we describe the so-called L^2 theory, the most elegant part of the subject. In Part 5, we prove the existence of $[M]$ when M is continuous. This proof is inextricably linked to the formula (23.2). We then round off the theory of stochastic integrals relative to continuous semimartingales. Part 6 presents applications.

One advantage of this approach is that we avoid having to repeat the L^2 theory in Chapter VI. We believe that it also clarifies for this present chapter what really matters.

24. Stable spaces of $\mathcal{M}_0^2$, c$\mathcal{M}_0^2$, d$\mathcal{M}_0^2$. Let $\mathcal{M}^2$ be the space of martingales bounded in L^2. Thus, an element M of $\mathcal{M}^2$ is an R-martingale such that

$$\sup_t \mathbf{E}(M_t^2) < \infty.$$

The space $\mathcal{M}^2$ is in one-to-one correspondence with the Hilbert space $L^2(\mathcal{F}_\infty)$, the bijection being given by:

(24.1(i)) $$\{M_t: t \geqslant 0\} \in \mathcal{M}^2 \mapsto M_\infty \equiv \lim M_t \in L^2(\mathcal{F}_\infty),$$

(24.1(ii)) $$M_\infty \in L^2(\mathcal{F}_\infty) \mapsto M_t \equiv \mathbf{E}(M_\infty \mid \mathcal{F}_t) \in \mathcal{M}^2.$$

We shall often identify an element M of $\mathcal{M}^2$ simply by specifying its limiting value M_∞ in $L^2(\mathcal{F}_\infty)$. The space $\mathcal{M}^2$ inherits a Hilbert space structure from that of $L^2(\mathcal{F}_\infty)$. Note that, for M in $\mathcal{M}^2$,

$$\sup_t \mathbf{E}(M_t^2) = \|M_\infty\|^2 = \mathbf{E}(M_\infty^2).$$

Recall Doob's L^2 inequality. For this, we need the standard notations

(24.2) $$M_t^* = \sup_{s \leqslant t} |M_s|, \quad M_\infty^* \equiv \sup_{s \geqslant 0} |M_s|.$$

(*24.3*) LEMMA (Doob's L^2 inequality, II.70.2).

$$\mathbf{E}\{(M_\infty^*)^2\} \leqslant 4\mathbf{E}(M_\infty^2).$$

(*24.4*) *Convergence in $\mathcal{M}^2$.* Suppose that $\{M^{(n)}\}$ is a Cauchy sequence in $\mathcal{M}^2$, equivalently, $\{M_\infty^{(n)}\}$ is Cauchy in $L^2(\mathcal{F}_\infty)$. Then $M_\infty^{(n)} \to M_\infty$ in $L^2(\mathcal{F}_\infty)$ for some M_∞, and if $M_t \equiv \mathbf{E}(M_\infty | \mathcal{F}_t)$, then

$$M^{(n)} \to M \text{ uniformly in } L^2$$

in that

$$\sup_t |M^{(n)}(t) - M(t)| \to 0 \text{ in } L^2.$$

In this case, there is a subsequence $(n(k))$ such that, almost surely,

(24.5) $$M^{n(k)} \to M \text{ uniformly over } [0, \infty].$$

In particular, we have the following lemma:

(*24.6*) LEMMA. *The space $c\mathcal{M}^2$ of continuous martingales in $\mathcal{M}^2$ is closed in the topology of $\mathcal{M}^2$.*

Proof. A uniform limit of a sequence of continuous functions is continuous. □

The space $\mathcal{M}_0^2$ of elements of $\mathcal{M}^2$ null at 0 is obviously also a closed subspace, as is $c\mathcal{M}_0^2$.

It is convenient to collect here some important further definitions, even though it will be some time before we shall use them.

(*24.7*) DEFINITION ('Weak' and strong orthogonality in $\mathscr{M}_0^2$.) *Let M and N be elements of* $\mathscr{M}_0^2$. *Then*:

(i) *M and N are called* weakly orthogonal *if M and N are orthogonal in the usual Hilbert space sense*: $\mathbf{E}(M_\infty N_\infty)=0$.

(ii) *M and N are called* strongly orthogonal *if for every (finite or infinite) stopping time T*, $\mathbf{E}(M_T N_T)=0$.

(*24.8*) LEMMA. *M and N are strongly orthogonal if and only if MN is a UI martingale.*

Proof. Because of Lemma II.77.6, this is trivial. □

(*24.9*) DEFINITION (Stable subspaces of $\mathscr{M}_0^2$.) *A subspace* $\mathscr{N}_0$ *of* $\mathscr{M}_0^2$ *is called* stable *if*:

(i) $\mathscr{N}_0$ *is a closed subspace*;

(ii) $\mathscr{N}_0$ *is stable under stopping, so that if* $N\in\mathscr{N}_0$, *then* N^T *is in* $\mathscr{N}_0$ *for every stopping time T.*

Example. $c\mathscr{M}_0^2$ is a stable subspace of $\mathscr{M}_0^2$.

We are going to decompose $\mathscr{M}_0^2$ as the ('strong') direct sum of its continuous part $c\mathscr{M}_0^2$ and its purely discontinuous part $d\mathscr{M}_0^2$. This will be a consequence of the following theorem.

(*24.10*) THEOREM. *Let* $\mathscr{N}_0$ *be a stable subspace of* $\mathscr{M}_0^2$. *Let* $\mathscr{N}_0^\perp$ *be the set of elements Z of* $\mathscr{M}_0^2$ *which are weakly orthogonal to every element N of* $\mathscr{N}_0$. *Then* $\mathscr{N}_0^\perp$ *is a stable subspace of* $\mathscr{M}_0^2$, *and every Z in* $\mathscr{N}_0^\perp$ *is strongly orthogonal to every element N in* $\mathscr{N}_0$. *Finally, every M in* $\mathscr{M}_0^2$ *decomposes uniquely as*

$$M=N+Z, \tag{24.11}$$

where $N\in\mathscr{N}_0$ *and* $Z\in\mathscr{N}_0^\perp$.

Proof. That $\mathscr{N}_0^\perp$ is closed is trivial.

Let $Z\in\mathscr{N}_0^\perp$. We must show that, for a stopping time T, we have $Z^T\in\mathscr{N}_0^\perp$; in other words, that for $N\in\mathscr{N}_0$, $\mathbf{E}(Z_T N_\infty)=0$. Now, since $N^T\in\mathscr{N}_0$, we have $\mathbf{E}(Z_\infty N_T)=0$. But

$$\begin{aligned}\mathbf{E}(Z_\infty N_T)&=\mathbf{E}\mathbf{E}[Z_\infty N_T|\mathscr{F}_T]=\mathbf{E}(N_T\mathbf{E}[Z_\infty|\mathscr{F}_T])\\&=\mathbf{E}(N_T Z_T)=\ \dots\ =\mathbf{E}(N_\infty Z_T).\end{aligned}$$

We have proved both that $\mathscr{N}_0^\perp$ is stable under stopping and that each Z in $\mathscr{N}_0^\perp$ is strongly orthogonal to each N in $\mathscr{N}_0$.

The decomposition (24.11) is trivial. □

(*24.12*) DEFINITION (purely discontinuous martingale, $d\mathscr{M}_0^2$). *We define*:

$$d\mathscr{M}_0^2=(c\mathscr{M}_0^2)^\perp.$$

We say that an element of $\mathrm{d}\mathcal{M}_0^2$ *is* purely discontinuous.

Each M in $\mathcal{M}_0^2$ *decomposes uniquely as:*

$$M = M^c + M^d, \quad M^c \in \mathrm{c}\mathcal{M}_0^2, \quad M^d \in \mathrm{d}\mathcal{M}_0^2;$$

and for U in $\mathrm{c}\mathcal{M}_0^2$ *and V in* $\mathrm{d}\mathcal{M}_0^2$, UV *is a UI martingale.*

(*24.13*) **Exercise.** Prove that for M in $\mathcal{M}_0^2$, and T a stopping time,

$$(M^c)^T = (M^T)^c.$$

25. Elementary stochastic integrals relative to M in $\mathcal{M}_0^2$. Let $M \in \mathcal{M}_0^2$. Let $H \in \mathrm{b}\mathcal{E}$ (recall §6), so that H has the form

(25.1) $$H = \sum_{i=1}^{n} Z_{i-1}(T_{i-1}, T_i],$$

where (T_k) is a finite sequence of stopping times with $0 \leqslant T_0 \leqslant T_1 \leqslant \ldots \leqslant T_n$, and $Z_k \in \mathrm{b}\mathcal{F}(T_k)$. Then we know from (6.7) that the obvious definition

(25.2) $$(H \bullet M)(t) \equiv \sum Z_{i-1}\{M(T_i \wedge t) - M(T_{i-1} \wedge t)\}$$

gives a uniformly integrable martingale $H \bullet M$. We can now show that $H \bullet M \in \mathcal{M}^2$. Indeed, if we write

$$\Delta_i \equiv M(T_i) - M(T_{i-1}),$$

then

$$\mathbf{E}\{(H \bullet M)_\infty^2\} = \mathbf{E}\left\{\left(\sum Z_{i-1}\Delta_i\right)^2\right\} = \mathbf{E}\sum Z_{i-1}^2 \Delta_i^2,$$

because, for $i<j$,

$$\mathbf{E}\{Z_{i-1}Z_{j-1}\Delta_i\Delta_j \mid \mathcal{F}(T_{j-1})\} = Z_{i-1}Z_{j-1}\Delta_i \mathbf{E}\{\Delta_j \mid \mathcal{F}(T_{j-1})\} = 0.$$

Let us summarize.

(*25.3*) LEMMA. *Let H be a bounded elementary process as at* (*25.1*). *Let* $M \in \mathcal{M}_0^2$. *Then* $H \bullet M$ *is in* $\mathcal{M}_0^2$, *and*

(25.4) $$\mathbf{E}\{(H \bullet M)_\infty^2\} = \mathbf{E}\sum Z_{i-1}^2\{M(T_i) - M(T_{i-1})\}^2.$$

Our essential need is to express the right-hand side of (25.4) in a more tractable form. This is exactly what the introduction of $[M]$ allows us to do.

Before moving on to the next section, we record the following extension of Lemma 25.3, the proof of which is left as an exercise.

(*25.5*) LEMMA. *Let H be a uniformly bounded simple process of the form:*

$$H = \sum_{i \geqslant 1} Z_{i-1}(T_{i-1}, T_i],$$

where (T_k) *is an infinite sequence of stopping times with* $0 \leqslant T_0 \leqslant T_1 \leqslant \ldots$, $Z_k \in \mathrm{b}\mathscr{F}(T_k)$, *and* $\sup|Z_k| \leqslant c$. *Define* $H \bullet M$ *by the obvious analogue of (25.2). Then* $H \bullet M$ *is in* $\mathscr{M}_0^2$, *the obvious analogue of (25.4) holds, and hence*:

$$\mathbf{E}\{(H \bullet M)_\infty^2\} \leqslant c^2 \mathbf{E}(M_\infty^2).$$

26. The processes $[M]$ **and** $[M, N]$**.** Here is the fundamental theorem which we assume for now.

For its statement, we adopt a notation used previously: for an R-function f on $[0, \infty)$, Δf is the function on $(0,\infty)$ with

$$(\Delta f)(t) \equiv f(t) - f(t-).$$

(*26*) THEOREM (Meyer). *Let* $M \in \mathscr{M}_0^2$. *Then there exists a unique increasing process* $[M]$ *null at* 0 *such that*:

(i) $M^2 - [M]$ *is a uniformly integrable martingale*;

(ii) $\Delta[M] = (\Delta M)^2$ *on* $(0,\infty)$.

This is the first draft of a better theorem (see §VI.36.6), but it is exactly what we need for now. The intuitive significance of $[M]$ is described at (23.1).

For M and N in $\mathscr{M}_0^2$, *introduce the process* $[M, N]$ *by polarization*:

$$[M, N] \equiv \tfrac{1}{4}([M+N] - [M-N]). \tag{26.1}$$

Then we have the following continuation of Theorem 26.

(*26.2*) THEOREM (continuation of Theorem 26, Kunita and Watanabe). *The process* $[M, N]$ *is the unique FV process null at* 0 *such that*:

(i) $MN - [M,N]$ *is a uniformly integrable martingale*;

(ii) $\Delta[M, N] = (\Delta M)(\Delta N)$ *on* $(0, \infty)$.

Note. Of course, $[M] = [M, M]$.

(*26.3*) DEFINITION (quadratic variation, quadratic covariation). *Let M and N be elements of* $\mathscr{M}_0^2$. *Then* $[M]$ *is called the* quadratic-variation process *of M, and* $[M, N]$ *is called the* quadratic-covariation process *of M and N.*

You can check that $[M, N]$ is a bilinear form. You can also check that for a stopping time T,

$$[M^T, N^T] = [M, N]^T.$$

Hence, we may localize Theorem 26.2 as follows:

(*26.4*) PRE-THEOREM. Let $M, N \in \mathscr{M}_{0,\mathrm{loc}}^2$. *Then there exists a unique FV process* $[M, N]$ *null at* 0 *such that*:

(i) $MN - [M, N]$ *is a local martingale*;

(ii) $\Delta[M, N] = (\Delta M)(\Delta N)$ *on* $(0, \infty)$.

However, this result is true for *any* local martingales M and N, and not merely for local martingales M and N which are locally bounded in L^2. This explains why we referred to Theorem 26 as a 'first draft' and why (26.4) is only a 'pre-theorem'.

27. Constructing stochastic integrals in L^2. Let $M \in \mathscr{M}_0^2$. Lemma (26.3) shows that if $H = \sum Z_{i-1}(T_{i-1}, T_i]$ is an element of $\mathrm{b}\mathscr{E}$, and we define $H \bullet M$ in the obvious way:

(27.1) $$(H \bullet M)_t \equiv \sum Z_{i-1}\{M(T_i \wedge t) - M(T_{i-1} \wedge t)\},$$

then $H \bullet M$ is again an L^2-bounded martingale; indeed,

(27.2) $$\mathbf{E}(H \bullet M)_\infty^2 = \mathbf{E}\sum Z_{i-1}^2(M(T_i) - M(T_{i-1}))^2.$$

Now, since $M_t^2 - [M]_t$ is a uniformly integrable martingale, then for any stopping times S and T with $S \leqslant T$,

$$\begin{aligned}\mathbf{E}[(M_T - M_S)^2 | \mathscr{F}_S] &= \mathbf{E}[M_T^2 - M_S^2 | \mathscr{F}_S] \\ &= \mathbf{E}([M]_T - [M]_S | \mathscr{F}_S).\end{aligned}$$

So (27.2) yields the crucial identity

(27.3) $$\begin{aligned}\mathbf{E}(H \bullet M)_\infty^2 &= \mathbf{E}\sum Z_{i-1}^2([M]_{T_i} - [M]_{T_{i-1}}) \\ &= \mathbf{E}\int_0^\infty H_s^2 d[M]_s.\end{aligned}$$

For any previsible process, we now define the norm

$$\|H\|_M \equiv \left(\mathbf{E}\int_0^\infty H_s^2 d[M]_s\right)^{1/2}.$$

Then the space

$$L^2(M) \equiv \{\text{previsible processes } H \text{ such that } \|H\|_M < \infty\}$$

(strictly speaking, this is of course a space of certain equivalence classes) is a Hilbert space, containing $\mathrm{b}\mathscr{E}$. Lemma 13.8 shows that all evanescent processes are previsible; they clearly belong to the same 'null' equivalence class. The stochastic integral $H \bullet M$ is already defined for $H \in \mathrm{b}\mathscr{E}$, *and* (27.3) *shows that the map*

$$I: L^2(M) \cap \mathrm{b}\mathscr{E} \to L^2(\mathscr{F}_\infty)$$

defined by

$$I(H) = (H \bullet M)_\infty$$

is an isometry. Thus I extends uniquely to an isometry on the closure $\overline{\mathscr{U}}$ in $L^2(M)$ of the subspace $\mathscr{U} = L^2(M) \cap \mathrm{b}\mathscr{E}$. But Lemma 6.5 implies that $\overline{\mathscr{U}}$ contains all bounded previsible processes, and so $\overline{\mathscr{U}} = L^2(M)$.

We summarize the construction of the stochastic integral in the following theorem. (It is time to emphasize once again the basic bijection (24.1) between $\mathscr{M}^2$

and $L^2(\mathscr{F}_\infty)$, so that in the theorem, the element $H \bullet M$ of $\mathscr{M}_0^2$ is identified by its limiting value.)

(*27.4*) DEFINITION, THEOREM (Kunita and Watanabe). *Let $M \in \mathscr{M}_0^2$ and $H \in L^2(M)$. Then the (Itô) integral $H \bullet M$ of H with respect to M is the image of H under the extension of the isometry I to the whole of $L^2(M)$: in particular:*

$$\mathbf{E}(H \bullet M)_\infty^2 = \mathbf{E}\int_0^\infty H_s^2\, d[M]_s. \tag{27.5}$$

Since $Z \bullet M$ is clearly evanescent if Z is evanescent, we may (and shall) think of $H \bullet M$ as a (representative) process rather than as an equivalence class.

The next result provides the key to extending the stochastic integral by localization.

(*27.6*) THEOREM. *Fix $M \in \mathscr{M}_0^2$. Then $\mathrm{b}\mathscr{P} \subset L^2(M)$, and for any $H \in L^2(M)$ and for any stopping time T:*

(i) $(H \bullet M)^T = H(0, T] \bullet M = H \bullet M^T$.

(ii) $(H \bullet M)_t^2 - \int_0^t H_s^2 d[M]_s$ *is a uniformly integrable martingale.*

(iii) *If $H, K \in \mathrm{b}\mathscr{P}$, then $H \bullet (K \bullet M) = (HK) \bullet M$.*

(iv) $\Delta(H \bullet M) = H \Delta M$.

(v) *If M is continuous, then $H \bullet M$ is also continuous.*

Proof. (i) The linear maps $f_1: L^2(\mathscr{F}_\infty) \to L^2(\mathscr{F}_\infty)$, $f_2: L^2(M) \to L^2(M)$, and $f_3: L^2(M) \to L^2(M^T)$ defined by:

$$f_1(Y) \equiv \mathbf{E}(Y \mid \mathscr{F}_T), \quad f_2(H) \equiv H(0, T], \quad f_3(H) \equiv H,$$

are all contractions. Hence, if $I^{(T)}$ denotes the stochastic-integral map $I^{(T)}(H) \equiv (H \bullet M^T)_\infty$ associated with M^T, then the maps

$$f_1 \circ I, \quad I \circ f_2, \quad I^{(T)} \circ f_3$$

are continuous linear maps on $L^2(M)$; and since these maps agree on the dense subspace $\mathrm{b}\mathscr{S}$ (as is easily verified directly), they agree on $L^2(M)$. It only remains to notice that, since $H \bullet M$ is a uniformly integrable martingale, $(f_1 \circ I)(H) = (H \bullet M)_\infty^T$.

(ii) Let:

$$N_t \equiv (H \bullet M)_t^2 - \int_0^t H_s^2\, d[M]_s.$$

Then N is certainly uniformly integrable; indeed, by Doob's L^2 inequality, $(H \bullet M)_t^2$ is dominated by the integrable random variable $(H \bullet M)_\infty^{*2}$, and the increasing process $\int H_s^2 d[M]_s$ is integrable. Thus by Lemma II.77.6, it is sufficient to prove that for any stopping time T,

$$0=\mathbf{E}N_T=\mathbf{E}(H\bullet M)_T^2-\mathbf{E}\int_0^T H_s^2\,d[M]_s.$$

But $(H\bullet M)_T=((H\bullet M)^T)_\infty$, and $\int_0^T H_s^2\,d[M]_s=\int_0^\infty (H(0,T])_s^2\,d[M]_s$; so the result follows from the property (i) and (27.5).

(iii) The identity $H\bullet(K\bullet M)=(HK)\bullet M$ is immediate for $H, K\in b\mathscr{E}$. For $K\in b\mathscr{P}$, find a uniformly bounded sequence (K^n) of elements of $b\mathscr{E}$ such that $K^n\to K$ in $L^2(M)$. Then for $H\in b\mathscr{E}$,

$$\begin{aligned}\mathbf{E}(\{H\bullet(K\bullet M)-H\bullet(K^n\bullet M)\}_\infty^2)&=\mathbf{E}(\{H\bullet((K-K^n)\bullet M)\}_\infty^2)\\&=\mathbf{E}\int_0^\infty H_s^2\,d[(K-K^n)\bullet M]_s\\&=\mathbf{E}\int_0^\infty H_s^2(K_s-K_s^n)^2\,d[M]_s,\quad\text{by (ii)},\\&\to 0\quad\text{as}\quad n\to\infty.\end{aligned}$$

Clearly $HK^n\to HK$ in $L^2(M)$, so the identity extends to $H\in b\mathscr{E}$, $K\in b\mathscr{P}$. The extension to $H\in b\mathscr{P}$ is similar but easier.

(iv) Property (iv) is immediate if $H\in b\mathscr{E}$. Next, suppose that $H\in L^2(M)$, and that (H^n) is a sequence in $b\mathscr{E}$ with $H^n\to H$ in $L^2(M)$. Then $H^n\bullet M\to H\bullet M$ in $\mathscr{M}_0^2$. In a suitable subsequence $(n(k))$, and except for ω in some null set, we have both

(27.7) $$\int_0^\infty (H_s^{n(k)}-H_s)^2\,d[M]_s\to 0$$

and (see (24.5))

(27.8) $$H^{n(k)}\bullet M\to H\bullet M \text{ uniformly on } [0,\infty].$$

Now, at any point s at which M is continuous, $H^{n(k)}\bullet M$ is continuous by definition, so that, from (27.8), $H\bullet M$ is continuous at s. If $\Delta M_s\neq 0$, then $\Delta[M]_s>0$, so that, from (27.7), $H_s^{n(k)}\to H_s$, and thus $\Delta(H\bullet M)_s=H\Delta M_s$.

(v) This follows from (iv). □

(*27.9*) LEMMA. *Because of Theorem 27.6 (i), we can now extend the definition of $H\bullet M$ to the case when $H\in lb\mathscr{P}$ and $M\in\mathscr{M}^2_{0,\mathrm{loc}}$, and then $H\bullet M\in\mathscr{M}^2_{0,\mathrm{loc}}$. Moreover, for $H\in lb\mathscr{P}$ and $M\in c\mathscr{M}^2_{0,\mathrm{loc}}$, we have $H\bullet M\in c\mathscr{M}^2_{0,\mathrm{loc}}$.*

(*27.10*) *Agreement of L^2 and FV integrals.* Suppose that $H\in lb\mathscr{P}$, $M\in\mathscr{M}^2_{0,\mathrm{loc}}$, and that M has paths of finite variation. We certainly want $H\bullet M$ to agree with the pathwise Stieltjes integral, and indeed it does. (See § VI.37.9. The proof given there will be accessible to you once you have read the next section IV.28, so you can read it then if you are particularly anxious so to do. We do not actually need to establish the agreement of the two integrals until Chapter VI.)

28. The Kunita–Watanabe inequalities. The definition of the stochastic integral afforded by (27.4) is nowadays preferred, because it provides a good starting point for the stochastic calculus. But, in terms of elegance, it falls far short of the original Kunita–Watanabe definition provided by the following theorem.

(*28.1*) THEOREM (Kunita and Watanabe). *Let* $M \in \mathcal{M}_0^2$ *and* $H \in L^2(M)$. *Then* $(H \bullet M)_\infty$ *is the unique element of* $L^2(\mathcal{F}_\infty)$ *such that for every* $N \in \mathcal{M}_0^2$,

$$\mathbf{E}\{(H \bullet M)_\infty N_\infty\} = \mathbf{E}\{(H \bullet [M, N])_\infty\}. \tag{28.2}$$

Moreover, we have (up to evanescence) the process identity:

$$[H \bullet M, N] = H \bullet [M, N] \equiv \int H_s d[M, N]_s. \tag{28.3}$$

Do look at that astounding paper, Kunita and S. Watanabe [1], in which you will find this and so much else which spurred the later work.

The proof rests on the following inequalities.

(*28.4*) THEOREM (Kunita and Watanabe). *Let* $M, N \in \mathcal{M}_0^2$, *and suppose that* H *and* K *are measurable processes. Then:*

$$\left(\int_0^\infty |H_s K_s| . |d[M, N]_s|\right)^2 \leqslant \int_0^\infty H_s^2 \, d[M]_s . \int_0^\infty K_s^2 \, d[N]_s \quad \text{a.s.} \tag{28.5}$$

Hence:

$$\mathbf{E}\left(\int_0^\infty |H_s K_s| . |d[M, N]_s|\right) \leqslant \left(\mathbf{E}\int_0^\infty H_s^2 \, d[M]_s\right)^{1/2} \left(\mathbf{E}\int_0^\infty K_s^2 \, d[N]_s\right)^{1/2}. \tag{28.6}$$

Remarks. Results (28.5) and (28.6) are called the *Kunita–Watanabe inequalities.* The second follows from the first by an application of the Schwarz inequality. The first, and more general, inequality is in fact a later discovery due to Courrège.

Proof. To begin with, we can find a measurable process σ with values in $\{1, -1\}$ such that $|d[M, N]_t| = \sigma_t d[M, N]_t$ (see the amplification below), so by replacing H_s by $H_s \sigma_s \operatorname{sgn}(H_s K_s)$, it is enough to prove

$$\left(\int H_s K_s \, d[M, N]_s\right)^2 \leqslant \int H_s^2 \, d[M]_s . \int K_s^2 \, d[N]_s, \quad \text{a.s.} \tag{28.7}$$

By monotone-class arguments, it is enough to prove for H and K of the form

$$H \equiv \sum U_i (t_i, t_{i+1}], \quad K \equiv \sum V_i (t_i, t_{i+1}],$$

where $0=t_0<t_1<\ldots<t_{n+1}$ are reals, and U_i and V_i are bounded random variables. You can show by the Schwarz inequality that (28.7) holds for such H and K if

$$\Delta_i[M,N]\leqslant\Delta_i[M]^{1/2}\cdot\Delta_i[N]^{1/2},\quad \text{a.s.,} \tag{28.8}$$

where

$$\Delta_i[M,N]\equiv[M,N]_{t_{i+1}}-[M,N]_{t_i}.$$

But for each $\lambda\in\mathbb{Q}$, $[M+\lambda N]$ is an increasing process, so

$$\Delta_i[M+\lambda N]=\Delta_i[M]+2\lambda\Delta_i[M,N]+\lambda^2\Delta_i[N]\geqslant 0,\quad \text{a.s.;}$$

whence (28.8), from which (28.7) follows. □

Amplification. An IV_0 process A induces a measure on $((0,\infty)\times\Omega,\ \mathscr{B}(0,\infty)\times\mathscr{F})$ by the recipe

$$F\mapsto \mathbf{E}\int_0^\infty I_F(s,\omega)\,dA_s(\omega).$$

Clearly the measures induced by $[M,N]$ and $\int_0^\bullet |d[M,N]_s|$ are absolutely continuous with respect to the measure induced by $[M+N]+[M-N]$, and σ is constructed from the densities in the obvious way.

Proof of equality (28.2). Fix M and N. We know that $H\mapsto H\bullet M$ is an isometry of $L^2(M)$ to $L^2(\mathscr{F}_\infty)$, so that the left-hand side of (28.2) is a continuous function of H in $L^2(M)$. On taking $K=1$ in (28.6), we see that

$$|\mathbf{E}(H\bullet[M,N])_\infty|\leqslant\|N\|_2\|H\|_M,$$

so that the right-hand side of (28.2) is also a continuous function of H in $L^2(M)$. But we can prove (28.2) for H in $\mathrm{b}\mathscr{E}$ by an argument similar to that used to prove (27.3). Since $\mathrm{b}\mathscr{E}$ is dense in $L^2(M)$, it follows that (28.2) is true for all H in $L^2(M)$. □

It is standard Hilbert-space theory that the element $(H\bullet M)_\infty$ of $L^2(\mathscr{F}_\infty)$ is determined by the values of the left-hand side of (28.2) for all N in $L^2(\mathscr{F}_\infty)$.

Proof of identity (28.3). In the same way that we deduced (27.6(ii)) from (27.6(i)), we can show by stopping that *$(H\bullet M)N-H\bullet[M,N]$ is a uniformly integrable martingale.* Moreover, from (26.2(ii)) and (27.6(iv)),

$$\begin{aligned}\Delta[(H\bullet M),N]&=\Delta(H\bullet M)\Delta N=H\Delta M\Delta N\\&=H\Delta[M,N]=\Delta(H\bullet[M,N]).\end{aligned}$$

Thus (28.3) holds. □

5. STOCHASTIC INTEGRALS WITH RESPECT TO CONTINUOUS SEMIMARTINGALES

29. Orientation. We now turn to the theory of stochastic integrals with respect to continuous semimartingales, which remains the most important part of the theory because it supports the theory of stochastic integrals for diffusions developed in Chapter V.

After Part 4, we can now unashamedly exploit the continuity assumption without caring about the generalizations which will feature in Chapter VI.

One nice special feature of continuous local martingales null at 0 is that they localize in a very strong sense: to bounded martingales.

(*29.1*) LEMMA. *Let* $M \in c\mathscr{M}_{0,\,\mathrm{loc}}$. *Fix* $n \in \mathbb{N}$. *Define* $T(n) \equiv \inf\{t: |M_t| > n\}$. *Then* $M^{T(n)}$ *is a (bounded) martingale.*

Proof. If (S_k) is a reducing sequence for M, and $s < t$, then

$$M(T_n \wedge S_k \wedge s) = \mathbf{E}\{M(T_n \wedge S_k \wedge t) | \mathscr{F}_s\}.$$

Now let $k \uparrow \infty$, and use the L^1 continuity of conditional expectations to conclude that

$$M(T_n \wedge s) = \mathbf{E}\{M(T_n \wedge t) | \mathscr{F}_s\}. \qquad \square$$

In particular, then,

$$c\mathscr{M}_{0,\,\mathrm{loc}} = c\mathscr{M}^2_{0,\,\mathrm{loc}}, \tag{29.2}$$

so that, once we have proved Theorems 26 and 26.2 for the case when M and N are continuous, we shall have at our disposal the stochastic integral $H \bullet M$ for locally bounded previsible H and continuous local martingales M null at 0.

Another nice feature of continuous local martingales is that if $M \in c\mathscr{M}_{0,\,\mathrm{loc}}$, then it is possible to tell exactly for which previsible integrands H the integral $H \bullet M$ may be defined. See Theorem 30.7 and the remark which follows it.

Having defined the integral with respect to continuous local martingales, we extend the range of integrators to continuous semimartingales in §31. Again the continuity assumption simplifies the theory because a continuous semimartingale has a *unique* decomposition

$$X = X_0 + M + A,$$

where $M \in c\mathscr{M}_{0,\,\mathrm{loc}}$ and $A \in \mathrm{cFV}_0$, the notation cFV_0 being self-explanatory.

The final section of Part 5 is devoted to the proof of Itô's formula for continuous semimartingales which, once established, dominates the remainder of this chapter and all of the next.

30. Quadratic variation for continuous local martingales. Let us restate Theorem 26 for the continuous case.

(*30.1*) THEOREM. *Let $M \in c\mathcal{M}_0^2$, so that M is a continuous martingale null at 0 and bounded in L^2. Then there exists a unique continuous increasing process $[M]$ null at 0 such that $M^2 - [M]$ is a uniformly integrable martingale.*

Proof. By localization, we may, and shall, assume that M is bounded.

For each $n \in \mathbb{N}$, define the stopping times

$$T_0^n \equiv 0, \quad T_{k+1}^n \equiv \inf\{t > T_k^n : |M(t) - M(T_k^n)| > 2^{-n}\} \quad (k \geqslant 0),$$

and abbreviate $t \wedge T_k^n$ to t_k^n. Then, by the summation-by-parts formula,

$$(30.2) \qquad M_t^2 = 2\sum_{k \geqslant 1} M(t_{k-1}^n)\{M(t_k^n) - M(t_{k-1}^n)\} + \sum_{k \geqslant 1}\{M(t_k^n) - M(t_{k-1}^n)\}^2.$$

If H^n denotes the simple process (see Lemma 25.5)

$$H^n \equiv \sum_{k \geqslant 1} M(T_{k-1}^n)(T_{k-1}^n, T_k^n],$$

then the first term on the right of (30.2) is the L^2-bounded martingale $2H^n \bullet M$. Hence, if we put

$$A_t^n \equiv \sum_{k \geqslant 1}\{M(t_k^n) - M(t_{k-1}^n)\}^2,$$

then (30.2) becomes

$$M_t^2 = 2(H^n \bullet M)_t + A_t^n.$$

Notice the following obvious features:

$$(30.3(\text{i})) \qquad \sup_t |H_t^n - H_t^{n+1}| \leqslant 2^{-n-1} \text{ and } \sup_t |H_t^n - M_t| \leqslant 2^{-n};$$

$$(30.3(\text{ii})) \qquad J_n(\omega) \equiv \{T_k^n(\omega);\ k \geqslant 0\} \subset J_{n+1}(\omega);$$

$$(30.3(\text{iii})) \qquad A^n(T_k^n) \leqslant A^n(T_{k+1}^n).$$

By Lemma 25.5,

$$\mathbf{E}\{(H^n \bullet M - H^{n+1} \bullet M)_\infty^2\} \leqslant 4^{-n-1}\mathbf{E}\{M_\infty^2\}.$$

The continuous martingales $H^n \bullet M$ therefore converge uniformly a.s. to a continuous martingale N (say). Consequently, the processes A^n converge uniformly a.s. to some continuous process A, and

$$M_t^2 = 2N_t + A_t.$$

But by (30.3(ii)–(iii)),

$$A(T_k^n) \leqslant A(T_{k+1}^n)$$

for all n and k, so that A is increasing, at least on the closure of $J(\omega) \equiv \bigcup_n J_n(\omega)$. But if I is an open interval in the complement of J, then no T_k^n lies in I, so that M must be constant throughout I, and hence the same must be true of A (look at the definition of A^n).

Thus we have constructed a continuous increasing process A, null at 0, such that $M_t^2 - A_t = 2N_t$, a uniformly integrable martingale. It remains only to prove the uniqueness assertion, which is a consequence of the following exceptionally useful result.

(*30.4*) THEOREM. *Suppose that M is a continuous local martingale null at* 0. *If M has paths of finite variation, then $M_t = 0$ for all t.*

Proof. By stopping M at the stopping time $\inf\{t: V_M(t) > n\}$, where V_M denotes the total variation process of M, we may suppose M is a bounded continuous martingale with paths of bounded variation. But now (compare with Lemma 2.16!) for the processes A^n defined above, we have

$$\begin{aligned} A_t^n &\equiv \sum_{k \geqslant 1} \{M(t_k^n) - M(t_{k-1}^n)\}^2 \\ &\leqslant 2^{-n} \sum |M(t_k^n) - M(t_{k-1}^n)| \\ &\leqslant 2^{-n} V_M(t) \\ &\to 0 \text{ as } n \to \infty. \end{aligned}$$

Thus the limit process A is identically zero, and M^2 is a martingale. But $M_0 = 0$, implying that $M_t = 0$ for all t. □

Proof of Theorem (*30.1*) concluded. If A and A' are continuous increasing processes null at zero such that $M^2 - A$ and $M^2 - A'$ are both uniformly integrable martingales, then $A - A'$ is a continuous finite variation martingale null at zero, and is therefore identically zero by Theorem 30.4. □

(*30.5*) *Remark.* The fact that the paths of continuous local martingales have finite (and generally non-zero) quadratic variation gives them an amazing 'rigidity' of structure for which there is no parallel in the theory of smooth functions. Watch for reflections of this in the Cameron–Martin–Girsanov theorem for Brownian motion, and, for example, in the fact that 2-dimensional Brownian motion started away from 0 may be reconstructed from its angular part.

We can now round off the theory of the construction of stochastic integrals with respect to continuous local martingales.

(*30.6*) LEMMA, DEFINITION (quadratic variation, quadratic covariation). *If M is a continuous local martingale null at zero, then there exists a unique continuous increasing process $[M]$ such that $M^2 - [M]$ is a continuous local martingale null at zero.*

The process $[M]$ is called the quadratic variation process of M, *and for* *continuous local martingales M and N null at zero, the process:*

$$[M, N] \equiv \tfrac{1}{4}\{[M+N]-[M-N]\}$$

is called the quadratic covariation process *of M and N.*

Proof. If $T_n \equiv \inf\{t: |M_t|>n\}$, then we define $[M]$ by

$$[M]_t = [M^{T(n)}]_t \quad \text{for} \quad 0 \leqslant t \leqslant T_n;$$

this definition is consistent, since

$$[M^{T(n)}]^{T(n-1)} = [M^{T(n-1)}],$$

and, since $T_n \uparrow \infty$, $[M]_t$ is defined for all t. The uniqueness follows from Theorem 30.4. □

Now it is an easy matter to define the stochastic integral with respect to a continuous local martingale M.

(*30.7*) THEOREM. *Let M be a continuous local martingale null at* 0, *and let H be a previsible process such that, with probability* 1,

$$\int_0^t H_s^2\, d[M]_s < \infty, \quad \forall t.$$

Then the stochastic integral $H \bullet M$ may be defined, and $H \bullet M$ is a continuous local martingale.

Proof. Let

$$U(n) \equiv \inf\{t: |M_t|>n\},$$

$$V(n) \equiv \inf\left\{t: \int_0^t H_s^2\, d[M]_s > n\right\}.$$

Let $T(n) \equiv U(n) \wedge V(n)$. Then,

$$M^{T(n)} \in c\mathscr{M}_0^2,$$

$$H^{T(n)} \in L^2(M^{T(n)}),$$

so that the integral

$$H^{T(n)} \bullet M^{T(n)}$$

exists and is a continuous L^2-bounded martingale. Hence, we can define the stochastic integral $H \bullet M$ such that

$$(H \bullet M)^{T(n)} = H^{T(n)} \bullet M^{T(n)},$$

and $H \bullet M$ is a continuous local martingale. □

We shall see later (Remark 34.14) that this is the maximal possible extension. The various relevant properties of L^2 stochastic integrals carry over; in particular, we have by extension of (28.3),

(30.8) $$[H \bullet M, K \bullet N] = (HK) \bullet [M, N],$$

an identity of which we shall make very frequent use.

(*30.9*) *Time transformation of stochastic integrals.* At a number of points in the book, we need to consider the effect of time transformations on stochastic integrals. The following result covers the cases we shall study. (If it is necessary, a previsible process H will be extended to $[0, \infty)$ by setting $H_0 = 0$.)

(*30.10*) PROPOSITION. *Let* $(\Omega, \mathcal{F}, \{\mathcal{F}_t\}, \mathbf{P})$ *satisfy the usual conditions, let* φ *be a continuous (adapted) increasing process null at* 0, *and let* τ *be its right-continuous inverse:*

$$\tau_t \equiv \inf\{u: \varphi_u > t\}.$$

Let $\tilde{\mathcal{F}}_t \equiv \mathcal{F}(\tau_t)$. *Then the following results hold:*

(i) *If* T *is an* $\{\mathcal{F}_t\}$*-stopping time, then* φ_T *is an* $\{\tilde{\mathcal{F}}_t\}$*-stopping time.*
(ii) *If* S *is an* $\{\tilde{\mathcal{F}}_t\}$*-stopping time, then* τ_S *is an* $\{\mathcal{F}_t\}$*-stopping time.*
(iii) *If* H *is an* $\{\tilde{\mathcal{F}}_t\}$*-previsible process, then* $H \circ \varphi$ *is an* $\{\mathcal{F}_t\}$*-previsible process.*
(iv) *Suppose that* M *is a continuous* $\{\mathcal{F}_t\}$*-local martingale such that* $\tilde{M}_t \equiv M(\tau_t)$ *is a continuous process and the following integrability condition holds: for some sequence* $S_n \uparrow \infty$ *of* $\{\tilde{\mathcal{F}}_t\}$*-stopping times, each* $M^{\tau(S_n)}$ *is a uniformly integrable martingale. Then* $\tilde{M}_t \equiv M(\tau_t)$ *is an* $\{\tilde{\mathcal{F}}_t\}$*-local martingale reduced by* $\{S_n\}$*, and for any locally bounded* $\{\tilde{\mathcal{F}}_t\}$*-previsible process* H,

(30.11) $$\int_{(0,t]} H_s \, d\tilde{M}_s = \int_{(0,\tau_t]} H(\varphi_s) \, dM_s.$$

Proof. It is easily checked that the filtration $\{\tilde{\mathcal{F}}_t\}$ satisfies the usual conditions.

(i) The continuity of φ implies that

$$\{\varphi_T \leqslant u\} = \{T \leqslant \tau_u\} \in \mathcal{F}(\tau_u) = \tilde{\mathcal{F}}_u \quad \text{for all } u.$$

(ii) $$\{\tau_S < t\} = \{\varphi_t > S\} = \bigcup_{p \in \mathbb{Q}} (\{\varphi_t > p\} \cap \{p \geqslant S\}) = \bigcup_{p \in \mathbb{Q}} (\{t > \tau_p\} \cap \{p \geqslant S\}).$$

But $\{p \geqslant S\} \in \tilde{\mathcal{F}}_p$, so $\{t > \tau_p\} \cap \{p \geqslant S\} \in \mathcal{F}_t$.

(iii) It is enough to check the result for H in some generating set. But if S is an $\{\tilde{\mathcal{F}}_t\}$-stopping time and $H = (S, \infty)$, then $H \circ \varphi = (\tau(S), \infty)$ is previsible.

(iv) It is easy to prove that $\tilde{M}$ is an $\{\tilde{\mathcal{F}}_t\}$-local martingale. The process $H(\varphi_\cdot)$ is a locally bounded $\{\mathcal{F}_t\}$-previsible process by (iii), so that both sides of (30.11) are well defined. Taking H of the form (S, ∞), the left-hand side is

$$\tilde{M}(t) - \tilde{M}(S \wedge t) = M(\tau_t) - M(\tau_S \wedge \tau_t)$$

which is the right-hand side. Thus (30.11) is valid for all bounded elementary processes, and extends by the now-familiar use of the monotone-class theorem to bounded previsible H, and thence to locally bounded previsible H; we leave the details to you. □

(*30.12*) *Remarks.* (i) It is not generally true that $M(\tau_t)$ is an $\{\tilde{\mathscr{F}}_t\}$-local martingale whenever M is an $\{\mathscr{F}_t\}$-local martingale. Why? (*Hint.* Take B to be a BM($\mathbb{R}$), and

$$\varphi_t = \text{meas}\,\{s \leqslant t : B_s \leqslant 0\}.)$$

(ii) Proposition 30.10 remains true without the continuity assumptions in (iv). By the time you have read Part 7 of Chapter VI, this extension will be clear to you.

31. Canonical decomposition of a continuous semimartingale. For the purposes of this chapter and the next, it is best to regard a continuous semimartingale X as defined to be a process which may be written:

$$X = X_0 + M + A, \tag{31.1}$$

where:

X_0 is $\mathscr{F}_0$ measurable,
M is a continuous local martingale null at 0,
and A is an continuous FV process null at 0.
(Recall that an FV process is *adapted.*)

Remarks. The perspicacious reader will have spotted the gap in the argument here. For there can exist a continuous process X of the form

$$X = X_0 + L + G, \tag{31.2}$$

(say) where X_0 is $\mathscr{F}_0$ measurable, L is a local martingale null at 0, where G is an FV_0 process, where neither L nor G is continuous, but where of necessity

$$\Delta L = -\Delta G.$$

The jumps of L are thus exactly cancelled by the jumps of G. According to our previous definition, this process X at (31.2) must be regarded as a semimartingale. But there is no contradiction: X can also be written in the form (31.1) with *continuous* M and A. We will have to assume this until §VI.24. (This seems to represent a falling away from our promise to give the full theory for continuous semimartingales, but we are doing all that is possible within the bounds of that theory!) □

Let X be as at (31.1). Then M and A are uniquely determined. The reason is the familiar one. If

$$X = X_0 + M' + A'$$

is another decomposition of the same type (with M' and A' continuous), then the process

$$N = M' - M = A - A'$$

is a continuous local martingale null at 0 and with paths of finite variation. By Theorem 30.4, $N=0$, $M'=M$ and $A'=A$.

(*31.3*) DEFINITION (canonical decomposition, X^{cm}). *The unique decomposition (31.1) of the continuous semimartingale X is called the* canonical decomposition of X. *The local martingale M is called the* continuous local martingale part *of X and written $M = X^{\text{cm}}$. We define*

$$[X, X] = [M, M] = [X^{\text{cm}}, X^{\text{cm}}].$$

Similarly, for two continuous semimartingales X and Y, we define

$$[X, Y] = [X^{\text{cm}}, Y^{\text{cm}}].$$

(*31.4*) DEFINITION. *For X as at* (31.1) *and $H \in \mathrm{lb}\mathscr{P}$, we define*

$$H \bullet X = \int_0^{\bullet} H\,dX = H \bullet M + H \bullet A.$$

We now know how $H \bullet M$ and $H \bullet A$ are defined. We see that $H \bullet X$ is a continuous semimartingale null at 0 and that

$$H \bullet X = H \bullet M + H \bullet A$$

is the canonical decomposition of $H \bullet X$. Thus,

$$(H \bullet X)^{\text{cm}} = H \bullet X^{\text{cm}},$$

and, from (30.8),

$$[H \bullet X, K \bullet Y] = (HK) \bullet [X, Y]. \tag{31.5}$$

32. Itô's formula for continuous semimartingales. The Itô integral is defined by a limiting procedure far too clumsy to provide an effective means of handling such integrals in practice. The Lebesgue integral is also defined by means of a limiting procedure, but in practice when faced with a calculation of, say, the Lebesgue integral of $f(x) = x^2$ from 0 to 1, we do not begin by approximating f by simple functions; the Lebesgue integral is useful precisely because of its *properties* and the *rules by which it can be manipulated*. The same is true of the Itô integral. In the previous section we explored some of its properties; here, we shall develop the rules by which it can be manipulated, commonly called the *stochastic calculus*. As before, we start with stochastic calculus for bounded continuous processes, and build up via localization.

(*32.1*) LEMMA. *Let M be a bounded continuous martingale null at* 0, *and let V be a continuous adapted process of bounded variation null at* 0. *Then*:

(32.2)
$$M_t^2 = \int_0^t 2M_s dM_s + [M]_t;$$

(32.3)
$$M_t V_t = \int_0^t M_s dV_s + \int_0^t V_s dM_s.$$

Proof. In Theorem 30.1, we constructed $[M]$ as the uniform limit in L^2 of processes
$$A_t^n \equiv M_t^2 - 2(H^n \bullet M)_t,$$
where each H^n was a bounded previsible process, such that $\sup_t |H_t^n - M_t| \leqslant 2^{-n}$. Thus $H^n \to M$ in $L^2(M)$ and so the martingales $H^n \bullet M$ converge to $M \bullet M$ uniformly in L^2, with (32.2) as an immediate consequence.

For (32.3), fixing n and setting $t_i^n \equiv (i2^{-n}) \wedge t$, by summation-by-parts,

$$\begin{aligned} M_t V_t &= \sum_{i \geqslant 1} M(t_i^n)\{V(t_i^n) - V(t_{i-1}^n)\} + \sum_{i \geqslant 1} V(t_{i-1}^n)\{M(t_i^n) - M(t_{i-1}^n)\} \\ &= \sum_{i \geqslant 1} M(t_i^n)\{V(t_i^n) - V(t_{i-1}^n)\} + \int_0^t H_s^n dM_s \end{aligned}$$

where H^n is the bounded previsible simple process
$$\sum V(t_{i-1}^n)(t_{i-1}^n, t_i^n].$$
The H^n are bounded and converge to V, by the continuity of V, so as $n \to \infty$, the second term tends to $\int_0^t V_s dM_s$. As for the first, simply use dominated convergence to deduce that it converges to $\int_0^t M_s dV_s$ as $n \to \infty$, yielding (32.3). □

The next result is the cornerstone of stochastic calculus.

(32.4) THEOREM. (Integration by parts formula). *Let X and Y be continuous semimartingales. Then*:

(32.5)
$$X_t Y_t - X_0 Y_0 = \int_0^t X_s dY_s + \int_0^t Y_s dX_s + [X, Y]_t.$$

Proof. We know that $X, Y \in \mathrm{lb}\mathscr{P}$, so that the stochastic integrals in (32.5) are well defined. Without loss of generality, we may suppose that $X_0 = Y_0 = 0$. Indeed, if we know the result for such X and Y, then for the general case, letting $\tilde{X} \equiv X - X_0$, $\tilde{Y} \equiv Y - Y_0$, we find that

$$\begin{aligned} X_t Y_t - X_0 Y_0 &= \tilde{X}_t \tilde{Y}_t + X_0 \tilde{Y}_t + Y_0 \tilde{X}_t \\ &= \int \tilde{X}_s d\tilde{Y}_s + \int \tilde{Y}_s d\tilde{X}_s + [\tilde{X}, \tilde{Y}]_t + X_0 \tilde{Y}_t + Y_0 \tilde{X}_t \\ &= \int \tilde{X}_s dY_s + \int \tilde{Y}_s dX_s + [X, Y]_t + \int X_0 dY_s + \int Y_0 dX_s \\ &= \int X_s dY_s + \int Y_s dX_s + [X, Y]_t. \end{aligned}$$

If $X = M + A$, $Y = N + B$, $M, N \in c\mathcal{M}_{0,\text{loc}}$, A, $B \in \text{FV}_0$, we may (by localization) assume that M and N are bounded martingales, and A and B are of bounded variation. The identity (32.2) polarizes to give

$$M_t N_t = \int_0^t M_s dN_s + \int_0^t N_s dM_s + [M, N]_t, \tag{32.6}$$

and (finite-variation) calculus gives

$$A_t B_t = \int_0^t A_s dB_s + \int_0^t B_s \, dA_s. \tag{32.7}$$

Assembling (32.3), (32.6), (32.7) yields (32.5). □

Just as in the case of finite-variation stochastic calculus, where we derived the change of variables formula (Itô's formula) from the integration-by-parts formula, so here we repeat the exercise (and the proof is essentially the same!) to derive the powerful Itô's formula for continuous semimartingales.

(*32.8*) THEOREM **(Itô's formula).** *Let* $f: \mathbb{R}^n \to \mathbb{R}$ *be* C^2, *and let* $X_t \equiv (X_t^1, \ldots, X_t^n)$ *be a continuous semimartingale in* $\mathbb{R}^n$ *(that is, each* X^i *is a continuous semimartingale). Then:*

$$f(X_t) - f(X_0) = \int_0^t D_i f(X_s) \, dX_s^i + \tfrac{1}{2} \int_0^t D_{ij} f(X_s) \, d[X^i, X^j]_s. \tag{32.9}$$

Remarks. We use the summation convention (summation over repeated indices is understood), and $D_i = \partial/\partial x^i$, $D_{ij} = D_i D_j$. Notice that $D_i f(X_s)$, as a continuous adapted process, is in lb $\mathcal{P}$ and therefore the stochastic integral is well defined.

(*32.10*) COROLLARY. *Under the conditions of Theorem 32.8,* $f(X)$ *is a continuous semimartingale. If* $X = X_0 + M + A$ *is the canonical decomposition of* X, *then:*

$$f(X_t)=f(X_0)+\int_0^t D_i f(X_s)\,dM_s^i$$
$$+\left\{\int_0^t D_i f(X_s)\,dA_s^i+\tfrac{1}{2}\int_0^t D_{ij}f(X_s)\,d[M^i,M^j]_s\right\}$$

is the canonical decomposition of $f(X)$.

Proof of the theorem. Convention dictates that Itô's formula should only be proved for $n=1$, the general case being left as an exercise, amid bland assurances that only the notation is any more difficult.

As in the finite-variation case (§18), we let $\mathscr{A}$ denote the collection of C^2 functions $f:\mathbb{R}\to\mathbb{R}$ for which Itô's formula holds. Plainly $\mathscr{A}$ is a vector space, but $\mathscr{A}$ is also an algebra. Indeed, if f and g are in $\mathscr{A}$, we can apply the integration-by-parts formula to the semimartingales $F_t\equiv f(X_t)$, $G_t\equiv g(X_t)$. By definition,

$$[F,G]_t\equiv[F^{\mathrm{cm}},G^{\mathrm{cm}}]_t$$
$$=\left[\int_0^\bullet f'(X_s)\,dM,\int_0^\bullet g'(X_s)\,dM\right]_t,$$

since the continuous local martingale part of F is $\int f'(X_s)\,dM$, Itô's formula holding for f by hypothesis. Hence,

$$[F,G]_t=\int_0^t f'(X_s)g'(X_s)\,d[M,M]_s,$$

by (31.5). Thus the integration-by-parts formula tells us that

$$F_tG_t-F_0G_0=\int_0^t F_s\,dG_s+\int_0^t G_s\,dF_s+\int_0^t f'g'(X_s)\,d[M]_s$$
$$=\int_0^t\{F_sg'(X_s)+f'(X_s)G_s\}\,dX_s$$
$$+\tfrac{1}{2}\int_0^t\{F_sg''(X_s)+2f'g'(X_s)+f''(X_s)G_s\}\,d[M]_s.$$

But this is exactly what Itô's formula states for the product fg. Thus, $\mathscr{A}$ is an algebra. Since $\mathscr{A}$ obviously contains $f(x)=x$, $\mathscr{A}$ contains all polynomials.

In order to complete the proof, define:

$$U_n\equiv\inf\{t:|X_t|+[M]_t>n\}.$$

Then (U_n) is a sequence of stopping times with $U_n\uparrow\infty$, and it suffices to prove Itô's formula for arbitrary f in C^2 on the interval $[0,U_n]$. By analogy with the previous argument (§18), choose polynomials f_k such that, for $r=0, 1, 2$, the rth derivative of f_k converges to that of f uniformly on compacts. Since Itô's formula holds for f_k, we have

$$f_k(X_{t\wedge U_n})-f_k(X_0)=I_{1,k}(t)+I_{2,k}(t)+I_{3,k}(t), \tag{32.11}$$

where

$$I_{1,k}(t)\equiv\int_0^t f'_k(X_s)I_{(0,U_n]}dM_s,$$

$$I_{2,k}(t)\equiv\int_0^t f'_k(X_s)I_{(0,U_n]}dA_s,$$

$$I_{3,k}(t)\equiv\int_0^t \tfrac{1}{2}f''_k(X_s)I_{(0,U_n]}d[M]_s.$$

Define $I_1(t)$, $I_2(t)$ and $I_3(t)$ analogously with f replacing f_k.

Now, on $(0, U_n]$, $|X|$ is bounded by n; and, for $r=0, 1, 2$:

$$_kQ_r\equiv\sup_{|x|\leqslant n}|f_k^{(r)}(x)-f^{(r)}(x)|\to 0, \quad (k\to\infty).$$

Since A and $[M]$ are of finite variation, it is true for each that

$$I_{2,k}(t)\to I_2(t), \quad I_{3,k}(t)\to I_3(t).$$

But, by the usual Hilbert space isometry property,

$$\mathbf{E}(\{I_{1,k}(t)-I_1(t)\}^2)\leqslant {_kQ_1^2}E[M]_{t\wedge U_n}\leqslant {_kQ_1^2}n.$$

Indeed, by combining this estimate with the submartingale property in the now-familiar way, we can show that as $k\to\infty$ in a subsequence, each term in (32.11) converges to the corresponding term in Itô's formula for f, uniformly over compact t-intervals.

The theorem is proved. □

(*32.12*) *Differential notation.* We can regard (32.9) as a Taylor series expansion where we set

$$dX^i dX^j\equiv d[X^i, X^j],$$

$$dX^i dX^j dX^k\equiv 0.$$

We shall make heavy use of this type of differential notation. Particularly important is the fact that if X is a continuous 1-dimensional semimartingale and V is a continuous 1-dimensional finite-variation process, then

$$dX\,dV=0.$$

Of course, this reflects results (32.3) and (32.7). Hence, if X_1, X_2, Y_1 and Y_2 are 1-dimensional continuous semimartingales such that both $X_1 - X_2$ and $Y_1 - Y_2$ are of finite variation, then

$$dX_1 dY_1 = dX_2 dY_2.$$

With this differential notation, we view (31.5) as follows. If X and Y are continuous semimartingales in $\mathbb{R}$, and H and K are previsible $\mathbb{R}$-valued processes, then

$$d(H \bullet X) = H dX,$$

so that

$$\begin{aligned} d[H \bullet X, K \bullet Y] &= d(H \bullet X) d(K \bullet Y) = H(dX) K(dY) \\ &= HK(dX)(dY) = HK d[X, Y]. \end{aligned}$$

6. APPLICATIONS OF ITÔ'S FORMULA

33. Lévy's theorem. Many of the most important and useful results of probability, originally proved by very complicated methods, become elegant and simple when we use Itô's formula. We begin with what is everyone's favourite proof, the Kunita–Watanabe proof of Lévy's theorem.

We remark that Lévy's theorem is the fundamental *martingale characterization* result, and thus it can be seen as motivation for the *martingale-problem approach* to diffusion theory. The result is of central importance in other areas too, notably in the theory of *filtering*. The 1-dimensional case of the theorem was stated in § I.2.

(*33.1*) THEOREM (Lévy, Kunita–Watanabe). *Let* $\{B_t^i : t \geqslant 0\}$ $(i = 1, \ldots, n)$ *be continuous local martingales such that* $B_0^i = 0$ *for each* i *and*

$$[B^i, B^j]_t = \delta_{ij} t.$$

Then $B_t \equiv (B_t^1, \ldots, B_t^n)$ *is a Brownian motion relative to* $(\Omega, \mathcal{F}, (\mathcal{F}_t), \mathbf{P})$. *In other words, for* $s < t$, *the increment* $B_t - B_s$ *is independent of* $\mathcal{F}_s$ *and has the normal distribution of zero mean and covariance matrix* $(t-s) I$, I *being the identity* $n \times n$ *matrix.*

Proof. All that we need to notice is that, for fixed θ in $\mathbb{R}^n$,

$$M_t^\theta \equiv f(B_t, t)$$

is a continuous martingale, where

$$f(x, t) \equiv \exp\{i(\theta, x) + \tfrac{1}{2}|\theta|^2 t\},$$

(θ, x) signifying the inner product. This follows immediately from Itô's formula applied to the smooth function f; in differential form,

$$\begin{aligned} d(f(B_t, t)) &= \frac{\partial f}{\partial x^j}(B_t, t)\, dB_t^j + \frac{\partial f}{\partial t}(B_t, t)\, dt \\ &\quad + \tfrac{1}{2}\frac{\partial^2 f}{\partial x^j \partial x^k}(B_t, t)\, d[B^j, B^k]_t \\ &= i\theta_j f(B_t, t)\, dB_t^j + \tfrac{1}{2}|\theta|^2 f(B_t, t)\, dt \\ &\quad + \tfrac{1}{2}(i\theta_j)(i\theta_k) f(B_t, t)\, \delta_{jk}\, dt \\ &= i\theta_j f(B_t, t)\, dB_t^j. \end{aligned}$$

Hence, since M is the sum of stochastic integrals with respect to continuous local martingales, M is itself a continuous local martingale. But, since $|M_t^\theta| = \exp(\frac{1}{2}|\theta|^2 t) < \infty$ for each t, M^θ is actually a *martingale*. Hence, for $s < t$,

$$\mathbf{E}[\exp\{i(\theta, B_t - B_s)\} \,|\, \mathscr{F}_s] = \exp(-\tfrac{1}{2}(t-s)\,|\theta|^2) \quad \text{a.s.},$$

and the theorem follows. □

34. Continuous local martingales as time-changes of Brownian motion. A striking application of Lévy's theorem shows that, modulo time-transformations, Brownian motion is the most general continuous local martingale. More specifically, let M be a continuous local martingale, and regard $[M]$ as the intrinsic clock carried by M. Then M has delusions of grandeur: it thinks it is a Brownian motion!

This result has many very important consequences. It also poses one of the most tantalizing of unsolved problems.

(*34.1*) THEOREM (Dubins and Schwarz [1]). *Let M be a continuous local martingale null at* 0 *such that $[M]_t \uparrow \infty$ as $t \uparrow \infty$. For $t \geqslant 0$, define the stopping time*

$$\tau_t \equiv \inf\{u \colon [M]_u > t\},$$

and set $\mathscr{G}_t \equiv \mathscr{F}(\tau_t)$. Then $B_t \equiv M(\tau_t)$ defines a Brownian motion B relative to the $\{\mathscr{G}_t\}$ filtration. Moreover, for each fixed t, $[M]_t$ is a $\{\mathscr{G}_t\}$ stopping time, and

$$M_t = B([M]_t), \quad \forall t. \tag{34.2}$$

Proof. We shall appeal to Lemma II.73.10 that *a right-continuous adapted process is progressive.*

Since M is $\{\mathscr{F}_t\}$-progressive (by the result just quoted), B is $\{\mathscr{G}_t\}$-adapted by Lemma II.73.11.

We now prove that

(34.3) *B is continuous.*

To establish this, we must show that (for almost every ω) M is constant on any interval of constancy of $[M]$. Localization allows us to pretend for a moment that $M \in \mathcal{M}_0^2$. It is clearly enough to show that if q is any nonnegative rational, and

$$S_q \equiv \inf\{t > q: [M]_t > [M]_q\},$$

then M is constant on $[q, S_q)$. But

$$\mathbf{E}[M(S_q)^2 - [M](S_q) | \mathcal{F}_q] = M_q^2 - [M]_q,$$

and, since $[M](S_q) = [M]_q$,

$$\mathbf{E}[\{M(S_q) - M_q\}^2 | \mathcal{F}_q] = 0.$$

Thus, (34.3) holds.

Next, we prove that

(34.4) *B and $B^2 - t$ are $\{\mathcal{G}_t\}$ local martingales.*

Set:

$$T(n) \equiv \inf\{t: |M_t| > n\}, \quad U(n) \equiv [M](T(n)).$$

You should check (carefully!) that, for each t,

$$\tau_{t \wedge U(n)} = T(n) \wedge \tau_t,$$

so that

$$B_{t \wedge U(n)} = M_{\tau(t)}^{T(n)}.$$

Let us check that each $U(n)$ is a $\{\mathcal{G}_t\}$ stopping time. For each fixed t,

$$\Lambda \equiv \{U(n) \leqslant t\} = \{T(n) \leqslant \tau_t\}$$

is in $\mathcal{F}(\tau_t) = \mathcal{G}_t$, by the easy result that $\{S \leqslant T\} \in \mathcal{F}_S \cap \mathcal{F}_T$ for stopping times S and T.

On applying the Optional Stopping/Sampling Theorem II.77.5 to the UI martingale $M^{T(n)}$, we therefore obtain for $s < t$,

$$\mathbf{E}[B_{t \wedge U(n)} | \mathcal{G}_s] = E[M_{\tau(t)}^{T(n)} | \mathcal{G}_s] = M_{\tau(s)}^{T(n)} = B_{s \wedge U(n)},$$

so that B is a $\{\mathcal{G}_t\}$ local martingale with $(U(n))$ as reducing sequence. Similarly, we can apply Theorem II.77.5 to the UI martingale $(M^2 - [M])^{T(n)}$ to obtain

$$\mathbf{E}[B_{t \wedge U(n)}^2 - t \wedge U(n) | \mathcal{G}_s] = B_{s \wedge U(n)}^2 - (s \wedge U(n)),$$

so that $B^2 - t$ is a $\{\mathcal{G}_t\}$ local martingale.

Because of Lévy's theorem, the proof of Theorem 34.1 is now complete. □

(*34.5*) *The 'pure local martingale' problem* (Dubins and Schwarz [1]). We have seen that $M = B([M])$, where, for each t, $[M]_t$ is a $\{\mathcal{G}_t\}$ stopping time. Let $\{\mathcal{B}_t\}$ be the augmented filtration generated by B, so that, for each u, $\mathcal{B}_u \subset \mathcal{G}_u$.

(*34.6*) DEFINITION (pure local martingale). *The (continuous) local martingale M is called* pure *if each $[M]_t$ is a $\{\mathscr{B}_t\}$ stopping time; then, M itself is adapted to the $\{\mathscr{B}_t\}$ filtration.*

The tantalizing problem referred to at the start of this section is the following:

(*34.7*) PROBLEM: *Characterize pure local martingales.*

Stroock and Yor [1] have some attractive results on this problem. In the next section, we shall look at a particularly fascinating case. □

We now explain how Theorem 34.1 may be generalized to allow us to drop the restriction that $[M]_t \uparrow \infty$ as $t \uparrow \infty$. This generalization forces us to enrich the given set-up $(\Omega, \mathscr{F}, \{\mathscr{F}_t\}, \mathbf{P})$. We do not give the formal definition of enrichment; for that, see Section II.7 of Ikeda and Watanabe [1]. We content ourselves with explaining the idea. Intuitively, things are very simple. The technicalities are simple too, compared with those of introducing 'killing' for Markov processes as was done in §III.18.

We shall need an easy lemma.

(*34.8*) LEMMA. *On the set $\{[M]_\infty < \infty\}$, the limit $M_\infty \equiv \lim_{t \uparrow \infty} M_t$ exists almost surely.*

Proof. You can easily show (by introducing a reducing sequence) that if, for $K > 0$, we define $S(K)$ as follows:

$$S(K) \equiv \inf\{t: [M]_t > K\},$$

then $M^{S(K)}$ is an L^2-bounded martingale, so that

$$\lim_{t \uparrow \infty} M_{t \wedge S(K)} \text{ exists, } \quad \text{a.s.}$$

The lemma follows. □

Now we introduce a Brownian motion W independent of the set-up $(\Omega, \mathscr{F}, \{\mathscr{F}_t\}, \mathbf{P}, M)$. The details are given in a moment. Then define

$$B_t = \begin{cases} M(\tau_t) & \text{if } t < [M]_\infty, \\ M_\infty + W(t) - W([M]_\infty) & \text{if } t \geqslant [M]_\infty. \end{cases}$$

Then B is a Brownian motion such that

$$M_t = B([M]_t). \tag{34.9}$$

To do this formally, introduce a new set-up $(\Omega^*, \mathscr{F}^*, \{\mathscr{F}_t^*\}, \mathbf{P}^*)$ carrying a Brownian motion W^*. Define

$$\tilde{\Omega} \equiv \Omega \times \Omega^*, \quad \tilde{\mathscr{F}}_t^\circ \equiv \mathscr{F}_t \times \mathscr{F}_t^*, \quad \tilde{\mathbf{P}} \equiv \mathbf{P} \times \mathbf{P}^*,$$

and, for $\tilde{\omega}=(\omega, \omega^*)$ in $\tilde{\Omega}$, define

$$\begin{aligned}\tilde{B}_t(\tilde{\omega}) &\equiv M(\tau_t(\omega)) \quad \text{if } t<[M]_\infty(\omega),\\ &\equiv M_\infty(\omega)+W^*(t, \omega^*)-W^*([M]_\infty(\omega), \omega^*).\end{aligned}$$

The process $\tilde{M}$ with

$$\tilde{M}_t(\tilde{\omega}) \equiv M_t(\omega), \quad [\tilde{M}]_t(\tilde{\omega}) \equiv [M]_t(\omega)$$

is a natural 'enrichment' of M, and we can—and do—regard it as essentially being M. It is now easy to check that $\tilde{B}$ is a Brownian motion on $(\tilde{\Omega}, \tilde{\mathcal{F}}, \{\tilde{\mathcal{F}}_t^\circ\}, \mathbf{P})$, and that

(34.10) $$\tilde{M}_t(\tilde{\omega})=\tilde{B}([\tilde{M}]_t(\tilde{\omega}), \tilde{\omega}).$$

This gives the precise meaning of (34.9).

In the following summary result, we allow a final further generalization to creep in. This will not cause you any problems.

(*34.11*) THEOREM. *Let* $\{M_t: 0 \leqslant t < \zeta\}$ *be a continuous process with lifetime* ζ *(finite or infinite) carried by the set-up* $(\Omega, \mathcal{F}, \{\mathcal{F}_t\}, \mathbf{P})$ *such that there exists a sequence* $(T(n))$ *of stopping times with* $T(n) \uparrow\uparrow \zeta$ *such that each process*

$$\{M_{t \wedge T(n)}: 0 \leqslant t < \infty\}$$

is a martingale. The process $\{[M]_t: 0 \leqslant t < \zeta\}$ *is defined in the obvious way. Then there exists a Brownian motion B (perhaps on some enriched set-up) such that, with the interpretation given before the statement of this theorem,*

(34.12) $$M_t=B([M]_t), \quad 0 \leqslant t < \zeta.$$

(*34.13*) COROLLARY. *On the set* $\{[M]_{\zeta-}<\infty\}$, *the limit* $M_{\zeta-} \equiv \lim_{t \uparrow\uparrow \zeta} M_t$ *exists almost surely. On the set* $\{[M]_{\zeta-}=\infty\}$,

$$\limsup_{t \uparrow\uparrow \zeta} M_t=+\infty, \quad \liminf_{t \uparrow\uparrow \zeta} M_t=-\infty.$$

This is an immediate consequence of (34.12).

(*34.14*) *Remark.* Let $M=\{M_t: 0 \leqslant t<\infty\}$ be a continuous local martingale. Let H be a previsible process. Then the stochastic integral $H \bullet M$ is a local martingale with

$$[H \bullet M]_t=\int_0^t H_s^2 \, d[M]_s.$$

It is now clear from (34.13) that the restriction (30.7) that H satisfy

$$\mathbf{P}\left[\int_0^t H_s^2\, d[M]_s < \infty, \forall t\right] = 1$$

cannot be dropped. □

(*34.15*) *Knight's Theorem.* There is an n-dimensional generalization of Theorem 34.1 due to Knight.

(*34.16*) THEOREM (Knight). *Let $M^1, M^2, \ldots, M^n$ be continuous local martingales such that, for each i, $[M^i]_t \uparrow \infty$ as $t \uparrow \infty$, and, for $i \neq j$, $[M^i, M^j] = 0$. Then there is a Brownian motion B on $\mathbb{R}^n$ such that, for each i,*

$$M_t^i = B^i([M^i]_t), \quad \forall t, \forall i.$$

See Knight [2], Meyer [5] and Cocozza and Yor [1]. □

(*34.17*) *Conformal martingales.* Suppose that

$$Z = X + iY$$

is a Brownian on $\mathbb{C}$, and that f is an analytic function on $\mathbb{C}$. On combining Itô's formula with the Cauchy–Riemann equations, we find that

$$df(Z) = f'(Z)\, dZ, \tag{34.18}$$

and that if we write $f(Z) = U + iV$, where U and V are real local martingales, then

$$[U] = [V], \quad \text{and} \quad [U, V] = 0. \tag{34.19}$$

In this case,

$$[U] = [V] = \int |f'(Z)|^2\, ds. \tag{34.20}$$

(34.21) DEFINITION (conformal local martingale). *A local martingale*

$$W = U + iV$$

in $\mathbb{C}$ (U and V are real) is said to be conformal *if* (34.19) *holds.*

Conformal local martingales were first defined, and their rôle in connection with a number of important inequalities developed, by Getoor and Sharpe [3].

(*34.22*) **Exercise.** Prove that if W is a conformal martingale, then

$$W = B([U])$$

for some Brownian motion B in $\mathbb{C}$.

The fact that analytic functions preserve Brownian tracks was already known to Lévy. In recent years, this fact has been cleverly exploited to give new insights into results on analytic functions and on other classes of functions on analytic manifolds. See B. Davis [1] for a proof of Picard's 'little' theorem on entire functions, and W. S. Kendall [2] for extensions.

(*34.23*) *An elementary application.* We can prove that a Brownian motion on $\mathbb{C}$ cannot hit 0 at a positive time by considering

$$W = \exp(Z),$$

where Z is a BM($\mathbb{C}$). Then W has Brownian tracks, $[W]_\infty = \infty$, and W cannot hit 0.

35. Bessel processes; skew products; etc.

Let X be a Brownian motion in $\mathbb{R}^n$. Let $r = |X|$. Suppose that $r_0 = a > 0$. Then

$$r^2 = X \cdot X,$$

so that by Itô's formula,

$$d(r^2) = 2r\,db + n\,dt, \tag{35.1}$$

where

$$db = r^{-1} X \cdot dX.$$

By Lévy's theorem (33.1), the (1-dimensional) process b is a Brownian motion. Let $v = r^2$. Then, we have

$$dv = 2v^{1/2}\,db + n\,dt, \quad v_0 = a^2. \tag{35.2}$$

The SDE (35.2) for v (just!) falls within the scope of the (essentially best possible) theorem of Yamada and Watanabe which is given in § V.40. From that theorem, we can conclude that for $\alpha > 0$, $a \geqslant 0$, and a given Brownian motion b, the SDE

$$dv = 2(v^+)^{1/2}\,db + \alpha\,dt, \quad v_0 = a^2, \tag{35.3}$$

(here, $v^+ \equiv \max(v, 0)$) has a unique solution v adapted to the augmented filtration $\{\mathcal{F}_t^b\}$ generated by b. Moreover, $v_t \geqslant 0$, $\forall t$. Then $r \equiv v^{1/2}$ is called an *'α-dimensional' Bessel process* BES(α), and v is called a BES^2(α) process.

We can prove that

(35.4) *for* $\alpha \geqslant 2$, BES^2(α) *never hits* 0 *at a positive time.*

Proof of (*35.4*) *when* $\alpha = 2$. This is the hardest case. Start v at $a^2 > 0$. From (35.3) we have, for $t<\zeta\equiv\sup\{s: |\log v_s|<\infty\}$,

$$d[v]=4v\,dt,$$
$$d(\log v)=v^{-1}\,dv-\tfrac{1}{2}v^{-2}\,d[v]=2v^{-1/2}\,db.$$

Thus, $M\equiv\log v$ defines a local martingale M on $[0,\zeta)$ in the sense of Theorem 34.11. (We can take $T(n)=\inf\{t: |\log v_t|>n\}$.) From Corollary 34.13, we see that on the set $\{\zeta<\infty\}$, we shall have in particular,

$$\limsup_{t\uparrow\uparrow\zeta} M_t=\infty, \quad \text{so that } \limsup_{t\uparrow\uparrow\zeta} v_t=\infty,$$

and this is impossible because the Yamada–Watanabe Theorem says that v is finite for all t.

This way of utilizing the fact that an exploding local martingale must explode to both $+\infty$ and $-\infty$ is often useful. □

(*35.5*) **Exercise.** Prove that, for $\alpha>2$, $\mathrm{BES}^2(\alpha)$ never hits 0 at a positive time. (*Hint.* Suppose that v starts at $a^2>0$. Prove that

$$v^{1-\alpha/2}=r^{2-\alpha}$$

is a local martingale, and, since it is nonnegative, hence a supermartingale by the Fatou Lemma 14.3.) □

We can derive (35.5) from (35.4) via the following result.

(*35.6*) THEOREM (Pythagoras, Shiga and Watanabe [1]). *Suppose that V_1 is a $\mathrm{BES}^2(\alpha_1)$ and that V_2 is a $\mathrm{BES}^2(\alpha_2)$ independent of V_1. Then $V=V_1+V_2$ is a $\mathrm{BES}^2(\alpha_1+\alpha_2)$.*

Proof. We have, for $i=1,2$,

$$dV_i=2V_i^{1/2}\,db_i+\alpha_i\,dt,$$

where b_1 and b_2 are independent Brownian motions. Then

$$dV=2V^{1/2}\,db+(\alpha_1+\alpha_2)\,dt,$$

where

$$db=(V_1/V)^{1/2}\,db_1+(V_2/V)^{1/2}\,db_2.$$

By Lévy's Theorem, b is a Brownian motion. □

Suppose now that $\alpha\geqslant 2$ and that v satisfies (35.3) with $a>0$. Since $v\mapsto v^{1/2}$ is smooth away from 0, and since v never hits 0, we can apply Itô's formula to find that $r=v^{1/2}$ satisfies

(35.7) $$dr=db+\tfrac{1}{2}(\alpha-1)r^{-1}\,dt, \quad r_0=a.$$

If $\alpha = n(\geqslant 2)$ is an integer, then the fact that r satisfies (35.7) corresponds to the fact that r is a Markov process with generator equal to the radial part

$$\tfrac{1}{2}d^2/dr^2 + \tfrac{1}{2}(n-1)r^{-1}d/dr \tag{35.8}$$

of $\tfrac{1}{2}\Delta$, where Δ is the Laplace operator for $\mathbb{R}^n$.

McKean observed that, for $\alpha \geqslant 2$, the (pathwise) uniqueness of the solution of (35.7) is trivial. For suppose that, for $i = 1, 2$,

$$dr_i = db + \tfrac{1}{2}(\alpha - 1)r_i^{-1}\,dt, \quad r_i = a.$$

Then,

$$r_1(t) - r_2(t) = \tfrac{1}{2}(\alpha - 1)\int_0^t [r_1(s)^{-1} - r_2(s)^{-1}]\,ds.$$

Thus, $r_1 - r_2$ decreases if $r_1 - r_2 > 0$ and increases if $r_1 - r_2 < 0$. This is impossible (Why?) unless $r_1(t) = r_2(t)$ for all t. □

For the most interesting case, surprisingly that in which $\alpha = 1$, see § 43 below. We shall find Bessel processes cropping up all over the place.

(*35.9*) *An example of a pure local martingale* (Stroock and Yor [1]). Let n be an odd positive integer. Let β be a Brownian motion, and let

$$M \equiv \int \beta^n\,d\beta, \quad [M] = \int \beta^{2n}\,ds, \quad \tau_t \equiv \inf\{u\colon [M]_u > t\}.$$

On differentiating $[M](\tau_t) = t$, we obtain

$$\beta_{\tau(t)}^{2n}\,\tau'(t) = 1. \tag{35.10}$$

Now, if $q \equiv \beta^{n+1}/(n+1)$, then, because n is odd, q is nonnegative and

$$dq = dM + \tfrac{1}{2}n\beta^{n-1}\,dt,$$

so that, if $V = q^2$, then,

$$dV_t = 2V^{1/2}\,dM_t + \alpha\,d[M]_t,$$

where $\alpha = (2n+1)/(n+1)$. If $v_t \equiv V(\tau_t)$ and $B_t \equiv M(\tau_t)$ as usual, then (see (30.10)) one can justify the formal conclusion:

$$dv = 2v^{1/2}\,dB + \alpha\,dt.$$

This is the equation (35.3) for a $\mathrm{BES}^2(\alpha)$, and we shall see in § V.48 that v is adapted to the augmented filtration $\{\mathscr{B}_t\}$ generated by B. Equation (35.10) now shows that the τ process is adapted to $\{\mathscr{B}_t\}$. Thus,

$$[M]_t = \inf\{u\colon \tau(u) > t\}$$

is a $\{\mathscr{B}_t\}$ stopping time, and M is adapted to the $\{\mathscr{B}_t\}$ filtration. The local martingale M is therefore pure. □

(*35.11*) PROBLEM. Suppose that n is a positive *even* integer. *Is the local martingale*

$$\int \beta^n \, d\beta$$

pure? This problem remains unsolved. □

(*35.12*) *Skew-product representation for* BM($\mathbb{C}$). Suppose that a Brownian motion

$$Z = (X, Y) = (r, \theta)$$

starts at a point of $\mathbb{C}$ away from 0. Then, since Z never hits 0, there exists a continuous determination of θ in $\mathbb{R}$. One way to think of this is that

$$Z \mapsto \log Z = \log r + i\theta$$

is a conformal map from $\mathbb{C} \backslash \{0\}$ to the Riemann surface for the log function. Since $(d/dz)\log z = 1/z$, we have, from our study of conformal martingales at (34.17),

$$(\log r_t, \theta_t) = B\left(\int_0^t r^{-2}\, ds\right), \tag{35.13}$$

where $B = (B_1, B_2)$ is a Brownian motion on $\mathbb{C}$. The facts:

(35.14(i)) *r is a* BES(2);

(35.14(ii)) $\theta_t = B_2\left(\int_0^t r^{-2}\, ds\right)$, *where B_2 is a Brownian motion independent of r*;

constitute the so-called *skew-product representation* of BM($\mathbb{C}$).

Why is B_2 independent of r? One has to be a little careful. What we certainly know is that B_2 is independent of B_1. Hence, if we prove that the augmented filtration of r is contained in that of B_1 (in fact, it is equal to it), then we have finished. This exactly amounts to proving that

(35.15) $M \equiv \log r$ *is a pure local martingale.*

Proof of (35.15). We have, from (35.7),

$$dr = d\beta + \tfrac{1}{2} r^{-1}\, dt$$

for some Brownian motion β. Hence,

$$dM = r^{-1}\, d\beta, \quad d[M] = r^{-2}\, dt.$$

Let $\tau(t) \equiv \inf\{s: [M]_s > t\}$. Then $B_1(t) = M(\tau_t)$. We have

$$r_{\tau(t)}^{-2}\, \tau'(t) = 1,$$

so that, as usual, we need only prove that $r_{\tau(t)}$ is adapted to the filtration of B_1. But this is trivial:

$$r_{\tau(t)} = \exp(B_1(t)).$$ □

We shall see later that

(35.16) *any 1-dimensional diffusion 'in natural scale' is a pure local martingale.*

(*35.17*) *Notes on the filtration of* BM($\mathbb{C}$). Again, suppose that Z is a BM($\mathbb{C}$) started away from 0. Then $\{\theta_t: t \geqslant 0\}$ generates the same augmented filtration as Z. This is because, from (35.13),

$$d[\theta]_t = r_t^{-2}\, dt,$$

so that the r process is adapted to the θ filtration.

On the one hand, it seems slightly surprising that the 1-dimensional process θ carries all the information about Z. On the other, if you are sitting at the origin watching the Brownian particle through a telescope, then, when it gets close to you, you will soon wake up to the fact!

It is easy to believe that if $Z_0 = 0$, then, for any fixed t, θ_t is (now considered modulo 2π) uniformly distributed on the unit circle and independent of the r-process. See Stroock and Yor [1]. □

(*35.18*) *Windings.* The last four years or so have seen many very remarkable papers on the windings of Brownian motion in $\mathbb{C}$. See, for example, Lyons and McKean [1] and Pitman and Yor [3,4]. □

(*35.19*) *Skew-product representation for* BM($\mathbb{R}^n$), $n \geqslant 2$. The skew product representation for a BM($\mathbb{R}^n$) started away from 0 is as follows:

(35.20(i)) $|X|$ is a BES(n);
(35.20(ii)) $X/|X| = b(\int r^{-2}\, ds)$;

where b is a Brownian motion on the unit sphere S^{n-1} independent of $|X|$.

See §V.31 for more information.

36. Brownian martingale representation. As already remarked in §21, there is an important general principle to the effect that a martingale representation result (such as Lévy's theorem) implies a martingale representation result (such as Theorem 36.1 below). This principle will be explained in §V.25, and references will be given there. For now, the following proof is very instructive.

(*36.1*) THEOREM (Itô). *Let $\{B_t: t \geqslant 0\}$ be Brownian motion in $\mathbb{R}^n$ on some filtered probability space, and let $\{\mathscr{G}_t\}$ be the filtration generated by B, augmented in the usual way (see §II.67). If $Y \in L^2(\mathscr{G}_\infty)$, then there exists a $\{\mathscr{G}_t\}$ previsible $\mathbb{R}^n$-valued process H with $\mathbf{E}\int_0^\infty |H_s|^2\, ds < \infty$ such that:*

$$\mathbf{E}[Y|\mathscr{G}_t] = \mathbf{E}Y + \int_0^t H_s\, dB_s, \tag{36.2}$$

and H is uniquely determined modulo Leb × **P**-*null sets.*

Proof. The zero-one law shows that

$$\mathbf{E}[Y|\mathscr{G}_0]=\mathbf{E}Y.$$

It is now clear that it is enough to prove the theorem under the assumption that $\mathbf{E}Y=0$. Moreover, it is obviously enough to establish the result in the case when (as we now also assume) $Y\in L^2(\mathscr{G}_T)$ for some constant time $T>0$.

For a $\{\mathscr{G}_t\}$ previsible process H with values in $\mathbb{R}^n$, define

$$\|H\|_{2,T}\equiv\left(\mathbf{E}\left(\int_0^T |H_s|^2\,ds\right)\right)^{1/2}.$$

Then

$$L_T^2(B)\equiv\{H\colon H \text{ is } \{\mathscr{G}_t\} \text{ previsible and } \|H\|_{2,T}<\infty\}$$

is a Hilbert space, and the (stochastic integral) map $I\colon L_T^2(B)\to L^2(\mathscr{G}_T)$ defined by

$$I(H)\equiv\int_0^T H_s\,dB_s$$

is an isometry. Thus $V\equiv I(L_T^2(B))\subset L^2(\mathscr{G}_T)$ is a complete (and therefore closed) subspace of $L_0^2(\mathscr{G}_T)\equiv\{Z\in L^2(\mathscr{G}_T)\colon \mathbf{E}Z=0\}$. The aim is to prove that $V=L_0^2(\mathscr{G}_T)$.

Suppose that Z is an element of the orthogonal complement of V in $L_0^2(\mathscr{G}_T)$. We must prove that $Z=0$. Let $Z_t\equiv\mathbf{E}(Z|\mathscr{G}_t)$, an L^2-bounded martingale. Note that $Z_0=0$.

Let $H\in L_T^2(B)$, and define $N_T=I(H)$, $N_t\equiv\mathbf{E}(N_T|\mathscr{G}_t)$ for $0\leqslant t\leqslant T$. Then $N_T\in V$, and, for any stopping time $S\leqslant T$, $N_S=I(H(0,S])$ is also in V. Thus, $\mathbf{E}(ZN_S)=\mathbf{E}(Z_SN_S)=0$, and, since Z_T and N_T are in L^2, it follows by Lemma II.77.6 that Z_tN_t is a uniformly integrable martingale.

Now, we saw in the proof of Theorem 33.1 that each martingale M_t^θ occurring there is the stochastic integral with respect to B of the process $H=i\theta M_t^\theta$ in $L_T^2(B)$. (Since $H\notin L_\infty^2(B)$, we have been forced to stop at finite T.) Thus, $Z_tM_t^\theta$ is a martingale, so that, for $0\leqslant s\leqslant t\leqslant T$,

$$\mathbf{E}[Z_t\exp\{i(\theta,B_t-B_s)\}\,|\mathscr{G}_s]=Z_s\exp(-\tfrac{1}{2}(t-s)|\theta|^2). \tag{36.3}$$

Taking $0<t_1<\ldots<t_m\leqslant T$, and writing $\Delta_k\equiv B(t_k)-B(t_{k-1})$, we find that

$$\begin{aligned}\mathbf{E}\left[Z_T\exp\left\{i\sum_{j=1}^m(\theta^j,\Delta_j)\right\}\right] &=\mathbf{E}\left[Z_0\exp\left\{-\tfrac{1}{2}\sum(t_j-t_{j-1})|\theta^j|^2\right\}\right]\\ &=0,\end{aligned} \tag{36.4}$$

by conditioning successively on the $\mathscr{G}_{t_j}$ and using (36.3). Since Z_T is $\mathscr{G}_T$-measurable, and (36.4) holds for every θ^j in $\mathbb{R}^n$, we conclude from the

Stone–Weierstrass theorem and monotone-class arguments that $Z_T=0$. The uniqueness assertion is trivial. □

This important result generalizes easily to local martingales.

(*36.5*) THEOREM. *Every $\{\mathscr{G}_t\}$-local martingale is continuous, and is the stochastic integral with respect to B of a previsible process H such that:*

$$\mathbf{P}\left[\int_{(0,t)} |H_s|^2\, ds < \infty, \forall t\right] = 1.$$

Proof. It is clearly enough to prove the result for $M \in \mathrm{UI}\mathscr{M}_0$, since every $M \in \mathscr{M}_{0,\mathrm{loc}}$ can be reduced to this case. So we suppose that M is UI: $M_t = \mathbf{E}(M_\infty | \mathscr{F}_t)$. Then there exist bounded M^n such that $\|M_\infty - M_\infty^n\|_1 \leqslant 3^{-n}$, so, by Doob's submartingale maximal inequality II.70.1,

$$\mathbf{P}(\sup_t |M_t^n - M_t| > 2^{-n}) \leqslant 2^n \mathbf{E}|M_\infty - M_\infty^n| \leqslant (\tfrac{2}{3})^n,$$

where $M_t^n \equiv \mathbf{E}(M_\infty^n | \mathscr{F}_t)$. Hence, by the Borel–Cantelli lemma, the martingales M^n converge uniformly a.s. to M; but since $M_\infty^n \in L^\infty \subseteq L^2$, the martingales M^n are continuous (by Theorem 36.1), and so the uniform limit M is also continuous.

Since M is continuous, we may take a stopping time T such that M^T is bounded, and then by Theorem 36.1,

$$M^T = \int_0^T H_s\, dB_s$$

for some previsible process H unique modulo Leb × $\mathbf{P}$ null sets. □

Shortly, we shall consider what can be said about the explicit form of the integrand in the stochastic integral representation of a given martingale. Note that we can in a sense give an 'explicit' description of H at (36.2) as follows:

(36.6) $$H_t = (d/dt)[Y, B]_t,$$

where Y is the martingale $Y_t \equiv \mathbf{E}[Y | \mathscr{G}_t]$.

However, Clark's description (41.3) is the one that matters in practice.

37. Exponential semimartingales; estimates. The next application of Itô's formula is of central importance because it is the fundamental result for the Cameron–Martin–Girsanov change-of-measure theorem. It yields a rare example of a stochastic differential equation with an explicit solution. Because the solution is explicit, we can discuss it outside the framework of the general theory of stochastic differential equations.

(*37.1*) THEOREM (Doléans). *Let X be a continuous semimartingale, with $X_0=0$, and suppose that Z_0 is some $\mathscr{F}_0$-measurable random variable. Then there* *exists a unique (continuous) semimartingale Z such that:*

(37.2)
$$Z_t = Z_0 + \int_{(0,t]} Z_s \, dX_s.$$

The unique solution is given explicitly by:

(37.3)
$$Z_t = Z_0 \mathscr{E}(X)_t,$$

where
$$\mathscr{E}(X)_t \equiv \exp(X_t - \tfrac{1}{2}[X]_t).$$

Proof. On applying Itô's formula, we find that

$$\begin{aligned} d(\mathscr{E}(X)_t) &= \mathscr{E}(X)_t dX_t - \tfrac{1}{2}\mathscr{E}(X)_t d[X]_t + \tfrac{1}{2}\mathscr{E}(X)_t d[X]_t \\ &= \mathscr{E}(X)_t dX_t, \end{aligned}$$

so that $\mathscr{E}(X)$ is a solution of (37.2) with $Z_0 = 1$. The fact that (37.3) implies (37.2) for a general $\mathscr{F}_0$-measurable Z_0 is an immediate extension.

To prove uniqueness, define $Y_t \equiv \exp(-X_t + \tfrac{1}{2}[X]_t)$ and notice that by Ito's formula,

(37.4)
$$dY_t = -Y_t dX_t + Y_t d[X]_t.$$

Now let Z be any solution of (37.2). Then, by the integration-by-parts formula,

$$\begin{aligned} d(Z_t Y_t) &= Z_t dY_t + Y_t dZ_t + d[Y, Z]_t \\ &= Z_t(-Y_t dX_t + Y_t d[X]_t) + Y_t(Z_t dX_t) + (-Y_t Z_t) d[X]_t, \end{aligned}$$

the last term arising by (31.5). Thus,

$$d(Z_t Y_t) = 0,$$

which implies that $Z_t Y_t$ is constant, and so $Z_t = Z_0 \mathscr{E}(X)_t$. □

Remarks. The stochastic differential equation (37.2) is called an *exponential stochastic differential equation*, and the solution (37.3) is called an *exponential semimartingale*. (Compare §19). The martingales M used in the proof of Theorem 33.1 are examples of exponential semimartingales, when the 'driving' semimartingale X is $i(\theta, B_t)$.

(*37.5*) **Exercise.** Suppose more generally that X_t and Y_t are given continuous semimartingales. Prove that there is a unique continuous semimartingale Z_t such that

$$Z_t = Y_t + \int_{(0,t]} Z_s \, dX_s,$$

given by

$$Z_t = \mathscr{E}(X)_t \left\{ Y_0 + \int_0^t \mathscr{E}(X)_s^{-1} (dY_s - d[X, Y]_s) \right\}.$$

(*37.6*) *Some important estimates.* We saw a fine indication of the power of estimates based on exponential martingales as early as in McKean's proof of the law of the iterated logarithm in §I.16. We now derive some elementary, but exceedingly useful, estimates which generalize the inequality used there.

Let M be a continuous local martingale null at 0. Then

$$\mathscr{E}(\theta M)=\exp(\theta M-\tfrac{1}{2}\theta^2[M])$$

is a nonnegative local martingale, and hence, by the Fatou Lemma 14.3, is a supermartingale. In particular,

$$\mathbf{E}\exp(\theta M-\tfrac{1}{2}\theta^2[M])\leqslant 1, \quad \forall t. \tag{37.7}$$

(*37.8*) THEOREM. *Suppose that, for each t, there exists a constant K_t such that:*

$$[M]_t<K_t \quad \text{a.s.}$$

Then, for every t, and every $y>0$,

$$\mathbf{P}\left[\max_{s\leqslant t} M_s>y\right]\leqslant\exp(-y^2/2K_t), \tag{37.9}$$

and each $\mathscr{E}(\theta M)$ is a true martingale (*so that equality holds at* (37.7)).

Note. McKean [1] and Stroock and Varadhan [1] make brilliant use of this kind of estimate.

Proof. Since $\mathscr{E}(\theta M)$ is a non-negative supermartingale such that $\mathscr{E}(\theta M)_0=1$, we see that for $\theta>0$, $y>0$,

$$\mathbf{P}\left[\max_{s\leqslant t} M_s>y\right]\leqslant\mathbf{P}\left[\max_{s\leqslant t}\mathscr{E}(\theta M)_s>\exp(\theta y-\tfrac{1}{2}\theta^2K_t)\right]$$
$$\leqslant\exp(-\theta y+\tfrac{1}{2}\theta^2K_t).$$

Choose $\theta=y/K_t$ to obtain (37.9).

Let $Z(t)\equiv\max\{M_s: s\leqslant t\}$. Then, for any $\theta>0$,

$$\mathbf{E}\,e^{\theta Z(t)}=\int e^{\theta y}\mathbf{P}[Z\in dy]$$
$$=1+\theta\int e^{\theta y}\mathbf{P}[Z>y]\,dy$$
$$\leqslant 1+\theta\int e^{\theta y}\exp(-y^2/2K_t)\,dy<\infty.$$

We know that, for $\theta>0$,

$$N_s \equiv \mathscr{E}(\theta M)_s = \exp(\theta M_s - \tfrac{1}{2}\theta^2[M]_s)$$

defines a local martingale. Thus, there exists a sequence of stopping times $T(n)$ with $T(n)\uparrow\infty$ such that, for $s\leqslant t$,

$$\mathbf{E}[N_{t\wedge T(n)}\,|\,\mathscr{F}_s] = N_{s\wedge T(n)}. \tag{37.10}$$

But each variable $N_{t\wedge T(n)}$ is dominated by $\exp(\theta Z(t))$. Hence, we can let $n\uparrow\infty$ in (37.10) to obtain

$$\mathbf{E}[N_t\,|\,\mathscr{F}_s] = N_s,$$

so that $N=\mathscr{E}(\theta M)$ is a true martingale.

We deal with $\mathscr{E}(\theta M)$ for $\theta<0$ by considering θM as $(-\theta)(-M)$. □

The following two corollaries are very frequently used.

(*37.11*) COROLLARY. *Let B be a Brownian motion in $\mathbb{R}^n$ and let H be a bounded previsible process in $\mathbb{R}^n$. Then:*

$$\exp\left[\int H\,dB - \tfrac{1}{2}\int |H_s|^2\,ds\right]$$

is a true martingale.

Proof. This is obvious. Take $\theta=1$ and $M=\int H\,dB$. □

(*37.12*) COROLLARY. *Let M be a continuous local martingale null at 0. Then for positive t, y and K, the following estimates hold:*

$$\mathbf{P}\left[\max_{s\leqslant t} M_s \geqslant y;\ [M]_t \leqslant K\right] \leqslant \exp(-y^2/2K),$$

$$\mathbf{P}\left[\max_{s\leqslant t} |M_s| \geqslant y;\ [M]_t \leqslant K\right] \leqslant 2\exp(-y^2/2K).$$

Proof. Let $T=\inf\{u: [M]_u \geqslant K\}$. Then,

$$\mathbf{P}\left[\max_{s\leqslant t} M_s \geqslant y;\ [M]_t \leqslant K\right] \leqslant \mathbf{P}\left[\max_{s\leqslant t} M_s^T \geqslant y\right] \leqslant \exp(-y^2/2K),$$

by applying (37.9) to M^T. □

(*37.13*) *Remarks.* Suppose that M is a continuous local martingale. Deciding when $\mathscr{E}(M)$ is a true martingale can be a tricky problem. For Novikov's result that

$$\mathbf{E}\exp(+\tfrac{1}{2}[M]_t) < \infty, \quad \forall t,$$

is a sufficient condition, and for an illuminating example of what can go wrong, see (13.34) of Elliott [1].

38. Cameron–Martin–Girsanov change of measure. Suppose we are given a set-up $(\Omega, \mathscr{F}, \{\mathscr{F}_t\}, \mathbf{P})$ satisfying the usual conditions, and a measure $\mathbf{Q}$ on $(\Omega, \mathscr{F})$ equivalent to $\mathbf{P}$.

It was remarked in §16 that if X is an integrator under $\mathbf{P}$, then X is obviously an integrator under $\mathbf{Q}$. Because of the fundamental Theorem 16.4 that integrators are the same as semimartingales, it follows that (as stated at (16.5)):

X is a semimartingale under $\mathbf{P}$ *if and only if X is a semimartingale under* $\mathbf{Q}$.

We now give the classical proof of this result. This proof proceeds by explicit decomposition of a $\mathbf{P}$ semimartingale as a $\mathbf{Q}$ local martingale plus a finite-variation process.

This result is of particular importance. There are many circumstances in which the 'drift' or finite-variation part of a process is of considerable nuisance value, and it is often advantageous to 'remove the drift' via a change of measure. Illustrations will be given in the next two sections, and it will become clear that the theory now being developed can be seen as a generalization of its paradigm example, the *Doob h-transform* for Markov processes.

Recently, the C–M–G theorem has acquired still greater importance in the version of the *Malliavin calculus* developed by Bismut. We take a first look at this technique in §41.

We begin by recalling Theorem 17.1. According to that result,

$$Z_t \equiv \left.\frac{d\mathbf{Q}}{d\mathbf{P}}\right|_{\mathscr{F}_t}$$

defines a strictly positive uniformly integrable martingale. Moreover, a process N is a $\mathbf{Q}$ local martingale if and only if NZ is a $\mathbf{P}$ local martingale.

Because we do not yet have the general Itô's formula (for possibly discontinuous processes), we have to work for the time being in a restricted situation in which N and Z are continuous, and we should emphasize that this covers incomparably the most important cases. (But, of course, our work on likelihood ratios for chains in §22 is closely related.) So, for now, we restrict attention to the case covered by:

(*38.1*) (ASSUMPTION) *Z is a continuous martingale.*

Since Z is strictly positive and continuous, the process Z^{-1} is locally bounded and previsible. Hence, we can define a continuous local martingale X as follows:

(38.2) $$X_t \equiv \int_0^t Z_s^{-1}\,dZ_s.$$

Then:

(38.3) $$Z_t = Z_0 + \int_0^t Z_s\,dX_s;$$

that is, Z is the solution

$$Z = Z_0 \mathscr{E}(X)$$

of the exponential SDE (37.2) driven by the local martingale X!

(*38.4*) THEOREM. *Let M be a continuous* **P** *local martingale. Then:*

$$N_t \equiv M_t - [M, X]_t = M_t - \int_0^t Z_s^{-1}\,d[M, Z]_s$$

defines a continuous **Q** *local martingale N, and*

$$[N]_t = [M]_t.$$

Proof. We need only check that $N_t Z_t$ is a **P** local martingale. By the integration-by-parts formula,

$$d(M_t Z_t) = M_t\,dZ_t + Z_t\,dM_t + d[M, Z]_t.$$

Also,

$$\begin{aligned} d([M, X]_t Z_t) &= [M, X]_t\,dZ_t + Z_t\,d[M, X]_t \\ &= [M, X]_t\,dZ_t + Z_t(Z_t^{-1}\,d[M, Z]_t), \end{aligned}$$

there being no Itô correction term here because $[M, X]$ is a finite variation process. Subtracting, we see NZ exhibited as a sum

$$NZ = (NZ)_0 + \int N\,dZ + \int Z\,dM$$

of terms each of which is a stochastic integral with respect to a **P** local martingale, whence NZ is itself a **P** local martingale.

One can prove that $[N] = [M]$ similarly, but it is more elegant to note that (at least when M is bounded) $[M]_t$ is obtained as a limit (almost surely (**P** or **Q**!) down a fast subsequence) of processes of the form

$$A_t \equiv \sum \{M(t_k) - M(t_{k-1})\}^2$$

(as we saw in § 30). Changing M to N only adds a continuous finite-variation process to M whose influence on the sum vanishes in the limit as the mesh of the dissection tends to zero. *Note.* By now, you should have no difficulty in making this kind of argument rigorous. If $\Delta_j f \equiv f(t_j) - f(t_{j-1})$, then, for a continuous finite-variation process F,

$$\sum(\Delta_j(M+F))^2 = \sum(\Delta_j M)^2 + \sum(\Delta_j F)\,[2\Delta_j M + \Delta_j F].$$

The last sum is bounded by $\mathrm{Var}(F)\cdot\sup|2\Delta_j M + \Delta_j F|$, and the supremum tends to 0 by continuity—you know the idea! □

The impact of a change of measure is even more striking if we put it in a specific context. Take $\Omega = C(\mathbb{R}^+, \mathbb{R}^n)$, the Polish space of continuous $\mathbb{R}^n$-valued functions on $\mathbb{R}^+$, let

$$X_t(\omega) \equiv \omega(t) \quad \text{for } \omega \in \Omega, \quad t \geqslant 0,$$

let $\mathscr{F}_t^\circ \equiv \sigma(\{X_s: s \leqslant t\})$, and take for the probability $\mathbf{P}$ the law of n-dimensional Brownian motion started at 0. Let $\mathscr{F}_t$ be the usual $\mathbf{P}$-augmentation of $\mathscr{F}_t^\circ$. Then we have the following important result.

(*38.5*) THEOREM (Cameron, Martin, Girsanov, . . .) (i) *Let* $\mathbf{Q}$ *be a law on* $(\Omega, \mathscr{F})$ *which is equivalent to* $\mathbf{P}$. *Then there exists a previsible* $\mathbb{R}^n$*-valued process* c_t *such that*:

$$Z_t \equiv \left.\frac{d\mathbf{Q}}{d\mathbf{P}}\right|_{\mathscr{F}_t} = \exp\left(\int_0^t c_s\,dX_s - \tfrac{1}{2}\int_0^t |c_s|^2\,ds\right) \tag{38.6}$$

and, under $\mathbf{Q}$,

$$\tilde{X}_t \equiv X_t - \int_0^t c_s\,ds \tag{38.7}$$

is a Brownian motion.

(ii) *Let* γ_t *be a previsible* $\mathbb{R}^n$*-valued process such that*:

$$\zeta_t \equiv \exp\left(\int_0^t \gamma_s\,dX_s - \tfrac{1}{2}\int_0^t |\gamma_s|^2\,ds\right)$$

defines a uniformly integrable $\mathbf{P}$ *martingale* ζ. *Define a measure* $\mathbf{Q}$ *on* $(\Omega, \mathscr{F})$ *by*:

$$\frac{d\mathbf{Q}}{d\mathbf{P}} = \zeta_\infty. \tag{38.8}$$

Then, under $\mathbf{Q}$,

$$X_t - \int_0^t \gamma_s\,ds \text{ is a Brownian motion.}$$

Proof. (i) The process Z is a $\mathbf{P}$ martingale, so, by (36.5), is continuous. Moreover, by the zero-one law (see §§I.12 and III.9), $Z_0 = 1$. By (36.5) again, the continuous **P** local martingale $Y_t \equiv \int_0^t Z_s^{-1}\, dZ_s$ has an integral representation

$$Y_t = \int_0^t c_s\, dX_s,$$

and Z solves the exponential SDE

$$Z_t = 1 + \int_0^t Z_s dY_s$$

whose unique solution $\mathscr{E}(Y)$ is simply the right-hand side of (38.6). By Theorem 38.4,

$$\tilde{X}_t^i = X_t^i - [X^i, Y]_t = X_t^i - \int_0^t c_s^i\, ds$$

defines a continuous **Q** local martingale $\tilde{X}^i$, and since

$$[\tilde{X}^i, \tilde{X}^j]_t = [X^i, X^j]_t = \delta_{ij} t,$$

Lévy's Theorem 33.1 tells us that X is a Brownian motion under **Q**.

(ii) Since ζ is a uniformly integrable martingale, we shall have $d\mathbf{Q}/d\mathbf{P} = \zeta_t$ on $\mathscr{F}_t$. The remaining assertions follow from part (i). □

Now, Theorem 38.5 is not the appropriate result for applications. In practice, we are most commonly concerned with situations in which, for our canonical Brownian motion $(\Omega, \mathscr{F}^\circ, \{\mathscr{F}_t^\circ\}, \mathbf{P})$, we are interested in a measure **Q** which is equivalent to **P** on each $\mathscr{F}_{t+}^\circ$, but not equivalent to **P** on $\mathscr{F}^\circ$. In this situation, the idea of the usual conditions is a nonsense. See the Example below.

(*38.9*) THEOREM. *Let* $(\Omega, \mathscr{F}^\circ, \{\mathscr{F}_t^\circ\}, \mathbf{P})$ *relate to* CBM($\mathbb{R}^n$). *Suppose that* γ *is an* $\{\mathscr{F}_{t+}^\circ\}$ *previsible* $\mathbb{R}^n$*-valued process such that:*

$$\zeta_t \equiv \exp\left(\int_0^t \gamma_s\, dX_s - \tfrac{1}{2} \int_0^t |\gamma_s|^2\, ds \right) \tag{38.10}$$

defines a martingale ζ. *In particular, Corollary* 37.11 *shows that this will be true when* γ *is a bounded process. Then there exists a unique measure* **Q** *on* $(\Omega, \mathscr{F}^\circ)$ *such that:*

$$\left.\frac{d\mathbf{Q}}{d\mathbf{P}}\right|_{\mathscr{F}_{t+}^\circ} = \zeta_t, \quad \forall t, \tag{38.11}$$

and, under **Q**,

$$X_t - \int_0^t \gamma_s ds$$

defines a Brownian motion relative to $\{\mathscr{F}_{t+}^\circ\}$.

Remark. We do, of course, know that X is a **P** Brownian motion relative to $\{\mathscr{F}^\circ_{t+}\}$. See §II.67.10. Using $\{\mathscr{F}^\circ_{t+}\}$ instead of $\{\mathscr{F}^\circ_t\}$ allows us more freedom.

Proof. In effect, replacing γ by $\gamma I_{[0,t]}$ for the various t-values reduces the problem to that covered by Theorem 38.5. The only point to watch is the existence of the measure **Q** such that (38.11) holds for every t. However, this is an immediate consequence of the Daniell–Kolmogorov theorem. □

Example. As we shall see in the next section, an important illustration of this theorem will be the case when **Q** is the law of Brownian motion on $\mathbb{R}$ with constant non-zero drift μ. Let

$$\Lambda \equiv \{\omega \in \Omega: \lim t^{-1}\omega(t) = \mu\}.$$

Then Λ is of **Q** measure 1 and of **P** measure 0. Thus Λ is in every $\mathscr{F}^{\mathbf{P}}_t$ for the usual **P** augmentation $\{\mathscr{F}^{\mathbf{P}}_t\}$ of $\{\mathscr{F}^\circ_t\}$. We cannot expect to make $X_t - \mu t$ a **Q** Brownian motion relative to $\{\mathscr{F}^{\mathbf{P}}_t\}$ because every $\mathscr{F}^{\mathbf{P}}_t$ contains the whole future of X under **Q**!

39. First applications: Doob *h*-transforms, hitting of spheres, etc. We now look at several examples to get the idea of how the Cameron–Martin–Girsanov change of measure works.

(*39.1*) *Doob h-transforms.* We study Doob h-transforms of space-time Brownian motion. Let $(X_t, \Omega, \mathscr{F}^\circ, \mathscr{F}^\circ_t, \mathbf{P})$ be canonical Brownian motion on $\mathbb{R}$, started at 0. A strictly positive h: $\mathbb{R}^+ \times \mathbb{R} \to (0, \infty)$ is called *space-time regular* if, for $s > 0$, $t \geqslant 0$, $x \in \mathbb{R}$,

$$h(t,x) = \int p(s; x, y) h(t+s, y)\, dy \tag{39.2}$$

where, of course, $p(\cdot\,;\cdot,\cdot)$ is the Brownian transition density. Assume without loss of generality that $h(0, 0) = 1$. It is not hard to show that h is a C^∞ function, and that a measure **Q** on $(\Omega, \mathscr{F}^\circ)$ may be defined so that

$$\left.\frac{d\mathbf{Q}}{d\mathbf{P}}\right|_{\mathscr{F}^\circ_{t+}} = Z_t \equiv h(t, X_t).$$

Under **Q**, the coordinate process X is a non-homogeneous Markov process with transition density

$$p^h(t, t+s; x, y) = h(t, x)^{-1} p(s; x, y) h(t+s, y).$$

Applying Itô's formula to the **P** martingale Z_t gives

$$dZ_t = h'(t, X_t)dX + \{\dot{h}(t, X_t) + \tfrac{1}{2}h''(t, X_t)\}dt, \tag{39.3}$$

where a dot denotes a derivative with respect to t, and a prime denotes a derivative with respect to x. But the curly-bracketed term is identically zero, either directly from (39.2), or because the left-hand side of (39.3) is a **P** local martingale, as is the first term on the right, so the second term, being of finite variation, must vanish. Hence:

$$dZ_t = Z_t c_t dX_t$$

where $c_t = h'(t, X_t)h(t, X_t)^{-1}$, so that $Z = \mathscr{E}(c \bullet X)$, and by Theorem 38.9,

(39.4) $\quad X_t = X_t - \int_0^t h'(s, X_s)h(s, X_s)^{-1}\, ds$ *is a* **Q**-*Brownian motion.*

(39.5) Brownian motion with constant drift, hitting spheres. If $\mu \in \mathbb{R}$, then:

$$h(t, x) = \exp\{\mu x - \tfrac{1}{2}\mu^2 t\}$$

is a space-time regular function for Brownian motion and $h'(t, x)/h(t, x) = \mu$ for all t, x. So, under **Q**, the h-transformed law, the canonical process X is Brownian motion with drift μ.

Generalizing to $\mathbb{R}^n$, it is easy to show that if $\mu \in \mathbb{R}^n$, then h-transforming Wiener measure for $\mathbb{R}^n$ via

$$h(t, x) = \exp\{(\mu, X) - \tfrac{1}{2}|\mu|^2 t\}$$

gives the law of Brownian motion with drift μ started at zero. Let this law be denoted by $\mathbf{Q}^\mu$, and the associated expectation by $\mathbf{E}^\mu$.

As a first example where a change of measure transforms a seemingly difficult problem into an easy one, we prove the following result. This was first obtained by Reuter via explicit solution of certain partial differential equations. The following proof was discovered by one of us, and reported in Kent [1].

(39.6) THEOREM (Reuter). *Let* $(X_t, \mathbf{Q}^\mu)$ *be Brownian motion on* $\mathbb{R}^n$ *with constant drift* μ *and started at zero. Let* $T \equiv \inf\{t: |X_t| = 1\}$. Then X_T *and* T *are independent under the law* $\mathbf{Q}^\mu$.

Proof. Let $f: S^{n-1} \to [0, 1]$ and $g: \mathbb{R}^+ \to [0, 1]$ be measurable. Then if $\mu = 0$, it is obvious that X_T and T are independent, so:

(39.7) $$\mathbf{E}^0 f(X_T) g(T) = \mathbf{E}^0 f(X_T) \mathbf{E}^0 g(T).$$

For $\mu \neq 0$, using the fact that $h(t \wedge T, X_{t \wedge T})$ is a uniformly integrable martingale, and hence that

$$d\mathbf{Q}^\mu / d\mathbf{P} = h(T, X_T) \text{ on } \mathscr{F}^\circ_{T+},$$

we calculate

$$\begin{aligned}
\mathbf{E}^\mu[f(X_T)g(T)] &= \mathbf{E}^0[f(X_T)g(T)\exp\{(\mu, X_T) - \tfrac{1}{2}|\mu|^2 T\}] \\
&= \mathbf{E}^0[f(X_T)e^{(\mu, X_T)}]\mathbf{E}^0[g(T)e^{-\frac{1}{2}|\mu|^2 T}] \qquad \text{(by 39.7))} \\
&= \frac{\{\mathbf{E}^0[f(X_T)e^{(\mu, X_T)}]\mathbf{E}^0[e^{-\frac{1}{2}|\mu|^2 T}]\mathbf{E}^0[g(T)e^{-\frac{1}{2}|\mu|^2 T}]\mathbf{E}^0[e^{(\mu, X_T)}]\}}{\mathbf{E}^0[e^{-\frac{1}{2}|\mu|^2 T}]\mathbf{E}^0[e^{(\mu, X_T)}]} \\
&= \mathbf{E}^\mu[f(X_T)]\, \mathbf{E}^\mu[g(T)],
\end{aligned}$$

combining terms in the numerator and denominator using (39.7). □

(*39.8*) *The von Mises distribution.* On taking $g \equiv 1$ in the above calculations, we see that

$$\mathbf{E}^{\mu}[f(X_T)\} = C_\mu \mathbf{E}^0[f(X_T)e^{(\mu, X_T)}],$$

where

$$C_\mu = \mathbf{E}^0[e^{-\frac{1}{2}|\mu|^2 T}].$$

Hence X_T has density $C_\mu \exp[(\mu, \bullet)]$ relative to the normalized 'area' measure on S^{n-1}. We say that X_T *has the von Mises distribution with pole at* $\mu/|\mu|$ *and concentration parameter* $|\mu|$. See Kent [2], and Pitman and Yor [1] for more information on its probabilistic properties, and D. G. Kendall [1] and Mardia [1] for statistical applications. The distribution will feature again in § V.14.

(*39.9*) *Brownian motion conditioned to hit a particular boundary point.* We saw an interesting example of a Doob h-transform for Brownian motion in § III.31. This was the example of Brownian motion in the open unit ball D in $\mathbb{R}^3$ conditioned to hit the boundary sphere ∂D at the unit vector ξ. We found that the conditioned process is the Doob h-transform of BM(D) where

$$h \equiv \kappa(x, \xi) \equiv (1 - |x|^2)/|\xi - x|^3,$$

and has infinitesimal generator

$$\tfrac{1}{2}h^{-1}\Delta h = \tfrac{1}{2}\Delta + h^{-1}(\text{grad h})\cdot\text{grad}.$$

To understand the significance of the type of drift which arises here, let us consider the 2-dimensional situation. We work with the upper half-plane H rather than the disc D. (*Exercise.* Go from one to the other via a conformal map.)

By (I.19.7), the probability density function h of hitting the real axis at 0 for a BM($\mathbb{R}^2$) started at (x,y) with $y>0$ has the Cauchy form:

$$h(x,y) = \pi^{-1} y r^{-2}, \qquad (r^2 = x^2 + y^2).$$

Thus, the generator of Brownian motion on H conditioned to hit the real axis at 0 is

(39.10) $$\tfrac{1}{2}h^{-1}\Delta h = \tfrac{1}{2}\Delta - 2xr^{-2}\frac{\partial}{\partial x} + (x^2 - y^2)r^{-2}y^{-1}\frac{\partial}{\partial y}.$$

The stochastic differential equations for the associated process (x_t, y_t) are

(39.11 (i)) $$dx_t = db_t - 2x_t r_t^{-2}\,dt,$$
(39.11 (ii)) $$dy_t = dw_t + (x_t^2 - y_t^2)r_t^{-2}y_t^{-1}\,dt,$$

where b and w are independent Brownian motions.

The integral curves of the vector field described by the first-order term in (39.10) are semicircles though the origin with centres on the x-axis—no surprise!

If we invert the picture in the unit circle, so that

$$(x, y) \mapsto (X, Y) \equiv r^{-2}(x, y),$$

the integral curves transform to straight lines parallel to the y-axis. We find that (see Exercise 39.13 below)

$$(39.12) \qquad (X, Y)(t) = (\xi, \eta)\left(\int_0^t R^4 ds\right),$$

where $R^2 = X^2 + Y^2$, and

(i) ξ is a Brownian motion,
(ii) η is (what else?!) a BES(3),
(iii) ξ and η are independent.

This gives a simple picture of the (x, y) motion which fits in perfectly with our intuition.

(*39.13*) **Exercise.** Prove that

$$dX = R^2 dB, \quad dY = R^2 dW + R^4 Y^{-1} dt,$$

where (with $x = r\cos\theta$, $y = r\sin\theta$),

$$dB = -\cos 2\theta\, db - \sin 2\theta\, dw,$$

$$dW = -\sin 2\theta\, db + \cos 2\theta\, dw.$$

Use Lévy's theorem to show that (B, W) is a BM ($\mathbb{R}^2$). Complete the proof of (39.12).

(*39.14*) *References*: Durrett [1] and Cranston [1].

40. Further applications: bridges, excursions, etc. We now use the methods of the last two sections to study Brownian bridges and the scaled Brownian excursion.

(*40.1*) *The Brownian bridge.* Let $\{B_t : t \geq 0\}$ denote a Brownian motion in $\mathbb{R}^n$, $B_0 = 0$. Roughly speaking, the Brownian bridge from 0 to some fixed point a of $\mathbb{R}^n$ is B 'conditioned to be at a at time 1', though some care is needed in interpreting this conditioning on an event of probability zero. However, there are more robust definitions which are more useful in practice.

(*40.2*) DEFINITION (Brownian bridge). *Fix* $a \in \mathbb{R}^n$. *A continuous Gaussian process* $\{X_t : 0 \leq t \leq 1\}$ *such that*:

$$\mathbf{E}X_t = at, \ \operatorname{cov}(X_t, X_s) = \{(s \wedge t) - st\} I \qquad (0 \leq s, t \leq 1)$$

is called the Brownian bridge from 0 to a (*if* $a = 0$, *it is simply called the* Brownian bridge). Here, I is the $n \times n$ identity matrix.

(*40.3*) THEOREM. *For fixed* $a \in \mathbb{R}^n$, *the following are equivalent definitions of the (law of) Brownian bridge from* 0 *to* a.

(i) *For some Brownian motion B,*

$$X_t = B_t + t(a - B_1) \qquad (0 \leqslant t \leqslant 1).$$

(ii) *For some Brownian motion B,*

$$X_t = at + (1-t)B(t/[1-t]) \qquad (0 \leqslant t \leqslant 1).$$

(iii) $\{X_t : 0 \leqslant t \leqslant 1\}$ *is a continuous process such that*

$$X_t - at + \int_0^t (1-s)^{-1}(X_s - as)ds \text{ is a Brownian motion.}$$

(iv) *The law of X is the h-transform of Wiener measure* $\mathbf{P}$ *on* $C([0,1], \mathbb{R}^n)$, *where:*

$$h(t,x) = (2\pi(1-t))^{-n/2}\exp\{-|x-a|^2/2(1-t)\}.$$

Note. The significance of the h-transform as inducing the required conditioning should be clear to you: intuitively,

$$h(t,x) = \text{`}\mathbf{P}[B_1 = a \mid B_t = x]\text{'}.$$

Proof. It is trivial to verify that (i) and (ii) are equivalent to Definition (40.2).

If B is a Brownian motion in $\mathbb{R}^n$, then

$$M_t \equiv \int_0^t (1-s)^{-1}dB_s \qquad (0 \leqslant t < 1)$$

is a continuous martingale, and

$$[M^i, M^j]_t = \delta_{ij}\int_0^t (1-s)^{-2}ds = \delta_{ij}t/(1-t).$$

Then by an immediate extension of Theorem 34.1, $\{M_t : 0 \leqslant t < 1\}$ has the same distribution as $\{B(t/[1-t]) : 0 \leqslant t < 1\}$. Hence, if X is defined by (ii), then X has the same distribution as

$$\tilde{X}_t \equiv at + (1-t)\int_0^t (1-s)^{-1}dB_s$$
$$= at + B_t - \int_0^t (1-s)^{-1}(\tilde{X}_s - as)\,ds,$$

by Itô's formula. By reversing the argument, any process X satisfying (iii) satisfies (ii).

The h-transform description (iv) is the most interesting. For each $\varepsilon \in (0,1)$ there is a measure $\mathbf{Q}^\varepsilon$ on Ω such that

$$\left.\frac{d\mathbf{Q}^\varepsilon}{d\mathbf{P}}\right|_{\mathscr{F}_t^\circ} = M(t \wedge (1-\varepsilon))$$

where M_t is the martingale $h(t, X_t)/h(0, 0)$. On applying (39.4), we see that under $\mathbf{Q}$,

$$\tilde{X}_t^\varepsilon \equiv X_t - at + \int_0^{t\wedge(1-\varepsilon)} (1-s)^{-1}(X_s - as)ds$$

is a Brownian motion on $[0, 1]$. Hence there is a measure $\mathbf{Q}$ on $C([0, 1), \mathbb{R}^n)$ such that $\mathbf{Q}|_{\mathscr{F}^\circ(1-\varepsilon)} = \mathbf{Q}^\varepsilon|_{\mathscr{F}^\circ(1-\varepsilon)}$ for all $\varepsilon > 0$, and on $[0, 1)$,

$$\tilde{X}_t \equiv X_t - at + \int_0^t (1-s)^{-1}(X_s - as)\,ds$$

is a Brownian motion. But then, by (iii), $\{X_t : 0 \leqslant t < 1\}$ is a Brownian bridge under $\mathbf{Q}$, $\lim X_t = a$, $\mathbf{Q}$-a.s., and $\mathbf{Q}$ extends to a measure on $C([0, 1], \mathbb{R}^n)$, the law of a Brownian bridge from 0 to a. □

The principal interest of the Brownian bridge is the following. For each $a \in \mathbb{R}^n$, let $\mathbf{P}(a, \bullet)$ be the law of the Brownian bridge from 0 to a:

$$(\mathbf{P}f)\,(a) \equiv \int_\Omega \mathbf{P}(a, d\omega) f(\omega) = \mathbf{E}[f(B_\cdot + \bullet(a - B_1))]$$

for all bounded continuous $f: \Omega \equiv C([0, 1], \mathbb{R}^n) \to \mathbb{R}$. Then the kernel $\mathbf{P}$ from $\mathbb{R}^n$ to Ω is continuous ($\mathbf{P}f$ is obviously continuous if f is) and is a regular conditional distribution for Wiener measure given B_1. This follows from the easily verified fact that:

$X_t \equiv B_t + t(Y - B_1)$ *is a Brownian motion*

if Y is a normal random variable with zero mean and identity covariance independent of B. For an application of this to proving smoothness of transition densities for one-dimensional diffusions, see Rogers [3].

(40.4) The scaled Brownian excursion. Consider a process r on $(0, \infty)$ which is 'Brownian motion conditioned to hit 0 for the first time at time 1'. This process will be the (*formal*) h-transform of Brownian motion where

$$h(t, x) = \text{`}\mathbf{P}[H_0 = 1 \mid B_t = x]\text{'}, \quad 0 < t < 1, \quad x > 0.$$

Thus, $h(t, x)$ is the value of the density at time $1-t$ of the $\mathbf{P}^x$ law of H_0:

$$\begin{aligned} h(t, x) &= \mathbf{P}^x[H_0 \in 1 - dt]/dt \\ &= (2\pi(1-t)^3)^{-1/2} x \exp[-x^2/2(1-t)] \end{aligned}$$

(see § I.9), and

$$h'/h = x^{-1} - x/(1-t).$$

So, we have an SDE for r of the form:

(40.5) $$dr = db + r^{-1}dt - r\,dt/(1-t),$$

where b is a Brownian motion. For this process, 0 acts as an entrance boundary similar to that for BES(3). The process r started at 0 is called the *scaled Brownian excursion*.

Now let X be the 3-dimensional Brownian bridge from 0 to 0 in time 1. Then from part (iii) of Theorem 40.3,

$$dX = dW - X\,dt/(1-t)$$

where W is a 3-dimensional Brownian motion. Itô's formula shows that $R \equiv |X|$ satisfies

(40.6) $$dR = dB + R^{-1}dt - R\,dt/(1-t),$$

where $dB = (R^{-1}X)\cdot dW$, so that B is a (1-dimensional) Brownian motion.

If we take for granted certain uniqueness theorems for SDEs studied in the next chapter, we can conclude from (40.5) and (40.6) that:

(40.7) *the scaled Brownian excursion is identical in law to the 3-dimensional Bessel bridge* BES(3)BR, *that is, to the radial part of a 3-dimensional Brownian bridge from* 0 *to* 0.

This result was proved by Markovian methods in Williams [7]. (Well, actually, it was left as an exercise for the reader!)

41. Explicit Brownian martingale representation. The Cameron–Martin–Girsanov formula for changing the drift (equivalently, changing the measure) of Brownian motion leads us naturally to a *stochastic 'calculus of variations'*, one of whose earliest dividends is the following method (due to Bismut[1]) of obtaining Clark's explicit description of the stochastic integrand in the Brownian martingale representation (Theorem 36.1).

We restrict attention to the time interval [0, 1]. So let $(\Omega, \mathscr{F}, \{\mathscr{F}_t\}, \mathbf{P})$ relate to canonical Brownian motion on the time-parameter interval [0, 1]. Thus $\Omega = C[0,1]$, $\mathbf{P}$ is Wiener measure (for the restricted time-interval), $X(t,\omega) = \omega(t)$, etc.

(41.1) Invariance principle for the Malliavin–Bismut calculus. Let u be a bounded previsible process, and let

$$\varphi_t = \int_0^t u_s\,ds.$$

For $\varepsilon \in \mathbb{R}$, define

$$Z^\varepsilon \equiv \exp\left[\varepsilon \int_0^1 u_s\,dX_s - \tfrac{1}{2}\varepsilon^2 \int_0^1 u_s^2\,ds\right].$$

Define a measure $\mathbf{Q}^\varepsilon$on $(\Omega, \mathscr{F})$ by $d\mathbf{Q}^\varepsilon/d\mathbf{P} = d\mathbf{P} = Z^\varepsilon$ on $(\Omega, \mathscr{F})$. Then, under $\mathbf{Q}$, $X - \varepsilon\varphi$ *is a Brownian motion*, so that, for any functional F of the paths,

$$\mathbf{E}[F(X)] = \mathbf{E}^\varepsilon[F(X - \varepsilon\varphi)],$$

where $\mathbf{E}^\varepsilon$ denotes the expectation associated with $\mathbf{Q}^\varepsilon$. In other words, we have the *fundamental invariance principle*:

$$\text{(41.2)} \qquad \mathbf{E}[F(X)] = \mathbf{E}\left[F(X-\varepsilon\varphi)\exp\left\{\varepsilon\int_0^1 u_s\,dX_s - \tfrac{1}{2}\varepsilon^2\int_0^1 u_s\,ds\right\}\right].$$

This principle underlies Bismut's approach to the Malliavin calculus. The formula (41.2), less sophisticated than Bismut's original, appeared in the attempt by one of us (Williams [13]) to introduce people to the first steps in the Malliavin calculus.

The calculus of variations arises because, under suitable conditions, we can differentiate (41.2) with respect to ε at $\varepsilon=0$. For our applications, we shall insist that

u is a bounded continuous adapted process.

Let us rewrite (41.2) as follows:

$$\text{(41.3)} \qquad \begin{aligned} 0 &= \mathbf{E}[\varepsilon^{-1}\{F(X-\varepsilon\varphi)-F(X)\}\,\{Z^\varepsilon-1\}] \\ &\quad + \mathbf{E}[\varepsilon^{-1}\{F(X-\varepsilon\varphi)-F(X)\}] \\ &\quad + \mathbf{E}[F(X)\varepsilon^{-1}\{Z^\varepsilon-1\}]. \end{aligned}$$

Suppose now that F satisfies the following three conditions:

(41.4 (i)) $\mathbf{E}[F(X)^2] < \infty$.

(41.4 (ii)) There exists a constant γ such that for $\varphi\in\Omega$,

$$|F(X+\varphi)-F(X)| \leqslant \gamma\|\varphi\| \quad (X\in\Omega).$$

(41.4 (iii)) There is a kernel F' from Ω to $[0, 1]$ such that for each φ in $C^1[0, 1]$,

$$\lim_{\varepsilon\to 0}\varepsilon^{-1}\{F(X+\varepsilon\varphi)-F(X)\} = \int_0^1 F'(X, dt)\,\varphi(t)$$

for almost all $X\in\Omega$.

(*41.5*) *Remarks.* (i) Condition (ii) does in fact imply condition (i).

(ii) The conditions (41.4) are satisfied if F is differentiable with bounded derivative, for then

$$F(X+\varepsilon\varphi)-F(X) = \varepsilon F'(X)(\varphi) + o(|\varepsilon|\;\|\varphi\|),$$

where $F'(X)$ is a bounded linear functional on Ω, that is, a measure. As we shall soon see, there are interesting functionals F where differentiability fails, but the conditions (41.4) hold. □

Now consider what happens when we let $\varepsilon\to 0$ in (41.3). The first term on the right-hand side is bounded by $\gamma\;\|\varphi\|\,\mathbf{E}|Z^\varepsilon-1|$, and so tends to 0. Conditions (ii) and (iii) of (41.4) allows us to use the Dominated Convergence Theorem on the second. Finally, we have

$$\varepsilon^{-1}(Z^\varepsilon - 1) = \int_0^1 Z_s^\varepsilon u_s \, dX_s,$$

so that

$$\varepsilon^{-1}(Z^\varepsilon - 1) \to \int_0^1 u_s \, dX_s \text{ in } L^2.$$

Hence:

(41.6)
$$0 = \mathbf{E}\left[-\int_0^1 F'(X, dt)\varphi(t) + F(X)\int_0^1 u_s \, dX_s \right].$$

By the martingale-representation result (36.1), we have, for some C in $L^2(X)$,

$$F(X) = \mathbf{E}(F(X)) + \int_0^1 C_s \, dX_s$$

and so

$$\mathbf{E}\left[F(X)\int_0^1 u_s \, dX_s\right] = \mathbf{E}\left[\int_0^1 C_s dX_s \cdot \int_0^1 u_s \, dX_s\right]$$
$$= \mathbf{E}\left[\int_0^1 u_s C_s \, ds\right].$$

On substituting into (41.6), rearranging and integrating by parts, we obtain

(41.7)
$$\mathbf{E}\left[\int_0^1 u_s C_s \, ds\right] = \mathbf{E}\left[\int_0^1 u_s H_s \, ds\right],$$

where

$$H_s \equiv F'(X_s, (s, 1]).$$

Now, we shall see in Chapter VI that there a previsible process G, called the previsible projection of H, such that in particular,

$$\mathbf{E}(H_s \mid \mathcal{F}_s) = G_s,$$

so that

$$\mathbf{E}(u_s H_s) = \mathbf{E}(u_s G_s).$$

Hence,

$$\mathbf{E}\int_0^1 u_s C_s \, ds = \mathbf{E}\int_0^1 u_s G_s \, ds$$

for every bounded continuous adapted process u. But (see Exercise 41.8) the bounded continuous adapted processes are dense in $L^2(X)$, whence it is almost surely true that

$$C_s = G_s \textit{ for almost all } s.$$

(*41.8*) **Exercise.** Let S and T be stopping times with $S<T$. Find bounded continuous adapted processes (Y_n) such that

$$Y_n \to 1(S, T].$$

Deduce that the bounded continuous adapted processes are dense in $L^2(X)$. □

We have proved the following theorem.

(*41.9*) THEOREM (Clark [1]). *Let* $F: C[0, 1] \to \mathbb{R}$ *be a functional satisfying conditions* (41.4 (*i*)–(*iii*)). *Then*:

$$F(X) = \mathbf{E}(F(X)) + \int_0^1 C_t dX_t, \tag{41.10}$$

where C is the previsible projection of the process

$$H_t \equiv F'(X, (t, 1]).$$

(*41.11*) *Example.* Take $F(X) = \int_0^1 X_t dt$. Then F is a bounded linear map from Ω to $\mathbb{R}$, and $F'(X)(\varphi) = \int_0^1 \varphi_t dt$ for all $\varphi \in \Omega$. Thus

$$H_t \equiv F'(X, (t, 1]) = 1 - t$$

which is previsible, so $C_t = H_t$. By symmetry, $\mathbf{E}F(X) = 0$, so the representation (41.10) tells us that

$$F(X) \equiv \int_0^1 X_t dt = \int_0^1 (1-t) dX_t.$$

(*41.12*) **Exercise.** Confirm this directly. □

(41.13) *Example.* Define the increasing process $M_t \equiv \sup\{X_s : 0 \leqslant s \leqslant t\}$ and suppose $F(X) = M_1$. Now F is not a differentiable function, but it satisfies the conditions (41.4 (i)–(iii)). The verification of the first two is immediate, and as for the third, for almost all $X \in \Omega$, the maximum M_1 is attained at a unique time $\sigma(X)$. Moreover, for almost all X,

$$X(t+\sigma) - X(\sigma) \leqslant -|t|^{1/2+\delta}$$

in some neighbourhood of $t=0$, for each $\delta > 0$, and this implies

$$\lim \varepsilon^{-1}\{F(X+\varepsilon\varphi) - F(X)\} = \varphi(\sigma(X))$$

for almost all X if φ is smooth.

In this case, $H_t = I_{\{\sigma > t\}}$, and so

$$C_t = \mathbf{P}[\sigma > t \mid \mathcal{F}_t]$$
$$= \mathbf{P}(\tau_\xi < 1 - t)$$

where $\tau_a \equiv \inf\{t\colon X_t = a\}$, and $\xi = M_t - X_t$. Simple calculations then give (see §I.13)

$$C_t = 2\Phi\left[\frac{M_t - X_t}{(1-t)^{1/2}}\right]$$

where Φ is the tail of the Gaussian distribution with zero mean and unit variance. Since $\mathbf{E}M_1 = (2/\pi)^{1/2}$, Theorem 41.9 tells us that

$$M_1 = \left(\frac{2}{\pi}\right)^{1/2} + \int_0^1 2\Phi\left[\frac{M_t - X_t}{(1-t)^{1/2}}\right] dX_t.$$

(*41.14*) **Exercise.** Obtain this directly! □

(*41.15*) *Remarks.* The explicit martingale representation generalizes to Brownian motion in $\mathbb{R}^n$. For much more substantial generalizations of importance in control theory, see Haussmann [1], Bismut [2], Ocone [1], and the references contained therein.

42. Burkholder–Davis–Gundy inequalities. Martingale inequalities of diverse and wondrous forms have been the subject of intensive investigation in recent years, and it would be impossible to attempt a detailed coverage. Several references are given at the end of this section.

The Burkholder–Davis–Gundy inequalities have joined those of Doob amongst the classic results of the subject. We shall need to use them in our work on local time, and in our study of SDEs.

We remark that in many contexts, inequality (37.12) is extremely powerful, improving indeed on the B–D–G results.

We say that a function $F\colon \mathbb{R}^+ \to \mathbb{R}^+$ is *moderate* if it is continuous and increasing, $F(x) = 0$ if and only if $x = 0$, and for some (and then for every) $\alpha > 1$,

$$\sup_{x>0} \frac{F(\alpha x)}{F(x)} < \infty.$$

The most obvious examples of moderate F are the powers $F(x) = x^p$, where $p > 0$.

(*42.1*) THEOREM. *Let F be moderate. Then there exist universal constants $c_F \leqslant C_F$ such that, for every continuous local martingale M null at zero,*

$$c_F \mathbf{E}\, F([M]_\infty^{1/2}) \leqslant \mathbf{E}\, F(M_\infty^*) \leqslant C_F \mathbf{E}\, F([M]_\infty^{1/2}). \tag{42.2}$$

Here, as usual, $M_t^* \equiv \sup_{s \leqslant t} |M_s|$.

Remarks. For the case $F(x)=x^p$, Getoor and Sharpe [3] give a nice proof using stochastic calculus. We follow Burkholder's proof which relies on the 'good λ inequality'.

(*42.3*) LEMMA (good λ inequality). *Let X and Y be nonnegative random variables, and suppose that there exists $\beta>1$ such that, for all $\lambda>0$, $\delta>0$,*

(42.4) $$\mathbf{P}(X>\beta\lambda,\ Y\leqslant\delta\lambda)\leqslant\psi(\delta)\mathbf{P}(X>\lambda),$$

where $\psi(\delta)\downarrow 0$ as $\delta\downarrow 0$. Then, for each moderate function F, there exists a constant C depending only on β, ψ, and F such that

(42.5) $$\mathbf{E}[F(X)]\leqslant C\mathbf{E}[F(Y)].$$

Proof of Lemma. Without loss of generality, we may assume that $\mathbf{E}F(Y)<\infty$. Moreover, if the good λ inequality (42.4) holds, it holds with X replaced by $X\wedge a$ for any $a>0$, so that we may also suppose that $\mathbf{E}F(X)<\infty$. Pick a constant γ such that, for all x, $F(x/\beta)\geqslant\gamma F(x)$. Now integrate both sides of (42.4) with respect to $F(d\lambda)$. On the right we obtain $\psi(\delta)\mathbf{E}F(X)$, so that

$$\begin{aligned}\psi(\delta)\mathbf{E}F(X)&\geqslant\int_0^\infty F(d\lambda)\mathbf{E}I_{\{Y/\delta\leqslant\lambda<X/\beta\}}\\&=\mathbf{E}\left[\left(\int_0^{X/\beta}F(d\lambda)-\int_0^{Y/\delta}F(d\lambda)\right)^+\right]\\&\geqslant\mathbf{E}F(X/\beta)-\mathbf{E}F(Y/\delta)\\&\geqslant\gamma\mathbf{E}F(X)-\mathbf{E}F(Y/\delta).\end{aligned}$$

On rearranging, we have

$$\{\gamma-\psi(\delta)\}\mathbf{E}F(X)\leqslant\mathbf{E}F(\delta^{-1}Y).$$

Take $\delta<1$ so small that $\gamma-\psi(\delta)>\gamma/2$, and suppose that μ is such that $F(\delta^{-1}x)\leqslant\mu F(x)$ for all x. This gives inequality (42.5) with $C=2\mu/\gamma$, the same constant whatever the laws of X and Y. □

Proof of the theorem. We establish the good λ inequalities for $X=M^*_\infty$ and $Y=[M]^{1/2}_\infty$. If N is a continuous local martingale null at zero, if $a<0<b$, and if $\tau_x\equiv\inf\{t: N_t=x\}$, it is an easy exercise to prove that

(42.6) $$\mathbf{P}(\tau_b<\tau_a)\leqslant -a/(b-a).$$

Now take $\beta>1$ and $\lambda>0$, and let $\tau\equiv\inf\{u: |M_u|>\lambda\}$. Take $0<\delta<(\beta-1)$. Then $N_t\equiv(M_{\tau+t}-M_\tau)^2-([M]_{\tau+t}-[M]_\tau)$ is a continuous local martingale relative to $\mathcal{F}_t\equiv\mathcal{F}_{\tau+t}$, and on the event $\{M^*_\infty>\beta\lambda,\ [M]^{1/2}_\infty\leqslant\delta\lambda\}$, N reaches $(\beta-1)^2\lambda^2-\delta^2\lambda^2>0$ before $-\delta^2\lambda^2$, so by the above estimate (42.6),

$$\mathbf{P}(M^*_\infty>\beta\lambda,\ [M]^{1/2}_\infty\leqslant\delta\lambda\,|\,\mathcal{F}_\tau)\leqslant\delta^2/(\beta-1)^2.$$

Thus:

$$\begin{aligned}\mathbf{P}(M_\infty^* > \beta\lambda, [M]_\infty^{1/2} \leqslant \delta\lambda) &= \mathbf{P}(M_\infty^* > \beta\lambda, [M]_\infty^{1/2} \leqslant \delta\lambda, \tau < \infty) \\ &= \mathbf{E}(\mathbf{P}(M_\infty^* > \beta\lambda, [M]_\infty^{1/2} \leqslant \delta\lambda \mid \mathscr{F}_\tau) I_{\{\tau<\infty\}}) \\ &\leqslant \frac{\delta^2}{(\beta-1)^2}\, \mathbf{P}(\tau < \infty).\end{aligned}$$

The proof of the good λ inequality with the roles of M_∞^* and $[M]_\infty^{1/2}$ interchanged is accomplished by interchanging the roles of M^* and $[M]_\infty^{1/2}$ in the above proof! □

(*42.7*) *References to the literature.* An exhaustive survey of recent progress in the area of martingale inequalities cannot be attempted here; but as an indication of the intensity of activity in this field we refer you to Barlow and Yor [1, 2], Barlow, Jacka and Yor [1], Burkholder [1], Garsia [1], Lenglart, Lépingle and Pratelli [1], and (for an especially striking result) Jacka [1], and to the references which these papers contain.

43. Semimartingale local time; Tanaka's formula. The applications of Itô's formula we have seen so far have been valid for continuous semimartingales with values in $\mathbb{R}^n$ for any $n \geqslant 1$. For 1-dimensional continuous semimartingales, we have the added great advantage of having the concept of local time. Itô's formula and local time together supply simple and beautiful existence, uniqueness, and comparison results for 1-dimensional SDEs, and the combination of local time and Itô excursion theory can be applied very effectively to 1-dimensional diffusions. All this will be explained in Chapters V and VI.

Physicists like to think of the Brownian local time l_t at 0 as given by the expression,

(43.1) $$l_t = \int_0^t \delta(B_s)\, ds,$$

where δ is the Dirac delta function at 0. Now, if $f(x) = |x|$, then $f'(x) = \operatorname{sgn}(x)$ and $\frac{1}{2}f''(x) = \delta(x)$. Thus, formally, Itô's formula gives

(43.2) $$|B_t| - |B_0| = \int_0^t \operatorname{sgn}(B_s)\, dB_s + l_t;$$

and we can now make (43.1) precise by using (43.2) as a definition of l_t. The fundamental Theorem 43.3 carries through this idea for a general continuous semimartingale X.

Note. Comments on the different normalizations of local time, and on the relation between semimartingale local time and Markovian local time, are collected at the end of this section. Do not worry about them now.

(*43.3*) THEOREM (Tanaka's formula). *Let X be a continuous semimartingale. Then there exists a continuous increasing adapted process $\{l_t: t \geqslant 0\}$ such that:*

(43.4) $$|X_t| - |X_0| = \int_0^t \operatorname{sgn}(X_s)\, dX_s + l_t,$$

where $\operatorname{sgn}(x) = -1$ *if* $x < 0$; $\operatorname{sgn}(x) = 1$ *if* $x > 0$, *and (please note)* $\operatorname{sgn}(0) = -1$.

The process l is called the (semimartingale) local time of X at zero. It grows only when $X = 0$:

(43.5) $$\int_0^t I_{\{X_s \neq 0\}}\, dl_s = 0.$$

Remarks. (i) The identity (43.4), which can be taken as the definition of local time, is called Tanaka's formula. It can be generalized to semimartingales which are not continuous; see Meyer [1]. We follow Meyer's development, taking advantage of the short-cuts opened up by the continuity assumption.

(ii) In many applications, the definition of sgn (0) is irrelevant; here, though, it is essential for the semimartingale equivalent of Trotter's theorem that sgn is left continuous.

(iii) The alternative forms of Tanaka's formula

(43.6) $$X_t^+ - X_0^+ = \int_0^t I_{\{X_s > 0\}}\, dX_s + \tfrac{1}{2} l_t,$$

(43.7) $$X_t - X_0 = -\int_0^t I_{\{X_s \leqslant 0\}}\, dX_s + \tfrac{1}{2} l_t$$

are often useful. They are immediate consequences of (43.4).

Proof. The obvious idea is to approximate the function $f(x) = |x|$ by C^2 functions.

Let φ be a C^∞ increasing function $\varphi: \mathbb{R} \to [-1, 1]$ such that $\varphi(x) = -1$ for $x \leqslant 0$, and $\varphi(x) = 1$ for $x \geqslant 1$. Define C^∞ functions f_n by

$$f_n(x) = -x \quad (x \leqslant 0), \quad f_n'(x) = \varphi(nx) \quad \text{for all } x.$$

Then the f_n converge uniformly to f, and $f_n'(x) \uparrow \operatorname{sgn}(x)$. By Itô's formula,

(43.8) $$\begin{aligned} f_n(X_t) - f_n(X_0) &= \int_0^t f_n'(X_s)\, dX_s + \tfrac{1}{2} \int_0^t f_n''(X_s)\, d[X]_s \\ &= \int_0^t f_n'(X_s)\, dX_s + C_t^n, \end{aligned}$$

say. Since $f_n'' \geqslant 0$, C^n is a continuous increasing process and, since $f_n''(x) = 0$ for $|x| \geqslant n^{-1}$, we have

(43.9) $$\int_0^t I_{\{|X_s| > 1/n\}}\, dC_s^n = 0.$$

Let $X_t = X_0 + M + A$ be the canonical decomposition of X. Localization allows us to reduce the problem to the case when M is bounded and A is of bounded variation. Then:

$$\left\| \int_0^\infty (\operatorname{sgn}(X_s) - f_n'(X_s))\, dM_s \right\|_2^2 = \mathbf{E} \int_0^\infty (\operatorname{sgn}(X_s) - f_n'(X_s))^2 \, d[M]_s$$

$$\to 0 \quad \text{as } n \to \infty.$$

By Doob's L^2 martingale inequality, we may suppose that (on passing to a subsequence if necessary)

$$\sup_t \left| \int_0^t (\operatorname{sgn}(X_s) - f_n'(X_s)) dM_s \right| \to 0$$

almost surely and in L^2. Also,

$$\left| \int_0^t (\operatorname{sgn}(X_s) - f_n'(X_s)) dA_s \right| \leqslant \int_0^t |\operatorname{sgn}(X_s) - f_n'(X_s)| . |dA_s|$$

$$= \int_0^t \{\operatorname{sgn}(X_s) - f_n'(X_s)\} . |dA_s|.$$

The monotone-convergence theorem allows us to conclude that

$$\int_0^t f_n'(X_s) dA_s \to \int_0^t \operatorname{sgn}(X_s) dA_s,$$

the convergence being uniform in t. Thus the left-hand side of (43.8) converges uniformly to the left-hand side of (43.4), and the stochastic integral on the right of (43.8) converges uniformly a.s. to the stochastic integral on the right of (43.4). Hence C^n also converges uniformly a.s., to a limit l which is continuous and increasing, and (43.4) holds.

If we regard C^n as the distribution function of a measure, then C^n converges weakly to l, so, by (I.42.3), it follows from (43.9), that, for each n,

$$\int_0^t I_{\{|X_s| > 1/n\}} dl_s = 0;$$

(43.5) follows immediately. □

(*43.9*) *Markovian local time.* We made a first study of Markovian local time in §III.16, and will study this concept again in our treatments of 1-dimensional diffusions in Chapter V and of excursion theory in Chapter VI. In Markov process theory, we are interested in local time as a continuous additive functional generated by a certain λ-potential.

Let us now check that, for Brownian motion, the Markovian and semimartingale local times agree (modulo normalization). (Yor [2] points out a case in which a process X is both a Markov process and a semimartingale in which the two types of local time are fundamentally different.)

Let $(\Omega, \mathscr{F}, \mathscr{F}_t, \mathbf{P}, B)$ relate to CBM(R). We saw in §§ III.16.6 and III.17.5 that Brownian local time at 0 has the property that

$$\mathbf{E}^x \int_0^\infty \exp(-\tfrac{1}{2}s)dA_s = \exp(-|x|)$$

(in other words, the $\frac{1}{2}$-potential of A is $f(x) = \exp(-|x|)$), and that

$$C_t = \int_0^t \exp(-\tfrac{1}{2}s)\,dA_s$$

is a continuous increasing adapted process such that

(43.10) $$\exp(-\tfrac{1}{2}t)f(B_t) = \exp(-\tfrac{1}{2}t - |B_t|) = M_t - C_t,$$

where M is a martingale. As remarked in § III.17, this is the Meyer decomposition of the regular class (D) potential on the left-hand side of (43.10)—see § VI.31. But

$$Y \equiv \exp(-\tfrac{1}{2}t - |B|) = \mathscr{E}(-|B|),$$

so that

(43.11) $$dY = -Y d|B| = -Y \operatorname{sgn}(B) dB - Y dl,$$

where l is semimartingale local time at 0 for B. Thus, on comparing (43.10) with (43.11), we see that

$$C_t - \int_0^t Y_s dl_s$$

is a continuous local martingale of finite variation null at 0, and so is identically zero. Moreover, since l grows only when $B=0$, that is, when $Y_s = \exp(-\frac{1}{2}s)$, we have

$$\int_0^t \exp(-\tfrac{1}{2}s)dA_s = \int_0^t \exp(-\tfrac{1}{2}s)dl_s,$$

so that

$$l = A.$$

(43.12) Remarks on the vexed question of normalization. The normalization of semimartingale local time has now become standard. It is unfortunate that, for Brownian motion, it proves to be twice the 'standard local time' of Itô and McKean.

We shall always use the symbol l to denote the semimartingale Brownian local time, which, as we shall prove in §45, is the occupation density with respect to Lebesgue measure.

44. Study of joint continuity. Though we have so far defined local time of a continuous semimartingale X only at zero, there is nothing to prevent our defining analogously the local time at a by:

$$l_t^a \equiv |X_t - a| - |X_0 - a| - \int_0^t \operatorname{sgn}(X_s - a)\, dX_s. \tag{44.1}$$

This immediately raises questions about the measurability of the 2-parameter process $\{l_t^a : a \in \mathbb{R},\, t \geqslant 0\}$; the following result deals with them.

(*44.2*) THEOREM (Yor [2]). *There exists a version of* $\{l_t^a : a \in \mathbb{R},\, t \geqslant 0\}$ *which is jointly continuous in t and right-continuous with left limits in a*:

$$\lim_{\substack{s \to t \\ b \downarrow a}} l_s^b = l_t^a,$$

$$\lim_{\substack{s \to t \\ b \uparrow\uparrow a}} l_s^b = l_t^{a-} \text{ exists.}$$

Moreover,

$$l_t^a - l_t^{a-} = 2\int_0^t I_{\{X_s = a\}}\, dA_s, \tag{44.3}$$

where $X_t = X_0 + M_t + A_t$ *is the canonical decomposition of* X.

Proof. Because of the usual localization arguments, we may suppose that M and the variation of A are both bounded by some constant K. Define

$$\xi_1(a, t) \equiv \int_0^t \operatorname{sgn}(X_s - a)\, dM_s,$$

$$\xi_2(a, t) \equiv \int_0^t \operatorname{sgn}(X_s - a)\, dA_s,$$

$$\xi_3(a, t) \equiv |X_t - a| - |X_0 - a|.$$

Then ξ_3 is jointly continuous, ξ_2 is continuous in t and right continuous with left limits in a, and

$$\lim_{\substack{s \to t \\ b \uparrow\uparrow a}} \xi_2(b, s) = \int_0^t (1 - 2I_{\{X_s < a\}})\, dA_s$$

$$= \xi_2(a-, t).$$

Since

$$\xi_2(a, t) - \xi_2(a-, t) = -2\int_0^t I_{\{X_s = a\}}\, dA_s,$$

all we have to do is prove that ξ_1 has a jointly continuous version. Now there is a classical and very powerful tool for proving the existence of continuous versions of stochastic processes:

(*44.4*) LEMMA (Kolmogorov). *Let $\{\xi(x)\colon x \in \mathbb{R}^n\}$ be a stochastic process with values in some complete metric space (S, ρ). If there exist positive constants α, β, ε such that:*

$$\mathbf{E}[\rho(\xi(x), \xi(y))^\alpha] \leqslant \beta |x-y|^{n+\varepsilon}$$

for all $x, y \in \mathbb{R}^n$, then a continuous version of ξ exists.

Proof. See Meyer [7]. □

We apply this to the case where $n = 1$, $S = C([0, \infty], \mathbb{R})$ with the uniform norm, and $\xi(x) \equiv (t \mapsto \xi_1(x, t))$. We have for $a < b$,

$$\text{(44.5)} \qquad \mathbf{E}[\sup_t |\xi_1(a, t) - \xi_1(b, t)|^p] = 2^p \mathbf{E}\left[\sup_t \left|\int_0^t I_{\{a < X_s \leqslant b\}}\, dM_s\right|^p\right]$$

$$\leqslant C_p \mathbf{E}\left[\left|\int_0^\infty I_{\{a < X_s \leqslant b\}}\, d[M]_s\right|^{p/2}\right]$$

by the Burkholder–Davis–Gundy inequalities. Now take some C^2 function $f\colon \mathbb{R} \to \mathbb{R}$ with the following properties: $0 \leqslant f''(x) \leqslant 1$, $f''(x) = 0$ for $x \notin (-1, 2)$, $f''(x) = 1$ for $x \in [0, 1]$, $f'(x) = 0$ for $x \leqslant -1$. Letting $\varphi(x) \equiv f(\delta^{-1}(x-a))$, where $\delta \equiv b - a$, we have, using the fact that $0 \leqslant f'(x) \leqslant 3$ at the fifth step below,

$$0 \leqslant \tfrac{1}{2}\int_0^t I_{\{a < X_s \leqslant b\}}\, d[X]_s$$

$$\leqslant \tfrac{1}{2}\delta^2 \int_0^t \varphi''(X_s)\, d[X]_s$$

$$= \delta^2 \left\{\varphi(X_t) - \varphi(X_0) - \int_0^t \varphi'(X_s)\, dX_s\right\}$$

$$\leqslant \delta^2 |\varphi(X_t) - \varphi(X_0)| + \delta^2 \left|\int_0^t \varphi'(X_s)\, dM_s\right| + \delta^2 \int_0^t |\varphi'(X_s)| . |dA_s|$$

$$\leqslant 3\delta |X_t - X_0| + \delta \left|\int_0^t f'(\delta^{-1}(X_s - a))\, dM_s\right| + 3\delta \int_0^t |dA_s|$$

$$\leqslant 9K\delta + \delta \sup_t \left|\int_0^t f'(\delta^{-1}(X_s - a))\, dM_s\right|.$$

Hence by the Burkholder–Davis–Gundy inequalities again,

$$\text{(44.6)} \qquad \mathbf{E}\left\{\left|\int_0^\infty I_{\{a<X_s\leqslant b\}}\, d[X]\right|^{p/2}\right\}$$
$$\leqslant \delta^{p/2} C_p'\left(1+\mathbf{E}\left|\int_0^\infty f'(\delta^{-1}(X_s-a))^2\, d[M]_s\right|^{p/2}\right)$$
$$\leqslant \delta^{p/2} C_p'(1+3^p\mathbf{E}([M]_\infty^{p/2})).$$

Now since $\mathbf{E}([M]_\infty^{p/2})\leqslant C_p\mathbf{E}((M_\infty^*)^p)\leqslant K^pC_p''<\infty$, we can assemble (44.5) and (44.6) to give for some $\beta_p<\infty$,

$$\mathbf{E}\left[\sup_t|\xi_1(a,t)-\xi_1(b,t)|^p\right]\leqslant \beta_p|a-b|^{p/2}.$$

It remains only to take $p>2$ and use Kolmogorov's Lemma. □

Remarks. (i) For the extension to semimartingales which are not necessarily continuous, see Yor [2].

(ii) *In the case of a continuous local martingale, $A=0$, so from (44.3) we see that l is actually jointly continuous; for Brownian motion, this is part of Trotter's celebrated theorem* (see § I.5).

(iii) In general, the local time is not jointly continuous. Let B be a Brownian motion, and let $X=|B|$. Then, with an obvious notation,

$$dX=d|B|=\text{sgn}(B)dB+dl^B(t,0);$$

and, by definition of l^X and because $X=|X|$ and sgn (0) is defined to equal -1,

$$X_t-X_0=\int_0^t \text{sgn}(X_s)dX_s+l^X(t,0)$$
$$=\int_0^t [1-2I_{\{0\}}(X_s)]\,dX_s+l^X(t,0)$$
$$=X_t-X_0-2\int_0^t I_{\{0\}}(X_s)\text{sgn}(B_s)\,dB_s$$
$$-2\int_0^t I_{\{0\}}(B_s)\,dl^B(t,0)+l^X(t,0)$$
$$=X_t-X_0-2l^B(t,0)+l^X(t,0).$$

Thus,

$$l^X(t,0)=2l^B(t,0)$$

agreeing with the fact that

$$l^X(t,0)=l^X(t,0)-l^X(t,0-)=2\int_0^t I_{\{0\}}(X_s)\,dA,$$

where $A_t=l^B(t,0)$, the finite-variation part of X.

(iv) Return to the general case. If M is bounded, then obviously

$$L^2\text{-lim}\int_0^{\cdot} I_{\{a-2^{-n}<X_s\leqslant a\}}\,dM_s=\int_0^{\cdot} I_{\{X_s=a\}}\,dM_s.$$

The continuity of ξ_1 immediately tells us that a.s. for all a, t

$$\int_0^t I_{\{X_s=a\}}\,dM_s=0,$$

or equivalently

$$\int_0^t I_{\{X_s=a\}}\,d[M]_s=\int_0^t I_{\{X_s=a\}}\,d[X]_s=0.$$

(*44.7*) **Exercise.** Use Kolmogorov's Lemma to provide a quick proof of the existence of a continuous version of Brownian motion.

45. Local time as an occupation density; generalized Itô–Tanaka formula. Trotter's theorem I.5 is still only partly proved; it remains to show that local time is the occupation density for Brownian motion. We will deduce this from the extension of Itô's formula to convex functions provided by the next theorem. Theorem 45.1 (in the generality in which it is presented) is the result on which we base our treatment of 1-dimensional diffusion theory in Part 7 of Chapter V.

Recall that a function f on $\mathbb{R}$ is called *convex* if, whenever $x<y<z$, then

$$(z-x)f(y)\leqslant(z-y)f(x)+(y-x)f(z),$$

or, equivalently,

$$(z-y)\{f(z)-f(x)\}\leqslant(z-x)\{f(z)-f(y)\}.$$

Thus, for $x<y<z$,

$$(z-x)^{-1}\{f(z)-f(x)\}\leqslant(z-y)^{-1}\{f(z)-f(y)\},$$

so the left-hand derivative

$$(D_-f)(z)\equiv\uparrow\lim_{y\uparrow\uparrow z}(z-y)^{-1}\{f(z)-f(y)\}$$

exists. The monotone-convergence theorem now yields

$$\int_a^b (D_-f)(y)\,dy=f(b)-f(a).$$

It is easily verified that D_-f is left-continuous and non-decreasing, so that we can define a measure μ on $\mathscr{B}(\mathbb{R})$ via

$$\mu[a, b)=(D_-f)(b)-(D_-f)(a).$$

For $g \in C_K^\infty$,

$$\int g'' f\,dx = \int g\,d\mu,$$

and we say that $f''=\mu$ as a measure (or positive Schwartz distribution).

(*45.1*) THEOREM (Trotter, Meyer). *Let X be a continuous semimartingale with local time process $\{l_t^a : a \in \mathbb{R}, t \geqslant 0\}$ (assumed continuous in t, right continuous with left limits in a). If $f\colon \mathbb{R} \to \mathbb{R}$ is a convex function, with $f''=\mu$ as a measure, and D_-f is the left derivative of f, then:*

$$f(X_t)-f(X_0)=\int_0^t D_-f(X_s)\,dX_s+\tfrac{1}{2}\int_{\mathbb{R}} l_t^a \mu(da). \tag{45.2}$$

Moreover, for any bounded measurable φ,

$$\int_0^t \varphi(X_s)\,d[X]_s=\int_{\mathbb{R}} l_t^a \varphi(a)\,da. \tag{45.3}$$

Remarks. (i) Of course, (45.2) is valid when f is the difference of two convex functions. If $f(x)=|x|$, then (45.2) is simply Tanaka's formula. Note that the use of the left-hand derivative tallies with our current convention that sgn $(0)=-1$.

(ii) For Brownian motion, $[X]_t = t$ and (45.3) is the occupation identity of Trotter's theorem I.5.

(iii) The theorem appeared (in more general form) in Meyer [1].

Proof. By localization, we may suppose that X is bounded; we can also suppose that μ is supported in $[-N, N]$ and we lose no generality in supposing that $\mu(\mathbb{R})=1$. Further, if (45.2) is true for some f, it is also true for $f(x)+cx$ for any c, so we can assume that $D_-f=0$ for $x \leqslant -N$.

As remarked above, if the support of μ is a singleton, then (45.2) is essentially just Tanaka's formula, and by linearity, (45.2) is still valid if the support of μ is a finite set.

Suppose that Y is a random variable with law μ and distribution function F. Define $h_n\colon \mathbb{R} \to \mathbb{R}$ by

$$h_n(x)=j2^{-n} \quad \text{if} \quad (j-1)2^{-n}<x \leqslant j2^{-n},$$

so that $x \leqslant h_n(x) < x+2^{-n}$, and let $Y_n \equiv h_n(Y)$. Let F_n be the distribution function of Y_n; F_n is the distribution function of a measure μ_n whose support is finite, and the linear extension of Tanaka's formula yields

$$f_n(X_t)-f_n(X_0)=\int_0^t D_-f_n(X_s)\,dX_s+\tfrac{1}{2}\int_{\mathbb{R}} l_t^a \mu_n(da),$$

where

$$f_n(x) = \int_{-\infty}^{x} F_n(y)\,dy.$$

The following facts are elementary:

(i) If ψ is bounded and right-continuous,

$$\mathbf{E}\psi(Y_n) = \int \psi(a)\mu_n(da) \to \mathbf{E}\psi(Y) = \int \psi(a)\mu(da)$$

(ii) $F_n(x-) \uparrow F(x-)$ as $n \to \infty$.

(iii) $f_n(x) \uparrow f(x) = \int_{-\infty}^{x} F(y)\,dy.$

Using (i) and the right continuity in a of l,

$$\int_{\mathbb{R}} l_t^a \mu_n(da) \to \int_{\mathbb{R}} l_t^a \mu(da).$$

Using (ii), standard arguments show that

$$\int_0^t F_n(X_s-)\,dX_s \to \int_0^t F(X_s-)\,dX_s = \int_0^t D_- f(X_s)\,dX_s$$

uniformly in L^2 over $[0, \infty)$, and finally, using (iii) gives us

$$f_n(X_t) - f_n(X_0) \to f(X_t) - f(X_0).$$

Result (45.2) is therefore proved. Result (45.3) follows for φ in C^2 on comparing (45.2) with the classical Itô formula, and then extends to measurable φ by monotone-class arguments. □

The following extension is often useful.

(*45.4*) (CONTINUATION OF THEOREM 45.1). *If g is a positive Borel function on $[0, \infty) \times \mathbb{R}$, then:*

$$\int_0^t g(s, X_s)\,d[X]_s = \int_{x \in \mathbb{R}} dx \int_{s=0}^{t} g(s, x)\,d_s l^x.$$

Proof. You should have no difficulty in establishing this result by first considering the case in which g has the form

$$g(t, x) = I_{[u, v]}(t)\varphi(x)$$

and then using monotone-class arguments. □

(*45.5*) *Berman's 'Fourier' approach.* A very attractive and important series of results on local time as occupation density for Gaussian and other processes, obtained by Fourier techniques, may be found in Berman [1, 2, 3].

(*45.6*) *Markovian proof of the occupation density result for Brownian motion.* Once we know from the last section that, for Brownian motion B, l^x is jointly continuous in t and x, we can prove that

$$\int_a^b l_t^x\,dx = \int_0^t I_{[a,b]}(B_s)\,ds \tag{45.7}$$

as follows. First, we need to argue that the left-hand side of (45.7) defines a PCHAF of B, and then we prove that (for $\lambda > 0$) the two sides of (45.7) have the same λ-potential. Then we can appeal to the Volkonskii–Šur–Meyer theorem (see §III.16.7). You can carry through this approach as an exercise.

In Chapter VI, we shall take up the matter raised in §III.16, of the relationship between the V–Š–M theorem and the Meyer decomposition theorem for supermartingales.

(*45.8*) *A different relaxation of the C^2 condition for Itô's formula.* For certain applications, particularly those in control theory, we need to apply Itô's formula to functions which are neither C^2 nor differences of convex functions.

We just give a simple result of this type which we shall need in §V.28.

(*45.9*) LEMMA. *Let X be a continuous semimartingale in $\mathbb{R}$, with $X_0 = 0$. Let f be a C^1 function on $\mathbb{R}$ such that, in the sense of Schwarz distributions, f'' is a function h: thus, h is a measurable function integrable on each interval $[-a, a]$ such that, for $x < y$,*

$$f'(y) - f'(x) = \int_x^y h(z)\,dz.$$

Then Itô's formula holds for f:

$$f(X_t) - f(X_0) = \int_0^t f'(X_s)\,dX_s + \tfrac{1}{2}\int_0^t h(X_s)\,d[X]_s. \tag{45.10}$$

Proof. It is enough to fix $a > 0$ and establish (45.10) for $t < \tau_a$, where

$$\tau_a \equiv \inf\{t : |X_t| = a\}.$$

Let (h_n) be a sequence of continuous functions on $\mathbb{R}$ such that

$$\varepsilon_n \equiv \int_{-a}^{a} |h_n(x) - h(x)|\,dx \to 0.$$

Define functions f_n on $\mathbb{R}$ via

$$f_n'' = h_n, \quad f_n'(0) = f'(0), \quad f_n(0) = f(0).$$

Then:

$$\text{on } [-a, a], \quad |f_n' - f'| < \varepsilon_n, \quad \text{and} \quad |f_n - f| < a\varepsilon_n. \tag{45.11}$$

By Itô's formula applied to the C^2 function f_n,

$$f_n(X_t)-f_n(X_0)=\int_0^t f'_n(X_s)\,dX_s+\tfrac{1}{2}\int_0^t h_n(X_s)\,d[X]_s. \tag{45.12}$$

Consider what happens as $n\to\infty$. By (45.11), the left-hand side of (45.12) converges to the left-hand side of (45.10), and it is now well within your powers to prove that the first integral on the right-hand side of (45.12) also does the correct thing.

It only remains to notice that, for $t<\tau_a$,

$$\left|\int_0^t \{h_n(X_s)-h(X_s)\}\,d[X]_s\right|=\left|\int_{-a}^a \{h_n(x)-h(x)\}\,l_t^x\,dx\right|$$
$$\leqslant \varepsilon_n \sup\{l_t^x : x\in[-a,a]\},$$

whence, since $x\mapsto l_t^x$ is an R-function and is therefore bounded on $[-a,a]$, the result follows. □

Note: The result follows immediately from Theorem 45.1 because f is the difference of two convex functions!

For a general result for $f(B)$ where B is a BM($\mathbb{R}^n$), see Theorem 2.10.1 of Krylov [1].

46. The Stratonovich calculus. Let X and Y be continuous semimartingales. The Stratonovich integral $S=\int Y\partial X$ is defined by

$$S_t=\int_0^t Y\partial X\equiv\int_0^t Y\,dX+\tfrac{1}{2}[Y,X]_t. \tag{46.1}$$

Throughout the book, we always use:

d to denote the Itô differential,
∂ to denote the Stratonovich differential.

The following notations are therefore used to describe (46.1):

$$\partial S = Y\partial X, \tag{46.2}$$

$$dS = Y\,dX+\tfrac{1}{2}d[Y,X] = Y\,dX+\tfrac{1}{2}dY\,dX. \tag{46.3}$$

(*46.4*) THEOREM (rules for Stratonovich calculus). *The Stratonovich calculus obeys the same rules as the Newton–Leibnitz calculus in the following sense. If X and Y are continuous semimartingales, and $f\in C^3$, then:*

$$X_tY_t-X_0Y_0=\int_0^t X_s\partial Y_s+\int_0^t Y_s\partial X_s, \tag{46.5}$$

(46.6) $$f(X_t)-f(X_0)=\int_0^t f'(X_s)\partial X_s.$$

Proof. Result (46.5) is obvious because of the Itô rule:

$$\begin{aligned} d(XY) &= XdY+YdX+dXdY \\ &= XdY+\tfrac{1}{2}dXdY+YdX+\tfrac{1}{2}dYdX. \end{aligned}$$

Next, for $f\in C^3$, we have

(46.7) $$d\{f(X)\}=f'(X)dX+\tfrac{1}{2}f''(X)d[X],$$

(46.8) $$d\{f'(X)\}=f''(X)dX+\tfrac{1}{2}f'''(X)d[X].$$

From (46.8) and the Kunita–Watanabe result (31.5), we have

(46.9) $$d[f'(X),X]=f''(X)d[X].$$

Equation (46.6) is immediate from (46.8) and (46.9). □

(*46.10*) *Notation.* According to the notation (46.2), we write (46.6) in differential form:

(46.11) $$\partial f(X)=f'(X)\partial X.$$ □

(*46.12*) *There is much more to it!* The relation between the Itô and Stratonovich integrals at (46.1) may be written:

(46.13) $$S_t=\int_0^t Y\partial X=\left\{\int_0^t YdM\right\}+\left\{\int_0^t YdA+\tfrac{1}{2}[Y,X]_t\right\},$$

where $X=X_0+M+A$ is the canonical decomposition of X. Now, the right-hand side of (46.13) is the canonical decomposition of the semimartingale S into its local-martingale and finite-variation parts.

It was explained at 15.8 that, while the concept of a semimartingale X on a manifold Σ has an obvious definition (namely: $h(X)$ is a semimartingale on $\mathbb{R}$ for all h in $C^\infty(\Sigma)$), it is not at all obvious how to define a local martingale on a manifold. Indeed, *to define a local martingale on a manifold, we need an extra structure which is exactly that of a linear connection on* Σ. This extremely nice link between martingale theory and differential geometry was made by Bismut and Meyer—see Meyer [6, 7] and §V.33.

Recall that, as we have repeatedly emphasized, the central property of Itô integrals is their preservation of the local-martingale property.

What now emerges—see Meyer [6, 7]—is that the Stratonovich integral (of 1-forms along semimartingale paths or over more general 'chains') may be defined on any smooth manifold, while the Itô integral may be defined only on manifolds with a connection (and on which, therefore, the local-martingale property is meaningful).

47. Riemann-sum approximation to Itô and Stratonovich integrals; simulation. Throughout this section, we use the familiar notation:

$$t_i^n = t \wedge (i2^{-n}).$$

(*47.1*) LEMMA (the Itô case). *Let C be a locally bounded adapted* L*-process, and let X be a continuous semimartingale. Put* $C(0)=0$, *and set:*

$$I^n(t) \equiv \sum_{i \geqslant 1} C(t_{i-1}^n)\{X(t_i^n) - X(t_{i-1}^n)\},$$

$$I(t) \equiv \int_0^t C\,dX.$$

Then $I^n \to I$ uniformly in probability on compact intervals: for $K>0$ and $\varepsilon>0$,

$$\mathbf{P}\left\{\sup_{t \leqslant K} |I^n(t) - I(t)| > \varepsilon\right\} \to 0 \quad \text{as } n \to \infty.$$

Proof. Let

$$C^n(t) \equiv C(t_{i-1}^n) \quad \text{on } (t_{i-1}^n, t_i^n].$$

Then $C^n(t, \omega) \to C(t, \omega)$, $\forall t$, $\forall \omega$. Also,

$$I^n(t) = \int_0^t C^n\,dX = (C^n \bullet X)(t).$$

Let $X = X_0 + M + A$ be the canonical decomposition of X. Then, by classical Lebesgue theory, for each ω,

$$C^n \bullet A \to C \bullet A \textit{ uniformly on compact intervals;}$$

and it remains only to prove that

$$C^n \bullet M \to C \bullet M \quad \text{(u.p.c.)} \tag{47.2}$$

where 'u.p.c.' signifies 'uniformly in probability on compact intervals'.

It is clearly enough to prove that, for a sequence $(T(k))$ of stopping times with $T(k) \uparrow \infty$,

$$(C^n \bullet M)^{T(k)} \to (C \bullet M)^{T(k)} \textit{ uniformly in probability.}$$

But, by (10.2) and (29.1), this reduces the problem to showing that if C is bounded and $M \in \mathcal{M}_0^2$, then

$$C^n \bullet M \to C \bullet M \textit{ uniformly in probability,}$$

and this is obvious from the L^2 isometry property of the stochastic integral and Doob's L^2 inequality. □

(*47.3*) LEMMA (the Stratonovich case). *Let X and Y be continuous semimartingales. Define*:

$$S^n(t) \equiv \sum_{i \geq 1} \tfrac{1}{2}\{Y(t_i) + Y(t_{i-1})\}\{X(t_i) - X(t_{i-1})\},$$

$$S(t) \equiv \int_0^t Y \partial X.$$

Then $S^n \to S$ *(u.p.c.).*

Proof. Put $C = Y$ in the previous lemma. Then

$$S^n(t) = I^n(t) + \tfrac{1}{2}\sum_i \Delta Y(t_i^n)\Delta X(t_i^n),$$

where $\Delta Z(t_i^n) \equiv Z(t_i^n) - Z(t_{i-1}^n)$. Lemma 47.1 and polarization reduce the problem to showing that

$$\sum_i \{\Delta X(t_i^n)\}^2 \to [X](t) \quad \text{(u.p.c.)}.$$

But, by Theorem 30.1,

$$\sum_i \{\Delta M(t_i^n)\}^2 \to [X](t) \equiv [M](t) \quad \text{(u.p.c.)}.$$

Also, if Z denotes either M or A, then

$$\left|\sum_i \Delta Z(t_i^n)\Delta A(t_i^n)\right| \leq \sup|\Delta Z(t_i^n)| \operatorname{Var}_A(t) \to 0 \quad \text{(u.p.c)},$$

since Z is uniformly continuous on compact intervals. The result follows.

(*47.4*) *Simulation.* The two lemmas just given are clearly relevant to the problem of simulating processes described by stochastic differential equations.

The most important simulation technique is that due to Mihlstein [1]. See Clark [3], Pardoux and Talay [1].

CHAPTER V

Stochastic Differential Equations and Diffusions

1. INTRODUCTION

1. What is a diffusion in $\mathbb{R}^n$? One of the main uses of the theory of stochastic differential equations (habitually abbreviated to SDEs) is to construct and study diffusions in $\mathbb{R}^n$. So what is a diffusion in $\mathbb{R}^n$? Everybody knows intuitively what a diffusion is, yet mathematical texts featuring the word 'diffusion' in the title have been written without any precise definition of the term appearing! The problem is that the obvious definition, as a continuous time-homogeneous $\mathbb{R}^n$-valued strong Markov process, is too wide; it embraces examples with very unruly behaviour. (The process (X, Y) on $\mathbb{R}^2$, where X and Y are independent Feller–McKean processes (see §III.23), is an indication of the type of bizarre process which can arise.)

Let us then be guided in the framing of our (current!) definition by the *intuitive* notion that a diffusion is a continuous process $X = (X^1, \ldots, X^n)$ such that

(1.1(i)) $$\mathbf{E}[X^i_{t+h} - X^i_t \mid \mathscr{X}_t] = b^i(X_t)h + o(h),$$

(1.1(ii)) $$\mathbf{E}[(X^i_{t+h} - X^i_t - hb^i(X_t))(X^j_{t+h} - X^j_t - hb^j(X_t)) \mid \mathscr{X}_t] = a^{ij}(X_t)h + o(h)$$

for $i, j = 1, \ldots, n$, $t \geqslant 0$ and $h > 0$. Here, $\mathscr{X}_t \equiv \sigma(\{X_s : s \leqslant t\})$, and b^i and a^{ij} are measurable functions. Notice that $a(x) \equiv (a^{ij}(x))$ is a nonnegative definite symmetric matrix for each x.

Chapter IV affords us a language to translate this intuition into rigorous mathematics.

(*1.2*) DEFINITION (diffusion; covariance; drift). *Let a and b be measurable functions such that a: $\mathbb{R}^n \to S_n^+$ and b: $\mathbb{R}^n \to \mathbb{R}^n$, where*:

(1.3) $$S_n^+ \equiv \{\text{real } n \times n \text{ nonnegative-definite symmetric matrices}\}.$$

A diffusion *in $\mathbb{R}^n$ with* covariance *a and* drift *b is a continuous $\mathbb{R}^n$-valued semimartingale $X = (X^1, \ldots, X^n)$ (defined on some filtered probability space* *satisfying the usual conditions) such that, for each* $i=1,\ldots,n$,

(1.4)
$$M_t^i \equiv X_t^i - X_0^i - \int_0^t b^i(X_s)\,ds$$

is a continuous local martingale, for which

(1.5)
$$[M^i, M^j]_t = \int_0^t a^{ij}(X_s)\,ds.$$

For brevity, we refer to a diffusion with covariance a and drift b as an (a, b)*-diffusion.*

Do notice that there is *a priori* no reason why an (a, b)-diffusion X should be a Markov process. We shall see later, though, that *if a and b satisfy certain conditions which will hold in nearly every practical situation, then any* (a, b)*-diffusion is a strong Markov process.* For the time being, we concentrate on diffusions as semimartingales, and exploit the theory developed in Chapter IV. In particular, we have Itô's formula at our disposal; a simple (and very important) application is to show that, *if* X *is an* (a, b)*-diffusion, then:*

(1.6) *for each* $f \in C^\infty$, *the process*
$$C_t^f \equiv f(X_t) - f(X_0) - \int_0^t \mathscr{L}f(X_s)\,ds$$
is a local martingale, where $\mathscr{L}$ *is the operator*

(1.7)
$$\mathscr{L} = \tfrac{1}{2}\sum_{i,j=1}^n a^{ij}(\cdot)D_iD_j + \sum_{i=1}^n b^i(\cdot)D_i.$$

(Here, $D_i \equiv \partial/\partial x^i$.) Indeed, by Itô's formula,
$$C_t^f = \sum_{i=1}^n \int_0^t D_i f(X_s)\,dM_s^i.$$

If, additionally, a and b are locally bounded, then

(1.8) *for each* $f \in C_K^\infty$, C^f *is a martingale.*

In fact, each of (1.6) and (1.8) is an alternative definition of an (a, b)-diffusion.

(*1.9*) LEMMA. *Let* X *be a continuous semimartingale in* $\mathbb{R}^n$, *and suppose that* $a\colon \mathbb{R}^n \to S_n^+$ *and* $b\colon \mathbb{R}^n \to \mathbb{R}^n$ *are locally bounded. Then the following are equivalent:*

(i) X *is an* (a, b)*-diffusion;*
(ii) X *satisfies* (1.6);
(iii) X *satisfies* (1.8).

Proof. That (i) $\Rightarrow$ (ii) is just Itô's formula, and that (ii) $\Rightarrow$ (i) follows by taking firstly $f(x) = x^i$, and then $f(x) = x^i x^j$. If $f \in C_K^\infty$, then both f and $\mathscr{L}f$ are bounded, and so the fact that (ii) $\Rightarrow$ (iii) is immediate. To show that (iii) $\Rightarrow$ (ii), stop X at

$T_k \equiv \inf\{t : |X_t| > k\}$, and notice that

$$C^f(T_k \wedge \cdot) = C^{f_k}(T_k \wedge \cdot),$$

where f_k in C_K^∞ is such that $f_k(x) = f(x)$ if $|x| \leqslant k$. Thus C^f is a local martingale. □

(*1.10*) *Matrix differential notation.* Recall the convention that for continuous real-valued semimartingales u and v,

$$du\,dv \equiv d[u, v].$$

Again, if U and V are continuous semimartingales taking values in the spaces of $k \times m$ and $m \times n$ matrices respectively, then we define

$$(dU dV)_{ij} \equiv \sum_{q=1}^{m} dU_{iq} dV_{qj}.$$

With this convention, *we can reformulate the fact that X is an (a, b)-diffusion by stating that X is a continuous $\mathbb{R}^n$-valued semimartingale such that:*

(1.11(i)) *$dM \equiv dx - b(X)dt$ defines a local martingale M in $\mathbb{R}^n$,*

(1.11(ii)) *$(dM)\,(dM^T) = a(X)dt$.*

We shall find this notation extremely convenient.

(*1.12*) *Summation convention.* Throughout the chapter, we employ the summation convention so that *summation is implied over any index which appears precisely twice in any product, once as an upper index and once as a lower.* Thus, we can write (1.7) as follows:

$$\mathscr{L} = \tfrac{1}{2} a^{ij}(x) D_i D_j + b^i(x) D_i.$$

We shall sometimes write the summation sign explicitly, for example, when we wish to emphasize the range of summation.

(*1.13*) *Remark.* We should mention that interest is now being shown—for certain applications to physics, as well as for intrinsic reasons—in 'diffusions' defined via *Dirichlet forms* (see Fukushima [1], Silverstein [1, 2] for the theory). Such diffusions need not be semimartingales.

2. FD diffusions recalled. Recall (see §III.13) that an FD diffusion is an FD process in $\mathbb{R}^n$ with continuous sample paths and a generator whose domain contains C_K^∞. We shall restrict attention to honest FD diffusions, for which the lifetime is infinite $\mathbf{P}^x$-a.s. for each $x \in \mathbb{R}^n$. In this case, Theorem III.13.3 says that, restricted to C_K^∞, the generator is a second-order operator of the form (1.7), with a and b continuous. If now X is the canonical process on the path space

$\Omega = C(\mathbb{R}^+, \mathbb{R}^n)$, then (see § III.10.13)

(2.1) *for* $f \in C_K^\infty$ *and* $x \in \mathbb{R}^n$, C^f *is a* $\mathbf{P}^x$*-martingale.*

Hence, by Lemma 1.9, for each $x \in \mathbb{R}^n$,

(2.2) X *is an* (a, b)*-diffusion under* $\mathbf{P}^x$.

Thus our definition of an (a, b)-diffusion is consistent with our earlier use of the term in the FD set-up (and we shall even see (§ 22) that, conversely, when a and b satisfy mild conditions, an (a, b)-diffusion is an FD diffusion). But do notice the shortcomings of the FD setting; given continuous $a: \mathbb{R}^n \to S_n^+$ and $b: \mathbb{R}^n \to \mathbb{R}^n$, then, at present, *only by appealing to deep theorems on partial differential equations* (see Chapter 3 of Stroock and Varadhan [1]) *would we be able to decide*:

(i) *whether there exists any FD diffusion whose generator restricts to* $\mathscr{L}$ *on* C_K^∞;

(ii) *if such a diffusion does exist, whether it is unique.*

($\mathscr{L}$ is defined in terms of a and b by (1.7), of course). The whole of § III.13 is concerned with the *properties* of FD diffusions, but does nothing to help us with the problem of their *existence*.

A direct means of constructing diffusions would clearly be invaluable.

3. SDEs as a means of constructing diffusions. *Itô's great idea was to construct the sample paths of a diffusion* X *directly from those of a Brownian motion* B, *describing* X *in terms of* B *via a stochastic differential equation.*

Suppose that for some positive integer d, we can find a (measurable) map

$$\sigma: \mathbb{R}^n \to L(\mathbb{R}^d, \mathbb{R}^n) = \{n \times d \text{ real matrices}\}$$

such that, for each x in $\mathbb{R}^n$, $\sigma(x)\sigma(x)^T = a(x)$. Often in practice, d will equal n and, for each x, $\sigma(x)$ will be the unique nonnegative definite square root of $a(x)$; but the flexibility which we are allowing ourselves is essential for some applications.

Finally, *suppose that we can find a set-up* $(\Omega, \{\mathscr{F}_t\}, \mathbf{P})$ *carrying both an* $\mathbb{R}^n$*-valued semimartingale* X *and an* $\mathbb{R}^d$*-valued Brownian motion* B *relative to* $(\{\mathscr{F}_t\}, \mathbf{P})$ *such that*:

(3.1) $$dX = \sigma(X_t)dB + b(X_t)dt.$$

In full, with $\sigma_q^i(x)$ denoting the (i, q)th component of $\sigma(x)$, equation (3.1) reads:

(3.2) $$X_t^i = X_0^i + \sum_{q=1}^d \int_0^t \sigma_q^i(X_s)\,dB_s^q + \int_0^t b^i(X_s)\,ds,$$

giving the canonical decomposition of X. Then:

$$dM_t \equiv dX_t - b(X_t)dt = \sigma(X_t)dB_t \text{ defines a local martingale } M,$$

and, since $dBdB^T = Idt$,

$$dMdM^T = \sigma(X)dBdB^T\sigma(X)^T = \sigma(X)\sigma(X)^T dt = a(X)dt.$$

Thus, by (1.11),

(3.3) X *is an* (a, b)*-diffusion.*

Our attention is therefore focussed on the problems of existence and uniqueness of solutions of SDEs such as (3.1).

Although this use of SDEs as a means of constructing diffusions is the main reason for studying them, we are now going to look at several examples which show that *SDEs often arise in their own right* rather than via association with a given operator $\mathscr{L}$.

4. Example: Brownian motion on a surface. Suppose that $\Sigma(\subseteq \mathbb{R}^3)$ is the level set

$$\Sigma = \{(x, y, z) \in \mathbb{R}^3 \colon F(x, y, z) = c\}$$

of a smooth function F on $\mathbb{R}^3$ whose gradient does not vanish on Σ. Then we can define the unit normal field in a neighbourhood of Σ by the equation

$$n \equiv (\text{grad } F)/|\text{grad } F|,$$

in terms of which $P(x) \equiv I - n(x)\,n(x)^T$ *is the orthogonal projection on to the tangent plane to* Σ *at* x. Let R_0 be a fixed point of Σ, and $\{R_t \colon t \geqslant 0\}$ solve the Stratonovich SDE:

(4.1) $$\partial R = P(R)\partial B,$$

where B is a BM($\mathbb{R}^3$). Since

$$\partial F(R) = (\text{grad } F(R)) \cdot \partial R = 0,$$

$R_t \in \Sigma$ for all t. We shall see in §31 that R *is a Brownian motion on* Σ. The Itô equation corresponding to (4.1) reads:

(4.2) $$dR = P(R)dB + c(R)n(R)dt,$$

where $c \equiv -\frac{1}{2}\text{div }(n)$ is the so-called mean curvature in the direction of n.

This method furnishes a concrete means of constructing a diffusion on the manifold Σ without recourse to patching or to sophisticated geometric concepts, provided that we can solve the SDE (4.2).

5. Examples: modelling noise in physical systems. SDEs are used in engineering and physics to describe how random factors ('noise') can be incorporated into classical dynamical equations. The random force is often represented by 'white noise', the formal derivative dB/dt of Brownian motion. There is no problem in making this rigorous. However, engineers are well aware that the use of white noise can provide only a rough guide, and use more realistic band-limited noise for more precise calculation.

(*5.1*) *The Ornstein–Uhlenbeck model for a Brownian particle.* For the x-coordinate of the motion of a Brownian pollen grain on the surface of a liquid, mathematical Brownian motion, which is non-differentiable and which therefore makes the concept of velocity meaningless, is not a good model. A better model, due to Ornstein and Uhlenbeck, is to make the *velocity* of the particle a diffusion process. We take a phase-space picture (U, V), U denoting the x-coordinate, and V the velocity in the x-direction. We take Newton's law in the form:

$$dU = V dt, \qquad U_0 = u, \tag{5.2(i)}$$

$$dV = -cV dt + \sigma dB, \qquad V_0 = v, \tag{5.2(ii)}$$

where u and v are fixed points of $\mathbb{R}$. Here, $(-cV)$ is the 'damping' or 'frictional' force due to viscosity, and $\sigma dB/dt$ is the formal white-noise force due to the buffeting of the particle by the molecules in the liquid. (The mass of the particle is normalized to 1.) Note that V, the O–U velocity process, is an autonomous diffusion. We can solve equation (5.2(ii)) explicitly:

$$d(e^{ct}V) = \sigma e^{ct} dB,$$

so that

$$e^{ct}V_t - v = \sigma \int_0^t e^{cs} dB_s,$$

whence

$$V_t = v e^{-ct} + \sigma \int_0^t e^{-c(t-s)} dB_s. \tag{5.3}$$

The processes V and (U, V) are Gaussian and Markov. The process U is not Markov.

The O–U process has a notable history in physics. See Nelson [1].

(*5.4*) *A random linear oscillator.* Much interest has been shown by engineers in the system

$$dU = V dt, \tag{5.5(i)}$$

$$dV = -2c\omega V dt - \omega^2 U dt + \sigma U dB. \tag{5.5(ii)}$$

Here, U represents the position and V the velocity of a particle under
(i) the simple harmonic restoring force $-\omega^2 U$,
(ii) the damping force $-2c\omega V$, $(c > 0, \omega > 0)$
(iii) the random white-noise force, now proportional to U.
For the deterministic system in which $\sigma = 0$, we have

$$U'' + 2c\omega U' + \omega^2 U = 0,$$

which is explicity solvable, and we see that

$$\lim_{t\to\infty}(U_t, V_t)=(0,0). \tag{5.6}$$

The fundamental problem is: *for which values σ does the stability property (5.6) remain true (with probability 1)?*

Khasminskii's general method (Khasminskii [1, 2]) for solving such problems has created an industry. But to carry through his method even for this apparently simple case is splendid fun, and was achieved by Kozin and Prodromou [1]. An introduction to Khasminskii's method is given in §37.

(*5.7*) *The Kalman–Bucy filter.* The ubiquitous Kalman–Bucy filter concerns the system

$$dY=Vdt+\alpha dW, \tag{5.8(i)}$$

$$dV=-cVdt+\sigma dB. \tag{5.8(ii)}$$

Here, V is the velocity process, and Y is a noisy observation of the position process subject to observation errors described by αdW, where W is Brownian motion independent of B.

The fundamental problem here is: how does one best estimate the present velocity V_t from the observation $\{Y_s: s\leqslant t\}$?

See §VI.9 for the solution.

(*5.9*) *Quantum fluctuations.* Nelson ([1], and especially [2]) found a fascinating link between diffusion processes and quantum theory, and between Newton's law and the Schrödinger equation. In some sense, for example, a particle in the ground state of a quantum harmonic oscillator associated with the Hamiltonian

$$\tfrac{1}{2}p^2/m+\tfrac{1}{2}m\omega^2q^2$$

can be regarded as if performing a diffusion process

$$dX=-\omega Xdt+\sigma dB,$$

where $\sigma=(\hbar/m)^{1/2}$, where $\hbar$ is Planck's constant divided by 2π, and where X is run under its invariant probability law which is normal with mean 0 and variance σ^2/ω. The process X is then identical to its time-reversal. If we define the mean forward velocity, DX, and mean 'backward' velocity, D_*X, as follows:

$$(DX)_t\equiv\lim_{h\downarrow 0}h^{-1}\mathbf{E}[X_{t+h}-X_t|X_t],$$

$$(D_*X)_t\equiv\lim_{h\downarrow 0}h^{-1}\mathbf{E}[X_t-X_{t-h}|X_t],$$

then $DX = -\omega X$, $D_* X = \omega X$, and the *mean acceleration*

$$A \equiv \tfrac{1}{2}(DD_* + D_* D)X$$

satisfies $A = -\omega^2 X$. Then we have the *stochastic Newton's law*:

$$\text{Force} = \text{mass} \times \text{mean acceleration}$$

because the force on a particle at position X in the classical oscillator is $-m\omega^2 X$. Now see Nelson [2] and the many references given there.

Whatever the philosophical interpretation, stochastic mechanics is a fine motivator of interesting problems in probability. You might be intrigued by Albeverio, Blanchard and Høegh-Krohn's paper [1] on the 'quantum mechanics' of the solar system, a nice reversal of pre-quantum theories of the atom!

6. Example: Skorokhod's equation. We discussed the SDE theory for BES(n), $n \geqslant 2$, in § IV.35. We now consider BES(1), that is, reflecting Brownian motion.

Recall that, if Z is a BM($\mathbb{R}$), then Tanaka's formula (IV.43.4) for the process $X \equiv |Z|$ reads:

$$X_t = X_0 + Y_t + l_t, \tag{6.1}$$

where

$$Y_t = \int_0^t \operatorname{sgn}(Z_s)\,dZ_s \text{ defines a Brownian motion } Y,$$

and l is the local time at 0 for X, a continuous increasing process null at 0 and satisfying

$$\int_0^t I_{\{X(s) \neq 0\}}\,dl_s = 0, \quad \forall t.$$

With this example in mind, we prove the following lemma.

(*6.2*) LEMMA (Skorokhod). *Let* $y: \mathbb{R}^+ \to \mathbb{R}$ *be a continuous function with* $y(0) \geqslant 0$. *Then there exists a unique pair* (x, a) *of continuous functions such that*

$$x(t) = y(t) + a(t), \tag{6.3(i)}$$

$$x(t) \geqslant 0, \quad a \text{ is increasing}, \ a(0) = 0, \tag{6.3(ii)}$$

$$\int_0^t I_{\{x(s) \neq 0\}}\,da_s = 0, \quad \forall t. \tag{6.3(iii)}$$

Moreover, the function a *is given explicitly by the formula*:

$$a(t) = \sup\{y(s)^- : 0 \leqslant s \leqslant t\}. \tag{6.4}$$

Of course, we are writing $y = y^+ - y^-$, $y^- \equiv -(0 \wedge y)$.

We momentarily postpone the proof to give an application.

(*6.5*) *Proof of Lévy's presentation of reflecting Brownian motion.* Note that in the context of (6.1) when $X(0)=0$, we have

$$X_t = Y_t - \min\{Y_s: s \leqslant t\}, \quad l_t = -\min\{Y_s: s \leqslant t\}.$$

Just reversing signs now yields Lévy's result that if B is a $\mathrm{BM}^0(\mathbb{R})$, then

$$(|B|, l^0) \sim (M - B, M),$$

where $M_t \equiv \max\{B_s: s \leqslant t\}$ and l^0 is local time at 0 for $|B|$.

(*6.6*) *Proof of Lemma 6.2: existence part.* Let y be a given continuous function satisfying $y(0) \geqslant 0$. Define the function a by (6.4), and set

$$x(t) \equiv y(t) + a(t) = y(t) - (0 \wedge \min\{y(s): s \leqslant t\}).$$

Then, clearly, $x(t) \geqslant 0$, $a(\cdot)$ is increasing, $a(0)=0$, and, since a grows only when y is at a new negative minimum (in which case $x=0$), it is clear that equation (6.3(iii)) holds. Thus conditions (6.3) are all satisfied.

Proof of uniqueness. Suppose that we have another solution:

$$\tilde{x}(t) = y(t) + \tilde{a}(t),$$

satisfying the conditions of the lemma. Suppose that $\tilde{x}(v) > x(v)$ for some v, and set

$$u \equiv \sup\{t < v: \tilde{x}(t) = x(t)\}.$$

Then, for $u < t \leqslant v$, we have $\tilde{x}(t) > x(t) \geqslant 0$, so that, by (6.3(iii)),

$$\tilde{a}(v) = \tilde{a}(u).$$

Thus,

$$\begin{aligned} 0 < \tilde{x}(v) - x(v) &= \tilde{a}(v) - a(v) \leqslant \tilde{a}(u) - a(u) \\ &= \tilde{x}(u) - x(u) = 0, \end{aligned}$$

a contradiction. Hence, $\tilde{x}(v) \leqslant x(v)$ for all v, and, similarly, $x(v) \leqslant \tilde{x}(v)$ for all v.

(*6.7*) *The general form of Skorokhod's equation.* The general 1-dimensional form of Skorokhod's equation reads:

$$X_t = X_0 + \int_0^t \sigma(X_s)\,dB_s + \int_0^t b(X_s)\,ds + A_t,$$

where A is a continuous increasing process null at 0 such that

$$\int_0^t I_{\{X(s) \neq 0\}}\,dA_s = 0, \quad \forall t.$$

The idea is that X will be a reflecting diffusion with generator

$$\mathscr{L} = \tfrac{1}{2}\sigma(x)^2 D^2 + b(x) D, \quad D \equiv d/dx,$$

acting on functions satisfying the reflecting (Neumann) boundary condition:

$$f'(0)=0.$$

An indication of how to study such equations is given in our treatment in § 13 of a control problem due to Jacka. See Chaleyat-Maurel and el Karoui [1] for a nice systematic treatment.

(*6.8*) *The multi-dimensional case.* For treatments of stochastic differential equations with boundary conditions, see Ikeda and Watanabe [1], and Stroock and Varadhan [3].

7. Examples: control problems. Stochastic control theory has great real-world importance. Many very fine books on the subject have appeared recently. Whittle's book [1] has a marvellous feel for the subject and for its scope. The stochastic-calculus approach is excellently covered in: Fleming and Rishel [1], Krylov [1].

All that we do in this volume is to explain briefly how the Itô calculus may be applied in some simple examples.

(*7.1*) *Whittle's 'flypaper' example.* The name of this example relates to the illuminating discussion in Whittle [1; § 20.3].

A particle is controlled to move within the closed interval $[0, z]$, the points 0 and z acting as absorbing boundaries. The process X, where X_t denotes the position of the particle at time t, is given by

$$\begin{aligned} X_t &= x + B_t + \int_0^t u_s\,ds, \quad t < T \equiv \inf\{r\colon X_r \in \{0, z\}\}, \\ &= X_T \quad (t \geqslant T). \end{aligned}$$

Here, x is the starting point, B is a Brownian motion, and u is the control process. We must determine u_t from the path $\{X_s\colon s \leqslant t\}$. The object is to minimize the expected cost, the cost associated with a path being

$$C = \int_0^T \tfrac{1}{2}(u_s^2 + 1)\,ds + K(X_T),$$

$K(0)$ and $K(z)$ signifying the terminal cost of hitting 0 and z respectively.

This is one of many situations in which it is possible to arrive at the answer by a heuristic argument. We give such an argument here, and justify it via the martingale principle of optimality in § 13 below.

Let $F(x)$ $(0 \leqslant x \leqslant z)$ be the minimal expected cost when the starting point for X is x. It is obvious from the nature of the problem that, conditional on $\mathscr{F}_t$, the minimal expected cost from time t on is $F(X_t)$. (It is all to do with 'terminal times' in the technical sense of Markov-process theory, but never mind.) It is further intuitively obvious that the control u_t to be applied at time t will depend only on X_t. Indeed, since we have made life easy by doing probability rather than statistics, how can extra information about the path $\{X_s: s \leqslant t\}$ affect our choice of u_t?

Start X at x. Following the idea of *Bellman's dynamic programming*, we split the cost into the cost over time-interval $[0, h]$ plus the cost over $[h, T]$. The cost over $[0, h]$ is $\frac{1}{2}(u^2+1)h+o(h)$. The minimum expected cost over $[h, T]$ is $F(X_t)$. So, a control u applied over $[0, h]$ followed by an optimal control on $[h, T]$ will have expected cost:

$$(7.2)\qquad \begin{aligned} &\tfrac{1}{2}(u^2+1)h+\mathbf{E}F(X_h)+o(h)\\ &=\tfrac{1}{2}(u^2+1)h+\mathbf{E}[F(x)+F'(x)(X_h-x)+\tfrac{1}{2}F''(x)(X_h-x)^2]+o(h)\\ &=\tfrac{1}{2}(u^2+1)h+F(x)+uhF'(x)+\tfrac{1}{2}hF''(x)+o(h)\\ &=F(x)+h\{\tfrac{1}{2}(u^2+1)+uF'(x)+\tfrac{1}{2}F''(x)\}+o(h). \end{aligned}$$

Note the way in which Itô's formula is implicit in the above calculation. We must choose u to minimize the expression (7.2), so

$$(7.3)\qquad u(x) = -F'(x).$$

Moreover, since the minimum value of the expression (7.2) must equal $F(x)$, we have

$$(7.4)\qquad \tfrac{1}{2}(u^2+1)+uF'(x)+\tfrac{1}{2}F''(x)=0.$$

On combining (7.3) and (7.4), we have:

$$(7.5)\qquad F''(x)-F'(x)^2+1=0, \quad F(0)=K(0), \quad F(z)=K(z).$$

You should give a little thought to the 'obvious' boundary conditions. The substitution $g=e^{-F}$ reduces (7.5) to the form

$$g''(x)=g(x),$$

so that equation (7.5) may be solved explicitly.

(7.6) *A controlled-variance problem.* In this example, the controlled process is of the form

$$X_t = x_0 + \int_0^t u_s\, dB_s,$$

where the control process u is previsible with respect to the natural filtration of X, and takes values in $[\delta, K]$, where δ and K are constants such that $0<\delta<K<\infty$. How would you control the process so as to maximize

$$\mathbf{E}\int_0^\infty e^{-\alpha s} I_{[-a,a]}(X_s)\, ds?$$

Here, α and a are fixed positive reals. The functional whose expectation one wishes to maximize grows only when $X_t \in [-a, a]$, so one would try to make the particle spend as much time as possible in $[-a, a]$. Thus, one would choose control δ when the particle is in $[-a, a]$; and when it is outside $[-a, a]$, one would attempt to get the particle into $[-a, a]$ as quickly as possible by using control K. So it is a plausible guess that the optimally controlled process should solve

(7.7)
$$X_t = x_0 + \int_0^t u(X_s)\, dB_s,$$

where $u(x) = \delta$ if $|x| \leqslant a$; $u(x) = K$ otherwise.

Can we make all this rigorous? See § 42 below.

For a remarkable paper on a much more difficult controlled-variance problem, see MacNamara [1].

(7.8) A finite-fuel control problem. Here is another control problem, this time with a finite-fuel constraint. It is solved by Jacka [1]. Given some increasing C^1 function σ: $\mathbb{R} \to (0, \infty)$ with bounded derivative, and a Brownian motion B, the controlled process X satisfies

$$X_t = x + \int_0^t \sigma(X_s)\, dB_s + u_t,$$

where u is some continuous increasing process null at 0 and adapted to the filtration of X. The choice of u is at our disposal; *the only constraint is that* $\sup u_t \leqslant C$, where C is the 'fuel limit'. *How should we choose u to minimize* $\mathbf{E}\, e^{-\alpha\tau(u,\, x)}$, *where* $\tau(u, x) \equiv \inf\{t\colon X_t < 0\}$? *Here, α is a given positive constant.* Again, intuition provides a plausible candidate for an optimal control. If X is at zero, we must apply some control to stop the process entering $(-\infty, 0)$. Away from zero, the quadratic variation of the stochastic integral is at least as bad as it is at zero, so we might as well save our control effort until X is at zero. This suggests the following recipe for constructing an optimal control; let (X, A) solve an equation essentially of the Skorokhod type:

(7.9)
$$\begin{cases} X_t = x + \int_0^t \sigma(X_s)\, dB_s + A_t; \\ X_t \geqslant 0, \quad A_t \text{ is continuous increasing, } A_0 = 0, \\ \int_0^\cdot I_{\{X_s \neq 0\}}\, dA_s = 0. \end{cases}$$

Then an optimal control should be given by $u_t = C \wedge A_t$. For a rigorous proof, see § 15.

For further examples of finite-fuel control problems with explicit solutions, see Beneš, Shepp and Witsenhausen [1].

The introduction of Skorokhod-type equations into stochastic-control theory was made by Bather in a series of papers. See, for example, §§ 3–6 of Chapter 38 of Whittle [1]. It is nice to see local-time models feature in this way.

2. PATHWISE UNIQUENESS, STRONG SDEs, AND FLOWS

8. Our general SDE; previsible path functionals; diffusion SDEs. We are going to restrict attention to SDEs of the form

$$X_t = \xi + \int_0^t \sigma(s, X.)\, dB_s + \int_0^t b(s, X.)\, ds \tag{8.1}$$

where B is a Brownian motion on some $(\Omega, \{\mathscr{F}_t\}, \mathbf{P})$, ξ is $\mathscr{F}_0$-measurable, and σ and b are suitable coefficients depending on the path of X. In order that the stochastic integral in (8.1) should be defined, we must ensure that $\sigma(s, X.)$ is previsible, and, since X is adapted, we must also have that $b(s, X.)$ is adapted. Both of these will happen if σ and b are *previsible path functionals*, which we define now.

(*8.2*) *Path-space* $(W, \{\mathscr{X}_t\})$. We let W (or, when essential, W^n) denote the Polish space $C(\mathbb{R}^+, \mathbb{R}^n)$ of continuous paths in $\mathbb{R}^n$, with canonical process $\{x_t\}$, so that

$$x_t(w) = w(t) \quad (t \geq 0, \quad w \in W).$$

We let $\mathscr{X}_t$ (or $\mathscr{X}_t^n$ if we have to) and $\mathscr{X}$ (or $\mathscr{X}^n$) denote the σ-algebras

$$\mathscr{X}_t \equiv \sigma\{x_s : s \leq t\}, \quad \mathscr{X} \equiv \sigma\{x_s : s < \infty\}.$$

Note that *the* $\mathscr{X}_t$ *are not completed in any way.*

(*8.3*) DEFINITION (previsible path functionals). *Let* $\mathscr{P}_W$ *be the previsible* σ*-algebra on* $(0, \infty) \times W$ *associated with the filtration* $\{\mathscr{X}_t\}$. *Thus,* $\mathscr{P}_W$ *is the smallest* σ*-algebra on* $\mathbb{R}^{++} \times W$ *such that every* $\{\mathscr{X}_t\}$*-adapted* L-process *is* $\mathscr{P}_W$*-measurable. A* $\mathscr{P}_W$*-measurable map on* $\mathbb{R}^{++} \times W$ *is called a* previsible path functional.

(*8.4*) *Important remark.* Because we are dealing with the *uncompleted* σ-algebras on the *canonical* path-space for *continuous* paths, several dramatic simplifications of the later theory in Chapter VI can be made. In fact (see Theorem IV.97 of Dellacherie and Meyer [1]), *for* $(W, \mathscr{X}_t)$, *the concepts*

previsible, optional, progressive and *non-anticipating*

(non-anticipating = measurable and adapted) *agree*, modulo silly things regarding conventions about time 0.

This information should reassure you when you move from book to book!

(8.5) A composition result. Now let X be a continuous $\mathbb{R}^n$-valued adapted process carried by some $(\Omega, \{\mathscr{F}_t\})$. For each ω in Ω, the corresponding path of X is the map

$$X_{\cdot}(\omega) \equiv (t \mapsto X_t(\omega)).$$

Thus, the path map $X_{\cdot}$ maps Ω to W.

It is easy to check that the map

$$i_X\colon \mathbb{R}^{++} \times \Omega \to \mathbb{R}^{++} \times W$$

$$(t, \omega) \mapsto (t, X_{\cdot}(\omega))$$

is $\mathscr{P}/\mathscr{P}_W$ measurable, where $\mathscr{P}$ is the previsible σ-algebra on $(0, \infty) \times \Omega$ associated with $\{\mathscr{F}_t\}$. Indeed, if $T\colon W \to \mathbb{R}^+$ is an $\{\mathscr{X}_t\}$-stopping time, then $T(X_{\cdot})$ is an $\{\mathscr{F}_t\}$-stopping time.

It now follows by composition that

(8.6) *if α is a previsible path functional, then $\alpha(t, X_{\cdot})$ is previsible relative to $\{\mathscr{F}_t\}$.*

We now agree to the following convention.

(8.7) CONVENTION. *The functions:*

$$\sigma\colon \mathbb{R}^{++} \times W \to L(\mathbb{R}^d, \mathbb{R}^n),$$

$$b\colon \mathbb{R}^{++} \times W \to \mathbb{R}^n$$

are previsible ($\mathscr{P}_W$-measurable) path functionals.

Under the convention (8.7), the right-hand side of (8.1) is well-defined for any continuous adapted X, provided that (with $|\sigma|^2 \equiv \text{trace}\,(\sigma\sigma^T)$)

(8.8) $$\int_0^t \{|\sigma(s, X_{\cdot})|^2 + |b(s, X_{\cdot})|\}\, ds < \infty \quad \text{a.s.} \quad \text{for each } t.$$

In practice, we do not worry much about the finiteness condition (8.8); if it fails, we simply solve the SDE up to the explosion time $\zeta \equiv \inf\left\{t\colon \int_0^t \{|\sigma(s, X_{\cdot})|^2 + |b(s, X_{\cdot})|\}\, ds = \infty\right\}$ and say nothing about the solution after ζ. Routine localization reduces this case to the case where (8.8) holds, so for tidy statements of results, we shall always make (8.8) part of our definition of a solution.

An important special case of the SDE (8.1) deserves to be singled out.

(8.9) DEFINITION (diffusion SDE). *We say that the SDE* (8.1) *is a* diffusion SDE (*or an* SDE of diffusion type) *if*:

(8.10) $$\sigma(t, x_{\cdot}) = \sigma(x_t), \quad b(t, x_{\cdot}) = b(x_t),$$

for some measurable functions

$$\sigma\colon \mathbb{R}^n \to L(\mathbb{R}^d, \mathbb{R}^n), \quad b\colon \mathbb{R}^n \to \mathbb{R}^n.$$

If our SDE is of diffusion type, then condition (8.7) is automatically satisfied. Notice that our diffusion SDEs are time-homogeneous: the functions σ and b do not vary with time. The time-inhomogeneous case can formally be reduced to this case by considering the space-time process.

9. Pathwise uniqueness, exact SDEs. First, we make a convention.

(*9.1*) CONVENTION. *By a* set-up

$$(\Omega, \{\mathcal{F}_t\}, \mathbf{P}, \xi, B)$$

we understand that:

(i) $(\Omega, \{\mathcal{F}_t\}, \mathbf{P})$ *satisfies the usual conditions,*
(ii) B *is an* $(\{\mathcal{F}_t\}, \mathbf{P})$ *Brownian motion on* $\mathbb{R}^d$,
(iii) ξ *is an* $\mathcal{F}_0$ *measurable* $\mathbb{R}^n$*-valued random variable.*

(*9.2*) DEFINITION (pathwise uniqueness). *We say that* pathwise uniqueness *holds for equation* (*8.1*) *if the following statement is true. Given*:

(9.3(i)) *any set-up* $(\Omega, \{\mathcal{F}_t\}, \mathbf{P}, \xi, B)$, *and*
(9.3(ii)) *any two continuous semimartingales* X *and* X' *relative to* $(\{\mathcal{F}_t\}, \mathbf{P})$ *such that the finiteness condition* (*8.8*) *is satisfied for* X *and* X', *and*

$$X_t = \xi + \int_0^t \sigma(s, X.)\, dB_s + \int_0^t b(s, X.)\, ds,$$

$$X_t' = \xi + \int_0^t \sigma(s, X'.)\, dB_s + \int_0^t b(s, X'.)\, ds;$$

then:

$$\mathbf{P}[X_t = X_t', \quad \forall t] = 1.$$

It is essential to appreciate that *pathwise uniqueness refers simultaneously to all situations in which conditions* (*9.3*) *hold*: given any set-up $(\Omega, \{\mathcal{F}_t\}, \mathbf{P}, \xi, B)$, there is at most one semimartingale X satisfying (8.1). An SDE for which, given any set-up, there is *one and only one* semimartingale X satisfying (8.1) is obviously an important concept, for which we must have a name.

(*9.4*) DEFINITION (pathwise exact SDE). *We shall say that the SDE* (8.1) *is a* pathwise exact SDE *if, given any set-up* $(\Omega, \{\mathcal{F}_t\}, \mathbf{P}, \xi, B)$ *there is (modulo indistinguishability) exactly one semimartingale* X *such that* (*8.1*) *and* (*8.8*) *hold.*

Thus pathwise uniqueness holds for a pathwise exact SDE. We shall often drop the word 'pathwise', and refer simply to an *exact SDE*.

10. Relationship between exact SDEs and strong solutions. A (pathwise) exact SDE can be solved uniquely on any set-up, so, in particular, it can be solved on the 'minimal' set-up, generated by the initial condition ξ and the driving Brownian motion B. It turns out that the solution on this minimal set-up is *a deterministic function of* ξ *and* B; and even on more general set-ups, the solution is this *same* function of the initial condition and the driving Brownian motion. See Theorem 10.4. Thus conceptually the solution of an exact SDE is very clean; one simply takes ξ and B, puts them into some (measurable) function, and, hey presto, there is the solution process! Some little care is needed in the statement of this idea, though, and the precise form of it is not quite what one would hope. (*On a first reading, you should skip the rest of this section until you have read as far as* § 16.)

Fix some probability measure μ on $\mathbb{R}^n$, and define the *canonical set-up with initial law* μ as the set-up $(\Omega, \{\mathscr{F}_t\}, \mathbf{P}, \xi, B)$, where $\mathbf{Q}^d$ is Wiener measure on $(W^d, \mathscr{W}^d)$,

$$\Omega = \mathbb{R}^n \times W^d, \quad \mathbf{P} = \mu \times \mathbf{Q}^d,$$
$$\xi(y, w) = y, \quad B_t(y, w) = w(t) \quad \text{for} \quad (y, w) \in \Omega,$$
$$\mathscr{F}_t^\circ = (\{\xi, B_s: s \leqslant t\}),$$

and $\{\mathscr{F}_t\}$ is the usual $\mathbf{P}$-augmentation of $\{\mathscr{F}_t^\circ\}$. Let $\mathscr{P}^\circ$ denote the σ-algebra of $\{\mathscr{F}_t^\circ\}$-previsible sets, and let $\mathscr{P}$ denote the σ-algebra of $\{\mathscr{F}_t\}$-previsible sets.

Here is a technical result which we shall need before long, but which is of independent interest.

(*10.1*) LEMMA. *Suppose that* φ: $\mathbb{R}^{++} \times \Omega \to \mathbb{R}^d$ *is* $\mathscr{P}/\mathscr{B}(\mathbb{R}^d)$*-measurable, and suppose that*:

$$\mathbf{P}\left[\int_0^t |\varphi(s; \xi, B.)|^2 \, ds < \infty \text{ for all } t\right] = 1.$$

Then there exists a function Φ: $\mathbb{R}^n \times W^d \to W^1$ *which is* $\mathscr{B}(\mathbb{R}^n) \times \mathscr{W}^d/\mathscr{W}^1$*-measurable such that*:

$$\mathbf{P}\left[\int_0^\bullet \varphi(s; \xi, B.) \, dB_s = \Phi(\xi, B.)\right] = 1. \tag{10.2}$$

Moreover, on any set-up $(\tilde{\Omega}, \{\tilde{\mathscr{F}}_t\}, \tilde{\mathbf{P}}, \tilde{\xi}, \tilde{B})$, *where* $\tilde{\xi}$ *has law* μ,

$$\tilde{\mathbf{P}}\left[\int_0^\bullet \varphi(s; \tilde{\xi}, \tilde{B}.) \, d\tilde{B}_s = \Phi(\tilde{\xi}, B.)\right] = 1. \tag{10.3}$$

Proof. Since any $\{\mathscr{F}_t\}$-adapted L-process is indistinguishable from some $\{\mathscr{F}_t^\circ\}$-adapted L-process, φ is indistinguishable from some $\mathscr{P}^\circ$-measurable process (see Exercise 10.12). Thus we can assume that φ is $\mathscr{P}^\circ$-measurable. By localizing, we may further assume that $\varphi \in L^2(B)$; that is,

$$\|\varphi\|_B^2 \equiv \mathbf{E}\left[\int_0^\infty |\varphi(s;\xi, B.)|^2\, ds\right] < \infty.$$

If φ is an elementary integrand, the conclusions of the lemma are immediate. More generally, by monotone class arguments, there exist elementary φ_n such that

$$\|\varphi - \varphi_n\|_B \to 0 \quad \text{as} \quad n \to \infty,$$

and by passing to a fast subsequence if need be we can assume that

$$\int_0^\bullet \varphi_n(s;\xi, B.)\, dB_s \to \int_0^\bullet \varphi(s;\xi, B.)\, dB_s \quad \text{uniformly} \quad \text{a.s.}$$

If Φ_n is the measurable function such that

$$\int_0^\bullet \varphi_n(s;\xi, B.)\, dB_s = \Phi_n(\xi, B.) \quad \text{a.s.},$$

then we let $\Phi(\xi, B.) = \lim \Phi_n(\xi, B.)$ if the (uniform) limit exists; or 0 if it does not. Thus $\mathbf{P}$-a.s., $(\int(\varphi_n - \varphi)\, dB)_\infty^* \to 0$ and $\|\Phi_n(\xi, B.) - \Phi(\xi, B.)\| \to 0$, from which we obtain (10.2). The same approximation argument also yields (10.3). □

(*10.4*) THEOREM. *Suppose that on the canonical set-up with initial law μ the SDE (8.1) has a solution X:*

$$X_t = \xi + \int_0^t \sigma(s, X.)\, dB_s + \int_0^t b(s, X.)\, ds,$$

satisfying the finiteness condition (8.8). Then there exists a function:

(10.5) $$F_\mu\colon \Omega \equiv \mathbb{R}^n \times W^d \to W^n$$

such that for all $t \geq 0$

(10.6) $$F_\mu^{-1}(\mathscr{X}_t^n) \subseteq \mathscr{F}_t$$

and such that

(10.7) $$\mathbf{P}[X. = F_\mu(\xi, B.)] = 1.$$

Moreover, on any set-up $(\tilde\Omega, \{\tilde{\mathscr{F}}_t\}, \tilde{\mathbf{P}}, \tilde\xi, \tilde B)$ where the law of $\tilde\xi$ is μ, the process $\tilde X. \equiv F_\mu(\tilde\xi, B.)$ satisfies

(10.8) $$\tilde X_t \equiv \tilde\xi + \int_0^t \sigma(s, \tilde X.)\, dB_s + \int_0^t b(s, X.)\, ds.$$

Proof. Since $X.$ is $\mathscr{F}_\infty$-measurable, there is some $\mathscr{F}_\infty^\circ$-measurable $X'.$ such that $\mathbf{P}(X. = X'.) = 1$, and then (since W^n is Polish) there exists a $\mathscr{F}_\infty^\circ/\mathscr{X}^n$-measurable function $F_\mu : \Omega \to W^n$ such that $X'_{\cdot} = F_\mu(\xi, B_{\cdot})$. Thus (10.7) holds, and since $X_{\cdot}$ is adapted to $\{\mathscr{F}_t\}$, the measurability (10.6) follows immediately.

To prove the last assertion, we use Lemma 10.1. The function $\varphi^i : \mathbb{R}^{++} \times \Omega \to \mathbb{R}^d$ defined by

$$\varphi^i(s; \xi, B_{\cdot}) \equiv \sigma^i_{\cdot}(s, F_\mu(\xi, B_{\cdot}))$$

satisfies the conditions of the lemma (by (8.6)), and hence there exists Φ^i: $\mathbb{R}^n \times W^d \to W^1$ such that

$$\int_0^{\cdot} \varphi^i(s; \xi, B_{\cdot})\, dB_s = \Phi^i(\xi, B_{\cdot}).$$

Thus the statement (10.8) is, according to (10.3), the same as the statement that

$$F^i_\mu(\tilde{\xi}, \tilde{B}_{\cdot})_t = \tilde{\xi}^i + \Phi^i(\tilde{\xi}, \tilde{B}_{\cdot})_t + \int_0^t b^i(s, F_\mu(\tilde{\xi}, \tilde{B}_{\cdot}))\, ds \quad \text{for all } t$$

$\tilde{\mathbf{P}}$-a.s. But this is simply a statement about some functional of $(\tilde{\xi}, \tilde{B}_{\cdot})$, which has the same law as $(\xi, B_{\cdot})$, for which the statement is true. □

It follows immediately that *an SDE is exact if and only if pathwise uniqueness holds and, for every initial law μ, there is a solution on the canonical set-up with initial law μ.*

Of course, what we want is to say that, if $F_y(\cdot) \equiv F_{\delta_y}(y, \cdot)$, $y \in \mathbb{R}^n$, then for an exact SDE the process

$$X_{\cdot} \equiv F_\xi(B_{\cdot})$$

solves (8.1). The snag is that $(y, B_{\cdot}) \mapsto F_y(B_{\cdot})$ need not be jointly measurable. However, in many cases it is possible to prove that it is, in which case we say that F is a *strong solution* to the SDE.

(10.9) DEFINITION (strong solution). *We say that a function*

$$F : \mathbb{R}^n \times W^d \to W^n$$

is a strong solution *to the SDE (8.1) if*

(10.10(i)) $\quad F^{-1}(\mathscr{X}^n_t) \subseteq \mathscr{B}(\mathbb{R}^n) \times \bar{\mathscr{X}}^d_t$ *for all* $t \geq 0$;

(10.10(ii)) *on any set-up* $(\Omega, \{\mathscr{F}_t\}, \mathbf{P}, \xi, B)$, *the process*

$$X_{\cdot} \equiv F(\xi, B_{\cdot})$$

solves the SDE (8.1).

Here, $\{\bar{\mathscr{X}}^d_t\}$ *is the* $\mathbf{Q}^d$ *augmentation of* $\{\mathscr{X}^d_t\}$.

For example, we show (Theorem 13.1) that if σ and b are Lipschitz (see 11.1), then the SDE has a strong solution.

(*10.11*) *Remark.* Our strong solutions are subject to nicer regularity conditions than those of Ikeda and Watanabe [1], so that there may be situations in which a strong solution exists in their sense but not in ours.

(*10.12*) **Exercise.** Prove the statement made at the beginning of the proof of Lemma 10.1 that a $\mathscr{P}$-measurable process is indistinguishable from some $\mathscr{P}^{\circ}$-measurable process. (*Hint.* Consider the set $\mathscr{H}$ of those bounded $\mathscr{P}$-measurable processes which are indistinguishable from some $\mathscr{P}^{\circ}$-measurable process, and use Theorem II.3.2 with $\mathscr{C}$ the set of $\{\mathscr{F}_t\}$-adapted L-processes.)

11. The Itô existence and uniqueness result. Let us suppose that the coefficients σ and b are Lipschitz in the following sense. If f is a continuous function on $\mathbb{R}^+$ with values in some Euclidean space whose norm is denoted by $|\cdot|$, we define for each $t>0$, $f_t^* \equiv \sup\{|f(s)|: s \leqslant t\}$. Then we say that σ and b are Lipschitz if there is some $K<\infty$ such that:

$$\begin{aligned} &|\sigma(t, x.) - \sigma(t, y.)| \leqslant K(x-y)_t^* \\ &|b(t, x.) - b(t, y.)| \leqslant K(x-y)_t^* \end{aligned} \tag{11.1}$$

for all $t \geqslant 0$ and $x, y \in W^n$.

If $\sigma \equiv 0$, and b is Lipschitz, we are looking at an ordinary differential equation; the existence and uniqueness of a solution in this case is a beautiful application of the contraction mapping theorem. It was Itô's idea to extend the method to SDEs. It must be emphasized that *the one result, Theorem 11.2, is by far the most useful existence and uniqueness result in the subject*; for example, if $\sigma(t, x.)$ and $b(t, x.)$ are C^1 functions of x_t, then (subject to certain boundedness conditions) Theorem 11.2 holds, and there exists a unique solution to (8.1). Thus all the examples one meets in stochastic differential geometry are covered; we discuss a selection in Part 5. Protter [1], Doléans, Dade and Meyer [1], and Metivier and Pellaumail [1] discuss extensions of the method to R-processes, and to processes with values in a Banach space.

(*11.2*) THEOREM (Itô). *Suppose that the coefficients σ and b in (8.1) satisfy the Lipschitz condition (11.1) and that for each constant $T>0$ there is some C_T such that:*

$$|\sigma(s, 0)| + |b(s, 0)| \leqslant C_T \tag{11.3}$$

for all $s \leqslant T$. Then our SDE (8.1) is an exact SDE.

Recall that (8.1) reads, when written in integrated form:

$$X_t = \xi + \int_0^t \sigma(s, X.)\, dB_s + \int_0^t b(s, X.)\, ds. \tag{11.4}$$

The proof of this theorem rests upon a technical lemma. Though we shall use the lemma in this section only in the case when $p=2$ (when we could use Doob's L^2 inequality instead of Burkholder's), it is essential in §13 to have the result for $p>2$.

(*11.5*) LEMMA. *Fix $T>0$. Let $p\geqslant 2$. Suppose that a process X is defined on some $(\Omega,\{\mathcal{F}_t\},\mathbf{P})$ via the equation*

$$X_t=\xi+\int_0^t\sigma_s\,dB_s+\int_0^t b_s\,ds,$$

where B is Brownian motion in $\mathbb{R}^d$, $\xi\in L^p(\mathcal{F}_0)$, σ is a previsible $L(\mathbb{R}^d,\mathbb{R}^n)$-valued process, and b is a previsible $\mathbb{R}^n$-valued process. Then there exists a constant C depending only on p, T, and n such that for $0\leqslant t\leqslant T$,

$$\mathbf{E}(X_t^{*p})\leqslant C\{\mathbf{E}|\xi|^p+\mathbf{E}\int_0^t(|\sigma_s|^p+|b_s|^p)\,ds\}. \tag{11.6}$$

Proof. We have

$$X_t^*\leqslant|\xi|+\left(\int_0^\bullet\sigma\,dB\right)_t^*+\int_0^t|b_s|\,ds=F+G+H\ \text{(say)}.$$

By Jensen's inequality,

$$(X_t^*)^p\leqslant 3^{p-1}(F^p+G^p+H^p),$$

and

$$H^p\leqslant t^{p-1}\int_0^t|b_s|^p ds.$$

By the Burkholder–Davis–Gundy inequality (see §IV.42.1),

$$\mathbf{E}(G^p)\leqslant C_p\mathbf{E}\left[\left(\int_0^t|\sigma|^2ds\right)^{p/2}\right],$$

and, by Jensen's inequality again,

$$\left(\int_0^t|\sigma_s|^2ds\right)^{p/2}\leqslant t^{p/2-1}\int_0^t|\sigma_s|^p\,ds.$$

On assembling these pieces, we have a complete proof. □

As a corollary, we deduce the following result.

(*11.7*) COROLLARY. *Suppose that $p\geqslant 2$, that ξ and η are $\mathbb{R}^n$-valued random variables such that $\xi,\eta\in L^p(\mathcal{F}_0)$, and that previsible path functionals σ and b satisfy the Lipschitz condition (11.1). Let X and Y be continuous adapted $\mathbb{R}^n$-valued* *processes carried by some* $(\Omega, \{\mathscr{F}_t\}, \mathbf{P})$, *and define* $\tilde{X}$ *and* $\tilde{Y}$ *as follows:*

$$\tilde{X}_t \equiv \xi + \int_0^t \sigma(s, X.)\,dB_s + \int_0^t b(s, X.)\,ds$$

$$\tilde{Y}_t \equiv \eta + \int_0^t \sigma(s, Y.)\,dB_s + \int_0^t b(s, Y.)\,ds.$$

Then there exists a constant C (depending only on K, n, T and p) such that, for $0 \leqslant t \leqslant T$,

$$\mathbf{E}[(\tilde{X} - \tilde{Y})_t^{*p}] \leqslant C\left[\mathbf{E}|\xi - \eta|^p + \mathbf{E}\left(\int_0^t (X - Y)_s^{*p}\,ds\right)\right]. \tag{11.8}$$

Proof of the 'existence' part of Theorem 11.2. Suppose that σ and b satisfy the conditions of the theorem. Suppose given a set-up $(\Omega, \{\mathscr{F}_t\}, \mathbf{P}, \xi, B)$. We are going to give a 'Picard' algorithm for constructing a solution X of the equation (11.4). We can carry through this construction working in the set-up $(\Omega, \mathscr{G}, \{\mathscr{G}_t\}, \mathbf{P})$, where $\{\mathscr{G}_t\}$ is the augmented filtration generated by (ξ, B) defined precisely as follows. First, let

$$\mathscr{G}_t^\circ \equiv \sigma(\{B_s : s \leqslant t\} \cup \{\xi\}), \quad \mathscr{G}^\circ \equiv \bigvee_t \mathscr{G}_t^\circ,$$

and let $\mathscr{N}$ be the collection of $\mathbf{P}$-null sets in the $\mathbf{P}$-completion $\mathscr{G}$ of $\mathscr{G}^\circ$. Then the filtration $\mathscr{G}_t \equiv \sigma(\mathscr{G}_{t+}^\circ \cup \mathscr{N})$ satisfies the usual conditions.

We fix $T > 0$, and let L_T^2 be the space of continuous $\{\mathscr{G}_t\}$-adapted processes with norm $\mathbf{E}[(X_T^*)^2]^{1/2}$.

It is enough to prove the result when ξ is bounded; the general case follows from this by truncation.

We construct a sequence $\{X_t^{(n)} : 0 \leqslant t \leqslant T\}$ of approximations to a solution, whose limit X will solve (11.4) up to time T, and will be the unique solution. Since T is arbitrary, we shall have a unique solution for all time.

Define

$$X_t^{(0)} \equiv \xi \quad (0 \leqslant t \leqslant T);$$

$$\begin{aligned} X_t^{(n+1)} &\equiv (\mathscr{R} X^{(n)})_t \\ &\equiv \xi + \int_0^t \sigma(s, X_\cdot^{(n)})\,dB_s + \int_0^t b(s, X_\cdot^{(n)})\,ds \quad (0 \leqslant t \leqslant T) \end{aligned}$$

for each $n \in \mathbb{Z}^+$. Because of (11.3) and (11.6), both of $X^{(0)}$ and $X^{(1)}$ are in L_T^2, and $X^{(1)}$ is a continuous process adapted to $\{\mathscr{G}_t\}$. Because of (11.8), it follows inductively that each $X^{(n)}$ is in L_T^2, and is adapted to $\{\mathscr{G}_t\}$. Now define

$$\Delta_t^n \equiv \mathbf{E}(X^{(n+1)} - X^{(n)})_t^{*2} = \mathbf{E}(\mathscr{R}\,X^{(n)} - X^{(n)})_t^{*2}$$

and use (11.8); we find that

$$\Delta_t^{n+1} \leqslant C \int_0^t \Delta_s^n ds,$$

and, by induction, it follows that, for all $t \leqslant T$,

$$\Delta_t^n \leqslant \eta C^n t^n / n!,$$

where $\eta \equiv \Delta_T^0 < \infty$, the inequality being trivial for $n = 0$. Hence, on $[0, T]$, the sequence $X^{(n)}$ converges uniformly in L^2 to a limit X, which is also adapted to $\{\mathscr{G}_t\}$. Since

$$\mathbf{E}(X - X^{(n)})_T^{*2} \leqslant \eta \sum_{k \geqslant n} (CT)^k / k! \to 0 \quad \text{as } n \to \infty,$$

it follows from (11.8) that

$$\mathbf{E}(\mathscr{R} X - \mathscr{R} X^{(n)})_T^{*2} \equiv \mathbf{E}(\mathscr{R} X - X^{(n+1)})_T^{*2} \to 0 \quad \text{as } n \to \infty,$$

so that $X = \mathscr{R} X$, and X is a solution of (11.4).

Proof of pathwise uniqueness. If X and X' are two solutions of (11.4) associated with the same set-up $(\Omega, \{\mathscr{F}_t\}, \mathbf{P}, \xi, B)$, then estimate (11.8) implies that

$$0 \leqslant \mathbf{E}(X - X')_t^{*2} \leqslant C_T \int_0^t \mathbf{E}[(X - X')_s^{*2}]\, ds \tag{11.9}$$

which (see Exercise 11.11) implies that $\mathbf{E}(X - X')_t^{*2} = 0$ for $t \leqslant T$, and hence that $X = X'$ on $[0, T]$. □

A further simple application of (11.7) and (11.8) gives us a 'local' uniqueness result.

(*11.10*) COROLLARY. *Suppose that σ and b satisfy the hypotheses of Theorem 11.2 and that X is a solution of (11.4) associated with some set-up $(\Omega, \{\mathscr{F}_t\}, \mathbf{P}, \xi, B)$. Suppose that σ' and b' also satisfy the hypotheses of Theorem 11.2, and that:*

$$\sigma'(s, x.) = \sigma(s, x.)$$

$$b'(s, x.) = b(s, x.)$$

on $0 \leqslant s \leqslant \tau(x.)$, where $\tau \colon W^n \to \mathbb{R}^+$ is an $\{\mathscr{X}_t^n\}$-stopping time. If X' is the solution of

$$X'_t = \xi + \int_0^t \sigma'(s, X'_.)\, dB_s + \int_0^t b'(s, X'_.)\, ds,$$

then

$$\mathbf{P}[X_t = X'_t \quad \text{for all} \quad t < \tau(X')] = 1.$$

(*11.11*) **Exercise** (*Gronwall's Lemma*). Let ρ be a continuous function on $[0, T]$ such that

$$\rho_t \leqslant c + K\int_0^t \rho_s\, ds$$

for some positive constants c and K. Show that

$$(d/dt)\left[e^{-Kt}\int_0^t \rho_s\, ds\right] \leqslant ce^{-Kt}$$

and deduce that $\rho_t \leqslant ce^{Kt}$.

12. Locally Lipschitz SDEs; Lipschitz properties of $a^{1/2}$. The global Lipschitz conditions of Theorem 11.2 are stronger than is really necessary; the following variant, which is often useful, is a typical 'local' version of a 'global' result.

(*12.1*) THEOREM. *Suppose that the coefficients σ and b in the SDE* (*11.4*) *are such that for each N there is some K_N such that:*

$$\begin{aligned} |\sigma(s, x.) - \sigma(s, y.)| &\leqslant K_N (x-y)^*_s \\ |b(s, x.) - b(s, y.)| &\leqslant K_N (x-y)^*_s \end{aligned} \tag{12.2}$$

whenever $x^*_s \vee y^*_s \leqslant N$ and $0 \leqslant s \leqslant N$. Suppose also that for each constant $T > 0$, there is some C_T such that, for $0 \leqslant s \leqslant T$,

$$|\sigma(s, x.)| + |b(s, x.)| \leqslant C_T(1 + x^*_s). \tag{12.3}$$

Then the conclusions of Theorem 11.2 *remain valid.*

Proof. As before, we suppose without loss of generality that ξ is bounded. We build a solution X as follows. Define globally Lipschitz σ_N and b_N agreeing with σ and b on $\{(s, x); 0 \leqslant s \leqslant N, x^*_N \leqslant N\}$ such that (12.2) holds for all s, x and y, with a possibly larger K_N. Thus there is (by Theorem 11.2) a unique solution X^N to the SDE with coefficients σ_N and b_N. By Corollary 11.10, if $\tau_N \equiv \inf\{t: |X^{N+1}_t| \geqslant N\} \wedge N$, then X^N and X^{N+1} agree until τ_N. Hence, $\tau_N = \inf\{t: |X^N_t| \geqslant N\} \wedge N$, the τ_N increase, and we may define

$$X_t = X^N_t \quad \text{for} \quad 0 \leqslant t \leqslant \tau_N.$$

Now, clearly, for $0 \leqslant t \leqslant \tau_N$,

$$\begin{aligned} X_t = X^N_t &= \xi + \int_0^t \sigma_N(s, X^N_\cdot)\, dB_s + \int_0^t b_N(s, X^N_\cdot)\, ds \\ &= \xi + \int_0^t \sigma(s, X_\cdot)\, dB_s + \int_0^t b(s, X_\cdot)\, ds, \end{aligned}$$

so, provided we can prove that $\mathbf{P}(\sup \tau_N = \infty) = 1$, then X is a solution. But, for fixed $T > 0$, inequalities (11.6) and (12.3) show that there is a constant γ depending only on T and n such that, for $t \in [0, T]$,

$$\rho_t \equiv \mathbf{E}[X^*(t \wedge \tau_N)^2] \leqslant \gamma \left[1 + \mathbf{E} \int_0^{t \wedge \tau_N} X_s^{*2} \, ds \right]$$

which (by Gronwall's Lemma 11.11) implies that

$$\rho_t \leqslant \gamma e^{\gamma t}.$$

Thus

$$\mathbf{P}(\tau_N < T) \leqslant N^{-2} \rho_T \leqslant N^{-2} \gamma e^{\gamma T} \to 0 \quad \text{as} \quad N \to \infty,$$

and so $\mathbf{P}(\sup \tau_N = \infty) = 1$. Uniqueness of X is a consequence of Corollary 11.10, which shows that any solution must agree with X^N on $[0, \tau_N]$. □

(*12.4*) *Explosion.* Condition (12.3) is there to preclude explosion. Note that the 1-dimensional equation:

$$dX_t = (1 + X_t^2) \, dt, \quad X(0) = 0$$

has exploding solution $X(t) = \tan t$. One can prove that the equation

(12.5) $$dX = (1 + |X|^2) \, dB \text{ in } \mathbb{R}^3, \text{ where } B \text{ is a BM}(\mathbb{R}^3),$$

explodes in finite time. See § 51 and § 52 below.

There is a natural generalization of Theorem 12.1 for exploding solutions. For this, see Ikeda and Watanabe [1], § IV.2.

(*12.6*) *Discussion of previous examples.* The Brownian motion on a surface Σ described in § 4 clearly falls into the field of application of Theorem 12.1; and so do the examples in § 5. The theorem clearly allows us to construct a process corresponding to the conjectured optimal control for Whittle's example (7.1). By combining Theorem 12.1 with Skorokhod's lemma (6.2), we can construct a process corresponding to the conjectured optimal control for Jacka's example (7.8). As promised, we give a complete analysis of these control problems in § 15.

But Theorem 12.1 would not allow us to construct a process corresponding to the 'bang–bang' conjectured optimal control in Example 7.6. For that, we need a theorem of Nakao given later in § 41, or, for a 'weak solution' which is all that we really need, the 'change of time scale' method of § 26.

Also, we know that the Bessel processes lie outside the scope of Theorem 12.1, but that the Yamada–Watanabe theorem in § 40 below copes with them.

(*12.7*) *Lipschitz square roots.* We mentioned in § 3 that if we wish to construct a diffusion in $\mathbb{R}^n$ associated with a given operator

(12.8) $$\mathscr{L} = \tfrac{1}{2} a^{ij}(x) D_i D_j + b^i(x) D_i,$$

where $a\colon \mathbb{R}^n \to S_n^+$, then we consider the SDE

(12.9) $$dX_t = \sigma(X_t)\,dB_t + b(X_t)\,dt,$$

where the natural choice for σ is to take

(12.10) $$\sigma(x) \equiv a(x)^{1/2}$$

the unique nonnegative definite square root of $a(x)$. In order to be able to apply Theorem 12.1 to the equation (12.9), *we need to know that σ is (locally) Lipschitz* in that, for each N in $\mathbb{N}$, there exists some constant $C(N)$ such that

(12.11) $$\text{if } |x|,\ |y| \leqslant N, \quad \text{then} \quad |\sigma(x) - \sigma(y)| \leqslant C(N)|x-y|.$$

We can then use Theorem 12.1 to obtain a pathwise exact solution of (12.9) for any starting point X_0. We shall prove in § 22 that *X is then an FD diffusion with generator $\mathscr{L}$*. So we need to determine good sufficient conditions on the function a which will guarantee that the function $\sigma = a^{1/2}$ is locally Lipschitz.

(*12.12*) THEOREM (Phillips and Sarason). (i) *If the map $x \mapsto a(x)$ is C^1 and $a(x)$ is everywhere strictly positive definite, then the function $\sigma(\cdot) \equiv a(\cdot)^{1/2}$ is locally Lipschitz.*

(ii) *If $a(\cdot) \in C^2(\mathbb{R}^n, S_n^+)$, then the function $\sigma(\cdot) \equiv a(\cdot)^{1/2}$ is locally Lipschitz.*

Proof. We shall prove only part (ii). Part (i) is much easier to prove, and its proof will be self-evident from the following argument.

First we remark that if a is a fixed element of S_n^+ such that for some λ and μ with $0 < \lambda \leqslant \mu$,

$$\lambda|v|^2 \leqslant v^T a v \leqslant \mu|v|^2 \quad \text{for all} \quad v \in \mathbb{R}^n,$$

then we can write

$$a = \mu(I - S), \quad \text{where} \quad S = I - \mu^{-1}a,$$

and expand

$$a^{1/2} = \mu^{1/2}(I - S)^{1/2}$$

as a power series. This makes it clear that the map

(12.13) *$a \mapsto a^{1/2}$ is analytic on the set of strictly positive definite matrices.*

We now argue that to prove part (ii) of the theorem, it is enough to prove the following lemma.

(*12.14*) LEMMA. *Suppose that $t \mapsto a(t)$ is a C^2 map from $\mathbb{R}$ to S_n^+ such that:*

(12.15(i)) *for each t, $a(t)$ is strictly positive definite,*

(12.15(ii)) *for some constant K,*

$$|v^T \ddot{a}(t) v| \leqslant K|v|^2 \quad \textit{for all} \quad v \in \mathbb{R}^n,$$

where $\dot{a}(t) \equiv da(t)/dt$, $\ddot{a}(t) \equiv d^2a(t)/dt^2$, *etc. Let* $\sigma(t) \equiv a(t)^{1/2}$. *Then for* $1 \leqslant i, j \leqslant n$,

(12.16) $$|\sigma^{ij}(t) - \sigma^{ij}(s)| \leqslant C|t-s|,$$

where $C \equiv (1 + 2^{-1/2})K^{1/2}$.

Suppose that the lemma is established. Then we can immediately prove the lemma without condition (i) because we can replace $a(t)$ by $a_\varepsilon(t) \equiv a(t) + \varepsilon I$, which does not affect (12.15(ii)), apply the lemma to $a_\varepsilon(t)$, and then let $\varepsilon \downarrow 0$ in the 'ε version' of equation (12.16). If a: $\mathbb{R}^n \to S_n^+$ has all second-order derivatives bounded, we can apply the lemma to the map

$$t \mapsto a(y + tu)$$

where $y \in \mathbb{R}^n$ and u is a unit vector in $\mathbb{R}^n$ to deduce that σ is globally Lipschitz on $\mathbb{R}^n$. *Thus the Theorem follows.*

Proof of Lemma 12.14. Since each $a(t)$ is strictly positive definite, it follows from (12.13) that the map $t \mapsto \sigma(t)$ is C^2. We note that to prove (12.16), it is enough to prove that

(12.17) $$|\dot{\sigma}^{ij}(0)| \leqslant C$$

because there is nothing special about the value $t = 0$. Because we can replace $a(\bullet)$ by $H^T a(\bullet) H$, where $H^T a(0) H$ is diagonal and H is orthogonal, it does no harm to assume that

(12.18) $a(0)$ *is diagonal.*

Now,

$$a(t) = \sigma(t)\sigma(t), \quad \text{so that} \quad \dot{a}(t) = \dot{\sigma}(t)\sigma(t) + \sigma(t)\dot{\sigma}(t),$$

and, since $\sigma(0)$ is diagonal,

(12.19) $$\dot{a}^{ij}(0) = \dot{\sigma}^{ij}(0)\,(\sigma^{ii}(0) + \sigma^{jj}(0)).$$

For $v \in \mathbb{R}^n$, let

$$g_v(t) \equiv v^T a(t) v.$$

Then

$$\begin{aligned} 0 \leqslant g_v(t) &= g_v(0) + t\dot{g}_v(0) + \tfrac{1}{2}t^2\ddot{g}_v(\theta t) \\ &\leqslant g_v(0) + t\dot{g}_v(0) + \tfrac{1}{2}K|v|^2 t^2, \end{aligned}$$

so that as a quadratic form in t, the last expression has either no real zeros or coincident zeros. Thus,

$$\dot{g}_v(0)^2 \leqslant 2K|v|^2 g_v(0),$$

so that

(12.20) $$|\dot{g}_v(0)| \leqslant (2K)^{1/2}|v| g_v(0)^{1/2}.$$

Applying (12.20) to $v=\delta_i=(\delta_i^k: 1\leqslant k\leqslant n)$, and then to δ_j, and then to $\delta_i+\delta_j$, we obtain

$$|\dot{a}^{ii}(0)|\leqslant(2K)^{1/2}\sigma^{ii}(0),\quad |\dot{a}^{jj}(0)|\leqslant(2K)^{1/2}\sigma^{jj}(0),$$

$$|\dot{a}^{ii}(0)+2\dot{a}^{ij}(0)+\dot{a}^{jj}(0)|\leqslant 2K^{1/2}[a^{ii}(0)+a^{jj}(0)]^{1/2}$$

$$\leqslant 2K^{1/2}[\sigma^{ii}(0)+\sigma^{jj}(0)],$$

so that

$$|2\dot{a}^{ij}(0)|\leqslant(2+2^{1/2})K^{1/2}(\sigma^{ii}(0)+\sigma^{jj}(0)). \tag{12.21}$$

Result (12.17) now follows from (12.19) and (12.21). □

13. Flows; the diffeomorphism theorem; time-reversed flows. *Where the 'strong' or 'pathwise' approach of Itô's original theory of SDEs really comes into its own is in the theory of flows.* Flows are now very big business; and the martingale-problem approach, for all that it has other interesting things to say, cannot deal with them in any natural way.

We begin by proving an important refinement of Theorem 11.2.

(*13.1*) THEOREM (Blagoveščenkii, Freidlin). *Consider the equation*:

$$dX=\sigma(t, X.)dB+b(t, X.)dt, \tag{13.2}$$

where σ and b satisfy the hypotheses of Theorem 11.2. Then there exists a strong solution

$$F: \mathbb{R}^n\times W^d\to W^n$$

of (13.2), *with the extra regularity property that for each w in W^d, the map $x\mapsto F(x, w)$ is continuous from $\mathbb{R}^d$ into W^d.*

As a consequence of the definition (10.9) of a strong solution, whenever $(\Omega, \{\mathscr{F}_t\}, \mathbf{P}, B)$ is a set-up with B an $(\{\mathscr{F}_t\}, \mathbf{P})$-Brownian motion on $\mathbb{R}^d$, then, for each x, the process X^x, with

$$X_t^x\equiv F(x, B.)(t)$$

satisfies

$$X_t^x=x+\int_0^t\sigma(s, X_\cdot^x)\,dB_s+\int_0^t b(s, X_\cdot^x)\,ds. \tag{13.3}$$

Notation. Recall that $\mathbf{Q}^d$ denotes Wiener measure on $(W^d, \mathscr{X}^d)$, and $\{\bar{\mathscr{X}}_t^d\}$ denotes the usual $\mathbf{Q}^d$ augmentation of $\{\mathscr{X}_t^d\}$. Let

$$\mathbb{D}=\{j2^{-k}: j, k\in\mathbb{Z}\}$$

be the set of dyadic rationals in $\mathbb{R}$, and $\mathbb{D}^n=\mathbb{D}\times\ldots\times\mathbb{D}$ be the set of dyadic rationals in $\mathbb{R}^n$.

Proof. Theorem 11.2 shows that the SDE (13.2) is exact; and by Theorem 10.4 there exists a map $\tilde{F}^x$: $W^d \to W^n$, which is $\mathscr{X}_t^d/\mathscr{X}_t^n$ measurable for each t, such that for any set-up $(\Omega, \{\mathscr{F}_t\}, \mathbf{P}, B)$,

$$\tilde{X}_t^x \equiv \tilde{F}^x(B.)(t)$$

is the exact solution of (13.4). For the moment, think in terms of working with some arbitrary but fixed set-up $(\Omega, \{\mathscr{F}_t\}, \mathbf{P}, B)$.

Fix $T>0$, and let $p=n+1$. Estimate (11.8) shows that, for $x, y \in \mathbb{D}^n$ and $0 \leqslant t \leqslant T$, we have

$$\begin{aligned} 0 \leqslant \gamma_t &\equiv \mathbf{E}\{(\tilde{X}^x - \tilde{X}^y)_t^{*p}\} \\ &\leqslant C\left[|x-y|^p + \int_0^t \mathbf{E}\{(\tilde{X}^x - \tilde{X}^y)_s^{*p}\}\, ds\right] \\ &= C\left(|x-y|^p + \int_0^t \gamma_s\, ds\right). \end{aligned}$$

Hence, by Gronwall's Lemma (11.11),

$$0 \leqslant \gamma_t \leqslant C|x-y|^p e^{Ct} \leqslant C|x-y|^p e^{CT}.$$

Thus, we can use Kolmogorov's lemma (IV.44.4) to deduce that, with probability 1, the process $\{\tilde{X}_t^x\colon 0 \leqslant t \leqslant T, x \in \mathbb{D}^n\}$admits a continuous extension to a process $\{\tilde{X}_t^x\colon 0 \leqslant t \leqslant T, x \in \mathbb{R}^n\}$. If $N_T \subseteq \Omega$ is the set where no continuous extension is possible, then $\mathbf{Q}^d(N_T) = 0$ for all T. If $N \equiv \bigcup_{T \in \mathbb{N}} N_T$, we define for $x \in \mathbb{R}^n$,

$$F^x(w) = \begin{cases} \lim\limits_{\substack{y \to x \\ y \in \mathbb{D}^n}} \tilde{F}^y(w) = \lim \tilde{X}^y(w) & (w \notin N) \\ x, \quad \text{for all} \quad t \geqslant 0 & (w \in N). \end{cases}$$

The proof of Kolmogorov's lemma makes it clear that the function F can be defined in a way independent of the particular set-up $(\Omega, \{\mathscr{F}_t\}, \mathbf{P}, B)$.

Now consider solving (13.2) with initial condition ξ for some set-up $(\Omega, \{\mathscr{F}_t\}, \mathbf{P}, \xi, B)$. For $k \in \mathbb{N}$, let φ_k: $\mathbb{R}^n \to \mathbb{D}^n$ be a measurable map with $|\varphi_k(x) - x| \leqslant 2^{-k}$ for all x. Clearly, for each k,

$$X_\cdot^{(k)} \equiv F(\varphi_k(\xi), B.)$$

solves (13.2) with initial condition $\varphi_k(\xi)$. By estimate (11.8), the $X^{(k)}$ processes converge uniformly on compacts to the solution X of (13.2) with $X_0 = \xi$. But, by the continuity in x of $F(x, B.)$,

$$X_\cdot^{(k)} \to F(\xi, B.).$$

Therefore, $X = F(\xi, B.)$.

Finally, it is easy to conclude from the facts that

(i) $x \mapsto F(x, w)$ is continuous for each w,

(ii) $x \mapsto F(x, w)$ is $\bar{\mathscr{X}}^d_t/\mathscr{X}^n_t$-measurable for each x,

that:

(13.4) *for each t, F is $\mathscr{B}(\mathbb{R}^n) \times \bar{\mathscr{X}}^d_t/\mathscr{X}^n_t$-measurable.*

Hence F is a strong solution. □

(*13.5*) *Flow and Markov properties for the diffusion SDE case.* We have the following lemma.

(*13.6*) LEMMA (continuation of Theorem 13.1). *If the SDE* (13.2) *is of diffusion type, so that:*

$$\sigma(t, x.) = \sigma(x_t), \quad b(t, x.) = b(x_t)$$

for some Lipschitz functions $\sigma: \mathbb{R}^n \to L(\mathbb{R}^d, \mathbb{R}^n)$ and $b: \mathbb{R}^n \to \mathbb{R}^n$, then the strong solution F has the following flow property:

(13.7) $$\mathbf{P}[X^y_{u+\cdot}(w) = F(X^y_u, \theta_u w)] = 1, \quad \forall y, \forall u,$$

where $(\theta_u w)_s \equiv w_{u+s}$ as usual. For each y, the process $\{X^y_t : t \geq 0\}$ is Markovian.

Proof. For the strong solution X^y on the canonical path-space, we have:

$$X^y_{u+t} = X^y_u + \int_0^t \sigma(X^y_{u+s})\, d(\theta_u w)_s + \int_0^t b(X^y_{u+s})\, ds,$$

so that, for fixed y and u,

$$X^y_{u+\cdot} = F(X^y_u, \theta_u w) \quad \text{a.s.}.$$

For $h \in b\mathscr{B}(\mathbb{R}^n)$, define

$$P_t h(y) = \mathbf{E}h(X^y_t) = \mathbf{E}[h \circ F(y, w)(t)].$$

Because of (13.7), and because $\theta_t w$ is a Brownian motion independent of $\mathscr{X}_t$, we have

$$\mathbf{E}[h \circ X^y_{u+\cdot}(w) | \mathscr{X}_t] = \mathbf{E}[h \circ F(X^y_u, \theta_u w)(t) | \mathscr{X}_t]$$
$$= P_t h(X^y_u),$$

establishing the Markov property of X^y. See § III.7. □

In Part 4 of this chapter, we examine another way of thinking about the Markov property of SDEs which is more general and in some ways more fundamental. □

We now state the very important diffeomorphism theorem. The simplest proof is perhaps that to be found in Kunita [1]. You should certainly also read Carverhill and Elworthy [1].

(*13.8*) THEOREM (Diffeomorphism theorem, Part 1: Bismut, Elworthy, Kunita, Malliavin, . . .). *Let*

$$\sigma\colon \mathbb{R}^n \to L(\mathbb{R}^d, \mathbb{R}^n)$$

$$b\colon \mathbb{R}^n \to \mathbb{R}^n$$

be smooth (C^∞) *functions with globally bounded first derivatives. Then the strong solution F of the diffusion SDE*

(13.9)
$$dX_t = \sigma(X_t)\,dB_t + b(X_t)\,dt,$$

which is provided by the Blagoveščenskii–Freidlin Theorem 13.1 *has the further property that:*

for each t and each w, the map $x \mapsto X_t^x(w) \equiv F(x, w)(t)$ *is a diffeomorphism of* $\mathbb{R}^n$.

Recall that a diffeomorphism of $\mathbb{R}^n$ is a *one–one* C^∞ map of $\mathbb{R}^n$ *onto* $\mathbb{R}^n$ with C^∞ inverse map.

Let us write $Y(t, x)$ for the Jacobian map with (i, j)th component

(13.10)
$$Y(t, x)^i_j \equiv D_j X^i(t, x),$$

and, with the neat notation of Ikeda and Watanabe [1], $\sigma'_q(x)$ and $b'(x)$ for the $n \times n$ matrices

(13.11)
$$\sigma'_q(x)^i_j = D_j \sigma^i_q(x), \quad b'(x)^i_j = D_j b^i(x).$$

(*13.12*) THEOREM (Diffeomorphism theorem, Part 2). *The result of differentiating equation* (*13.7*) *formally with respect to the* jth *coordinate* x^j *of the starting point x is valid:*

(13.13)
$$Y(t, x) = I + \int_0^t \sigma'_q(X(s, x))\, Y(s, x)\, dB^q_s + \int_0^t b'(X(s, x))\, Y(s, x)\, ds.$$

(*13.14*) *Time-reversal of flows.* A very attractive method of establishing diffeomorphism properties of flows hinges on the idea of the time-reversed flow. The natural expression of this idea is the Stratonovich formulation. See § 31 for the general Stratonovich to Itô transformation rule.

(*13.15*) THEOREM (Kunita [2, 3]). *Suppose that σ and c are* C^∞ *functions with bounded first derivatives. Let*

$$(x, t) \mapsto X^x_t = X(x, t) \text{ from } \mathbb{R}^n \times [0, 1] \text{ to } \mathbb{R}^n$$

be the flow of the Stratonovich SDE

$$X_t^x = x + \int_0^t \sigma(X_s^x)\,\partial B_s + \int_0^t c(X_s)\,ds, \quad 0 \leqslant t \leqslant 1.$$

Let $\{B_u: 0 \leqslant u \leqslant 1\}$ *be the Brownian motion defined as follows:*

$$\hat{B}_u \equiv B_{1-u}.$$

Let

$$(z, u) \mapsto Z_u^z$$

be the flow of the SDE

$$Z_u^z = z + \int_0^u \sigma(Z_r)\,\partial\hat{B}_r - \int_0^u c(Z_r)\,dr, \quad 0 \leqslant u \leqslant 1. \tag{13.16}$$

Then, with probability 1,

$$Z_t^{X(x,1)} = X_{1-t}^x, \quad \forall (t, x) \in [0, 1] \times \mathbb{R}^n. \tag{13.17}$$

See Kunita [2, 3] for proof and for important applications.

(*13.18*) *A cautionary tale* (Elworthy). In § VIII.2C(c) of Elworthy [1], there is a very striking example of what can 'go wrong' with a flow. Consider the flow

$$Z_t^z = (z^{-1} + B_t)^{-1}, \quad Z_t^0 = 0, \tag{13.19}$$

where B is a BM($\mathbb{C}$), associated with the SDE

$$dZ = -Z^2 dB, \text{ or, equivalently, } \partial Z = -Z^2 \partial B. \tag{13.20}$$

(The Cauchy–Riemann equations force the agreement of the Itô and Stratonovich calculi, as was explained at (IV.34.18)). For each fixed starting point z in $\mathbb{C}$, it is almost surely true that $B(\bullet)$ never takes the value $-z^{-1}$, so that the solution $t \mapsto Z_t^z$ starting from z will almost surely never explode. However, for any path of B, the flow starting from any ('random') starting point z of the form $z = -B_t^{-1}$ for some t will explode (at time t, if not before).

(*13.21*) *Time-reversal of SDEs.* If we study time-reversal of a single SDE as opposed to time-reversal of a flow, matters become much more complicated. You can see the problem if you try to fix x and substitute $z = X(x, 1)$ in equation (13.16). Formally, we obtain

$$X_{1-u}^x = X_1^x + \int_0^u \sigma(X_{1-r}^x)\,\partial\hat{B}_r - \int_0^u c(X_{1-r}^x)\,dr, \quad 0 \leqslant u \leqslant 1.$$

If we try to interpret this, we must use the augmentation $\{\mathscr{G}_u\}$ of the filtration $\{\mathscr{G}_u^0\}$, where

$$\mathscr{G}_u^0 = \sigma(X_1^x) \vee \sigma(\hat{B}_r : r \leqslant u).$$

Then X_{1-u} is $\mathscr{G}_u$ measurable, but the point to notice is that, in general,

(13.22) $\{\hat{B}_u\colon u \leqslant 1\}$ is not a Brownian motion relative to $\{\mathscr{G}_u\}$.

As an exercise, you can prove this when $n=1$, $\sigma \equiv 1$, $b \equiv 0$. See Elliott and Anderson [1] for what to do.

We remark that *stochastic mechanics* (see § 5) relies on this type of time-reversal. See Zheng and Meyer [1].

We also remark that the theory of *grossissements* (see Jeulin [1], Jeulin and Yor [2] and Pardoux [2]) is very relevant to this problem.

14. Carverhill's noisy North–South flow on a circle. This very interesting example of a flow was presented in Carverhill [1].

The equation

$$\theta_t^\alpha = \alpha + \beta_t - K \int_0^t \sin \theta_s^\alpha \, ds, \tag{14.1}$$

where β is a BM ($\mathbb{R}$), represents the flow

$$(t, \alpha) \to \theta_t^\alpha$$

of the SDE

$$d\theta = d\beta - (K \sin \theta)\, dt. \tag{14.2}$$

Because $\sin\theta$ is periodic, we can regard the SDE (and the flow) as defined on the circle $S^1 = \mathbb{R}/2\pi\mathbb{Z}$. To make things agree with the Carverhill nomenclature, we have to agree that $\theta = 0$ represents the South Pole!

Let

$$\psi_t^\alpha = (\partial/\partial\alpha)\theta_t^\alpha. \tag{14.3}$$

Then, on differentiating (14.1) with respect to α, as we are allowed to do by Theorem 13.12, we find that

$$\psi_t^\alpha = 1 - K \int_0^t (\cos \theta_s^\alpha) \psi_s^\alpha \, ds,$$

so that

$$\psi_t^\alpha = \exp\left(-K \int_0^t \cos \theta_s^\alpha \, ds \right). \tag{14.4}$$

We now borrow a very elementary result from ergodic theory—see § 53. The diffusion $\{\theta_t^\alpha\colon t \geqslant 0\}$ has generator

$$\mathscr{L} = \tfrac{1}{2} D_\theta^2 - (K \sin \theta) D_\theta, \quad D_\theta \equiv \partial/\partial\theta,$$

and has invariant probability measure with density ρ satisfying

$$\mathscr{L}^* \rho = \tfrac{1}{2} D_\theta^2 \rho + D_\theta((K \sin \theta)\rho) = 0, \tag{14.5}$$

so that ρ is the von Mises density (see § IV.39)

$$\rho(\theta)=C_K\exp(2K\cos\theta). \tag{14.6}$$

Here, C_K is chosen so that

$$\int_0^{2\pi}\rho(\theta)\,d\theta=1.$$

Then, by the ergodic theorem,

$$t^{-1}\log\psi_t^\alpha=-Kt^{-1}\int_0^t\cos\theta_s^\alpha\,ds\to\lambda_K, \tag{14.7}$$

where

$$\lambda_K\equiv-K\int_0^{2\pi}(\cos\theta)\rho(\theta)\,d\theta=-KC_KI_K,$$

where

$$I_K\equiv\int_0^{2\pi}(\cos\theta)\exp(2K\cos\theta)\,d\theta.$$

Now, $I_0=0$, and

$$(d/dK)I_K=2\int_0^{2\pi}(\cos\theta)^2\exp(2K\cos\theta)\,d\theta>0.$$

Hence, $\lambda_K<0$, $K\neq0$.

Result (14.7) says that, if $K\neq0$, then, for each α, it is true with probability 1 that a little patch around α will get squashed together exponentially fast. The constant limit λ_K on the right of (14.7), which describes the rate of squashing-up, is called the Lyapunov exponent of the flow.

Of course, the flow cannot 'squash the whole circle' because it is not allowed to tear it apart. Figure V.1 shows a computer simulation by Peter Townsend and one of us (D.W.) of the flow for 19 initial θ-values and with time plotted radially. In other words, the picture is of paths $t\mapsto(t+c,\theta_t^\alpha)$ in polar coordinates, for 19 values of α. The various paths are pulled together, but the 'point of concentration' itself moves randomly.

A complete theoretical description of the flow was obtained by Carverhill. Here, we give a simpler derivation of this. The idea is to describe the function

$$\alpha\mapsto\theta_t^\alpha. \tag{14.8}$$

in terms of the process θ^0, which we regard as 'known'.

Let us write f' for df/dt. From (15.1), we obtain

$$(\theta_t^\alpha-\theta_t^0)'=-K(\sin\theta_t^\alpha-\sin\theta_t^0).$$

Figure V.1

Write $\mu_t^\alpha \equiv \frac{1}{2}(\theta_t^\alpha - \theta_t^0)$. Then

$$\begin{aligned}(\mu_t^\alpha)' &= -K(\sin\mu_t^\alpha)\cos\tfrac{1}{2}(\theta_t^\alpha + \theta_t^0)\\ &= -K(\sin\mu_t^\alpha)\cos(\mu_t^\alpha + \theta_t^0)\\ &= -K(\sin\mu_t^\alpha)(\cos\mu_t^\alpha)\cos\theta_t^0 + K(\sin\mu_t^\alpha)^2\sin\theta_t^0.\end{aligned}$$

Thus,

$$\begin{aligned}(\cot\mu_t^\alpha)' &= -(\operatorname{cosec}^2\mu_t^\alpha)(\mu_t^\alpha)'\\ &= (K\cot\mu_t^\alpha)\cos\theta_t^0 - K\sin\theta_t^0.\end{aligned}$$

Let

(14.9) $$a_t \equiv \exp\left(-K\int_0^t \cos\theta_s^0\,ds\right).$$

Then

$$(a_t\cot\mu_t^\alpha)' = -Ka_t\sin\theta_t^0.$$

Since $\mu_0^\alpha = \frac{1}{2}\alpha$, and $a_0 = 1$,

(14.10) $$a_t \cot \mu_t^\alpha = a_t \cot \tfrac{1}{2}(\theta_t^\alpha - \theta_t^0) = \cot \tfrac{1}{2}\alpha - g_t,$$

where

$$g_t = K \int_0^t a_s \sin \theta_s^0 \, ds.$$

Equation (14.10) provides the required description of the map (14.8) in terms of the θ^0 process.

(*14.11*) **Exercise:** *'gradient flow' on the circle* (see Carverhill, Chappell and Elworthy [1] and its Appendix by Elworthy and Stroock for more on gradient flows). Prove that the Lyapunov exponent for the flow on S^1 associated with the SDE

$$d\theta = (\cos\theta)\, d\beta_1 + (\sin\theta)\, d\beta_2,$$

where (β_1, β_2) is a BM $(\mathbb{R}^2)$, is equal to $-\frac{1}{2}$.

It is initially surprising that one has the same kind of 'squashing together' here as in Figure V.1. Peter Townsend and D.W. found that unless they quickly labelled computer printouts, they could not distinguish the cases (unless the time axis was so contracted as to obscure everything).

15. The martingale optimality principle in control. We saw in §7 that, in a number of control problems, it is possible to guess a good candidate for the control which minimises a given cost functional. There is a strikingly simple method for confirming one's guess (if it is indeed correct):

(*15.1*) (Martingale optimality principle): *Find a functional which is a martingale for the optimally controlled process, but a submartingale for any other control.*

(*15.2*) *Study of Whittle's 'flypaper' example* (*7.1*). There the model was the following:

$$X_t = x + B_t + \int_0^t u_s \, ds, \quad t < T \equiv \inf\{r : X(r) \in \{0, z\}\},$$
$$= X_T \quad (t \geqslant T);$$

and the cost C was given by

$$C = \int_0^T \tfrac{1}{2}(u_s^2 + 1)\, ds + K(X_T).$$

We conjectured that the minimal expected cost from starting point x in $[0, z]$ is given by $F(x)$, where F is the solution of the equation

(15.3) $$F''(x) - F'(x)^2 + 1 = 0, \quad F(0) = K(0), \quad F(z) = K(z),$$

and that the optimal control is given by

(15.4) $$u_t = -F'(X_t).$$

Now define F to be the solution of (15.3), define u by (15.4), and fix $x \in [0, z]$. If v is any possible control (previsible path functional) suppose that a process X on some $(\Omega, \{\mathcal{F}_t\}, \mathbf{P})$ satisfies

(15.5) $$X_t = x + B_t + \int_0^t v(s, X_\cdot)ds \quad (t \leqslant T)$$
$$= X_T \quad (t > T),$$

where $T \equiv \inf\{t: X_t \in \{0, z\}\}$, and B is a Brownian motion relative to $(\Omega, \{\mathcal{F}_t\}, \mathbf{P})$. Abbreviating $v(s, X_\cdot)$ to v_s, and defining

$$V_t \equiv F(X_{t \wedge T}) + \int_0^{T \wedge t} \tfrac{1}{2}(v_s^2 + 1)ds,$$

then the interpretation of V is that:

(*15.6*) *V_t is the conditional expectation (given $\mathcal{F}_t$) of the cost incurred by using control v up to time t, and thereafter using the optimal control*

(assuming that our conjecture about F is correct), and

$$V_\infty = \int_0^T \tfrac{1}{2}(v_s^2 + 1)ds + F(X_T) \equiv C,$$

the total cost incurred by using control v. In the light of this interpretation, it is not hard to convince yourself that *V must be a submartingale*; and it is even easier to prove this using Itô's formula! Indeed,

(15.7) $$dV_t = I_{\{t \leqslant T\}}[F'(X_t)dB_t + \{\tfrac{1}{2}F''(X_t) + v_t F'(X_t) + \tfrac{1}{2}(v_t^2 + 1)\}dt].$$

Now we are only going to be interested in controls for which $\mathbf{E}(C) < \infty$, and since $\mathbf{E}(T) \leqslant 2\mathbf{E}(C)$, we can assume that $\mathbf{E}(T) < \infty$ and so

$$\int_0^{T \wedge \cdot} F'(X_s)dB_s \textit{ is an } L^2\textit{-bounded martingale},$$

since F' is bounded on $[0, z]$. Since the quadratic form in v_t within the curly brackets $\{\cdot\}$ on the right-hand side of (15.7) is minimized to the value 0 when $v_t = -F'(X_t)$, we deduce from (15.7) that:

(15.8(i)) *V_t is a submartingale;*
(15.8(ii)) *V_t is an L^2-bounded martingale if $v_t \equiv v(t, X_\bullet) = -F'(X_t)$ for all t.*

Hence immediately

$$V_0 = F(x) \leqslant \mathbf{E}(C) \equiv \mathbf{E}(V_\infty),$$

with equality if $v=u$, illustrating the use of the martingale optimality principle (15.1); *the expected cost starting from x must be at least* $F(x)$, and all that remains is to confirm that this lower bound can be attained. But the obvious candidate for the optimally-controlled SDE

$$\bar{X}_t = x + B_{t\wedge T} - \int_0^{t\wedge T} F'(\bar{X}_s)\,ds \equiv x + B_{t\wedge T} + \int_0^{t\wedge T} u(s, \bar{X}.)\,ds$$

has Lipschitz coefficients, so is a pathwise exact SDE (Theorem 11.2) and has a pathwise unique strong solution. Thus (from 15.8(ii)), V is an L^2-bounded martingale, provided we can prove that $\mathbf{E}(T)<\infty$ for this control. But from (15.7) we know that V is a nonnegative local martingale, therefore a supermartingale, therefore $F(x)\geqslant \mathbf{E}(C)\geqslant \frac{1}{2}\mathbf{E}(T)$.

(*15.9*) *Study of Jacka's example* (*7.8*). Please re-read the earlier discussion of this example. We begin by considering the existence of a solution to equation (7.9).

Recall from §6 that for a given continuous function $y\colon \mathbb{R}^+\to\mathbb{R}$, $y(0)\geqslant 0$, there exists a unique pair (x, a) of continuous functions such that

(i) $x(t)=y(t)+a(t)$;

(ii) $x(t)\geqslant 0$, a is increasing, $a(0)=0$, and

$$\int_0^t I_{\{x(s)\neq 0\}}\,da_s = 0 \quad \text{for all } t,$$

and the solution is given explicitly by $a(t)\equiv\sup\{y(s)^-: 0\leqslant s\leqslant t\}$. This allows us to solve (7.9) by converting it into the SDE

$$Y_t = x + \int_0^t \sigma(Y_s + A_s(Y))\,dB_s \tag{15.10}$$

where $A(Y)$ is the functional of the path of Y defined by

$$A_t(Y)\equiv\sup\{Y_s^-;\, 0\leqslant s\leqslant t\}.$$

Now, σ has a derivative bounded by K, and it follows from the fact that $|\sigma(x)-\sigma(y)|\leqslant K|x-y|$ that the coefficient $\sigma\colon\mathbb{R}\times W^1\to\mathbb{R}$ defined by

$$\sigma(t, w)\equiv\sigma(w_t + A_t(w))$$

is Lipschitz. So, by Theorem 11.2, equation (15.10) has a strong solution Y, and $X_t = Y_t + A_t(Y)$ solves the original SDE (7.9).

We now use the martingale optimality principle to prove that the process X just constructed is indeed optimal. Let f be the unique decreasing solution of

$$\tfrac{1}{2}\sigma(x)^2 f''(x) = \alpha f(x), \quad f(0)=1, \quad f\geqslant 0. \tag{15.11}$$

(As we shall see in our study of 1-dimensional diffusions, $f(x)$ can be expressed explicitly:

$$f(x) = \mathbf{E}[e^{-\alpha H} | X_0 = x], \quad H \equiv \inf\{s: X_s = 0\}.)$$

Let $\rho(x) \equiv \sigma(x)^2/2\alpha$, an increasing C^1 function. Let

$$\Psi \equiv -f'(0) \geqslant 0. \tag{15.12}$$

We need one small result.

(*15.13*) PROPOSITION. $\varphi(x) \equiv f'(x) + \Psi f(x)$ *is nonnegative.*

Proof. By construction, $\varphi(0) = 0$, and $\lim_{x\to\infty} \varphi(x) \geqslant 0$, so if the proposition is false, there is some $x_0 > 0$ where φ attains its (negative) minimum, and there

$$\varphi(x_0) < 0, \quad \varphi'(x_0) = 0, \quad \varphi''(x_0) \geqslant 0;$$

that is,

$$f'(x_0) + \Psi f(x_0) < 0, \tag{15.14(i)}$$

$$f''(x_0) + \Psi f'(x_0) = 0, \tag{15.14(ii)}$$

$$f'''(x_0) + \Psi f''(x_0) \geqslant 0. \tag{15.14(iii)}$$

But $\rho f'' = f$, so that $\rho f''' + \rho' f'' = f'$. Since ρ is increasing and $f'' \geqslant 0$, it follows that $\rho f''' \leqslant f'$. Hence, (15.14(iii)) implies that

$$0 \leqslant f'(x_0) + \Psi \rho(x_0) f''(x_0) = f'(x_0) + \Psi f(x_0),$$

contradicting (15.14(i)). □

Now suppose that v is some admissible control, and let

$$Z_t = x + \int_0^t \sigma(Z_s)\, dB_s + v_t.$$

Define $\tau = \tau(x, v) \equiv \inf\{s: Z_s < 0\}$, define the semimartingale $\tilde{V}$ by

$$\tilde{V}_t = f(Z_t)\exp\{-\alpha t - (C - v_t)\Psi\}, \tag{15.15}$$

and let

$$V_t \equiv \tilde{V}(t \wedge \tau). \tag{15.16}$$

Then, because of (15.11), Itô's formula yields

$$\begin{aligned} dV_t &= I_{\{t \leqslant \tau\}}[f'(Z_t)dZ_t + \Psi f(Z_t) dv_t]\exp\{-\alpha t - (C - v_t)\Psi\} \\ &= [f'(Z_t)\sigma(Z_t)dB_t + \varphi(Z_t)dv_t] I_{\{t \leqslant \tau\}} \exp\{-\alpha t - (C - v_t)\Psi\}. \end{aligned}$$

This represents the bounded semimartingale V as the sum of a local martingale and a finite variation process which is both increasing (because $\varphi \geqslant 0$ and v is increasing) and bounded (because φ is bounded and $v_t \leqslant C$ for all t). Thus the local martingale is bounded, and so is a martingale; *the process V is a bounded submartingale*, and so

$$f(x)e^{-\Psi C}=V_0\leqslant \mathbf{E}V_{\tau(x,v)}\leqslant \mathbf{E}e^{-\alpha\tau(x,v)}. \tag{15.17}$$

However, *if the process Z is the solution X of the SDE* (7.9), so that $v=u$, then because u grows only when $Z=0$ and $\varphi(0)=0$, we see that *V is a bounded martingale.* Moreover, $u_{\tau(x,u)}=C$. Thus

$$f(x)e^{-\Psi C}=V_0=\mathbf{E}V_{\tau(x,u)}=\mathbf{E}e^{-\alpha\tau(x,u)},$$

whence from (15.17),

$$\mathbf{E}e^{-\alpha\tau(x,u)}=f(x)e^{-\Psi C}\leqslant \mathbf{E}e^{-\alpha\tau(x,v)}, \tag{15.18}$$

proving that u is an optimal control, and exhibiting the minimal expected 'cost' as

$$\inf_v \mathbf{E}e^{-\alpha\tau(x,v)}=f(x)e^{-\Psi C}.$$

(15.19) Discussion. For the Markov process X, $A_t\equiv A_t(Y)$ is local time at 0. Set

$$\gamma(t)\equiv\inf\{s: A_s>t\},$$

so that γ is inverse local time at 0 for X. Since $X(\gamma(t))=0$, the strong Markov theorem makes it clear (compare (VI.43.3)) that $\gamma(\bullet)$ is a subordinator, and, by the Lévy–Itô theorem of §II.64,

$$\mathbf{E}^0 e^{-\alpha\gamma(t)}=e^{-t\Psi(\alpha)}$$

where $\Psi(\bullet)$ is the 'Lévy exponent function'. We have, under $\mathbf{P}^x$,

$$\tau(x,u)=\gamma(C)=H+\gamma(C)\circ\theta_H,$$

where θ is the time-shift operator. Hence, by the strong Markov property applied at time H,

$$\begin{aligned}\mathbf{E}e^{-\alpha\tau(x,u)}&=\mathbf{E}^x e^{-\alpha\gamma(C)}\\&=\mathbf{E}^x[e^{-\alpha H}]\mathbf{E}^0[e^{-\alpha\gamma(C)}]\\&=f(x)e^{-C\Psi(\alpha)}\end{aligned}$$

so that the content of (15.12) is the result (Itô and McKean [1; §6.2]) that

$$\Psi(\alpha)=-f'(0). \tag{15.20}$$

Recall that f is defined from α via equation (15.11). Of course, the argument due to Jacka which we have given proves (15.20).

Note that it is now clear that the definition of V given by (15.15) and (15.16) again has the right intuitive interpretation: *V_t is the conditional expectation (given $\mathcal{F}_t$) of the cost of utilizing control v up to time t and the optimal control thereafter.* The factor $e^{-\alpha t}$ arises because time t has already elapsed; and, because of (15.18),

$$f(Z_t)\exp\{-(C-v_t)\Psi\}$$

is the minimum cost if X starts at Z_t and there is an amount $(C-v_t)$ of fuel left.

(*15.21*) *Remark.* The main purpose which the stochastic-calculus technique is serving in the two examples just discussed is in proving that the 'Markov' controls are optimal. It is disappointing that not too many interesting examples of explicit 'non-Markov' solutions are known. Elliott gives some in Chapter 17 of his book [1].

3. WEAK SOLUTIONS, UNIQUENESS IN LAW

16. Weak solutions of SDEs; Tanaka's SDE. As we said earlier, one of the main purposes of studying SDEs is to be able to construct diffusions, and Theorem (11.2) gets us a long way towards this goal. It allows us to construct a $(\sigma\sigma^T, b)$-diffusion whenever σ and b are Lipschitz, which copes with the vast majority of cases arising in practice. But notice that Theorem 11.2 gives us far more than we really need; it delivers a strong solution, which implies that the SDE (8.1) has a solution on *every* set-up $(\Omega, \{\mathscr{F}_t\}, \mathbf{P}, \xi, B)$ (or equivalently, by Theorem 10.4, the SDE has a solution on the canonical set-up). In fact, as we saw in §3, to construct a $(\sigma\sigma^T, b)$-diffusion it is enough to find a solution to the SDE (8.1) on *some* set-up $(\Omega, \{\mathscr{F}_t\}, \mathbf{P}, \xi, B)$; in such cases we shall say that the SDE has a 'weak' solution. For similar reasons, the concept of weak solution is the appropriate one for control problems; our treatment in § 15 has allowed for this. There do exist SDEs with weak solutions but which have no solution on the canonical set-up (see § 16.5); the solution X cannot be constructed from ξ and B in such cases. Indeed, in many applications of the 'weak' solution method, it is B which is constructed from X, rather than the other way round!

We now formalize the definition of a weak solution, for our general SDE with previsible path functional coefficients σ and b.

(*16.1*) DEFINITION (Weak solution). *We say that the SDE*

$$X_t = X_0 + \int_0^t \sigma(s, X.)\,dB_s + \int_0^t b(s, X.)\,ds \tag{16.2}$$

has a weak solution with initial distribution μ *if there exists* $(\Omega, \{\mathscr{F}_t\}, \mathbf{P})$ *satisfying the usual conditions, together with continuous semimartingales X and B (with values in $\mathbb{R}^n$ and $\mathbb{R}^d$ respectively) such that:*

(i) *B is an $\{\mathscr{F}_t\}$-Brownian motion;*

(ii) *X_0 has law μ;*

(iii) $\displaystyle\int_0^t \{|\sigma(s, X.)|^2 + |b(s, X.)|\}\,ds < \infty$ a.s. *for all t;*

(iv) $\displaystyle X_t = X_0 + \int_0^t \sigma(s, X.)\,dB_s + \int_0^t b(s, X.)\,ds.$

If for each probability μ on $\mathbb{R}^n$ there is a (weak) solution to (16.2) *with μ as initial distribution, we say that* (16.2) *has a* weak solution.

The appropriate uniqueness concept for weak solutions is 'uniqueness in law'.

(*16.3*) DEFINITION (uniqueness in law). *We say that the solution of* (16.2) *is* unique in law *if whenever $\{X_t: t\geqslant 0\}$ and $\{X'_t: t\geqslant 0\}$ are two solutions (perhaps on different set-ups) such that the laws of X_0 and X'_0 are the same, then the laws of X and X' are the same.*

In an obvious way, we can speak about uniqueness in law in connection with a given initial distribution μ.

(*16.4*) *Remark.* If the coefficients σ and b satisfy the Lipschitz condition (11.1), then there exists a weak solution unique in law; indeed, Theorem 13.1 guarantees the existence of a strong solution, and since X is the image under the measurable map F of (X_0, B), its law is uniquely determined. We shall soon see that in complete generality, pathwise uniqueness implies uniqueness in law.

The difference between weak and strong solutions of SDEs is perfectly illustrated by a celebrated example, which we first encountered in our analysis of semimartingale local time (§IV.43).

(*16.5*) **Example:** *Tanaka's SDE.* Consider the (1-dimensional) SDE

$$X_t = X_0 + \int_0^t \operatorname{sgn}(X_s)\,dB_s. \tag{16.6}$$

Any solution X is a continuous local martingale with quadratic variation t, so, by Theorem IV.33.1, X is a Brownian motion. Thus the law of any solution is unique. Moreover, there do exist solutions; if X is a Brownian motion with initial law δ_0, and B is the Brownian motion

$$B_t = \int_0^t \operatorname{sgn}(X_s)\,dX_s, \tag{16.7}$$

then X solves (16.6). Thus the SDE (16.6) has a (weak) solution, which is unique in law.

Notice firstly that if $X_0=0$, and X is a solution of (16.6), then $-X$ is also a solution; this is no contradiction, because X and $-X$ are both Brownian motions and therefore have the same distribution, which is all that uniqueness in law says. The moral is that *even though the SDE may have a weak solution unique in law, there may be many 'pathwise' solutions on a fixed filtered probability space*; there is no pathwise uniqueness for the SDE (16.6).

Notice secondly that if $X_0=0$, then from Tanaka's formula IV.43.4, the Brownian motion B defined by (16.7) is $B=|X|-L$. In particular, B is adapted to the filtration of $|X|$, strictly smaller than the filtration of X. Thus *if we were given a Brownian motion B, and if $\mathscr{F}^\circ_t \equiv \sigma\{B_s: s\leqslant t\}$, and $\{\mathscr{F}_t\}$ the usual augmentation of*

$\{\mathcal{F}_t^\circ\}$, *then there is no semimartingale* X *on* $(\Omega, \{\mathcal{F}_t\}, \mathbf{P})$ *solving* (16.6)! The Tanaka SDE is an example of an SDE which has a weak solution, but does not have a strong solution. Tsirel'son's extraordinary illustration of the same phenomenon is given in § 18 below.

17. 'Exact equals weak plus pathwise unique'. We now come to a fundamental theorem of Yamada and Watanabe which helps sort things out.

(*17.1*) THEOREM (Yamada and Watanabe [1]). *Let* σ *and* b *be previsible path functionals, and consider the SDE*:

(17.2) $$dX = \sigma(t, X_\bullet)dB + b(t, X_\bullet)dt.$$

Then this SDE is exact if and only if the following two conditions hold:

(17.3(i)) *the SDE* (17.2) *has a weak solution,*

(17.3(ii)) *the SDE* (17.2) *has the pathwise-uniqueness property.*

Uniqueness in law then holds for equation (17.2).

It is clear from the statement of the theorem how it clarifies the structure of the subject. Its implication that

PATHWISE UNIQUENESS IMPLIES UNIQUENESS IN LAW

is very important. For example, it allows us to deduce that uniqueness in law holds for the 'Lipschitz' situations covered by Itô's theorems (11.2) and (12.1). Moreover, we shall see that, particularly in the 1-dimensional case and again thanks to Yamada and Watanabe, it is sometimes possible to establish properties (17.3) directly. As already mentioned, the Bessel processes provide a main application of this.

The proof of the theorem is very illuminating too but is best left until a second reading of this book. The idea is *to transfer the structure of a weak solution to a canonical set-up in which* W^n *becomes the sample space for* X *and* W^d *that for* B. This will allow two different weak solutions to be associated with Brownian motions which are not only the same in law but even pathwise identical. You should have sensed that this would be the crucial step in sorting things out.

What allows us to carry it through is the concept of *regular conditional probability*. This captures the idea of 'transferring the structure'.

In the following proof, we supply what we hope is about the right amount of detail, enough so that you could supply full details for yourself if forced to do so. On some of the steps of the proof, you will find all the details kindly and meticulously supplied in Chapter 8 of the Stroock and Varadhan [1].

Proof. It is a consequence of our definitions that if our SDE is exact then properties (17.3) hold. Now,

Assume that properties (*17.3*) *hold.*

During a first reading of the proof, take Lemmas 17.9 and 17.12 on trust.

Step 1: *proof of uniqueness in law.* Let μ be a given probability measure on $\mathbb{R}^n$, and suppose that (X, B) is a weak solution of (17.2) with initial distribution μ. Let π be the law on $\mathbb{R}^n \times W^d \times W^n$ of the triple (X_0, B, X), and let Q: $(\mathbb{R}^n \times W^d) \times \mathscr{B}(W^n) \to [0, 1]$ be a regular conditional distribution for X given (X_0, B). Similarly, if (X', B') is some other weak solution of (17.2) with initial distribution μ, let Q' be a regular conditional distribution for X' given (X_0, B'). Let S be the Polish space $S = \mathbb{R}^n \times W^d \times W^n \times W^n$; a typical member of S will be written (y, β, x, x'). With $\mathbf{Q}^d$ denoting Wiener measure on W^d, we define a probability $\bar{\mathbf{P}}$ on S by

$$\bar{\mathbf{P}}(dy, d\beta, dx, dx') \equiv \mu(dy)\mathbf{Q}^d(d\beta)Q((y, \beta), dx)Q'((y, \beta), dx').$$

The line of the proof should now be clear. Defining $\mathscr{G}_t^\circ$ to be $\sigma\{\beta_s, x_s, x_s' : s \leqslant t\}$, and $\{\mathscr{G}_t\}$ to be the usual $\bar{\mathbf{P}}$-augmentation of $\{\mathscr{G}_t^\circ\}$, we firstly prove (in Lemma 17.12 below) that $\{\beta_t : t \geqslant 0\}$ is a Brownian motion on $(S, \{\mathscr{G}_t\}, \bar{\mathbf{P}})$. Then, since the $\bar{\mathbf{P}}$-distribution of (y, β, x) is the same as the distribution of (X_0, B, X), it is not too hard to show as in the proof of Theorem 10.4 that

$$x_t = y + \int_0^t \sigma(s, x_\cdot)\,d\beta_s + \int_0^t b(s, x_\cdot)\,ds \tag{17.4}$$

and that the law of x is the law of X. Analogous reasoning holds for x', X', and so x, x' are two solutions of the SDE (17.2) on a common filtered probability space, with the same initial condition y and the same driving Brownian motion β; pathwise uniqueness implies

$$\bar{\mathbf{P}}[x_t = x_t' \quad \text{for all } t] = 1, \tag{17.5}$$

and so the laws of X and X' coincide, which completes the proof of uniqueness in law.

Step 2: *expressing x in terms of y and* β. From (17.5) we deduce that for $\mu \times \mathbf{Q}^d$-almost all (y, β), the product measure $Q((y, \beta), dx)Q'((y, \beta), dx')$ on $W^n \times W^n$ is concentrated on the diagonal. It is easy to see that this can only happen if the measure is degenerate:

$$Q((y, \beta), A) = Q'((y, \beta), A) = I_A(F_\mu'(y, \beta)),$$

where F_μ': $\mathbb{R}^n \times W^d \to W^n$ is measurable $(\mathscr{B}(\mathbb{R}^n) \times \mathscr{X}^d / \mathscr{X}^n)$. Hence immediately $\bar{\mathbf{P}}[x = F_\mu'(y, \beta)] = 1$, and there exists F_μ: $\mathbb{R}^n \times W^d \to W^n$ which is measurable $(\overline{\mathscr{B}(\mathbb{R}^n) \times \mathscr{X}^d} / \mathscr{X}^n)$ such that

$$x = F_\mu(y, \beta).$$

Here, we have used $\overline{\mathscr{B}(\mathbb{R}^n) \times \mathscr{X}^d}$ to denote the $\mu \times \mathbf{Q}^d$-completion of $\mathscr{B}(\mathbb{R}^n) \times \mathscr{X}^d$.

The aim now is to construct a solution on the canonical set-up with initial law μ (refer back to § 10 for the definition of the canonical set-up with initial law μ; we shall use the notation established there). Once we have done this, Theorem 10.4 tells us immediately that there is a solution with initial law μ on any set-up, and this together with pathwise uniqueness implies that the SDE (17.2) is exact.

The obvious candidate for a solution is

$$X_{\cdot} \equiv F_\mu(\xi, B_{\cdot}). \tag{17.6}$$

To show that this *is* a solution, we need to know that the map F_μ has the following stronger measurability property:

$$\text{for each } t \geqslant 0, \quad F_\mu^{-1}(\mathscr{X}_t^n) \subseteq \mathscr{F}_t$$

where $\mathscr{F}_t$ is the completion of $\mathscr{B}(\mathbb{R}^n) \times \mathscr{X}_t^d$ with all the $\mu \times \mathbf{Q}^d$-null subsets of sets in $\overline{\mathscr{B}(\mathbb{R}^n) \times \mathscr{X}^d}$. This is a consequence of the fact (proved in Lemma 17.9 below) that for $A \in \mathscr{X}_t^n$,

$$(y, \beta) \mapsto Q((y, \beta), A)$$

is measurable with respect to $\overline{\mathscr{B}(\mathbb{R}^n) \times \mathscr{X}_t^d}$.

Hence the map φ: $\mathbb{R}^{++} \times \Omega \to L(\mathbb{R}^d, \mathbb{R}^n)$ defined by

$$\varphi(s; \xi, B_{\cdot}) \equiv \sigma(s, F_\mu(\xi, B_{\cdot})) \tag{17.7}$$

is $\mathscr{P}/\mathscr{B}(L(\mathbb{R}^d, \mathbb{R}^n))$-measurable so, by Lemma 10.1, there exists measurable Φ: $\mathbb{R}^n \times W^d \to W^n$ such that, on any set-up $(\Omega', \{\mathscr{F}_t'\}, \mathbf{P}', \xi', B')$ where ξ' has law μ,

$$\mathbf{P}'\left[\int_0^{\cdot} \sigma(s, F_\mu(\xi', B'))\, dB_s' = \Phi(\xi', B')\right] = 1. \tag{17.8}$$

But $(S, \{\mathscr{G}_t\}, \bar{\mathbf{P}}, y, \beta)$ is one such set-up, for which $x = F_\mu(y, \beta_{\cdot})$ a.s. Thus in the light of (17.4), we conclude that

$$\bar{\mathbf{P}}\left[F_\mu(y, \beta_{\cdot}) = y + \Phi(y, \beta_{\cdot}) + \int_0^{\cdot} b(s, F_\mu(y, \beta_{\cdot}))\, ds\right] = 1.$$

Since the $\bar{\mathbf{P}}$-law of $(y, \beta_{\cdot})$ is the same as the $\mathbf{P}$-law of $(\xi, B_{\cdot})$, it follows immediately that

$$\begin{aligned} 1 &= \mathbf{P}\left[F_\mu(\xi, B_{\cdot}) = \xi + \Phi(\xi, B_{\cdot}) + \int_0^{\cdot} b(s, F_\mu(\xi, B_{\cdot}))ds\right] \\ &= \mathbf{P}\left[F_\mu(\xi, B_{\cdot}) = \xi + \int_0^{\cdot} \sigma(s, F_\mu(\xi, B_{\cdot}))\, dB_s + \int_0^{\cdot} b(s, F_\mu(\xi, B_{\cdot}))\, ds\right] \end{aligned}$$

and X defined by (17.6) is a solution of the SDE (17.2) on the canonical set-up with initial law μ. □

The proof of Theorem 17.1 is complete except for the lemmas. You should re-read Step 1 to set the scene.

(*17.9*) LEMMA. *For each* $t \geqslant 0$ *and each* $A \in \mathscr{X}_t^n$, *the map*

$$(y, \beta) \mapsto Q((y, \beta), A)$$

is measurable with respect to $\overline{\mathscr{B}(\mathbb{R}^n) \times \mathscr{X}_t^d}$.

Proof. Let g be a bounded measurable function on $\mathbb{R}^n$, and let Z be a random variable of the form $Z_1(\beta) Z_2(\beta)$, where Z_1 is a bounded functional of $\{\beta_s: s \leqslant t\}$, and Z_2 is a bounded functional of $\{\beta_u - \beta_t: u \geqslant t\}$. Then from the definition of Q,

$$\begin{aligned}
&\mathbf{E}[g(y) Z_1(\beta) Z_2(\beta) Q((y, \beta), A)] \\
&= \mathbf{E}[g(X_0) Z_1(B) Z_2(B) I_A(X)] \\
&= \mathbf{E}[g(X_0) Z_1(B) \mathbf{E}(Z_2(B) | \mathscr{F}_t) I_A(X)] \\
&= \mathbf{E}[g(X_0) Z_1(B) \mathbf{E}(Z_2(B)) I_A(X)] \\
&= \mathbf{E}[g(y) Z_1(\beta) \mathbf{E}(Z_2(\beta)) Q((y, \beta), A)].
\end{aligned}$$

Moreover, since $\{\beta_u - \beta_t: u \geqslant t\}$ is independent of $\{\beta_s: s \leqslant t\}$, we have

$$\mathbf{E}[g(y) Z_1(\beta) Z_2(\beta) | \overline{\mathscr{B}(\mathbb{R}^n) \times \mathscr{X}_t^d}] = g(y) Z_1(\beta) \mathbf{E} Z_2(\beta), \quad \text{a.s.}$$

Now, you may complete the argument by doing an exercise which we shall use frequently in Chapter VI. (It is proved as Lemma VI. 26.1.)

(*17.10*) **Exercise.** Let $(\Omega, \mathscr{F}, \mathbf{P})$ be a complete triple, and let $\mathscr{G}$ be a sub-σ-algebra of $\mathscr{F}$ containing all null sets in $\mathscr{F}$. Prove that if $\eta \in L^2(\Omega, \mathscr{F}, \mathbf{P})$, then $\eta \in L^2(\Omega, \mathscr{G}, \mathbf{P})$ if and only if for every ξ in $L^2(\Omega, \mathscr{F}, \mathbf{P})$ (or alternatively, for every ξ in a set whose linear span is dense in $L^2(\Omega, \mathscr{F}, \mathbf{P})$),

$$\mathbf{E}(\xi \eta) = \mathbf{E}(\mathbf{E}[\xi \mid \mathscr{G}] \eta). \tag{17.11}$$

Now apply this to the situation in which,

$$\Omega = \mathscr{R}^n \times W^d, \quad \mathbf{P} = \mu \times \mathbf{Q}^d, \quad \mathscr{F} = \overline{\mathscr{B}(\mathbb{R}^n) \times \mathscr{X}^d},$$

$$\mathscr{G} = \overline{\mathscr{B}(\mathbb{R}^n) \times \mathscr{X}_t^d}, \quad \eta = Q((y, \beta), A), \quad \xi = g(y) Z_1(\beta) Z_2(\beta).$$

Hint. Conditioning on $\mathscr{G}$ on the right of (17.11) shows that

$$\mathbf{E}(\xi \eta) = \mathbf{E}(\mathbf{E}[\xi \mid \mathscr{G}] \mathbf{E}[\eta \mid \mathscr{G}]) = \mathbf{E}(\xi \mathbf{E}[\eta \mid \mathscr{G}]). \qquad \square$$

In the following lemma, we return to the notation of Step 1.

(*17.12*) LEMMA. *The process* $\{\beta_t: t \geqslant 0\}$ *is a* $\{\mathscr{G}_t\}$*-Brownian motion.*

Proof. Take $0 \leqslant s \leqslant t$, $C \in \sigma\{\beta_r : r \leqslant s\}$, $A \in \sigma\{x_r : r \leqslant s\}$, and $A' \in \sigma\{x'_r : r \leqslant s\}$. Then by the definition of Q, Q',

$$\begin{aligned} &\mathbf{E}[\beta_t - \beta_s; C \cap A \cap A'] \\ &= \int \mu(dy)\, \mathbf{Q}^d(d\beta)\, I_C(\beta)\, Q\,((y,\beta), A) Q'((y,\beta), A')\, (\beta_t - \beta_s) \\ &= 0 \end{aligned}$$

since, by the above, $Q((y, \beta), A)$ is $\overline{\mathscr{B}(\mathbb{R}^n) \times \mathscr{X}^d_s}$- measurable, and since β is certainly a Brownian motion on $\mathbb{R}^n \times W^d$ under $\mu \times \mathbf{Q}^d$. Hence β is a continuous $\{\mathscr{G}_t\}$-martingale, and similarly $\beta^i \beta^j - \delta^{ij} t$ is an $\{\mathscr{G}_t\}$-martingale, implying that β is a $\{\mathscr{G}_t\}$-Brownian motion. □

18. Tsirel'son's example. This is a celebrated and mysterious example of a '1-dimensional' equation of type:

$$dX = dB + b(t, X.)dt, \quad X(0) = 0, \tag{18.1}$$

where b is a uniformly bounded previsible path functional. This equation automatically has a weak solution unique in law. For Tsirel'son's inspired choice of the functional b, X cannot be measurable on the filtration of B. This is an important counterexample in connection with the innovations problem in the theory of filtering. See § VI.8 for a brief indication of this.

First, note that, assuming only that b is bounded and previsible, we can construct a weak solution of (18.1) by the change-of-measure technique (given in more general form in § 27 below). Let $W = C([0, 1], \mathbb{R})$ be the space of continuous $\mathbb{R}$-valued paths, let $\mathbf{Q}$ be Wiener measure on $(W, \mathscr{X})$, and define a measure $\mathbf{P}$ on $(W, \mathscr{X})$ by the Cameron–Martin–Girsanov transformation:

$$d\mathbf{P}/d\mathbf{Q} = \exp\left[\int_0^t b(s, x.)\, ds - \tfrac{1}{2}\int_0^t b(s, x.)^2\, ds\right] \quad \text{on } \mathscr{X}_t. \tag{18.2}$$

By Theorem IV.38.9, under $\mathbf{P}$,

$$B_t \equiv B_t(x.) \equiv x_t - \int_0^t b(s, x.)\, ds$$

defines a Brownian motion relative to $(W, \{\mathscr{X}_t\}, \mathbf{P})$, so that

$$dx = dB + b(s, x.)\, dt,$$

and we have a weak solution of (18.1). (Uniqueness in law is not what really concerns us now, but it is easy to prove by using equation (18.2) in reverse to construct from the law $\mathbf{P}$ of a weak solution a law $\mathbf{Q}$ which must be Wiener measure. We look at this more systematically in § 27.)

Now we describe Tsirel'son's choice of b. Fix $t_0=1>t_{-1}>t_{-2}>\ldots$ such that $\inf t_n=0$, and define for $t\geqslant 0$, $x\in W$,

$$b(t,x.)=\begin{cases} f\left(\dfrac{x(t_j)-x(t_{j-1})}{t_j-t_{j-1}}\right) & \text{if } t_j<t\leqslant t_{j+1},\ -j\in\mathbb{Z}^+;\\ 0 & \text{if } t=0.\end{cases}$$

Here, $f(y)$ denotes the fractional part of y, a number in $[0, 1)$, and we define $t_1=\infty$. Thus the coefficient b is left continuous, bounded, and adapted. From now on, we assume that the functional b at (18.1) is as just described.

Let $\{\mathscr{B}_t: t\geqslant 0\}$ be the augmented filtration generated by B. Let $\delta_j=t_j-t_{j-1}$, *and let*

$$\mu_j\equiv f(\delta_j^{-1}(x(t_j)-x(t_{j-1})))=b(t_{j+1},x.)$$

denote the drift of the solution in $(t_j,t_{j+1}]$.

Here are some of the astonishing properties of the Tsirel'son example.

(*18.3*) THEOREM (Tsirel'son, Yor). (i) *For each* t,

$$\mathscr{X}_t=\mathscr{B}_t\vee\sigma(b(t,x.)).$$

(ii) *For each* j, μ_j *is uniformly distributed on* $[0, 1)$, *and is independent of* $\mathscr{B}_\infty$.
(iii) *The* σ*-algebra* $\mathscr{X}_{0+}$ *is trivial.*

Remarks. At any positive time t, the drift $b(t, x.)$ of the solution x is independent of the 'driving' Brownian motion B, and is uniform on $[0, 1)$; somehow, magically, this independent random variable has appeared from somewhere! Indeed, it really has appeared from thin air, because the third part of the theorem shows that it is not present at time 0!

Proof (Shiryaev, Yor). Abbreviate $b(t, x.)$ to b_t.

(i) It is trivial that $\mathscr{X}_t\supseteq\mathscr{B}_t\vee\sigma(b_t)$. We must prove the converse. If $t_j<t<t_{j+1}$, then, since

(18.4) $$x(t)-x(t_j)=B(t)-B(t_j)+(t-t_j)\mu_j\quad(t_j\leqslant t\leqslant t_{j+1})$$

it is clear that for each s in $[t_j, t)$, $x(s)-x(t_j)$ is measurable with respect to $\mathscr{B}_t\vee\sigma(b_t)$. Moreover, for some n,

$$\begin{aligned}x(t_j)-x(t_{j-1})&=(t_j-t_{j-1})(n+\mu_j)\\&=B(t_j)-B(t_{j-1})+(t_j-t_{j-1})\mu_{j-1}.\end{aligned}$$

Thus

(18.5) $$\mu_{j-1}=f(\mu_j-\delta_j^{-1}\{B(t_j)-B(t_{j-1})\})$$

is also measurable with respect to $\mathscr{B}_t\vee\sigma(b_t)$; inductively, every μ_k, $k\leqslant j$, is measurable with respect to $\mathscr{B}_t\vee\sigma(b_t)$, and hence

$$\mathscr{X}_t=\mathscr{B}_t\vee\sigma(b_t).$$

(ii) Writing $V_j = \delta_j^{-1}\{B(t_j) - B(t_{j-1})\}$, equation (18.5) implies that for all $n \in \mathbb{Z}$,

$$\exp\{i2n\pi\mu_{j-1}\} = \exp\{i2n\pi(\mu_j - V_j)\}.$$

But V_j is independent of $\mathscr{X}(t_{j-1})$; rearranging and conditioning on $\mathscr{X}(t_{j-1})$ yields

$$\begin{aligned}\mathbf{E}[\exp\{i2n\pi\mu_j\} | \mathscr{X}(t_{j-1})] &= \exp\{i2n\pi\mu_{j-1}\}\mathbf{E}[\exp\{i2n\pi V_j\}] \\ &= \exp(-2n^2\pi^2/\delta_j)\exp\{i2n\pi\mu_{j-1}\}.\end{aligned}$$

Thus

$$\begin{aligned}\mathbf{E}[\exp\{i2n\pi\mu_j\} | \mathscr{X}(t_{j-r})] &= \exp\left\{-2n^2\pi^2 \sum_{l=0}^{r-1} \delta_{j-l}^{-1} + i2n\pi\mu_{j-r}\right\} \\ &\to 0 \text{ as } r \to \infty,\end{aligned}$$

provided $n \neq 0$. Hence $\mathbf{E}\exp\{i2n\pi\mu_j\} = 0$ for $n \neq 0$, implying that μ_j is uniformly distributed on $[0, 1)$.

For $0 \leqslant u < v$, let $\mathscr{B}(u, v) \equiv \sigma\{B_s - B_u: u \leqslant s \leqslant v\}$. Then, since B is an $\{\mathscr{X}_t\}$ Brownian motion,

μ_j is independent of $\mathscr{B}(t_j, \infty)$.

Next, if $k < j$ and if Z is $\mathscr{B}(t_k, t_j)$-measurable and bounded, then, for $n \neq 0$,

$$\begin{aligned}\mathbf{E}[Z\exp(i2n\pi\mu_j)] &= \mathbf{E}\left[Z\exp\left\{i2n\pi \sum_{k+1}^{j} V_r\right\}\exp(i2n\pi\mu_k)\right] \\ &= \mathbf{E}\left[Z\exp\left\{i2n\pi \sum_{k+1}^{j} V_r\right\}\right]\mathbf{E}\exp(i2n\pi\mu_k) \\ &= 0\end{aligned}$$

by the independence of μ_k and $\mathscr{B}(t_k, t_j)$. Thus μ_j is independent of every $\mathscr{B}(t_k, t_j)$, and so, letting $k \to -\infty$,

μ_j is independent of $\mathscr{B}_{t_j}$.

Result (ii) of the theorem follows.

(iii) By Blumenthal's 0–1 law (Theorem III.9), $\mathscr{H}_{0+}$ is trivial for $\mathbf{Q}$. It is therefore trivial for $\mathbf{P}$ too. □

(*18.6*) *Important remark: Zvonkin's observation.* It is important to note that, in contrast to Tsirel'son's example, if b is a bounded measurable function on $\mathbb{R}$, then the equation

(18.7) $$dX = dB + b(X_t)dt,$$

in which the drift depends only on the current position X_t, is exact, so that X is measurable on the B filtration. See (28.8).

4. MARTINGALE PROBLEMS, MARKOV PROPERTY

19. Definition; orientation. We now investigate how the concepts of existence and uniqueness in law for weak solutions of our SDE

$$dX = \sigma(t, X_\cdot)\, dB + b(t, X_\cdot)\, dt \tag{19.1}$$

appear in an alternative formulation known as the *martingale problem* method. As we shall see, the two formulations are fully equivalent, and the martingale problem method has several advantages.

Suppose that X is a weak solution to (19.1) starting at $y \in \mathbb{R}^n$. Let $\mathbf{P}^y$ be the law of X; $\mathbf{P}^y$ is a probability measure on $(W^n, \{\mathscr{X}_t^n\})$. Then $\mathbf{P}^y$ has the following decisive properties:

(19.2(i)) $\mathbf{P}^y(x_0 = y) = 1$;
(19.2(ii)) under $\mathbf{P}^y$, for each $f \in C_K^\infty(\mathbb{R}^n)$,

$$C_t^f \equiv f(x_t) - f(x_0) - \int_0^t \mathscr{L} f(s, x_\cdot)\, ds$$

is an $\{\mathscr{X}_t^n\}$-martingale.

Here,

$$\mathscr{L} f(s, x_\cdot) \equiv \tfrac{1}{2} a^{ij}(s, x_\cdot) D_i D_j f(x_s) + b^i(s, x_\cdot) D_i f(x_s), \tag{19.3}$$

where

$$a^{ij}(s, x_\cdot) \equiv \sum_{q=1}^{d} \sigma_q^i(s, x_\cdot) \sigma_q^j(s, x_\cdot). \tag{19.4}$$

(In fact, we only proved this in § 3 assuming (19.1) was an SDE of diffusion type, but the calculations are just the same.)

(*19.5*) DEFINITION (solution of a martingale problem). *Suppose that* $a: \mathbb{R}^{++} \times W^n \to S_n^+$ *and* $b: \mathbb{R}^{++} \times W^n \to \mathbb{R}^n$ *are previsible path functionals. Then for* $y \in \mathbb{R}^n$, *we say that the probability* $\mathbf{P}^y$ *on* $(W^n, \{\mathscr{X}_t^n\})$ *is a* solution to the martingale problem for (a, b) started at y *if conditions* (19.2) *hold, where* $\mathscr{L}$ *is defined by* (19.3). *If there is only one such measure* $\mathbf{P}^y$, *we say that the solution to the martingale problem* (*for* (a, b) *starting at* y) *is* unique, *and if for every* y *there exists a unique solution to the martingale problem starting at* y, *then we say that the martingale problem is* well posed.

We shall often use the terminology 'martingale problem for $\mathscr{L}$' as an alternative to 'martingale problem for (a, b)'.

(*19.6*) *Remark.* We omit explicit reference to the coefficients (a, b) and the starting point y except where confusion might arise.

(*19.7*) *Equivalence of formulations.* We have seen that if X is a weak solution to the SDE (19.1) starting at y, then the law of X solves the martingale problem for

$(\sigma\sigma^T, b)$ starting at y. In the next section, we prove the converse: if $\mathbf{P}^y$ solves the martingale problem starting at y, then there exists a weak solution to the SDE (19.1) with law $\mathbf{P}^y$. Thus:

there exists a weak solution to the SDE (19.1) starting at y	$\Leftrightarrow$	there exists some solution to the martingale problem starting at y
uniqueness in law holds for the SDE (19.1) starting at y	$\Leftrightarrow$	there exists at most one solution to the martingale problem starting at y

The martingale-problem formulation of the SDE (19.1) is therefore completely equivalent to the weak-solution formulation.

(*19.8*) *The advantages of the martingale-problem formulation.* By focusing our attention on probability measures on the Polish space W, the martingale-problem method puts at our disposal three powerful techniques:

the theory of weak convergence;
the theory of regular conditional probabilities;
localization.

Moreover, the method is perfectly tailored to allow full exploitation of these techniques.

We shall see in §23 that the tightness criterion from the theory of weak convergence allows us to prove a very satisfying existence theorem for solutions of the martingale problem.

(*19.9*) *Markov property.* The theory of regular conditional probabilities allows us to prove in §21 that *if we are in the diffusion context*:

$$a(t, x.) = a(x_t), \quad b(t, x.) = b(x_t),$$

and the martingale problem is well posed, then $(x_t, \mathbf{P}^y)$ *is a strong Markov process.* In §IV.20, we saw a similar result in the much simpler context of finite Markov chains.

The theory of regular conditional probabilities may also be combined with some deep analysis to obtain the fundamental uniqueness theorem of Stroock and Varadhan (§24). The 'well-posedness' of the martingale problem for the 'diffusion' case is inextricably linked with the Markov property.

(*19.10*) *Martingale characterization implies martingale representation.* We have remarked before on the important general principle just italicized. See §25 for how it applies in the present context.

(*19.11*) *Martingale problems for processes on manifolds.* It is important that if $\mathscr{L}$ is a second-order elliptic operator on a manifold, then the concept of a solution of the martingale problem for $\mathscr{L}$ is coordinate-free.

(*19.12*) *Important note about the usual conditions.* The point made at IV.20.7 that the martingale problem is unaffected by augmentation applies in the present context. Thus we can apply all of the theory we have developed.

20. Equivalence of the martingale-problem and 'weak' formulations. The following result establishes the non-trivial part of the asserted equivalence.

(*20.1*) THEOREM. *Let* $\mathbf{P}^y$ *be a solution to the martingale problem for* $(\sigma\sigma^T, b)$ *started at* y. *Then there exists a weak solution to the SDE* (19.1) *starting at* y *whose law is* $\mathbf{P}^y$.

Remarks. We have to build a filtered probability space carrying a d-dimensional Brownian motion $\tilde{B}$ and a semimartingale X such that

(20.2) $$X_t = y + \int_0^t \sigma(s, X_\cdot)\, d\tilde{B}_s + \int_0^t b(s, X_\cdot)\, ds$$

and such that the law of X is $\mathbf{P}^y$. We do this by 'adding in' an independent Brownian motion to $(W^n, \mathbf{P}^y)$; some such procedure is evidently needed if $\sigma \equiv 0$, for then $\mathbf{P}^y$ is a degenerate measure, and $(W^n, \mathbf{P}^y)$ cannot support a Brownian motion. If σ is everywhere invertible, we can manufacture the Brownian motion from the canonical process x under law $\mathbf{P}^y$, but more generally we have to have an independent Brownian motion available to 'step in' whenever the rank of σ becomes deficient.

Proof. Let $\Omega \equiv W^n \times W^d$. For $(w, w') \in \Omega$, we define the canonical processes $X_t(w, w') \equiv w(t)$, $B_t(w, w') \equiv w'(t)$, and we let $\mathscr{F}_t^\circ$ be the filtration generated by (X, B), with the product measure $\mathbf{P} \equiv \mathbf{P}^y \times \mathbf{Q}^d$ (remember that $\mathbf{Q}^d$ is d-dimensional Wiener measure). Then it is immediate that:

$$\mathbf{P} \text{ solves the martingale problem for } \left(\begin{pmatrix} a & 0 \\ 0 & I \end{pmatrix}, \begin{pmatrix} b \\ 0 \end{pmatrix} \right) \text{ started at } (y, 0),$$

where $a \equiv \sigma\sigma^T$. Applying (19.2(ii)) to a C_K^∞ function equal to the ith coordinate map in some large neighbourhood of zero, we see that:

(20.3) $$M_t^i \equiv X_t^i - \int_0^t b^i(s, X_\cdot)\, ds \text{ is a local martingale,}$$

and, on taking a C_K^∞ function which is $x^i x^j$ in a neighbourhood of zero, we deduce similarly that

$$M_t^i M_t^j - \int_0^t a^{ij}(s, X_\cdot)\, ds \text{ is a local martingale.}$$

Equivalently,

(20.4) $$[M^i, M^j]_t = \int_0^t a^{ij}(s, X_\cdot)\, ds.$$

If ($d=n$ and) σ is everywhere invertible, we can define a local martingale

$$\tilde{B}_t = \int_0^t \sigma(s, X_\cdot)^{-1}\, dM_s \tag{20.5}$$

which is a Brownian motion by Lévy's Theorem IV.33.1. Then (20.2) follows immediately. But more generally, if σ is not invertible, then we define a continuous local martingale

$$\tilde{B}_t = \int_0^t \theta_s\, dM_s + \int_0^t \rho_s\, dB_s \tag{20.6}$$

where θ is some previsible $d \times n$ matrix process, ρ is a previsible $d \times d$ matrix process, chosen in such a way that B is Brownian motion, and (20.2) holds. The problem of finding such θ and ρ is a problem of elementary matrix algebra.

(*20.7*) LEMMA. *Let σ: $\mathbb{R}^d \to \mathbb{R}^n$ be linear, $V_0 \equiv \ker(\sigma)$, $U_1 \equiv \mathrm{Im}(\sigma)$, and let V_1 be the orthogonal complement of V_0. Then there exists linear θ: $\mathbb{R}^n \to \mathbb{R}^d$ with kernel $U_0 \equiv U_1^{\perp}$ such that*

(i) *$\sigma\theta$ (respectively $\theta\sigma$) is orthogonal projection on to U_1 (respectively V_1);*
(ii) *$\sigma\theta\sigma = \sigma$, $\theta\sigma\theta = \theta$.*

Proof. Exercise.

Now the quadratic variation process of $\tilde{B}$ as defined by (20.6) is

$$\begin{aligned}\frac{d[\tilde{B}^i, \tilde{B}^j]}{ds}s &= (\theta_s \rho_s)\begin{pmatrix} a_s & 0 \\ 0 & I \end{pmatrix}\begin{pmatrix} \theta_s^T \\ \rho_s^T \end{pmatrix} \\ &= \theta_s \sigma_s \sigma_s^T \theta_s^T + \rho_s \rho_s^T\end{aligned} \tag{20.8}$$

which must be I if B is Brownian motion (here, $\sigma_s \equiv \sigma(s, X_\cdot)$). Similarly, the quadratic variation process of the local martingale $M - \int_0^\cdot \sigma_s\, d\tilde{B}_s$ is

$$(\sigma_s\theta_s - I, \sigma_s\rho_s)\begin{pmatrix} a_s & 0 \\ 0 & I \end{pmatrix}(\sigma_s\theta_s - I, \sigma_s\rho_s)^T \tag{20.9}$$

which, should it vanish, proves that

$$M_t = \int_0^t \sigma_s\, d\tilde{B}_s,$$

which is our objective. But defining θ_s to be the generalized inverse of σ_s as given by Lemma (20.7), we see that $\pi_s \equiv \theta_s \sigma_s$ is an orthogonal projection, and so therefore is $\rho_s \equiv I - \pi_s$. Hence immediately the quadratic variation (20.8) is I, and the expression at (20.9) is

$$(\sigma_s\theta_s - I)\sigma_s\sigma_s^T(\sigma_s\theta_s - I)^T + \sigma_s\rho_s\rho_s^T\sigma_s^T = 0,$$

as required, because $(\sigma_s\theta_s - I)\sigma_s = 0$ and $\sigma_s\rho_s = 0$. □

21. Martingale problems and the strong Markov property. Here is the long-awaited result showing that an (a, b)-diffusion is a strong Markov process if the martingale problem for (a, b) is well-posed. (As is made clear by the statement of the following theorem, we associate path functionals a and b with functions a and b on $\mathbb{R}^n$ in the obvious way:

$$a(t, x_.) = a(x_t), \quad b(t, x_.) = b(x_t).)$$

(*21.1*) THEOREM. *Suppose that a: $\mathbb{R}^n \to S_n^+$ and b: $\mathbb{R}^n \to \mathbb{R}^n$ are bounded measurable functions such that the martingale problem for (a, b) is well-posed: for each $y \in \mathbb{R}^n$ there exists a unique probability measure $\mathbf{P}^y$ on W^n such that:*

(21.2(i)) $\mathbf{P}^y(x_0 = y) = 1$;

(21.2(ii)) *for each $f \in C_K^\infty$,*

$$C_t^f \equiv f(x_t) - f(x_0) - \int_0^t \mathscr{L}f(x_s)\,ds \text{ is a } \mathbf{P}^y\text{-martingale,}$$

where

$$\mathscr{L}f(x) \equiv \tfrac{1}{2}a^{ij}(x)D_iD_jf(x) + b^i(x)D_if(x).$$

Then under each $\mathbf{P}^y$, $\{x_t\colon t \geqslant 0\}$ is a time-homogeneous strong Markov process.

Proof. It is enough to consider bounded stopping times. Let $T \leqslant N$ be an $\{\mathscr{X}_t^n\}$-stopping time and define θ_T: $W \equiv W^n \to W$ by $(\theta_T x)(t) = x(t + T(x))$, $x \in W$. Then $\theta_T^{-1}(\mathscr{X}_s^n) \subseteq \mathscr{X}_{T+s}^n$.

We shall prove that if $(\mathbf{P}^y|\mathscr{X}_T)$: $W \times \mathscr{X} \to [0, 1]$ is a regular conditional distribution for $\mathbf{P}^y$ given $\mathscr{X}_T$, then for $(\mathbf{P}^y)$ almost every $x \in W$, the $(\mathbf{P}^y|\mathscr{X}_T)(x, \cdot)$ law of θ_T solves the martingale problem starting at $x(T)$, and therefore must be $\mathbf{P}^{x(T)}$. We agree to some definite version of $(\mathbf{P}^y|\mathscr{X}_T)(x, \cdot)$. What we must show is that except for some fixed $(\mathbf{P}^y)$ null set of x in W, it is true that whenever $s < t$, $A \in \mathscr{X}_s$, and $f \in C_K^\infty$, then

$$F(s, t, A, f)(x) = 0, \tag{21.3}$$

where

(21.4) $F(s, t, A, f)$ is the $(\mathbf{P}^y|\mathscr{X}_T)(x, \cdot)$ expectation of the random variable

$$\{(C_t^f - C_s^f)I_A\} \circ \theta_T = (C_{T+t}^f - C_{T+s}^f)I_H, \tag{21.5}$$

where $H = \theta_T^{-1}A \in \mathscr{X}_{T+s}$.

First, we establish that (21.3) holds for $(\mathbf{P}^y)$ almost all x when s, t, A and f (where $s < t$, $A \in \mathscr{X}_s$, and $f \in C_K^\infty$) are fixed. Now, $F(s, t, A, f)(x)$ is an $\mathscr{X}_T$-measurable function of x, and, by the definition of regular conditional probability (see § II.89), we have, for $G \in \mathscr{X}_T$,

$$\mathbf{E}[I_G F(s, t, A, f)] = \mathbf{E}[C_{T+t}^f - C_{T+s}^f;\ G \cap H]$$

which is 0 by the optional sampling theorem since C^f is a bounded martingale on $[0, N+t]$ and $G \cap H \in \mathscr{X}_{T+s}$. This establishes the desired result for fixed s, t, A and f.

We can prove that:

(21.6) *except for a certain* $(\mathbf{P}^y)$ *null set of* x *in* W, *equation (21.3) holds simultaneously for all* (s, t, A, f)

because we can find a 'countable dense subset' of (s, t, A, f) values in the following sense. We can restrict attention to pairs of rational (s, t) with $s<t$, and, for each rational s, we can restrict A to be an element of a countable determining class in $\mathscr{X}_s$. Moreover, because a and b are bounded, we can find a countable subset C' of C_K^∞ such that, for each f in C_K^∞, there exists a sequence (f_k) in C' such that $f_k \to f$ and $\mathscr{L}f_k \to \mathscr{L}f$ uniformly. Result (21.6) now follows from a simple continuity argument.

Thus, we have indeed proved that

$$(\mathbf{P}^y \circ \theta_T^{-1} \mid \mathscr{X}_T)(x, \cdot) = \mathbf{P}^{x(T)}(\cdot), \tag{21.7}$$

and we know from §II.90 that this expresses the strong Markov property. □

(*21.8*) *Note on measurability of* $y \mapsto \mathbf{P}^y$. Before equation (21.7) makes good sense, we need to know that the map $y \mapsto \mathbf{P}^y$ is measurable. See Exercise 6.7.4 of Stroock and Varadhan [1] for the proof of this in the general case; this relies on Kuratowski's theorem that a *one–one measurable map from a Borel subset of a Polish space onto another Polish space has Borel inverse.* In many important contexts, we can prove something much stronger than the stated measurability. The next section illustrates this very clearly.

(*21.9*) *Localizing the boundedness condition.* You can check that Theorem 21.1 remains true if a and b are locally bounded rather than bounded.

(*21.10*) *Important remark.* Certainly in the case of Lipschitz coefficients, this proof is more complicated than that of Lemma 13.7. However, the use of regular conditional probabilities in the proof becomes an essential step in the proof of the Stroock–Varadhan Theorem 24.1 below.

22. Appraisal and consolidation: where we have reached. Even though we have not yet made any serious use of the martingale problem method, it is worth pausing to emphasize the very substantial progress which is represented by the theorems already proved in this chapter.

Consider the SDE

$$dX = \sigma(X_t)\,dB + b(X_t)\,dt, \tag{22.1}$$

and the associated operator

(22.2) $$\mathscr{L} = \tfrac{1}{2}a^{ij}(x)D_iD_j + b^i(x)D_i,$$

where $a = \sigma\sigma^T$. *We suppose that we are in the Lipschitz case covered by Theorem 11.2.* It is important to realize that it could be that it is the operator $\mathscr{L}$ which is regarded as given, and that $\sigma(\cdot)$ is defined by $\sigma(\cdot) = a(\cdot)^{1/2}$.

By Itô's Theorem 11.2, we know that *the SDE (22.1) is exact.* By the Yamada–Watanabe Theorem 17.1, we know that *uniqueness in law holds for (22.1).* Thus, by Theorem 20.1, *the martingale problem associated with $\mathscr{L}$ is well-posed.* Let $\mathbf{P}^y$ denote the unique solution of the martingale problem for $\mathscr{L}$ with starting point y. By Theorem 21.1 and its 'locally bounded' extension, *the process*

(22.3) $$(W, \mathscr{X}, \{\mathscr{X}_t\}, \{x_t\}, \{\mathbf{P}^y\}),$$

where $W \equiv W^n$, etc., *is a strong Markov process.* Let $\{\mathbf{P}_t\}$ denote its transition function, so that

(22.4) $$(P_t f)(y) = \mathbf{E}^y f(x_t).$$

We are now going to show that:

(22.5) *the transition function $\{P_t\}$ is FD.*

Proof of (22.5). By the Blagoveščenskii–Freidlin Theorem 13.1, there is a map

$$(y, w) \mapsto F(y, w) \text{ from } \mathbb{R}^n \times W^d \text{ to } W^n,$$

continuous in y for each fixed w, such that for any set-up $(\Omega, \{\mathscr{F}_t\}, \mathbf{P}, B)$,

$$X^y_\cdot \equiv F(y, B_\cdot)$$

solves

$$X^y_t = y + \int_0^t \sigma(X^y_s)\,dB_s + \int_0^t b(X^y_s)\,ds.$$

Now,

$$P_t f(y) = \mathbf{E} f(X^y_t),$$

and, since the map $y \mapsto X^y_t$ is continuous from $\mathbb{R}^n$ to $\mathbb{R}^n$, it is immediate that

(22.6) $$P_t\colon C_b(\mathbb{R}^n) \to C_b(\mathbb{R}^n),$$

where $C_b(\mathbb{R}^n)$ is the space of bounded continuous functions on $\mathbb{R}^n$.

We must now prove that

(22.7) $$P_t\colon C_0(\mathbb{R}^n) \to C_0(\mathbb{R}^n).$$

Let X satisfy (22.1). We wish to find a convex decreasing function $\varphi\colon \mathbb{R}^+ \to \mathbb{R}^+$ tending to 0 at ∞ and such that

(22.8) $e^{-t}\varphi(V_t)$ *is a supermartingale, where* $V_t \equiv |X_t|^2$.

For then, if $\tau(N) \equiv \inf\{t: |X_t|^2 < N\}$, then

$$\mathbf{E}[e^{-\tau(N)}|X_0 = y] \leqslant \varphi(|y|^2)/\varphi(N) \to 0 \quad \text{as } |y| \to \infty,$$

and this fact (together with (22.6)) clearly implies (22.7). Now,

$$dV_t = 2X^T\sigma(X_t)\,dB + \{\text{trace}\, a(X_t) + 2X^Tb(X_t)\}\,dt,$$

so that, by Itô's formula,

$$d\{e^{-t}\varphi(V_t)\} = e^{-t}\psi(X_t)\,dt + d\text{ (local martingale)}, \tag{22.9}$$

where, with v denoting $|x|^2$,

$$\psi(x) = 2\varphi''(v)x^Ta(x)x + \varphi'(v)\{\text{trace}\, a(x) + 2x^Tb(x)\} - \varphi(v).$$

But, by the Lipschitz properties of σ and b, we have for some constant c,

$$\psi(x) \leqslant cv^2\varphi''(v) + cv|\varphi'(v)| - \varphi(v).$$

If we choose $\varphi(v) = v^{-\gamma}$, where $0 < \gamma c(\gamma + 2) < 1$, then:

$$\psi(x) \leqslant v^{-\gamma}[c\gamma(\gamma+1) + c\gamma - 1] \leqslant 0. \tag{22.10}$$

From (22.9), (22.10) and a familiar Fatou argument, we complete the proof of (22.7). □

Next, for $f \in C_K^\infty$, we have

$$\mathbf{E}\left\{f(X_t) - f(X_0) - \int_0^t \mathscr{L}f(X_s)\,ds\right\} = 0,$$

so that

$$(P_tf)(y) - f(y) - \int_0^t (P_s\mathscr{L}f)(y)\,ds = 0.$$

Now, as $s \downarrow 0$, $P_s\mathscr{L}f \to \mathscr{L}f$ uniformly on $\mathbb{R}^n$ (see equation (III.6.6(iv)*). Hence, for $f \in C_K^\infty$,

$$t^{-1}(P_tf - f) \to \mathscr{L}f \textit{ in the topology of } C_0.$$

Hence:

(22.11) *the infinitesimal generator of* $\{P_t\}$ *extends* $\mathscr{L}$ *on* C_K^∞.

We have in particular given a purely probabilistic proof of the following theorem which is purely analytic in statement.

(22.12) THEOREM. *Suppose that* $\sigma: \mathbb{R}^n \to L(\mathbb{R}^d, \mathbb{R}^n)$ *and* $b: \mathbb{R}^n \to \mathbb{R}^n$ *are globally Lipschitz.*

Let $a(x)=\sigma(x)\sigma(x)^T$, *and let*

$$\mathscr{L}=\tfrac{1}{2}a^{ij}(x)D_iD_j+b^i(x)D_i.$$

Then there exists a unique honest FD transition function $\{P_t\}$ *with generator extending* $\mathscr{L}$ *on* C_K^∞.

The uniqueness holds because if $\{\tilde{P}_t\}$ is another FD transition function with generator extending $\mathscr{L}$ on C_K, then the Markovian $\tilde{\mathbf{P}}^y$ measure induced by $\{P_t\}$ solves the martingale problem for (a, b) starting at y and must therefore equal $\mathbf{P}^y$.

Theorem 22.12 is a first illustration of how important analytic results may be obtained by probabilistic methods. More profound examples are provided by the Stroock–Varadhan Theorem 24.1 and by the Malliavin calculus (see § 38). Our main concern, however, is with the probability.

23. Existence of solutions of the martingale problem. To prove that a solution exists is usually easier than to prove that a solution is unique; in the next section, we shall discuss a deep uniqueness result of Stroock and Varadhan, but first we prove an existence result which dates back in essence to Skorokhod.

As already mentioned, this section shows how effectively the martingale-problem method combines with the theory of narrow convergence. *Indeed, showing this combination in action must be seen as the main point of this section.* The techniques used here can be applied to obtain important *approximation theorems*—see Chapter 11 of Stroock and Varadhan [1] and Ethier and Kurtz [1]—and are in no way limited to establishing the existence Theorem (23.5) below.

Recall Theorem II.85.3: if for $\delta>0$, $N\in\mathbb{N}$, $x\in W^n$ we define

$$\Delta(\delta, N; x)\equiv\sup\{|x(s)-x(t)|: s, t\in[0, N], |s-t|<\delta\},$$

then a subset H of $\Pr(W)\equiv\Pr(W^n)$ is conditionally compact if and only if

(23.1(i)) $$\lim_{a\uparrow\infty}\sup_{\mu\in H}\mu(\{x: |x(0)|>a\})=0;$$

(23.1(ii)) for each $\varepsilon>0$, and $N\in\mathbb{N}$,

$$\lim_{\delta\downarrow 0}\sup_{\mu\in H}\mu(\{x:\Delta(\delta, N; x)>\varepsilon\})=0.$$

We shall use the following estimate to establish conditional compactness.

(23.2) LEMMA. *Suppose that* $a:\mathbb{R}^{++}\times W\to S_n^+$, $b:\mathbb{R}^{++}\times W\to\mathbb{R}^n$ *are previsible path functionals, and that for some* K,

$$\sum_{i,j=1}^n\left|a^{ij}(t, x.)\right|+\sum_{i=1}^n\left|b^i(t, x.)\right|\leqslant K$$

for all t *and* x. *Let* $\mathbf{P}^y$ *be a solution to the martingale problem for* (a, b) *starting at* y. *Then there exists a constant* $c>0$ *such that for each* $y\in\mathbb{R}^n$, *for each* $\varepsilon>0$, m, $N\in\mathbb{N}$, *and* $m > K$,

(23.3) $$\mathbf{P}^y(\{x: \Delta(m^{-1}, N; x) > \varepsilon\}) \leqslant cNK^2/m\varepsilon^4.$$

Proof. If $\Delta(m^{-1}, N; x) > \varepsilon$, then for some $j = 0, 1, \ldots, Nm-1$, it must be that

$$\sup_{0 \leqslant t \leqslant m^{-1}} |x(jm^{-1}+t) - x(jm^{-1})| > \varepsilon/3.$$

Define the $\mathbf{P}^y$-local martingales

$$M_t^i \equiv x_t^i - \int_0^t b^i(s, x.)ds \quad (i = 1, \ldots, n);$$

as we have seen, $d[M^i, M^j]_t \equiv a^{ij}(t, x.)dt$. Now fix $0 \leqslant t_0 < t_1 \leqslant N$, and define

$$\tau \equiv \inf\{u > 0: |x(t_0+u) - x(t_0)| > \varepsilon/3\} \wedge (t_1 - t_0);$$

then $M(t_0 + (t \wedge \tau)) - M(t_0)$ is an $(\{\mathscr{F}_{t_0+t}\}, \mathbf{P}^y)$ martingale.

To simplify notation, take $t_0 = 0$. Then

$$x_{t\wedge\tau} - x_0 = M_{t\wedge\tau} - M_0 + \int_0^{t\wedge\tau} b(s, x.)ds$$

whence

$$|x_{t\wedge\tau} - x_0|^2 \leqslant 2|M_{t\wedge\tau} - M_0|^2 + 2K^2(t\wedge\tau)^2$$

so that

$$(x - x_0)^{*2}_{t\wedge\tau} \leqslant 2(M - M_0)^{*2}_{t\wedge\tau} + 2K^2\tau^2.$$

For any $p > 0$ we can use the Burkholder–Davis–Gundy inequalities and Jensen's inequality as we did in § 11 to conclude that there exists a constant C_p such that:

$$\begin{aligned}\mathbf{E}^y(x - x_0)^{*2p}_\tau &\leqslant C_p\mathbf{E}^y\left\{\left[\int_0^\tau \sum_{i=1}^n a^{ii}(s, x.)ds\right]^p + K^{2p}\tau^{2p}\right\}\\ &\leqslant C_p\mathbf{E}^y\{(K\tau)^p + (K\tau)^{2p}\}.\end{aligned}$$

Thus, 'restoring t_0', and setting $\eta = t_1 - t_0$,

$$(\varepsilon/3)^{2p}\mathbf{P}^y(\sup_{0\leqslant t\leqslant\eta} |x(t_0+t) - x(t_0)| > \varepsilon/3) \leqslant C_p(K\eta)^p + (K\eta)^{2p}.$$

Now take $p = 2$, and estimate, assuming $m > K$:

$$\begin{aligned}\mathbf{P}^y&\left(\bigcup_{j=0}^{Nm-1}\left\{\sup_{0\leqslant t\leqslant m^{-1}} |x(jm^{-1}+t) - x(jm^{-1})| > \varepsilon/3\right\}\right)\\ &\leqslant mN(\varepsilon/3)^{-4}C_2.2(K/m)^2\\ &= cNK^2/m\varepsilon^4\end{aligned}$$

where $c = 162C_2$. □

The martingale-problem approach is ideally suited to obtaining modulus of continuity results like (23.3); the proof was neither long nor devious! We need (23.3) to prove that a family of solutions to different martingale problems is conditionally compact and thence to draw out a convergent subsequence. But can we characterize the limit law as the solution to some martingale problem? Yes—again the martingale problem formulation is perfect! Before we prove the existence theorem we need a lemma.

(*23.4*) LEMMA. *Let* $\{g_k: k \geqslant 0\}$ *be a uniformly bounded sequence of measurable functions on a Polish space* S, *and suppose* $\mu_k \in \Pr(S)$, $\mu_k \Rightarrow \mu$. *If* $\{g_k: k \geqslant 0\}$ *is equicontinuous at each point of* S, *and* $g_k \to g$ *pointwise; then:*

$$\int_S g_k\, d\mu_k \to \int_S g\, d\mu.$$

Proof. It is well known (see, for example, Ikeda and Watanabe [1], Theorem I.2.7) that if $\mu_n \Rightarrow \mu$, then on some probability triple $(\Omega, \mathscr{F}, \mathbf{P})$ there exist S-valued random variables X_k and X such that

(i) $X_k \to X$ a.s.,

(ii) $\mu_k = \mathbf{P} \circ X_k^{-1}$, $\quad \mu = \mathbf{P} \circ X^{-1}$.

But if $\{g_k\}$ is equicontinuous at each point, and $g_k \to g$ pointwise, then $X_k(\omega) \to X(\omega)$ implies that $g_k(X_k(\omega)) \to g(X(\omega))$. Hence

$$\int g_k\, d\mu_k = \mathbf{E} g_k(X_k) \to \mathbf{E} g(X) = \int g\, d\mu. \qquad \square$$

(*23.5*) THEOREM. *Suppose that* $a: \mathbb{R}^{++} \times W^n \to S_n^+$, $b: \mathbb{R}^{++} \times W^n \to \mathbb{R}^n$ *are bounded previsible path functionals, and that* $a(t, \cdot)$ *and* $b(t, \cdot)$ *are continuous for each* $t \geqslant 0$. *Then for any* $\mu \in \Pr(\mathbb{R}^n)$, *there is a solution to the martingale problem for* (a, b) *with initial distribution* μ.

Proof. Define $\varphi_k: \mathbb{R}^{++} \times W^n \to \mathbb{R}^{++} \times W^n$ by $\varphi_k(t, x.) = ((t - k^{-1})^+, x.)$. Then, clearly, $a_k = a \circ \varphi_k$, $b_k \equiv b \circ \varphi_k$ *are previsible and* $a_k(t, \cdot)$ and $b_k(t, \cdot)$ are continuous for each $t \geqslant 0$. Moreover, if we define, for $f \in C_K^\infty(\mathbb{R}^n)$,

$$\mathscr{L}_k f(t, x.) \equiv \tfrac{1}{2} a_k^{ij}(t, x.) D_i D_j f(x_t) + b_k^i(t, x.) D_i f(x_t),$$

then, for each $t \geqslant 0$, $\mathscr{L}_k f(t, \cdot)$ is a continuous function on W^n, bounded uniformly in k.

Suppose given on some probability space an $\mathbb{R}^n$-valued random variable X_0 with law μ and an independent Brownian motion B in $\mathbb{R}^n$, and let σ denote the nonnegative definite square-root of a. Then, for each $k \in \mathbb{N}$, the SDE

$$X_t \equiv X_0 + \int_0^t \sigma((s - k^{-1})^+, X.)\, dB_s + \int_0^t b((s - k^{-1})^+, X.)\, ds \tag{23.6}$$

can obviously be solved pathwise. (Once the solution is known for $t \leqslant Nk^{-1}$, we can immediately find the solution up to $(N+1)k^{-1}$ because the coefficients in the integrals on the right of (23.6) are $\mathscr{F}(Nk^{-1})$-measurable for all $t\in[Nk^{-1},(N+1)k^{-1}]$). Let P_k be the law of the solution; P_k solves the martingale problem for (a_k, b_k) with initial law μ. From (23.1) (i)–(ii) and Lemma 23.2, the family $\{P_k: k\geqslant 1\}$ is conditionally compact, so by passing to a subsequence if need be, we may suppose that $P_k\Rightarrow P$. It remains only to show that P solves the martingale problem for (a, b) with initial law μ. But if we take $0<s<t, f\in C_K^\infty(\mathbb{R}^n)$, and any bounded continuous function ψ of the path $\{x_u: 0\leqslant u\leqslant s\}$, then the functions

$$g_k(x.)\equiv\left\{f(x_t)-f(x_s)-\int_s^t \mathscr{L}_k f(u, x.)du\right\}\psi(x.)$$

are uniformly bounded and are easily shown to be equicontinuous at each point of $W\equiv W^n$. So, by Lemma 23.4,

$$\int_W g_k dP_k\to\int_W g dP.$$

But P_k solves the martingale problem for (a_k, b_k), so $\int g_k dP_k=0$ for all k, implying

$$0=\int g dP=\mathbf{E}\left[\left\{f(x_t)-f(x_s)-\int_s^t \mathscr{L} f(u, x.)du\right\}\psi(x.)\right];$$

and since f and ψ are arbitrary, the result follows. □

Remarks. (i) This is the natural extension of a well-known result in ordinary differential equations which says that if the coefficients are bounded and continuous, then there exists some solution. Even then, though, the solution may not be unique; for example,

$$x_t=\int_0^t 2|x_s|^{1/2}ds$$

has solutions $x_t\equiv 0$ and $x_t=t^2$. In dimension greater than one, the analogous SDE

$$X_t=\int_0^t (|X_s|^\alpha\wedge 1)dB_s \tag{23.7}$$

has many solutions if $\alpha<1$ and only one if $\alpha\geqslant 1$ (which we already know since the coefficients are Lipschitz). In one dimension, though, things are different; (23.7) has a unique solution if and only if $\alpha\geqslant\frac{1}{2}$. We study the (Girsanov) SDE (23.7) in detail later, but it warns us that questions of uniqueness of solutions are likely to be more involved.

(ii) Note that the approximation used in Theorem 23.5 forces one to study non-diffusion SDEs.

(iii) If the martingale problem is of diffusion type and the coefficients are only assumed bounded and measurable, then provided there exists a $\delta>0$ such that, for all $t \geq 0$, λ, $x \in \mathbb{R}^n$,

$$\lambda^T a(t, x)\lambda \geq \delta|\lambda|^2,$$

the martingale problem has a solution. The simple approach of Theorem 23.5 breaks down, and much more sophisticated analysis is needed; see Krylov [1]. An interesting recent development is to use non-standard analysis to prove existence theorems of this type. See Cutland [1, 2], Albeverio, Fenstad, Høegh-Krohn and Lindström [1].

The criterion (23.3) for the conditional compactness of a family of solutions of a martingale problem has another important consequence.

(*23.8*) THEOREM. *Suppose that the martingale problem for* (a, b) *is well-posed, and that* $\{\mathbf{P}^y: y \in \mathbb{R}^n\}$ *is the unique family of solutions. If* a *and* b *are bounded and if* $a(t, \cdot)$ *and* $b(t, \cdot)$ *are continuous for each* $t \geq 0$, *then* $y(k) \to y$ *implies* $\mathbf{P}^{y(k)} \Rightarrow \mathbf{P}^y$.

Proof. By Lemma 23.2, the family $\{\mathbf{P}^{y(k)}: k \geq 0\}$ is conditionally compact. It is immediate to verify that any limit of a subsequence of $\mathbf{P}^{y(k)}$ solves the martingale problem starting from y, so must be $\mathbf{P}^y$. □

24. The Stroock–Varadhan uniqueness theorem. We already know that if σ and b are Lipschitz, then the martingale problem for $(\sigma\sigma^T, b)$ is well-posed. We have just seen that if a and b are bounded continuous, then there exists at least one solution to the martingale problem for each starting point, though in general, more than one. Deep results of Stroock and Varadhan say essentially that *continuity and uniform ellipticity of the coefficient a are enough to guarantee that the martingale problem is well posed*: we state here their main theorem for diffusion SDEs.

(*24.1*) THEOREM (Stroock–Varadhan). *Suppose* $a: \mathbb{R}^n \to S_n^+$, $b: \mathbb{R}^n \to \mathbb{R}^n$ *are measurable, and that*:

(i) *a is continuous;*
(ii) $a(x)$ *is strictly positive definite for each* x;
(iii) *for some constant* K, *for all* i, j *and* x,

$$|a^{ij}(x)| \leq K(1 + |x|^2), \quad |b^i(x)| \leq K(1 + |x|). \tag{24.2}$$

Then the martingale problem for (a, b) *is well-posed. The unique solutions* $\{\mathbf{P}^y: y \in \mathbb{R}^n\}$ *have the strong Markov property. Indeed, if we define*

$$P_t f(y) \equiv \mathbf{E}^y[f(x_t)] \qquad (t \geq 0, y \in \mathbb{R}^n),$$

then $\{P_t: t \geq 0\}$ *is the unique FD semigroup with generator extending* $\mathscr{L}$ *on* $C_K^\infty(\mathbb{R}^n)$. *Moreover, the semigroup is strong Feller; for each* $t > 0$,

$$P_t: b\mathscr{B}(\mathbb{R}^n) \to C(\mathbb{R}^n). \tag{24.3}$$

Proof. See Stroock and Varadhan ([3], Theorem 7.2.1, and Chapter 10).

Discussion of the proof. We do not present a proof of this result, but we give some clues to the main steps.

Step 1. The drift b does not create any serious problems: it is enough to establish the theorem when $b=0$ and then use the Cameron–Martin–Girsanov theorem to 'add in' b as described below. (Like several other statements made about this proof, this is a grotesque oversimplification. Extensive localization techniques are needed, and it is far from obvious how 'adding in b' affects the very subtle strong Feller property (24.3).)

Step 2. So we assume that $b=0$. Next, since $a(\cdot)$ is continuous, it is locally almost constant. Again, localization techniques make it sufficient to deal with the case where a is very close to a constant matrix, which we can take to be the identity:

(24.4) $$|c^{ij}(x)|<\eta \quad \forall i,j, \quad \forall x,$$

where $c(x)\equiv a(x)-I$, and η is some small enough constant.

Step 3. Fix y, and suppose that $\mathbf{P}^y$ is some solution to the martingale problem for $(a, 0)$ starting from y. Then for $\lambda>0$ and $f\in C_K^\infty$,

$$\int_0^\infty e^{-\lambda t}\mathbf{E}^y\left[f(x_t)-f(y)-\int_0^t \mathscr{L}f(x_s)ds\right]dt=0,$$

so that, after a familiar integration by parts,

(24.5) $$R_\lambda^y(\lambda-\mathscr{L})f=f(y),$$

where for $h\in C_K^\infty$ we define

(24.6) $$R_\lambda^y h\equiv \mathbf{E}^y\int_0^\infty e^{-\lambda t}h(x_t)dt.$$

Remember that we do not know that $\mathbf{P}^y$ is a Markovian measure, so the 'resolvent' notation, while it is meant to be suggestive, must not be allowed to mislead.

The important fact which must now be established is that,

(24.7) *for* $p>[\tfrac{1}{2}n]\vee 1$, *the map* $h\mapsto R_\lambda^y h$ *extends to a bounded linear functional on* $L^p(\mathbb{R}^n)$,

the L^p space with regard to Lebesgue measure. Perhaps the simplest method is that sketched in the proof of Theorem IV.3.3 of Ikeda and Watanabe [1]. The idea here is that we can, by (our) Theorem 23.5, find a weak solution of

$$X_t=y+\int_0^t \sigma(X_s)dB_s$$

with law $\mathbf{P}^y$. Define an mth approximation X^m to X by setting

$$\sigma_m(s, X.)=\sigma(X_r) \quad \text{if} \quad r=(k-1)2^{-m}<s\leqslant k2^{-m},$$

and

$$X^m(t) = y + \int_0^t \sigma_m(s, X^m_\cdot)\,dB_s.$$

Then X^m is (conditionally) piecewise Brownian motion with covariance matrix close to I, and we can do good estimates on it. For a detailed proof of (24.7), see Dellacherie, Doléans, Letta and Meyer [1].

Results such as (24.7) play an important role in control theory, and you will find a searching study in Krylov's book [1].

Step 4. We now wish to argue that result (24.5) determines the functional R^y_λ on $L^p(\mathbb{R}^n)$ because, if V_λ is the resolvent

$$V_\lambda = (\lambda - \tfrac{1}{2}\Delta)^{-1}$$

of Brownian motion, and $f = V_\lambda g$, then

$$\begin{aligned} f(y) &= R^y_\lambda(\lambda - \mathscr{L})f = R^y(\lambda - \tfrac{1}{2}\Delta - \tfrac{1}{2}c^{ij}D_iD_j)V_\lambda g \\ &= R^y_\lambda g - \tfrac{1}{2}R^y_\lambda(c^{ij}D_iD_jV_\lambda g). \end{aligned}$$

Now it is a result of Littlewood–Paley theory (for a fine probabilistic proof via the Burkholder–Davis–Gundy inequalities see Meyer [6]) that, for each i and j,

(24.8) *$D_iD_jV_\lambda$ is a bounded linear operator on L^p.*

If now $\tilde{R}^y_\lambda$ is the functional corresponding to R^y_λ for another solution $\tilde{\mathbf{P}}^y$ of the martingale problem for $(a, 0)$ starting at y, then it follows that

$$(R^y_\lambda - \tilde{R}^y_\lambda)g = \tfrac{1}{2}(R^y_\lambda - \tilde{R}^y_\lambda)(c^{ij}D_iD_jV_\lambda g)$$

so that we have the estimate for functionals and operators on L^p:

$$\|R^y - \tilde{R}^y\| \leqslant \tfrac{1}{2}\|R^y - \tilde{R}^y\|.\|c^{ij}D_iD_jV_\lambda\|.$$

By choosing η in (24.4) so small that

$$\tfrac{1}{2}\|c^{ij}D_iD_jV_\lambda\| < 1,$$

we force $R^y = \tilde{R}^y$. By inversion of Laplace transforms in (24.6), we see that:

(24.9) *for each t, the $\mathbf{P}^y$ law of x_t is uniquely determined.*

Step 5. But now things are straightforward. For we can use the regular conditional probability technique from the proof of Theorem 21.1 to show that $(\mathbf{P}^y \circ \theta_t^{-1} | \mathscr{X}_t)$ is ($\mathbf{P}^y$ almost surely) a solution of the martingale problem for $(a, 0)$ starting at x_t. But the $\mathbf{P}^{x(t)}$ law of x_u is uniquely determined by the argument of Step 4, so that

$$(\mathbf{P}^y \circ x_{t+u}^{-1} | \mathscr{X}_t) = \mathbf{P}^{x(t)} \circ x_u^{-1},$$

etc., etc.. We have met many such things. Thus $\mathbf{P}^y$ is the Markov law it has to be.

Step 6. Do the localization properly. See Ikeda and Watanabe [1] or Dellacherie, Doléans, Letta and Meyer [1].

Step 7. Read Stroock and Varadhan [1] to see how much more than this they achieved.

(*24.10*) *Other important references.* Jacod [2], Ethier and Kurtz [1], and Stroock [3].

(*24.11*) *Remark on the* 1-*dimensional case.* See, for example, Lemma 28.7 below for how things simplify for 1-dimensional problems.

25. Martingale representation. On several occasions, we have stressed that one of the most important consequences of the martingale problem's being well-posed is that one has martingale representation as a result. The charming method which we give here is due to Dellacherie, though without the simplification arising from the FD assumption, the method is a lot more difficult to use.

(*25.1*) THEOREM. *With the notation and assumptions of Theorem* 22.12, *for any* $y \in \mathbb{R}^n$ *and* $Y \in L^2 \equiv L^2(W^n, \mathscr{H}^n, \mathbf{P}^y)$, *there exist previsible* Φ^i $(i=1, \ldots, n)$ *such that, with* $\mathbf{E}$ *denoting* $\mathbf{E}^y$,

$$\mathbf{E}\left[\sum_{i,j=1}^{n} \int_0^\infty \Phi_s^i \Phi_s^j a^{ij}(x_s)\, ds\right] < \infty$$

and such that

$$Y = \mathbf{E}Y + \sum_{i=1}^{n} \int_0^\infty \Phi_s^i \, dM_s^i, \tag{25.2}$$

where

$$M_t^i \equiv x_t^i - x_0^i - \int_0^t b^i(x_s)\, ds.$$

Proof. Let $L^2(M)$ denote the Hilbert space of previsible $\Phi \equiv (\Phi^1, \ldots, \Phi^n)$: $\mathbb{R}^{++} \times W^n \to \mathbb{R}^n$ such that

$$\|\Phi\|_M^2 \equiv \mathbf{E}\left[\sum_{i,j=1}^{n} \int_0^\infty \Phi_s^i \Phi_s^j a^{ij}(x_s)\, ds\right] < \infty.$$

Then the map

$$I : \Phi \to \sum_{i=1}^{n} \int_0^\infty \Phi_s^i \, dM_s^i$$

is an isometry of $L^2(M)$ into L^2, whose image $I(L^2(M))$ is a closed subspace of L^2 (since $L^2(M)$ is complete), and is in fact a closed subspace of $L_0^2 \equiv \{Y \in L^2 \colon \mathbf{E}Y = 0\}$, since each $\Phi^i \bullet M^i$ is a martingale bounded in L^2. We remark that $\mathbf{E}Y = \mathbf{E}(Y \mid \mathscr{H}_0)$ by Blumenthal's 0–1 law (III.9).

The martingale representation result (25.2) is equivalent to the statement that $I(L^2(M)) = L_0^2$. Suppose that this is false. Then there exists non-zero $Z \in L_0^2$ which is orthogonal to $I(L^2(M))$. If we let $Z_t \equiv \mathbf{E}^y[Z|\mathscr{X}_t]$, then, *because x is a continuous FD process under* $\mathbf{P}^y$, by Theorem VI.15.1, $Z_{\cdot}$ is actually a *continuous* L^2-bounded martingale. Since $I(L^2(M))$ is a stable subspace of L^2 (see § IV.24.9) it follows that, for any $N \in I(L^2(M))$, and any stopping time T,

$$0 = \mathbf{E}[Z N_T] = \mathbf{E}[Z_T N],$$

so that Z_T is also orthogonal to $I(L^2(M))$ for any stopping time. Now choose

$$T \equiv \inf\{t>0\colon |Z_t| > \tfrac{1}{2}\}$$

so that, since $Z_0 = 0$ $\mathbf{P}^y$-a.s. and $Z_{\cdot}$ is continuous,

$$|Z_T| \leqslant \tfrac{1}{2}.$$

Since Z_T is orthogonal to the stable subspace $I(L^2(M))$, it follows from Theorem IV.24.10 and Lemma IV.24.8 that, for any $s>0$ and $f \in C_K^\infty(\mathbb{R}^n)$ the process

$$Z_{T\wedge t} C^f_{s\wedge t} \equiv Z_{T\wedge t}\{f(x_{s\wedge t}) - f(x_0) - \int_0^{s\wedge t} \mathscr{L} f(x_u)\,du\}$$

$$= Z_{T\wedge t} \int_0^{s\wedge t} D_i f(x_u)\,dM_u^i$$

is a UI martingale. Hence, if we define a new probability measure $\tilde{\mathbf{P}}^y$ by

(25.3) $$\frac{d\tilde{\mathbf{P}}^y}{d\mathbf{P}^y} \equiv 1 + Z_{T\wedge t} \quad \text{on } \mathscr{X}_t^n,$$

then, by Theorem IV.17.1 (iii), $C^f_{s\wedge\cdot}$ is also a $\tilde{\mathbf{P}}^y$ martingale. Thus $\tilde{\mathbf{P}}^y$ solves the martingale problem for (a, b) starting from y. But as the martingale problem is well-posed, $\tilde{\mathbf{P}}^y = \mathbf{P}^y$, and $Z = 0$, a contradiction. □

(*25.4*) *Remarks.* (i) The familiar Brownian martingale representation Theorem IV.36.1 is a special case of this result.

(ii) It is clear that Theorem 25.1 will also hold under the assumptions of Theorem 24.1.

(*25.5*) *References.* The classic reference on martingale representation is Jacod's book [2] following on from Jacod [1] and Jacod and Yor [1].

Transformation of SDEs

We have seen several existence and uniqueness results for SDEs, but many interesting examples are not covered by these results; in practice, we frequently transform an SDE into another which may be more amenable. There are three general methods for transforming an SDE: change of time scale, change of state space (both of which have analogues for ordinary differential equations), and change of measure.

26. Change of time scale; Girsanov's SDE. Suppose that $a\colon \mathbb{R}^n \to S_n^+$ and $b\colon \mathbb{R}^n \to \mathbb{R}^n$ are measurable and that $\mathbf{P}^y$ solves the martingale problem for (a, b) started at y. Suppose also that $\rho\colon \mathbb{R}^n \to (0, \infty)$ is a strictly positive measurable function, and define the strictly increasing adapted process φ by

$$\varphi_t \equiv \int_0^t \rho(x_s)\,ds.$$

We shall assume that

(26.1) $$\sup_t \varphi_t = \infty \quad \text{a.s.}.$$

We do not exclude the possibility that $\varphi_t = \infty$ for some finite t.

Then the time change

$$\tau_t \equiv \inf\{u\colon \varphi_u > t\}$$

is continuous and strictly increasing. Each τ_t is a stopping time. Hence, for each $f \in C_K^\infty(\mathbb{R}^n)$,

$$\tilde{C}_t^f \equiv C_{\tau_t}^f \equiv f(x_{\tau_t}) - f(x_0) - \int_0^{\tau_t} \mathscr{L} f(x_s)\,ds \text{ is an } \{\mathscr{F}(\tau_t)\}\text{-local martingale,}$$

where, as usual, $\mathscr{L} f(x) = \frac{1}{2} a^{ij}(x) D_i D_j f(x) + b^i(x) D_i f(x)$.

Now we simply change variable in the integral: writing $\tilde{x}_t \equiv x_{\tau_t}$, we find that

$$\tilde{C}_t^f = f(\tilde{x}_t) - f(\tilde{x}_0) - \int_0^t \rho(\tilde{x}_u)^{-1} \mathscr{L} f(\tilde{x}_u)\,du.$$

In particular, if ρ^{-1} is locally bounded, then $\tilde{C}^f$ is an $\{\mathscr{F}(\tau_t)\}$-martingale, and the law of x. is a solution to the martingale problem for $(\rho^{-1} a, \rho^{-1} b)$. *Thus we have transformed the martingale problem for* (a, b) *into the martingale problem for* $(\rho^{-1} a, \rho^{-1} b)$.

(*26.2*) *Girsanov's SDE in dimension 1.* As an example illustrating this, consider the 1-dimensional (Girsanov) SDE

(26.3) $$dX_t = \sigma_\alpha(X_t)\,dB_t, \; X_0 = 0,$$

where $\sigma_\alpha(x) \equiv |x|^\alpha \wedge 1$, and $\alpha > 0$. If $\alpha \geqslant 1$, σ_α is locally Lipschitz and, by Theorem 12.1, the SDE has the unique solution $X_t = 0$ for all t. The zero solution is of course always a solution, but for certain values of α, there are others.

Indeed, if we take $\rho_\alpha(x) \equiv |x|^{-2\alpha} \vee 1$, then, under Wiener measure on the path space W^1, for $f \in C_K^\infty(\mathbb{R})$,

$$f(x_t) - f(0) - \int_0^t \tfrac{1}{2} f''(x_s)\,ds \text{ is a martingale,}$$

and so, if we can time change by the inverse to the additive functional

$$\varphi_t = \int_0^t \rho_\alpha(x_s)\,ds, \tag{26.4}$$

we shall have the property:

$$f(\tilde{x}_t) - f(0) - \int_0^t \tfrac{1}{2}(|\tilde{x}_s|^{2\alpha} \wedge 1) f''(\tilde{x}_s)\,ds \text{ is an } \mathscr{X}(\tau_t)\text{-martingale}$$

and so $\tilde{x}$ will be a weak solution of the SDE (26.3). This will work provided that φ is strictly increasing, finite-valued, and unbounded; and the only difficulty is that φ may not be finite-valued. However, using the local time as the occupation density (Theorem IV.45.1) we have

$$\varphi_t \equiv \int_0^t \rho_\alpha(x_s)\,ds = \int_{-\infty}^{\infty} l_t^a \rho_\alpha(a)\,da = \int_{-\infty}^{\infty} l_t^a \frac{da}{|a|^{2\alpha} \wedge 1}$$

so that φ_t is finite if and only if $\alpha < \frac{1}{2}$.

Thus if $\alpha < \frac{1}{2}$, the Girsanov SDE has a non-zero weak solution obtained by changing the time scale. In particular then, *for $\alpha < \frac{1}{2}$, pathwise uniqueness fails for the Girsanov SDE* (otherwise there would be uniqueness in law—and $X_t \equiv 0$ is always a solution). We shall show in § 40 that, *for $\alpha \geqslant \frac{1}{2}$, pathwise uniqueness holds for the SDE (26.3),* and $X_t \equiv 0$ *is the only solution.*

(*26.5*) *Other solutions of the 1-dimensional Girsanov SDE.* If we replace the additive functional φ in (26.4) by

$$\varphi_t = \int_0^t \rho_\alpha(x_s)\,ds + c l_t^0,$$

where c is any nonnegative constant, then, after time transformation, we arrive at still more weak solutions of Girsanov's SDE. For a mysterious relation between this type of phenomenon and boundary theory for Markov chains, see Holley, Stroock and Williams [1].

(*26.6*) *Girsanov's SDE in dimension more than one.* In dimension greater than one, we can see that the locally Lipschitz result is essentially best possible by considering the following analogue of the 1-dimensional Girsanov SDE:

$$dX_t^i = \sigma_\alpha(|X_t|)\,dB_t^i, \quad X_0 = 0. \tag{26.7}$$

In this case, the additive functional $\varphi_t \equiv \int_0^t \rho_\alpha(|x_s|)\,ds$ is finite for any $\alpha<1$, because one can easily prove that $\mathbf{E}\,\varphi_t<\infty$ for each $t>0$, $\alpha<1$, and, as before, change of time scale furnishes a non-zero solution to (26.7).

27. Change of measure. The idea of changing the measure to change the drift of a Brownian motion is one of the most commonly used transformations of an SDE. We have already seen all the key ingredients in § IV.38 and in our study of the Tsirel'son example in § 18, and now there is little more to be said.

(*27.1*) THEOREM. *Suppose that* $\mathbf{P}^y$ *solves the martingale problem for* (a, b) *starting from* y, *suppose that* $\mu\colon \mathbb{R}^{++}\times W^n \to \mathbb{R}^n$ *is a bounded previsible path functional, and suppose that* a *and* b *are bounded. Then (by Theorem* IV. 37.8)

$$\zeta_t \equiv \exp\left[\int_0^t \mu(s, x.)^T dM_s - \tfrac{1}{2}\int_0^t \mu(s, x.)^T a(s, x.)\mu(s, x.)\,ds\right] \tag{27.2}$$

is a $\mathbf{P}^y$*-martingale, where*

$$M_t = x_t - \int_0^t b(s, x.)\,ds.$$

Define a measure $\tilde{\mathbf{P}}^y$ *on* $(W, \mathscr{X})$ *by*

$$d\tilde{\mathbf{P}}^y/d\mathbf{P}^y = \zeta_t \text{ on } \mathscr{X}_{t+}. \tag{27.3}$$

Then $\tilde{\mathbf{P}}^y$ *solves the martingale problem for* $(a, \tilde{b})$ *starting at* y, *where*

$$\tilde{b}^i(t, x.) = b^i(t, x.) + a^{ij}(t, x.)\mu_j(t, x.). \tag{27.4}$$

Proof. That (27.3) defines a measure on $(W, \mathscr{X})$ was proved in Theorem IV.38.9. Since ζ solves the exponential SDE

$$\zeta_t = 1 + \int_0^t \zeta_s\,dN_s,$$

where $N_t \equiv \int_0^t \mu(s, x.)^T dM_s$, then, by Theorem IV.38.4, under $\tilde{\mathbf{P}}^y$,

$$\begin{aligned}\tilde{C}_t^f &\equiv C_t^f - [C^f, N]_t\\ &= C_t^f - \int_0^t a^{ij}(s, x.)\mu_j(s, x.)D_i f(x_s)\,ds\end{aligned}$$

defines a continuous local martingale $\tilde{C}^f$. Boundedness of a and μ ensures that $\tilde{C}^f$ is a $\tilde{\mathbf{P}}^y$-martingale, and so $\tilde{\mathbf{P}}^y$ solves the martingale problem for (a, b) starting from y. □

(*27.5*) *Example.* Suppose that the coefficients (a, b) in the martingale problem are bounded measurable, and that, for some $\delta>0$ $(u, au)\geqslant\delta|u|^2$ for every $u\in\mathbb{R}^n$.

Then the martingale problem for (a, b) is well-posed if and only if the martingale problem for $(a, 0)$ is well-posed. Indeed, if α is the inverse to a, if $\mathbf{P}^y$ solves the martingale problem for (a, b), then by taking $\mu_i(s, x.) \equiv -\alpha_{ij}(s, x.)b^j(s, x.)$, Theorem (27.1) gives a solution $\tilde{\mathbf{P}}^y$ to the martingale problem for $(a, \tilde{b})$—but by the choice of μ, $\tilde{b}=0$. Similarly we can make a solution to the martingale problem for (a, b) from a solution for $(a, 0)$, establishing a one–one correspondence between solutions for the two problems. As already mentioned, this is a key step in the proof of the big Stroock–Varadhan result, Theorem 24.1.

28. Change of state–space; scale; Zvonkin's observation; the Doss–Sussmann method. Given an SDE, it is often helpful to consider some function $Y_t \equiv f(X_t)$ of the solution X. This is done either to exhibit Y as the solution of some simpler SDE or to establish some property of X which was not apparent from the original SDE. There are no theorems, nor even any general rules, but a few examples will serve to illustrate the technique.

(*28.1*) *Example: Bessel processes.* The need to make the transformation $V=R^2$ for Bessel processes R, which makes the singularity at 0 more amenable, has already been covered in §IV.35.

(*28.2*) *Scale function.* Let X be a solution of the 1-dimensional SDE

$$dX_t = \sigma(X_t)dB_t + b(X_t)dt, \tag{28.3}$$

where σ is continuous and b is measurable. If $Y=s(X)$, then

$$dy_t = (\sigma s')(X_t)\,dB_t + \{\tfrac{1}{2}\sigma^2 s'' + bs'\}(X_t)\,dt,$$

so that Y is a local martingale if and only if

$$\tfrac{1}{2}\sigma^2 s'' + bs' = 0.$$

(The fact that s is not necessarily of class C^2 is covered by (IV.45.9).) Assuming that $\sigma^{-2}b$ is locally integrable, this (ordinary) differential equation is solved by

$$s'(x) = \exp\left\{-\int^x 2b(u)\sigma(u)^{-2}du\right\}.$$

The function s (which is only determined to within multiples and additive constants) is called the *scale function* of the diffusion X. See §46. Notice that s is strictly increasing, and that Y solves the SDE

$$dY_t = g(Y_t)dB_t. \tag{28.4}$$

where

$$g(y) = (\sigma s')\circ s^{-1}(y). \tag{28.5}$$

Equation (28.4) has a unique weak solution if for some $0<\delta<K<\infty$,

(28.6) $$\delta \leqslant (\sigma s')(y) \leqslant K$$

for all y, by the Stroock–Varadhan Theorem 24.1.

However, uniqueness in law for solutions of 1-dimensional equation (28.4) may be established directly, and in greater generality.

(*28.7*) LEMMA. *If g is a measurable function such that $g(\cdot) \geqslant \delta$ for some $\delta>0$, then equation (28.4) has a weak solution Y unique in law. Moreover, Y is a pure local martingale in the sense of (IV.34.6).*

Proof. We know from Theorem IV.34.11 that if Y satisfies (28.4), then there exists a Brownian motion β such that

$$Y_t=\beta([Y]_t), \qquad [Y]_t=\int_0^t g(Y_s)^2\,ds.$$

Indeed,

$$\beta(t)=Y(\tau(t)), \quad \text{where} \quad \int_0^{\tau(t)} g(Y_s)^2\,ds=t.$$

We therefore have

$$g(Y(\tau_t))^2\tau'(t)=1=g(\beta_t)^2\tau'(t).$$

Hence,

$$Y_t=\beta([Y]_t),$$

where

$$[Y]_t=\inf\left\{u: \int_0^u g(\beta_s)^{-2}\,ds>t\right\}.$$

Since Y may be described explicitly in terms of the Brownian motion β, Y is pure, and the law of Y is uniquely determined. □

This sort of analysis plays a key role in the theory of 1-dimensional diffusion (§§ 44–54), where we shall see that the most general 1-dimensional diffusion (as a continuous strong Markov process) is obtained from Brownian motion by firstly changing the time scale, and then performing a deterministic transformation of the state space using the scale function. But beware: there are 1-dimensional diffusions which do not arise as a solution of an SDE (28.3).

(*28.8*) *Zvonkin's observation.* Zvonkin [1] observed that the transformation $Y=s(X)$ of equation (28.3) produces an equation (28.4) in which g is locally Lipschitz assuming only that σ is Lipschitz and bounded away from 0, and b is bounded and measurable. Under this assumption, it follows from Itô's theorem that equation (28.4) is exact, and therefore equation (28.3) is also exact.

In particular, if b is a bounded measurable function on $\mathbb{R}$, then the SDE

$$dX = dB + b(X_t)dt$$

is exact, in sharp contrast to the Tsirel'son example in which the drift depends on the history of X.

(*28.9*) *The Doss–Sussmann method.* Another (more sophisticated) application of change of state space is to the case of the (Stratonovich) SDE

$$X_t = x + \int_0^t \sigma(X_s)\,\partial B_s + \int_0^t b(X_s)\,\partial s, \tag{28.10}$$

where

(28.11(i)) $\sigma, b\colon \mathbb{R}^n \to \mathbb{R}^n$ are Lipschitz;
(28.11(ii)) σ is C^2;
(28.11(iii)) B is a 1-dimensional Brownian motion.

We know from Theorem 11.2 that the SDE (28.10) is exact. The idea is to express the solution in the form

$$X_t = F(B_t, Z_t)$$

where F is some deterministic function, and Z is the solution of some ordinary differential equation whose coefficients may depend on ω.

(*28.2*) THEOREM (Doss [1], Sussmann [1]). *Let $F\colon \mathbb{R} \times \mathbb{R}^n \to \mathbb{R}^n$ be the flow of the vector field σ; thus*

$$\frac{\partial F}{\partial t}(t, x) = \sigma(F(t, x)), \quad F(0, x) = x.$$

Then the process

$$X_t = F(B_t, Z_t) \tag{28.13}$$

is the pathwise unique strong solution of the SDE (28.10), where the continuous differentiable process Z is defined to be the solution to the ODE

$$\dot{Z}_t^j = K_i^j(B_t, Z_t) b^i(F(B_t, Z_t)), \quad Z_0 = x. \tag{28.14}$$

Here, the matrix $K(t, x)$ is the inverse to the Jacobian matrix

$$J_k^i(t, x) \equiv \frac{\partial F^i}{\partial x^k}(t, x).$$

Proof (*sketch*). Using the rules of Stratonovich calculus, we obtain

$$\begin{aligned}\partial X^i &= \frac{\partial F^i}{\partial t}(B, Z)\partial B + \frac{\partial F^i}{\partial x^j}(B, Z)\partial Z^j \\ &= \sigma^i(X)\partial B + \frac{\partial F^i}{\partial x^j}(B, Z)K^j_k(B, Z)b^k(X)\,dt \\ &= \sigma^i(X)\partial B + b^i(X)\,dt.\end{aligned}$$

Of course, one has to check that the Lipschitz and C^2 assumptions on σ ensure that the flow F exists, that F is sufficiently smooth, that the Jacobian is invertible and is sufficiently smooth for the ODE (28.14) to have a solution. All these points are straightforward, albeit lengthy; we refer you to Doss [1] and Sussmann [1] for further details. □

(*28.15*) *Remarks.* The ODE (28.14) can be solved *for each* ω, so that the solution (28.13) is itself defined *for each* ω. This is conceptually very attractive, and yields a number of dividends; one is that the solution X is obviously a strong solution, and another is that one may define a solution to (28.10) via (28.13), (28.14) for *any* continuous function B. The solution can be shown to depend continuously on x and B, and results of Wong and Zakai on approximations to the solution to (28.10) by replacing B by a smoothly mollified approximation to B follow immediately from this; see Sussmann [1].

(*28.16*) *Extensions, use in robust filtering, etc.* The following important papers develop extensions of the Doss–Sussmann technique and its application to non-linear filtering: Kunita [4], Davis [4, 5], Clark [3].

A brief introduction to filtering is given in §§ VI.8–11 below.

29. Krylov's example. Suppose that we solve the SDE

(29.1) $$dX_t = \sigma(X_t)dB_t, \quad X_0 \neq 0$$

in $\mathbb{R}^n$, $n \geq 2$, where σ is $n \times n$, and for some $0 < \delta < K < \infty$,

(29.2) $$\delta|u| \leq |\sigma(x)u| \leq K|u|$$

for all non-zero x, u in $\mathbb{R}^n$. The solution is then a local martingale with non-degenerate quadratic variation, and since Brownian motion in $\mathbb{R}^n$ never hits zero, one might think that the same would be true of X. It ain't necessarily so! Take

$$\sigma(x) = |x|^{-2}xx^T + C(I - |x|^{-2}xx^T) \quad (x \neq 0),$$

so that

$$\sigma(x) = \Pi_u + C(I - \Pi_u) \quad (u = x/|x|),$$

where $\Pi_u = uu^T$ is the orthogonal projection on to the space spanned by x. Then

$$\sigma(x)=\sigma(x)^T, \quad \text{and} \quad \sigma(x)^2=\Pi_u+C^2(I-\Pi_u),$$

so that

$$\operatorname{trace}(\sigma(x)\sigma(x)^T)=1+(n-1)C^2.$$

Although σ is not Lipschitz, it is Lipschitz outside any neighbourhood of 0 and an obvious localization allows us to construct a solution, at least up to the first time of reaching zero, which is all that interests us.

We have

$$d(|x|^2)=2|X|U\cdot dB+\{1+(n-1)C^2\}dt, \quad (U=X/|X|),$$

so that, if $R=|X|$, then

$$d(R^2)=2Rd\beta+\nu dt,$$

where β is the Brownian motion with $d\beta=U\cdot dB$, and

$$\nu=1+(n-1)C^2.$$

Thus,

$|X|$ *is a* BES(ν), *and* (see §48) X *will hit* 0 *if and only if* $\nu<2$, *that is, if and only if* $(n-1)C^2<1$.

5. OVERTURE TO STOCHASTIC DIFFERENTIAL GEOMETRY

30. Introduction; some key ideas; Stratonovich-to-Itô conversion. Differential geometry all began when Gauss, having first studied a surface Σ in $\mathbb{R}^3$ extrinsically (that is, utilizing its imbedding in $\mathbb{R}^3$), came to the profound realization that certain entities and concepts are intrinsic in that they depend only on the way in which distance is measured within Σ. Mathematicians, being what they are, seized on the intrinsic, and developed an immensely powerful subject whose sophistication can be something of a stumbling block to the novice. (Imbeddings became things which decent people had nothing to do with, and one kept very quiet about the fact that, for example, any smooth d-dimensional Riemannian manifold can be imbedded with preservation of metric in $\mathbb{R}^N$ for some N.) The intrinsic approach paid huge dividends in physics.

Recently the extrinsic approach is having something of a revival, especially for pedagogic purposes, thanks to fine papers such as that by Lewis [1].

Here, we attempt only to help get beginners (including ourselves) started, publishing the notes which we are making as we too struggle to learn. Our aim, like yours (and we are speaking only to novices), is to be able to read the splendid books by masters of the subject: notably, Bismut [1], Elworthy [1], Ikeda and Watanabe [1]. We hope that you will find that our 'overture' indicates that there are great things to be learnt. Of course, many excellent expository papers are to be found in the literature: Clark [2], the introduction to de Witt-Morette and Elworthy [1], Meyer [9, 10], and Pinsky [1, 2] form a sample of the more accessible ones. Historically, the insight of Itô [3, 4, 5] and Dynkin [3, 4] put them way ahead of their time.

CONVENTION. *Let us agree that throughout this Part of the Chapter, semimartingales and local martingales are understood to be continuous. The term 'smooth' means 'infinitely differentiable' or* C^∞.

(*30.1*) *Some key ideas.* We start by picking out three ideas which show that the extrinsic approach conforms perfectly to first intuition. We shall explain all terminology at appropriate stages, so we return to the conversational mode of learning the language with grammar supplied later.

Let Σ *be a* d*-dimensional submanifold of* $\mathbb{R}^N$. *For example,* Σ *might be an ordinary 2-dimensional surface in* $\mathbb{R}^3$. *For a point* x *of* Σ, *let* $P(x)$ *be the orthogonal projection on to a subspace through* 0 *parallel to the tangent space to* Σ *at* x.

(*30.2*) *What is a Brownian motion on* Σ? Let B be a Brownian motion in $\mathbb{R}^N$. Let X be a process in $\mathbb{R}^N$ with $X(0) \in \Sigma$ and satisfying the Stratonovich SDE

$$\partial X = P(X)\partial B. \tag{30.3}$$

Then X *stays on* Σ *and is a Brownian motion on* Σ. (You will have noticed that for this idea to make sense, $P(X)$ needs to be extended smoothly to some neighbourhood of Σ.)

The SDE (30.3) is the most natural candidate for an SDE whose solution remains on Σ; indeed, if B were differentiable, then one could interpret (30.3) as saying 'project the velocity of B onto the tangent space to Σ at X to obtain the velocity of X'. Thus the velocity of X would always be tangential to Σ, so that X would stay on Σ. Since the Stratonovich calculus obeys the same rules as Newton's, we can expect the same type of argument to apply; it does.

To prove that X is a Brownian motion on Σ, we must show that its generator is $\frac{1}{2}\Delta$ where Δ is the (intrinsic) *Laplace–Beltrami operator for* Σ. We shall explain one of the ways in which the Laplace–Beltrami operator can be introduced in geometry. But, for probabilists, there is no better way of introducing Δ than to define it to be twice the generator of the solution X of (30.3), and then show that it is intrinsic.

(*30.4*) *What is a* (*continuous*) *local martingale on* Σ? As mentioned in § IV.15, it is natural to define a *semimartingale* on Σ as a process X each component of which is a semimartingale on $\mathbb{R}$. Equivalently, we can require that $f(X)$ is a semimartingale for every smooth function f on Σ. The type of appeal to naive intuition which we have made in our discussion of (30.3) shows that it is natural to define a *local martingale on* Σ to be a semimartingale X on Σ such that

(30.5) $$\int P(X)\,dX \textit{ is a local martingale in } \mathbb{R}^N.$$

The integral in (30.5) is, of course, the Itô integral. *But is the definition which we have just made an intrinsic definition?* We shall see that it is.

(*30.6*) *What is horizontal Brownian motion on the orthonormal frame bundle* $O(\Sigma)$ *of* Σ? (We shall see at Theorem 34.86 below that this concept leads to a global intrinsic construction of Brownian motion on a general Riemannian manifold. It has other important applications for the advanced theory.) The *orthonormal frame bundle* $O(\Sigma)$ of Σ is the set of pairs (x, H), where x is a point of Σ and H is an orthonormal frame of tangent vectors to Σ at x. If we think of the elements of H as column vectors of an $N \times d$ matrix, then the properties of H are conveniently summarized as follows:

(30.7) $$P(x)H = H, \qquad H^T H = I_d,$$

where I_d is the identity $d \times d$ matrix. Consider the pair of Stratonovich SDEs:

(30.8(i)) $$\partial X = P(X)\partial B,$$

(30.8(ii)) $$\partial H = \partial(P(X))H,$$

where $(X(0), H(0))$ is a point of $O(\Sigma)$. Then *the pair* (X, H) *performs a 'horizontal' Brownian motion on* $O(\Sigma)$:

(30.9(i)) X *is* $BM(\Sigma)$, *as we have already seen;*

(30.9(ii)) H_t *is always an orthonormal frame at* X_t *in that equations* (*30.7*) *hold for all* t;

(30.9(iii)) *the column vectors of* H *(the vectors of the frame) move by parallel displacement along the curve* X.

We must explain what (30.9(iii)) means, and also decide the problem: *is the generator of the process* (X, H) *intrinsic*?

Several other important topics are discussed in Part 5. We have singled out a few ideas and presented them in a concrete way in the hope of convincing you that they are not at all unfriendly once you get to know them.

(*30.10*) *Switching from Stratonovich to Itô.* As we have seen, for formulating problems and results in the geometry, the Stratonovich version of the stochastic calculus is usually more appropriate than the Itô. But it is the Itô calculus which does all the work. You will therefore need (and will get!) plenty of practice in switching from the Stratonovich to the Itô versions. We shall often do the switching from first principles, but it is convenient to explain the general transformation rule now.

Recall from § IV.46 that in this book we use

d to signify the Itô differential,
∂ to signify the Stratonovich differential.

Recall too that, for two continuous semimartingales X and Y,

(30.11) $$X\partial Y = X\,dY + \tfrac{1}{2}dX\,dY,$$

where, as usual,

$$dX\,dY = d[X, Y].$$

In particular,

(30.12) *$X\partial Y$ and $X\,dY$ differ only by a finite-variation term.*

Consider a diffusion X in $\mathbb{R}^n$ driven by an m-dimensional Brownian motion B via the Stratonovich SDE

(30.13) $$\partial X^i = \sigma^i_q(X)\partial B^q + c^i(X)\partial t,$$

where σ^i_q $(1 \leqslant i \leqslant n,\ 1 \leqslant q \leqslant m)$ and c^i $(1 \leqslant i \leqslant n)$ are smooth functions.

(*30.14*) THEOREM (Transformation rule for Stratonovich SDEs). *Introduce the vector fields (first-order differential operators):*

(30.15) $$U_q \equiv \sigma^i_q(x)D_i, \quad V = c^i(x)D_i,$$

and the 'true drifts':

(30.16) $$b^i \equiv c^i + \tfrac{1}{2}\sum_{q=1}^{m} U_q(\sigma^i_q).$$

Then the Itô form of (30.13) reads:

(30.17) $$dX = \sigma(X)dB + b(X)dt.$$

Moreover, the differential generator

$$\mathscr{L} = \tfrac{1}{2}a^{ij}D_iD_j + b^iD_i,$$

where $a = \sigma\sigma^T$, may be calculated from the Stratonovich equation (30.13) as follows:

(30.18) $$\mathscr{L} = \tfrac{1}{2}\sum_{q=1}^{m} U_q^2 + V.$$

Proof. From (30.13) we obtain, for any smooth function f on $\mathbb{R}^n$,

(30.19) $$\partial f(X) = D_if(X)\partial X^i = U_qf(X)\partial B^q + Vf(X)\partial t.$$

On applying (30.19) to U_qf rather than f, we obtain

$$\partial(U_qf)(X) = U_rU_qf(X)\partial B^r + \partial G,$$

where G is of finite variation. Hence

$$d(U_q f)(X)dB^r = U_r U_q f(X)dt.$$

Thus, (30.19) takes the Itô form

$$df(X) = U_q f(X)dB^q + \mathscr{L}f(X)dt, \tag{30.20}$$

where $\mathscr{L}$ is given by (30.18). If we take $f(X) = X^i$, then (30.20) reduces to (30.17). □

31. Brownian motion on a submanifold of $\mathbb{R}^N$. As indicated at (30.2), we are going to give an explicit global construction of Brownian motion on a 'regular' d-dimensional C^∞ submanifold Σ imbedded in $\mathbb{R}^N$ $(N = n + d > d)$.

(*31.1*) DEFINITION (regular submanifold of $\mathbb{R}^N$). *A subset Σ of $\mathbb{R}^N$ is a* regular d-dimensional C^∞ submanifold of $\mathbb{R}^N$ (*usually abbreviated to* regular submanifold of $\mathbb{R}^N$) *if, for each x in Σ, there exist a subset G of Σ, open in the relative topology and containing x, and a C^∞ function F: $\mathbb{R}^d \to \mathbb{R}^n$ such that (after permuting the canonical coordinates of $\mathbb{R}^N$ if necessary)*

$$G = \{(y/F(y)) : y \in A\}$$

for some open subset A of $\mathbb{R}^d$. For typographical convenience, we have written $(y/F(y))$ for the column vector $\begin{pmatrix} y \\ F(y) \end{pmatrix}$.

Remarks. A regular submanifold of $\mathbb{R}^N$ 'locally looks like $\mathbb{R}^d$'. In particular, if

$$\pi(x^1, \ldots, x^N) \equiv (x^1, \ldots, x^d)$$

is the *canonical projection* of $\mathbb{R}^N$ onto $\mathbb{R}^d$, then

$$\pi\colon G \to A$$

is a one–one smooth map of G onto A with smooth inverse; we say that *π is a diffeomorphism of G with the open subset A of $\mathbb{R}^d$.*

We point out immediately that Definition 31.1 differs slightly from the usual definition of (an imbedded) submanifold of $\mathbb{R}^N$; *a regular submanifold carries the relative topology, while an imbedded submanifold need not.* We discuss the differences further after (34.100), but remark that the differences are irrelevant for our purpose, which is the construction of Brownian motion on an arbitrary smooth Riemannian manifold; we describe at (34.104) how the construction of Brownian motion on a smooth Riemannian manifold may be reduced to the construction of Brownian motion on a regular submanifold of $\mathbb{R}^N$, which we shall carry out in Theorem 31.11. At (34.86), we shall also give an intrinsic construction of Brownian motion on a Riemannian manifold.

(*31.2*) *Examples of regular submanifolds.* Suppose that f: $\mathbb{R}^N \to \mathbb{R}$ is C^∞, and that grad f does not vanish on $\Sigma \equiv f^{-1}(0)$. Then Σ is a regular $(N-1)$-dimensional submanifold of $\mathbb{R}^N$. As a particular case,

$$S^2 \equiv \{Y \in \mathbb{R}^3 : |Y|^2 - 1 = 0\}$$

is a regular submanifold of $\mathbb{R}^3$.

More generally, if $f: \mathbb{R}^N \to \mathbb{R}^p$ is C^∞, $p < N$, and the Jacobian matrix $(\partial f^m/\partial x^r)$ has full rank everywhere on $\Sigma \equiv f^{-1}(0)$, then Σ is a regular $(N-p)$-dimensional submanifold of $\mathbb{R}^N$. This follows from the implicit-function theorem.

We wish to introduce Christoffel symbols, etc., in the most concrete way possible. To do this, it will be helpful to make (for the moment) the following simplifying assumption.

(*31.3*) ASSUMPTION. *For simplicity, we shall assume for the moment that for some given C^∞ function $F: \mathbb{R}^d \to \mathbb{R}^n$,*

$$\Sigma = \{(y/F(y)) : y \in \mathbb{R}^d\}.$$

The point is that any regular submanifold is locally of this form. (Peep ahead to (31.15), where we explain the natural requirement on Σ.)

A smooth curve $t \mapsto y(t)$ in $\mathbb{R}^d$ lifts to a curve

$$t \mapsto x(t) = (y(t)/F(y(t)))$$

in Σ. The *tangent vector* $v(t) \equiv \dot{y}(t)$ to the curve (and therefore also to $\mathbb{R}^d$) at $y(t)$ lifts to the tangent vector

$$u(t) \equiv \dot{x}(t) = (I, J(y(t)))^T v(t) \text{ to } \Sigma \text{ at } x(t),$$

where $J(y)$ $(y \in \mathbb{R}^d)$ denotes the $d \times n$ Jacobian matrix with (i, r)th entry

(31.4) $$J_i^r(y) \equiv \partial F^r/\partial y^i \quad (1 \leqslant i \leqslant d, \quad d+1 \leqslant r \leqslant N).$$

In general, a tangent vector v to $\mathbb{R}^d$ at y lifts to a tangent vector

(31.5) $$u \equiv (I, J(y))^T v \text{ to } \Sigma \text{ at } x \equiv (y/F(y)).$$

We want to develop the idea that $\mathbb{R}^d$ coordinatizes Σ. Because π is a diffeomorphism, this coordinatization automatically preserves the differentiable structure. But we want also to make it metrically correct by *transferring the metric structure of Σ to $\mathbb{R}^d$*, giving $\mathbb{R}^d$ an extra structure making it into a *Riemannian manifold* $(\mathbb{R}^d, g)$ *with Riemannian metric* g. Then $(\mathbb{R}^d, g)$ will be a 'faithful' copy of Σ which we regard as indistinguishable from Σ because it carries all of the ('intrinsic') structure of Σ in which we are interested.

The square of the length of the tangent vector u at (31.5) is

(31.6) $$\|u\|^2 = u^T u = v^T g(y) v,$$

where $g(y)$ is the positive-definite symmetric matrix

(31.7) $$g(y) \equiv (I, J(y))(I, J(y))^T = I + J(y)J(y)^T.$$

If we define

$$\| v \|_g^2 = v^T g(y) v$$

for a tangent vector v to $\mathbb{R}^d$ at y, then the Riemannian metric g on $\mathbb{R}^d$ allows us to associate to the curve $\{y(r): 0 \leqslant r \leqslant t\}$ a length

$$\int_0^t \| v(r) \|_g dr = \int_0^t \{\dot{y}(r)^T g(y(r)) \dot{y}(r)\}^{1/2} dr$$

equal to the Euclidean length of the lifted curve $\{x(r): 0 \leqslant r \leqslant t\}$ on Σ.

We are now going to construct BM(Σ), Brownian motion on Σ. We shall see that the image of this process under π is a diffusion on Σ whose generator may be calculated from the function g. In this sense, BM(Σ) *is intrinsic: it depends only on the metric on Σ, and not on the way in which Σ is embedded in some $\mathbb{R}^N$*. There is another aspect of being intrinsic: intrinsic properties must remain invariant under change of coordinates. We examine this in § 34.

(*31.8*) *Constructing Brownian motion on Σ.* If $y \in \mathbb{R}^d$, and $x \equiv (y/F(y)) \in \mathbb{R}^N$, then any N-vector which is tangential to the submanifold Σ at x must lie in the subspace $T_x\Sigma$ of $\mathbb{R}^N$ spanned by the columns of the matrix $(I, J(y))^T$. Now it is easily confirmed that (the C^∞ function)

(31.9) $$P(y) \equiv (I, J(y))^T g(y)^{-1} (I, J(y))$$

is an orthogonal projection of $\mathbb{R}^N$ onto $T_x\Sigma$. With slight abuse of notation, let

(31.10) $$P(x) \equiv P(\pi(x))$$

for $x \in \mathbb{R}^N$. Then we have the following result.

(*31.11*) THEOREM. *Let B be Brownian motion in $\mathbb{R}^N$, and let X be the solution of the exact SDE*

(31.12) $$\partial X_t = P(X_t) \partial B_t, \quad X_0 \in \Sigma.$$

Then X takes values in Σ, and the differential generator $\mathscr{L}$ of the diffusion $\pi(X)$ on $(\mathbb{R}^d, g)$ may be written in terms of g as follows: $\mathscr{L} = \frac{1}{2}\Delta$, where

(31.13) $$\Delta = (\det g)^{-1/2} D_k (g^{ik} (\det g)^{1/2} D_i),$$

where $(g^{ik}(y))$ denotes the (i, k)th component of $g(y)^{-1}$.

(*31.14*) DEFINITION (Brownian motion on $(\mathbb{R}^d, g)$). *By a* Brownian motion on $(\mathbb{R}^d, g)$, *we mean a diffusion process the transition semigroup of which is the unique FD semigroup on $\mathbb{R}^d$ with generator extending $\frac{1}{2}\Delta$ on $C_K^\infty(\mathbb{R}^d)$.*

Remarks. (i) The naturalness of this definition is clear from the intrinsic property and the naïve geometric argument advanced in § 30.

(ii) Since P is a smooth function, and since P is bounded (P is an orthogonal projection on $\mathbb{R}^N$, whose norm is 1), the conditions of Theorem 12.1 are satisfied, and the SDE (31.12) is exact. The solution X is therefore Markov, and, for x in Σ, the law of $Y=\pi(X)$ when $X(0)=x$ is the unique solution of the martingale problem for $\mathscr{L}$ starting from $\pi(x)$.

(*31.15*) IMPORTANT OBSERVATION. *The description of X via the SDE (31.12) depends in no way on the assumed form (31.3) of the submanifold Σ; provided we can find a smooth function P defined in a neighbourhood of Σ such that $P(x)$ is orthogonal projection on to $T_x\Sigma$ for each $x \in \Sigma$, then (31.12) provides a global definition of Brownian motion on Σ. There is no need to 'patch'.*

Proof of Theorem 31.11. From the definition of P, it is trivial to verify that, for $r=d+1, \dots, N=d+n$,

$$\partial(X^r - F^r(X^1, \dots, X^d))=0,$$

so that

$$P(X_t=(Y_t/F(Y_t)) \text{ for all } t)=1,$$

where $Y_t \equiv \pi(X_t)$, and so X lives on Σ.

Remark Suppose that $\bar{P}(\cdot)$ is smooth in a neighbourhood of Σ and that $\bar{P}(\cdot)=P(\cdot)$ on Σ. Since the solution X of $\partial X=P(X)\partial B$ stays on Σ, we have $\partial X=\bar{P}(X)\partial B$, with X as the unique solution of the latter SDE. Local application of this principle justifies Observation (31.15).

We note that our equation $\partial X=P(X)\partial B$ implies the autonomous equation

$$\partial Y=g(Y)^{-1}(I, J(Y))\partial B \tag{31.16}$$

for the Y-diffusion on $\mathbb{R}^d$. We must now find the generator $\mathscr{L}$ of the Y-process. To do this, rewrite (31.16) as

$$\partial Y^j=(\sigma(Y)\partial B)^j=\sigma^{jq}(Y)\partial B_q, \tag{31.17}$$

where, with the standard convention that

$$g_{ij}(y)=(g(y))(i,j), \quad g^{ij}(y)=(g(y)^{-1})(i,j),$$

we have, for $1 \leqslant j \leqslant d$,

$$\begin{aligned} \sigma^{jk} &= g^{jk} && (1 \leqslant k \leqslant d) \\ \sigma^{js} &= g^{jl}J^s_l && (d+1 \leqslant s \leqslant N). \end{aligned}$$

It is immediate from (31.7) that $\sigma\sigma^T=g^{-1}$. Thus, by Theorem 30.14, we have

$$\mathscr{L}=\tfrac{1}{2}g^{jk}D_jD_k+b^iD_i, \quad D_i=\partial/\partial y^i,$$

where

$$b^i = \tfrac{1}{2}\sum_{k\leqslant d} U^k(\sigma^{ik}) + \tfrac{1}{2}\sum_{s>d} U^s(\sigma^{is}),$$

where

$$U^q \equiv \sigma^{jq} D_j \quad (1 \leqslant q \leqslant N).$$

Thus,

$$2b^i = \sum_{k\leqslant d} g^{jk} D_j(g^{ik}) + \sum_{s>d} g^{jk} J^s_k D_j(g^{il} J^s_l).$$

Now, from (31.7),

$$g^{-1} + g^{-1} J J^T = I,$$

so that, for each j, we have

$$D_j(g^{-1}) + D_j(g^{-1}J)J^T = -(g^{-1}J)D_j J^T. \tag{31.18}$$

Hence,

$$D_j g^{ik} + \sum_{s>d} D_j(g^{il} J^s_l) J^s_k = -\Gamma^i_{jk}, \tag{31.19}$$

where Γ^i_{jk} is the *Christoffel symbol*:

$$\Gamma^i_{jk} = g^{il} \sum_{r>d} J^r_l D_j J^r_k. \tag{31.20}$$

The significance of Christoffel symbols will be explained in subsequent sections.

We have therefore established the important formula:

$$2\mathscr{L} = g^{jk}(D_j D_k - \Gamma^i_{jk} D_i). \tag{31.21}$$

You can check that the right-hand sides of (31.21) and (31.13) agree. (Use cofactor expansion of the determinant.) However, as we shall see in the next two sections, the right-hand side of (31.21) is the natural expression of the Laplace–Beltrami operator for probabilists, and the fact that the Γ^i_{jk} are determined by the $g(\cdot)$ function is a fundamental property of Riemannian connections. So, formula (31.13) is not of primary importance to us. □

(*31.22*) *Brownian motion on a surface; martingale characterization.* As a special case, suppose that $N=3$ and $d=2$ (so that Σ is a surface in $\mathbb{R}^3$), and suppose that $f: \mathbb{R}^3 \to \mathbb{R}$ is C^∞, and that grad f does not vanish on $\Sigma \equiv f^{-1}(0)$. Then the normal vector

$$n(X) \equiv \operatorname{grad} f(X)/|\operatorname{grad} f(X)|$$

can be defined everywhere in a neighbourhood of Σ, and one can take

$$P(X) = I - n(X)n(X)^T.$$

In this case, the equation

$$\partial X_t = P(X_t)\partial B_t \tag{31.23}$$

has the Itô form (*Exercise!*)

$$dX = P(X_t)dB_t + c(X_t)n(X_t)dt, \tag{31.24}$$

where $c = -\frac{1}{2}\operatorname{div}(n)$ is the so-called *mean curvature*. Thus, $c(X)n(X)$ is the drift required to keep the particle on the surface.

It follows from (31.24) that

(31.25(i)) $dM = dX - c(X)n(X)dt$ defines a local martingale M in $\mathbb{R}^3$,

(31.25(ii)) $dMdM^T = (I - nn^T)dt, \quad n = n(X)$.

(*31.26*) THEOREM (van den Berg and Lewis). *Suppose that X is a semimartingale in $\mathbb{R}^3$ such that $X(0) \in \Sigma$ and equations (31.25) hold. Then X is a* BM(Σ).

Proof. Let $dN = \{I - P(X)\}dM$. Then it follows from (31.25(ii)) that

$$dNdN^T = \{I - P(X)\}P(X)\{I - P(X)\}dt = 0,$$

so that N is constant. Thus,

$$dM = P(X)dM = P(X)dX.$$

Now, let b be a BM($\mathbb{R}$) independent of X, and define

$$dB = dM + n(X)db.$$

Then B is a local martingale in $\mathbb{R}^3$, and

$$dBdB^T = (I - nn^T)dt + nn^T dt = Idt,$$

so that B is a BM($\mathbb{R}^3$). Finally,

$$dX = P(X)dB + c(X)n(X)dt,$$

or, in Stratonovich form,

$$\partial X = P(X)\partial B.$$ □

See Lewis [1] for the general martingale characterization result for submanifolds.

(*31.27*) **Exercise.** An alternative method of constructing a Brownian motion on a surface was announced in Price and Williams [1]: describe X as the solution of the Stratonovich SDE

$$\partial X_t = n(X_t) \times \partial B_t,$$

where $\times$ denotes the usual vector product in $\mathbb{R}^3$. Confirm that the generator of this process is the same as the generator of the solution of the SDE (31.23), and hence deduce that X is again a Brownian motion on Σ.

This idea arose in discussions with Aubrey Truman, and has since been developed in van den Berg and Lewis [1].

(*31.28*) *Example: skew-product representation of* BM($\mathbb{R}^3$). As an exercise, you can now check that if W is a BM($\mathbb{R}^3$) started away from 0, then $U \equiv W/|W|$ is a BM(S^2) run at rate $|W|^{-2}$ but otherwise independent of $|W|$. *Hint.* Calculate the generator of $(U_t, |W_t|)$ and show (as in § 26) that if $\tau_\cdot$ is inverse to $A_\cdot \equiv \int_0^\cdot |W_s|^{-2} ds$, then $U(\tau_\cdot)$ is BM(S^2) independent of $W(\cdot)$. Can you also prove that $P(A_\infty = \infty) = 1$?

(*31.29*) *Brownian motion on a surface of revolution.* A surface of revolution Σ is produced by rotating a smooth curve in the (x, z) plane about the z-axis. We parametrize the surface:

$$(x, y, z) = (r(\gamma)\cos\theta, r(\gamma)\sin\theta, z(\gamma)),$$

where γ is arc-length along the original curve, so that

$$\dot{r}(\gamma)^2 + \dot{z}(\gamma)^2 = 1. \tag{31.30}$$

The element of length ds on Σ is given by

$$(ds)^2 = (d\gamma)^2 + r(\gamma)^2(d\theta)^2 \quad \text{(non-stochastic!)},$$

so that

$$\begin{pmatrix} g_{\gamma\gamma} & g_{\gamma\theta} \\ g_{\theta\gamma} & g_{\theta\theta} \end{pmatrix} = \begin{pmatrix} 1 & 0 \\ 0 & r(\gamma)^2 \end{pmatrix}$$

Hence, from (31.13),

$$\tfrac{1}{2}\Delta = \tfrac{1}{2}\frac{\partial^2}{\partial\gamma^2} + \tfrac{1}{2}\frac{\dot{r}(\gamma)}{r(\gamma)}\frac{\partial}{\partial\gamma} + \frac{1}{r(\gamma)^2}\cdot\tfrac{1}{2}\frac{\partial^2}{\partial\theta^2}. \tag{31.31}$$

Thus BM(Σ), parametrized as (γ_t, θ_t), may be described as follows:

$$d\gamma = db + \tfrac{1}{2}\frac{\dot{r}(\gamma)}{r(\gamma)}dt, \quad \theta_t = \theta_0 + W\left(\int_0^t r(\gamma_u)^{-2} du\right), \tag{31.32}$$

where b and W are independent BM($\mathbb{R}$) processes.

You can of course derive this directly from the $\partial X = P(X)\partial B$ description.

The mean curvature at a point on our surface of revolution Σ is the mean of two terms which must be taken with appropriate signs: one is the curvature of the original plane curve C at the point; the other is the reciprocal of the distance from the point to the z-axis along the normal to the original curve.

(*31.33*) *Example: skew-product representation of* BM(S^2). For BM(S^2), where S^2 is the unit sphere in $\mathbb{R}^3$, in the coordinate system

$$(\sin\gamma\cos\theta, \ \sin\gamma\sin\theta, \ \cos\gamma),$$

there exist independent Brownian motions b and W such that

$$d\gamma = db + \tfrac{1}{2}(\cot\gamma)dt,$$

$$\theta_t = W\left(\int_0^t (\sin\gamma_u)^{-2}\,du\right).$$

(*31.34*) **Exercise.** Show that if Σ is produced by rotating the curve $x = \cosh z$ about the z-axis, then $\mathrm{BM}(\Sigma)$ is a local martingale in $\mathbb{R}^3$. The point is that the mean curvature is 0. See also Darling [1] and van den Berg and Lewis [1].

32. Parallel displacement; Riemannian connections. We return to the special extrinsic situation with which we began §31. Thus, as at (31.3),

$$\Sigma = \{(y/F(y)):\ y \in \mathbb{R}^d\}$$

and $\pi\colon \Sigma \to \mathbb{R}^d$ is the natural projection. We recall some further material from §31. A curve $y(\cdot)$ in $\mathbb{R}^d$ lifts to a curve $x(\cdot) \equiv (y(\cdot)/F(y(\cdot)))$ in Σ. The tangent vector $v(t) = \dot{y}(t)$ to $\mathbb{R}^d$ at $y(t)$ lifts to the tangent vector

$$u(t) = (I, J(y(t)))^T v(t) \quad \text{to } \Sigma \text{ at } x(t),$$

where J is the Jacobian matrix

$$J^r_i = \partial F^r/\partial y^i \quad (1 \leqslant i \leqslant d,\ d+1 \leqslant r \leqslant N = n+d).$$

Generally, a tangent vector v to $\mathbb{R}^d$ at y lifts to the tangent vector

$$u = (I, J(y))^T v \quad \text{to} \quad \Sigma \text{ at } x = (y/F(y)),$$

and

$$\|u\|^2 = v^T g(y) v,$$

where $g(\cdot)$ is the Riemannian inner product

$$g(y) = I + J(y)J(y)^T.$$

We think of $\mathbb{R}^d$ as endowed with the Riemannian metric g, so that $(\mathbb{R}^d, g)$ is a faithful copy of Σ.

(*32.1*) *Covariant differentiation.* Let $y(\cdot)$ be a smooth curve in $\mathbb{R}^d$, and suppose that, for each t, $w(t)$ is some tangent vector to $\mathbb{R}^d$ at $y(t)$, the function $t \mapsto w(t)$ being smooth. (Currently, you can think of $w(t)$ as the vector from $y(t)$ to $y(t) + w(t)$. For fancier ideas about tangent vectors, see §34. If you feel any insecurity about tangent vectors during your reading of this section, then peep ahead to §34.) Each vector $w(t)$ lifts to the vector

$$q(t) = (I, J(y(t)))^T w(t) \text{ tangent to } \Sigma \text{ at } x(t), \tag{32.2}$$

where $x(t) = (y(t)/F(y(t)))$ as usual. Now, $\dot{q}(t) = dq/dt$ represents the rate of change of $q(t)$ viewed in $\mathbb{R}^N$. But an observer living within Σ can observe only the tangential component of this rate of change, namely,

$$P(y(t))\dot{q}(t). \tag{32.3}$$

On using the fact (31.9) that

$$P(y)=(I, J(y)^T)g(y)^{-1}(I, J(y)),$$

we see that the expression at (32.3) is the lift of the vector

$$a(t)=g(y(t))^{-1}(I, J(y(t)))\dot{q}(t).$$

We are engaged in trying to transfer the structure of Σ to $\mathbb{R}^d$, and we see that $a(t)$ is the appropriate interpretation of the rate of change of $w(t)$ as it would be measured within $(\mathbb{R}^d, g)$. A simple calculation shows that

$$a^i(t)=\dot{w}^i(t)+\Gamma^i_{jk}(y(t))\dot{y}^j(t)w^k(t), \tag{32.4}$$

where, as at (31.20),

$$\Gamma^i_{jk}=g^{il}\sum_{r>d} J^r_l D_j J^r_k = g^{il}\sum_{r>d} J^r_l D_j D_k F^r. \tag{32.5}$$

We shall show at (32.15) that Γ^i_{jk} may be computed from the Riemannian metric tensor $g(\bullet)$.

(*32.6*) DEFINITION (covariant derivative). *The vector $a^i(t)$ is called the* covariant derivative *of $w(t)$ along the curve $y(\bullet)$ at time t (or along the vector $v(t)=\dot{y}(t)$) and is written*:

$$a(t)=\nabla_{v(t)}w(t).$$

(*32.7*) DEFINITION (parallel displacement). *We say that $w(\bullet)$ moves by* parallel displacement *along the curve $y(\bullet)$ if it has zero covariant derivative along the curve. Thus, we must have*

$$\dot{w}^i(t)+\Gamma^i_{jk}(y(t))\dot{y}^j(t)w^k(t)=0. \tag{32.8}$$

(*32.9*) *Symmetric (torsiòn-free) character of* Γ. We note from (32.5) that

$$\Gamma^i_{jk}=\Gamma^i_{kj}.$$

(*32.10*) *Parallel displacement preserves inner products.* Again consider our curve $y(\bullet)$ in $\mathbb{R}^d$. Suppose that, for each t, $w(t)$ and $\tilde{w}(t)$ are tangent to $\mathbb{R}^d$ at $y(t)$, and that both $w(\bullet)$ and $\tilde{w}(\bullet)$ move by parallel displacement along $y(\bullet)$. Then,

$$w(t)^T g(y(t))\tilde{w}(t) \text{ is independent of } t.$$

Proof. Define the lift $q(t)=(I, J(y(t)))^T w(t)$, and define $\tilde{q}(t)$ analogously. Then

$$w(t)^T g(y(t))\tilde{w}(t)=q(t)^T\tilde{q}(t), \tag{32.11}$$

the right-hand side being the inner product in $\mathbb{R}^N$. Since $w(\bullet)$ moves by parallel displacement, $\dot{q}(t)$ is orthogonal to the tangent plane $T_{x(t)}\Sigma$ to Σ at

$x(t) = (y(t)/F(y(t)))$. However, $\tilde{q}(t)$ lies within this tangent plane. It is now clear that the derivative of the right-hand side of (32.11) is zero. □

(*32.12*) *Connections.* By a connection Γ on $\mathbb{R}^d$, we now mean any collection $\Gamma = (\Gamma^i_{jk}(\cdot))$ of d^3 smooth functions on $\mathbb{R}^d$. With any connection, we can associate a covariant derivative, as is clear from (32.4) and (32.6), and hence we can introduce the idea of parallel displacement. Note that given the curve $y(\cdot)$ and $w(0)$, we can uniquely solve the first-order equation (32.8) for $w(\cdot)$.

(*32.13*) DEFINITION (Riemannian connection). *A connection Γ on the Riemannian metric space $(\mathbb{R}^d, g)$ is called* Riemannian *if*

(32.14(i)) *it is symmetric in that (32.9) holds;*
(32.14(ii)) *it is compatible with g in that parallel displacement under Γ preserves inner products.*

(*32.15*) THEOREM. *There is precisely one Riemannian connection $(\mathbb{R}^d, g)$ and it is given by*

$$\Gamma^i_{jk} = \tfrac{1}{2} g^{il}(D_j g_{lk} + D_k g_{lj} - D_l g_{jk}). \tag{32.16}$$

Since the connection introduced at (32.5) was proved to be Riemannian (32.9) and (32.10), it follows that the (extrinsic) formula (32.5) must agree with the (intrinsic) formula (32.16).

Proof of theorem. Let Γ be a Riemannian connection on $(\mathbb{R}^d, g)$. Suppose that for each t, $w(t)$ and $\tilde{w}(t)$ are tangents to $\mathbb{R}^d$ at $y(t)$. Suppose further that $w(\cdot)$ and $\tilde{w}(\cdot)$ move by parallel displacement along $y(\cdot)$, so that $w(\cdot)$ and $\tilde{w}(\cdot)$ satisfy (32.8). Then

$$(d/dt)\{w^i(t) g_{ij}(y(t)) \tilde{w}^j(t)\} = 0. \tag{32.17}$$

Evaluating the derivative on the left-hand side of (32.17), using (32.8), we have, on suppressing the t and y parameters and writing $v(t)$ for $\dot{y}(t)$,

$$\begin{aligned} 0 &= -\Gamma^i_{kl} v^k w^l g_{ij} \tilde{w}^j + w^i v^k (D_k g_{ij}) \tilde{w}^j - w^i g_{ij} \Gamma^j_{kl} v^k \tilde{w}^l \\ &= (-\Gamma^i_{kl} g_{ij} + D_k g_{lj} - g_{li} \Gamma^i_{kj}) w^l v^k \tilde{w}^j. \end{aligned}$$

For the 'preservation of the inner product' property, we must therefore have, using cyclic permutation for the second and third equations:

$$\Gamma^i_{kl} g_{ij} + \Gamma^i_{kj} g_{li} = D_k g_{lj}, \tag{32.18(i)}$$

$$\Gamma^i_{lj} g_{ik} + \Gamma^i_{lk} g_{ji} = D_l g_{jk}, \tag{32.18(ii)}$$

$$\Gamma^i_{jk} g_{il} + \Gamma^i_{jl} g_{ki} = D_j g_{kl}. \tag{32.18(iii)}$$

On adding the first and third of the above equations and subtracting the second, we obtain, on using the symmetry of g and the symmetry property (32.9) of Γ,

$$2\Gamma^i_{jk}g_{il}=(D_j g_{kl}+D_k g_{lj}-D_l g_{jk}),$$

so that (32.16) holds. □

(*32.19*) *Parallel transport of orthonormal frames.* We shall see in § 34 that the concept of parallel transport of frames is the key to obtaining an intrinsic SDE for Brownian motion on a Riemannian manifold. We prepare the way for this by working now with our faithful copy $(\mathbb{R}^d, g)$ of Σ. However, to be sure that everything is clear, we give at (32.23) below the treatment in terms of Σ imbedded in $\mathbb{R}^N$.

By an *orthonormal frame at* y in $(\mathbb{R}^d, g)$, we mean a collection

$$E=(e_1, e_2, \ldots, e_d)$$

of g-orthonormal vectors at y. We may regard E as the $d\times d$ matrix with columns $e_1, e_2, \ldots, e_d$, which satisfies

$$E^T g E=I. \tag{32.20}$$

Suppose that we are given a curve $\{y(t)\colon 0\leqslant t\leqslant 1\}$ in $(\mathbb{R}^d, g)$, and an orthonormal frame $E(0)$ at $y(0)$. Then we can uniquely determine vectors $e_j(t)$ $(1\leqslant j\leqslant d)$ at $y(t)$ such that each e_j moves by parallel displacement along $y(\cdot)$ (in the sense of the Riemannian connection Γ). Indeed, to obtain the matrix $E(t)$, we need only solve

$$\dot E(t)+\Lambda(t)E(t)=0, \tag{32.21}$$

where $\Lambda(t)$ has (i, k)th component

$$\Lambda^i_k(t)=\Gamma^i_{jk}(y(t))\dot y^j(t).$$

Because the Riemannian connection preserves inner products, it is true for each t that $E(t)$ is an orthonormal frame at $y(t)$, or, in other words an orthonormal basis for the tangent space to $(\mathbb{R}^d, g)$ at $y(t)$. An observer living in $(\mathbb{R}^d, g)$ and moving along the curvè $y(\cdot)$ can carry (or 'transport') with him the frame or basis $E(\cdot)$ which he regards as keeping fixed in its directions. In this way, he 'connects' the tangent spaces at $y(0)$ and $y(1)$. However, this 'connection' depends on the curve $y(\cdot)$: in particular, if $y(1)=y(0)$, then $E(1)$ need not equal $E(0)$. The standard example (Figure V.2) shows a frame being parallel transported around the boundary of an octant of a sphere, ending.up by being rotated by 90° from its original position.

(*32.22*) **Important exercise.** Suppose that $w(t)$ is a vector at $y(t)$. Our observer moving along $y(\cdot)$ with his frame $E(\cdot)$ will measure $w(t)$ as having coordinates $r(t)$, where

$$w(t)=E(t)r(t), \quad r(t)=E(t)^T g(y(t))w(t).$$

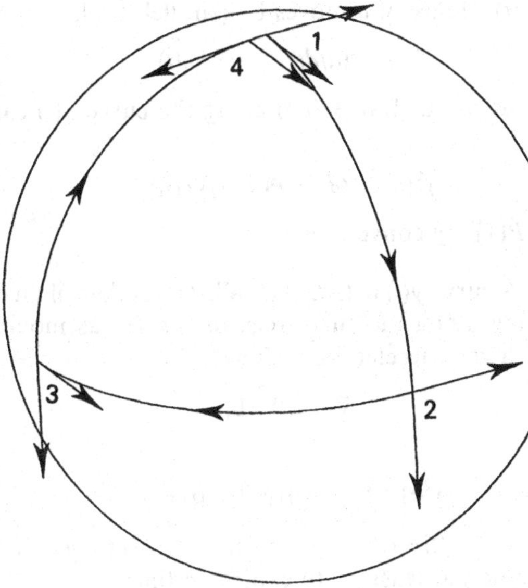

Figure V.2

To obtain the correct ('covariant') rate of change of $w(\cdot)$, we must transform the rate $\dot{r}(t)$ measured by the observer back to our coordinates:

$$\nabla_{v(t)} w(t) = E(t)\dot{r}(t),$$

where $v(t) = \dot{y}(t)$ as usual. Check this out!

(*32.23*) *The picture on* Σ. We continue to work with the notation of (32.21). The g-orthonormal frame E of tangent vectors at y in $\mathbb{R}^d$ lifts to the frame

$$H = (I, J(y))^T E$$

of tangent vectors to Σ at x which is orthonormal in the classical sense:

(32.24) $$H^T H = I_d, \text{ the unit } d \times d \text{ matrix.}$$

Of course, (32.24) is another expression of (32.20).

Recall the notation of (32.2) and (32.3). If $w(t)$ denotes a vector in $\mathbb{R}^d$ at the point $y(t)$ of our curve $y(\cdot)$ in $\mathbb{R}^d$, and

$$q(t) = (I, J(y(t)))^T w(t)$$

is the lift of $w(t)$ to $T_{x(t)}\Sigma$, then the tangential component of the rate of change of $q(t)$ is

$$P(y(t))\dot{q}(t) = \dot{q}(t) - \{(d/dt)P(y(t))\}\, q(t),$$

since $P(y(t))q(t)=q(t)$. Hence, if q moves by parallel displacement, then

$$\dot{q}(t)=\{(d/dt)P(y(t))\}q(t).$$

Parallel displacement of our frame $H(\cdot)$ along the curve $x(\cdot)$ can therefore be expressed as:

$$\dot{H}(t)=\{(d/dt)P(y(t))\}H(t). \tag{32.25}$$

Recall that '$P(x)=P(y)$' by convention.

(*32.26*) *Geodesics.* A curve $y(\cdot)$ in $(\mathbb{R}^d, g)$ is called a *geodesic* if an observer living in $(\mathbb{R}^d, g)$ and moving along the curve regards himself as moving at constant velocity, and so with zero acceleration. Thus

$$\nabla_{v(t)}v(t)=0, \tag{32.27}$$

or

$$\ddot{y}^i(t)+\Gamma^i_{jk}(y(t))\dot{y}^j(t)\dot{y}^k(t)=0. \tag{32.28}$$

If we are given $y(0)$ and $\dot{y}(0)=v(0)$, we can solve (32.28) to get a unique geodesic with these initial values, at least up to explosion time.

33. Extrinsic theory of BM^{hor}(O(Σ)); rolling without slipping; martingales on manifolds; etc. We now develop the extrinsic theory of *horizontal Brownian motion* BM^{hor}(O(Σ)) *on the orthonormal frame bundle* O(Σ) *of* Σ. *Again,* Σ *is assumed to be a smooth d-dimensional submanifold of* $\mathbb{R}^N$ *of the form (31.3).*

The set O(Σ) consists of all pairs (x, H), where x is a point of Σ and H is an orthonormal frame of tangent vectors to Σ at x. We think of H as the $N\times d$ matrix the columns of which are d orthonormal vectors in $\mathbb{R}^N$ spanning $T_x\Sigma$. Thus,

$$H^TH=I,\quad P(x)H=H.$$

The set O(Σ) has a natural manifold structure, as will be explained in § 34.

Construct a Brownian motion X on Σ with given starting point x as usual:

$$\partial X_t=P(X_t)\,\partial B_t,\quad X_0=x\in\Sigma. \tag{33.1}$$

(*33.2*) THEOREM. *Suppose that* $\{H_t: t\geqslant 0\}$ *is the solution to the SDE*

$$\partial H_t=\partial(P(X_t))H_t \tag{33.3}$$

with given initial value H_0, *an orthonormal frame for* $T_x\Sigma$. *Then, for every* $t\geqslant 0$, H_t *is an orthonormal frame for the tangent space to* Σ *at* X_t:

$$H_t^TH_t=I,\quad P(X_t)H_t=H_t. \tag{33.4}$$

The process (X, H) *therefore lives in* O(Σ). *Moreover,* H *is the parallel translate of* H_0 *in that*

$$P(X_t)\,\partial H_t=0. \tag{33.5}$$

Convention. We shall use shorthand notation in the usual way:
$\partial X = P(X)\partial B, \quad H^T H = I, \quad PH = H, \quad P\partial H = 0$, etc.

Remarks. (i) Obviously, the motivation for (33.3) is provided by (32.25).

(ii) One must regard $(X_\cdot, H_\cdot)$ as determined by the pair of equations (33.1) and (33.3), rewriting (33.3) as follows:

$$\partial H^i_q = (D_k P^i_r)(\partial X^k) H^r_q = (D_k P^i_r) H^r_q P^k_s \partial B^s,$$

whereupon the pair (33.1) and (33.3) forms a standard type of SDE for the process $(X_\cdot, H_\cdot)$. Since X stays on Σ and, as we shall see, H remains orthonormal, there is no chance of explosion.

Proof. We have:

$$\partial H = (\partial P) H = \partial (P^2) H = P(\partial P) H + (\partial P) PH$$
$$= P\partial H + (\partial P) PH.$$

But

$$\partial (PH) = P\partial H + (\partial P) H,$$

so that if $Z = (I - P) H$, then $Z_0 = 0$ and

$$\partial Z = (\partial P) PH - (\partial P) H = -(\partial P) Z. \tag{33.6}$$

Clearly, $(X_\cdot, Z_\cdot) = (X_\cdot, 0)$ is the unique solution to (33.1) and (33.6), so that $Z = 0$ and

$$PH = H \quad \text{for all } t. \tag{33.7}$$

Hence,

$$\partial H = P(\partial H) + (\partial P) H,$$

and (33.5) now follows, by (33.3).

Finally, since $P^T = P$, we have from (33.5) and its transposed version, and from (33.7),

$$H^T \partial H + (\partial H^T) H = \partial (H^T H) = \partial (H^T PH)$$
$$= (\partial H)^T PH + H^T (\partial P) H + H^T P\partial H$$
$$= 0 + H^T \partial H + 0 = H^T \partial H.$$

Thus, $(\partial H)^T H = 0$, $H^T \partial H = 0$, and $H^T H = \text{constant} = I$. □

(*33.8*) **Exercise.** Prove that

$$P = HH^T \quad \text{for all } t.$$

Hint. Either use elementary geometry, or else show that $Q \equiv P - HH^T$ satisfies

$$\partial Q = (\partial P) Q - Q(\partial P), \tag{33.9}$$

and deduce that $(X_t, Q_t) = (X_t, 0)$ for all t.

For reasons which will be explained in § 34, the process (X, H) is called the horizontal Brownian motion on $O(\Sigma)$. We now obtain what is for many purposes a better description of this process, using a driving Brownian motion W of the 'correct' dimension d.

(*33.10*) LEMMA. *Set W to be the process in $\mathbb{R}^d$ with $W_0=0$ and*

(33.11) $$\partial W = H^T \partial B.$$

Then equation (33.11) has the Itô form:

(33.12) $$dW = H^T dB,$$

and so W is a $\mathrm{BM}(\mathbb{R}^d)$. *We have*

(33.13) $$\partial X = H \partial W, \quad \partial H = \partial(P(X))H.$$

Comment. Though, as we shall see, (33.13) is a description of (X, H) with important consequences, it does not immediately imply that X is Markovian. But see the proof of Theorem 34.86.

Proof. It is immediate from (33.8) that if W is defined by (33.11), then (33.13) holds.

It remains to prove (33.12), which, in view of (33.11) amounts to showing that

(33.14) $$(\partial H^T)\partial B = 0.$$

Since $\partial X = P \partial B$, we have

$$(\partial X)\partial X^T = P(\partial B)(\partial B^T)P = P(\partial t I)P = P(\partial t) = (\partial X)\partial B^T.$$

Hence,

$$\begin{aligned}\{(\partial P)\partial X\}^i &= (D_k P^i_j)(\partial X^k)\partial X^j \\ &= (D_k P^i_j)(\partial X^k)\,\partial B^j \\ &= \{(\partial P)\partial B\}^i.\end{aligned}$$

Thus,

$$\begin{aligned}(\partial H^T)\partial B = H^T(\partial P)\partial B &= H^T(\partial P)\partial X \\ &= H^T(\partial P)P\partial B = (\partial H^T)P\partial B.\end{aligned}$$

But, by (33.5), $(\partial H^T)P = (P\partial H)^T = 0$. Thus, (33.14) holds, whence (33.11) holds, and W is a d-dimensional Brownian motion. □

(*33.15*) *Interpretation: rolling without slipping; stochastic development.* The equation

(33.16) $$\partial X = H \partial W$$

means that *an observer, moving in Σ according to the Brownian motion X and carrying the frame H which he regards as fixed in direction, will see X as a Brownian* *motion in* $\mathbb{R}^d$. This is because (33.16) says that ∂W is the vector in $\mathbb{R}^d$ giving the coordinates of the vector ∂X relative to H.

Let us now give the famous intuitive Cartan–Itô–Eells–Elworthy–Malliavin picture of how to construct X from W. We build a d-dimensional hyperplane L and mark a point of L as origin. At time 0, we place L tangential to Σ at x, the origin of L coinciding with the point x of Σ. Coordinate axes are marked and fixed in L lying along the vectors of the frame H_0. The process W is a Brownian motion in L relative to the coordinate axes fixed in L. We roll L without slipping on Σ so that at time t, L is tangential to Σ at the point W_t of L. The point X_t of Σ is then just the point of contact W_t of L interpreted within Σ, and the frame H_t is that formed by the marked coordinate-axis directions in L at time t. The 'without slipping' property corresponds to the 'parallel displacement' property that H_t remains fixed in direction according to an observer moving according to X within Σ. The curve X is called the *stochastic development* of the curve W in $\mathbb{R}^d$ onto Σ.

(*33.17*) *Example*: BM(S^2). Even though the case in which Σ is a sphere is very special, it is helpful to examine the way in which X may be developed from W in that situation. So suppose that Σ is the unit sphere S^2 in $\mathbb{R}^3$.

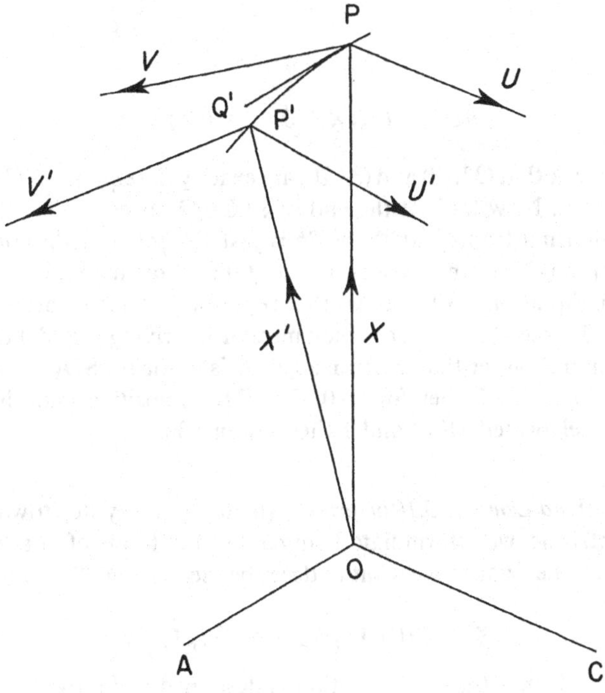

Figure V.3

Figure V.3 is meant to depict the situation. At time t, the plane L of the orthonormal vectors U, V touches the sphere at

$$P = W_t \,(\text{in } L) = X_t \,(\text{in } \Sigma).$$

The orthonormal vectors U and V represent the frame H at P at time t. The W increment is represented in the U, V plane by PQ′, with U, V coordinates $\partial\alpha$, $\partial\beta$, where (α, β) is a BM$(\mathbb{R}^2)$. Pretend that between times t and $t' = t + \partial t$, W moves in the straight line PQ′. Rolling the U, V plane without slipping will wrap PQ′ on to the great circle PP′ in the plane of OPQ′ (O is the centre of the sphere), PP′ having arc-length equal to the length of PQ′. Indeed, all we need to do is to rotate the set-up (X, U, V) to the new set-up (X', U', V') via 'infinitesimal rotation' about the OC axis corresponding to $X \times \partial X$. Then (X', U', V') will be the value of the (X, H) process at time t'. We have:

$$\partial X = (\partial\alpha)\, U + (\partial\beta)\, V \tag{33.18}$$

(correct to first, or should it be 'halfth', order, whatever that means!), and (see § 35 for a detailed study of infinitesimal rotations)

$$\partial(X\ U\ V) = (X\ U\ V) \times (\partial X \times X) = (X\ U\ V)\begin{pmatrix} 0 & -\partial\alpha & -\partial\beta \\ \partial\alpha & 0 & 0 \\ \partial\beta & 0 & 0 \end{pmatrix}, \tag{33.19}$$

so that

$$\partial U = -(\partial\alpha)\, X, \quad \partial V = -(\partial\beta)\, X. \tag{33.20}$$

You should check that (33.18) and (33.20) are exactly the equations (33.13) for the current situation. Now look at the end of § 4.6 of McKean [1].

The orthonormal frame bundle of S^2 is just the set of orthonormal triples $G = (X, U, V)$ in $\mathbb{R}^3$, in other words, the set O(3) of orthogonal 3×3 matrices: $O(S^2) = O(3)$. Equation (33.19) shows that G performs a left-invariant Brownian motion on O(3) – see § 35. The Brownian motion W driving G is 'deficient' in that it has dimension 2 rather than 3. Even so, if G_0 is in the set SO(3) of elements of O(3) of determinant $+1$, then for $t > 0$, G_t will have positive smooth density on SO(3) by the celebrated Hörmander theorem in § 38.

(33.21) Describing Lemma 33.10 in terms of $(\mathbb{R}^d, g)$. As a key step towards making everything intrinsic, we reformulate Lemma 33.10 in terms of our faithful copy $(\mathbb{R}^d, g)$ of Σ. In other words, we want to describe the motion of (Y, E) rather than (X, H), where

$$X = (Y/F(Y)) \in \Sigma, \quad H = (I, J)^T E.$$

You can check that the (Y, E) motion is described as follows:

$$\partial Y = E\, \partial W, \tag{33.22(i)}$$

(33.22(ii)) $$\partial E^i_m + \Gamma^i_{jr} E^r_m \partial Y^j = 0.$$

Of course, (33.22(ii)) is the stochastic analogue of (32.21).

(*33.23*) *Local martingales in manifolds.* Local martingales in manifolds can be used to provide beautifully clean probabilistic characterizations of various classes of maps between manifolds. See, for example, Darling [1].

A process X on our Σ imbedded in $\mathbb{R}^N$ is called a *semimartingale* if each coordinate process X^i is a semimartingale, or (intrinsically) *if $f(X)$ is a semimartingale for every smooth function f from Σ to $\mathbb{R}$.* We say that X is a *local martingale on Σ if the Itô integral*

$$\int P(X)\,dX$$

yields a local martingale in $\mathbb{R}^N$. The good sense of this is obvious.

As an exercise, you should prove that $X = (Y/F(Y))$ *is a local martingale on Σ if and only if Bismut's characterization* (which we shall see in § 35 is intrinsic) *holds:*

(33.24) $$dA^i + \tfrac{1}{2}\Gamma^i_{jk}\, d[M^j, M^k] = 0,$$

where

$$Y^i_t = Y^i_0 + M^i_t + A^i_t$$

is the canonical decomposition of the semimartingale Y.

One can show that X is a local martingale on Σ if and only if X can be obtained as the stochastic development of a local martingale in $\mathbb{R}^d$. This needs a study of SDEs with general driving local martingales, which we have not done.

(*33.25*) **Further exercise.** Prove that BM(Σ) is a local martingale in Σ. *Hint.* Look at formula (31.21) for Δ.

34. Intrinsic theory; normal coordinates; structural equations; diffusions on manifolds; etc.(!). We puzzled long over whether or not to present this 'intrinsic' section before applications to Brownian motions on matrix groups, Brownian motion of ellipses, and so on. *Many of the main results for those applications can be derived directly, as §§ 35, 36 will show.* However, as those paragraphs will also demonstrate, one's understanding of these applications is greatly enhanced by knowing something of the intrinsic theory. So, since this theory follows on naturally from what we have been doing in the last few sections, let's proceed with it now. To give an idea where we are headed, here is a fuller list of some of this section's topics than that provided by the section's title. The first block consists mainly of intrinsic versions of what we have already done extrinsically:

$\mathbb{R}^d$; diffeomorphisms; tangent vectors;

Manifold Σ, $C^\infty(\Sigma)$; diffeomorphisms; tangent vectors;

Vector field; invariance under a diffeomorphism;
Riemannian metric; invariance under a diffeomorphism;
Riemannian connection; parallel displacement; geodesics.

We then go on to new things, including:
Important example: the hyperbolic plane; geodesic polar coordinates;
Normal coordinates; volume element; Laplace–Beltrami operator;
Orthonormal vector fields; structural equations;
Horizontal vector fields on $O(\Sigma)$;
Diffusions on a manifold: global construction.

(*34.1*) $\mathbb{R}^d$; *smooth functions; diffeomorphisms; tangent vectors.* Let A be an open subset of $\mathbb{R}^d$. A function $f: A \to \mathbb{R}$ is called smooth, or C^∞, on A, and we write $f \in C^\infty(A)$, if f possesses continuous partial derivatives of all orders on A. More generally, a function $f: A \to \mathbb{R}^m$ is called smooth if $\pi^k \circ f$ is smooth for every k, where π^k is the map on $\mathbb{R}^m$ which gives the kth coordinate. Thus, $f: A \to B \subseteq \mathbb{R}^m$ is smooth if and only if $c \circ f$ is smooth for every c in $C^\infty(B)$.

Let A_1, A_2 be open subsets of $\mathbb{R}^d$. A map $\psi: A_1 \to A_2$ is called a *diffeomorphism* if:
(i) *ψ is a one–one map of A_1 onto A_2;*
(ii) *both ψ and ψ^{-1} are smooth.*

A tangent vector U at a point p of $\mathbb{R}^d$ is now considered not as a 'school' vector $(U^1, U^2, \dots, U^d)$ but rather as the associated derivative along U:

$$U = U^i D_i \quad \text{evaluated at } p. \tag{34.2}$$

(*34.3*) *A more elegant definition of tangent vector; derivations.* We define $C^\infty(p)$ to consist of real-valued functions f with open domain $\mathscr{D}(f)$ containing p, f being smooth on $\mathscr{D}(f)$. Of course, $f+g$ and fg will have domain $\mathscr{D}(f) \cap \mathscr{D}(g)$. A tangent vector U at p is a derivation of $C^\infty(p)$ at p:

$$U: C^\infty(p) \to \mathbb{R} \text{ is linear,} \tag{34.3(i)}$$

$$U(fg) = f(p)\,Ug + g(p)\,Uf, \quad f, g \in C^\infty(p). \tag{34.3(ii)}$$

It is clear that (34.2) implies (34.3(i)) and (34.3(ii)).

(*34.4*) **Exercise.** Prove that (34.3(i)) and (34.3(ii)) imply that U must have the form (34.2), where U^i is the value of U on any function agreeing in a neighbourhood of p with the function $q \mapsto (q-p)^i$.

Hints. (i) If f is constant on $\mathscr{D}(f)$, then $Uf = 0$.

(ii) Suppose that $f = 0$ in some neighbourhood N of p. Choose g with $\mathscr{D}(g) = \mathscr{D}(f)$ so that $g = 1$ at p and $g = 0$ off N. Then $fg = 0$ on $\mathscr{D}(fg) = \mathscr{D}(f)$, so that

$$0 = U(fg) = 0 \cdot Ug + 1 \cdot Uf = Uf.$$

(iii) Use the idea that if the line from p to q is contained in $\mathscr{D}(f)$, then

$$f(q)-f(p)=(q-p)^i h_i(q), \quad h_i(q)\equiv \int_0^1 (D_i f)(p+t(q-p))\,dt.$$

(*34.5*) *Manifold* Σ; $C^\infty(\Sigma)$; *diffeomorphisms.* The definition of a smooth d-dimensional manifold Σ makes precise the idea that:

(i) each point p of the manifold can be found on some 'chart' (G, φ)

(ii) the charts are compatible, together making an 'atlas' $\mathscr{A}$ for Σ.

The idea of finding p on a chart (G, φ) is that on some open subset G of Σ containing p, there is a 'local-coordinate' map

$$q \mapsto \varphi(q) = (x^1, x^2, \ldots, x^d) \in \mathbb{R}^d$$

giving local coordinates x of q. The compatibility requirement is that if we find p on some other chart $(\tilde{G}, \tilde{\phi})$ with associated local coordinates $(y^1, y^2, \ldots, y^d)$, then, roughly speaking, on $(G \cap \tilde{G})$ the y's are smooth functions of the x's, and conversely. To be precise:

(*34.6*) DEFINITION (smooth d-dimensional manifold). *A set* Σ *is called a* smooth d-dimensional manifold *if* Σ *is a* σ*-compact Hausdorff space supporting an atlas* $\mathscr{A}$ *of charts* $\{(G_\alpha, \varphi_\alpha): \alpha \in \mathscr{A}\}$ *in the following sense:*

(34.6(i)) *each* G_α *is an open subset of* Σ, *and* $\bigcup_\alpha G_\alpha = \Sigma$;

(34.6(ii)) φ_α *is a homeomorphism (continuous one–one onto map with continuous inverse) of* G *onto some open subset of* $\mathbb{R}^d$;

(34.6(iii)) *if* $G_\alpha \cap G_\beta \neq \varnothing$, *then* $\varphi_\beta \circ \varphi_\alpha^{-1}$ *is a diffeomorphism from* $\varphi_\alpha(G_\alpha \cap G_\beta)$ *onto* $\varphi_\beta(G_\alpha \cap G_\beta)$.

It follows easily from our definition that a smooth d-dimensional manifold is locally compact with a countable base for the topology. This would not be true if we only required that Σ be paracompact, as many geometers do.

The charts allow us to describe aspects of local behaviour (smoothness, etc.) in terms of corresponding concepts for $\mathbb{R}^d$. Thus, a function f from Σ to $\mathbb{R}$ is called smooth, and we write $f \in C^\infty(\Sigma)$, if and only if for each chart $(G_\alpha, \varphi_\alpha)$, $f \circ \varphi_\alpha^{-1}$ is smooth from the subset $\varphi_\alpha(G_\alpha)$ of $\mathbb{R}^d$ to $\mathbb{R}$. If Σ_1 and Σ_2 are smooth manifolds (possibly of different dimension), and $f: \Sigma_1 \to \Sigma_2$, then f is called smooth if and only if $c \circ f$ is smooth for every c in $C^\infty(\Sigma_2)$. A *diffeomorphism* between manifolds is a smooth one–one onto map with smooth inverse.

Exercise. Show that a smooth map must be continuous, and explain why two diffeomorphic manifolds must have the same dimension. (Though (34.8) below refers to a different situation, it should tell you the answer.)

(*34.7*) *More on local coordinates.* Suppose that (G, φ) and $(\tilde{G}, \tilde{\varphi})$ are two charts, and that for $q \in G \cap \tilde{G}$, we write local coordinates as

$$\varphi(q) = (x^1, x^2, \ldots, x^d), \quad \tilde{\varphi}(q) = (y^1, y^2, \ldots, y^d).$$

Let $f \in C^\infty(\Sigma)$. By abuse of notation, we shall write

$$D_{x(i)} f \quad \text{or} \quad \partial f / \partial x^i$$

when we should really write $(\partial/\partial x^i)(f \circ \varphi^{-1})$. We shall write

$$D_{x(i)}^{y(j)} \quad \text{or} \quad \partial y^j / \partial x^i$$

for what should be the impossibly clumsy

$$(\partial/\partial x^i)(\pi^j \circ \tilde{\varphi} \circ \varphi^{-1})$$

where π^j is the 'jth coordinate' map on $\mathbb{R}^d$. Note that the usual chain-rule argument gives

$$D_{y(r)}^{x(j)} D_{x(i)}^{y(r)} = \delta_i^j,$$

so that if D_x^y denotes the Jacobian matrix with (r, i)th entry $D_{x(i)}^{y(r)}$, then

$$D_x^y D_y^x = I. \tag{34.8}$$

(*34.9*) *Maximizing the atlas.* We wish to have the greatest possible flexibility in choosing local coordinates. The neat way to do this is to extend the atlas $\mathscr{A}$ to an atlas $\mathscr{A}^*$ containing every chart (G, φ) such that:

(i) G is open in Σ and $\varphi: G \to \varphi(G)$ is one–one onto an open subset of $\mathbb{R}^d$;

(ii) if $\alpha \in \mathscr{A}$ and $G_\alpha \cap G$ is non-empty, then $\varphi \circ \varphi_\alpha^{-1}$ is a diffeomorphism from $\varphi_\alpha(G_\alpha \cap G)$ to $\varphi(G_\alpha \cap G)$.

You can check that the thus-extended atlas has property (34.6(iii)). We may as well assume that $\mathscr{A}$ is already maximal: $\mathscr{A} = \mathscr{A}^*$.

(*34.10*) *Tangent vector at p; $T_p\Sigma$.* For $p \in \Sigma$, we can define $C(p)$ and a tangent vector U as a derivation at p in exact analogy with (34.3). Moreover, by using charts to 'transfer to $\mathbb{R}^d$', we can obviously conclude that in local coordinates,

$$U = U^{x(i)} D_{x(i)} \quad \text{evaluated at } p, \tag{34.11}$$

where $U^{x(i)}$ are the 'coordinates' of U associated with the coordinate system of x's.

What happens if we change coordinates to another system of 'y-coordinates' at p? By the chain rule, we have, for f smooth near p,

$$D_{x(i)} f = D_{x(i)}^{y(r)} D_{y(r)} f,$$

and we see that the coordinates of U transform as follows:

$$U = U^{y(r)} D_{y(r)}, \tag{34.12}$$

where

(34.13) $$U^{y(r)} = D^{y(r)}_{x(i)} U^{x(i)}.$$

This so-called *transformation rule for contravariant* 1-*tensors* is what guarantees the consistency of expressions (34.11) and (34.12) for the tangent vector U at p.

It is obvious from the derivation formulation (or from (34.11) that the tangent vectors at p form a vector space written $T_p\Sigma$. It is clear from (34.11) that this space is of dimension d.

(*34.14*) *Derivative of a map; velocity.* Suppose that $\Phi: \Sigma \to \tilde{\Sigma}$ is a smooth map from one manifold to another. The derivative of Φ at a point p of Σ is the linear map

$$U \mapsto U^{\Phi} \text{ of } T_p\Sigma \text{ to } T_{\Phi(p)}\tilde{\Sigma}$$

defined as follows: for $\tilde{f} \in C^{\infty}(\tilde{\Sigma})$,

(34.15) $$U^{\Phi}\tilde{f} = U(\tilde{f} \circ \Phi).$$

Suppose that $t \mapsto X(t)$ is a smooth function from $\mathbb{R}$ to Σ, $X(t)$ denoting the position of a particle at time t. Then the velocity of the particle at time t is the tangent vector

$$\dot{X} \equiv (d/dt)^X \text{ to } \Sigma \text{ at } X(t).$$

Thus,

(34.16) $$\dot{X}f = (d/dt)(f \circ X(t)).$$

(*34.17*) *Inner product on* $T_p\Sigma$. Stick with our fixed p on Σ with associated tangent space $T_p\Sigma$. Consider an inner product $g(\bullet, \bullet)$ on $T_p\Sigma$. In local x-coordinates, this will have coordinate expression

(34.18) $$g(U, V) = g_{x(i),\, x(j)} U^{x(i)} V^{x(j)}, \quad U, V \in T_p\Sigma,$$

where $g_{\cdot,\cdot}$ is a positive-definite symmetric $d \times d$ matrix. Rule (34.13) forces the following rule for coordinate changes:

(34.19) $$g_{y(r),\, y(s)} = D^{x(i)}_{y(r)} D^{x(j)}_{y(s)} g_{x(i),\, x(j)},$$

the transformation rule for covariant 2-*tensors*. By a convention with which we are already familiar, the inverse of $(g_{\cdot,\cdot})$ is written $(g^{\cdot,\cdot})$. It follows from (34.19) and (34.8) that

(34.20) $$g^{y(r),\, y(s)} = \mathrm{D}^{y(r)}_{x(i)} \mathrm{D}^{y(s)}_{x(j)} g^{x(i),\, x(j)},$$

so that $g^{\cdot,\cdot}$ *is a contravariant* 2-*tensor.*

(*34.21*) *Vector field.* Up until now in this section, we have been concerned with the tangent vectors at a single point p of Σ. For simplicity, we have denoted such a vector by U, but it would have been more proper to use a notation such as $U(p)$ which indicates the 'base point' of the vector.

We are now going to use the symbol U to signify a *vector field, that is, a smooth assignment*

$$p \mapsto U(p)$$

of a tangent vector at p to each point p of Σ. The smoothness property signifies that for $f \in C^\infty(\Sigma)$, the function Uf defined by

$$(Uf)(p) = U(p)f$$

is smooth. Thus, *we can regard the vector field U as a smooth first-order differential operator: U is a linear map*

$$U: C^\infty(\Sigma) \to C^\infty(\Sigma).$$

such that the derivation property

$$U(fg) = fUg + gUf$$

holds. In a chart, we shall have

$$U = U^{x(i)} D_{x(i)},$$

where $x \mapsto U^{x(i)}$ is now a smooth function. Note that this 'coordinate' definition of 'smooth' is, in fact, coordinate-free because of the transformation rule (34.13).

(*34.22*) *Invariance of a vector field under a diffeomorphism* ψ. This concept will play a key role in § 35 on Brownian motions on Lie groups.

Suppose that $\psi: \Sigma \to \Sigma$ is a diffeomorphism of Σ. A vector field U on Σ is called *invariant under* ψ if $U^\psi = U$ in the obvious sense:

$$U(p)(f \circ \psi) = U(\psi(p))f, \quad f \in C^\infty(\Sigma).$$

See (34.15).

(*34.23*) *Lie bracket* $[U, V]$ *of vector fields.* If we regard vector fields U and V as first-order differential operators, we can form their Lie bracket:

(34.24) $$[U, V] \equiv UV - VU.$$

You can check that $[U, V]$ *has the derivation property, and so is a vector field.* In local coordinates,

$$[U, V] = W^i D_i,$$

where

(34.25) $$W^i = U(V^i) - V(U^i).$$

(*34.26*) *Riemannian metric.* A Riemannian metric stands in the same relation to an inner product on $T_p\Sigma$ as does a vector field to a vector at p. In other words, *a Riemannian metric g is a smooth assignment*

$$p \mapsto \textit{inner product on } T_p\Sigma.$$

We write $g(U(p), V(p))$ for the inner product 'at p' of vectors $U(p)$ and $V(p)$ in $T_p\Sigma$. In particular, if U and V are vector fields, then $g(U, V)$ is a function on Σ, and the smoothness requirement on g is that $g(U, V) \in C^\infty(\Sigma)$ for all vector fields U and V on Σ. Another way to say this is that in a chart, each function

$$x \mapsto g_{x(i),\, x(j)}$$

(see (34.18)) is smooth; this can be seen to be coordinate-free because of (34.19).

(*34.27*) *Invariance of a Riemannian metric under a diffeomorphism.* Suppose that $\psi: \Sigma \to \Sigma$ is a diffeomorphism. We know that a vector field U induces a vector field U^ψ on Σ, via the obvious extension of (34.15) discussed at (34.22). The Riemannian metric g is called *invariant under* ψ if, for every p,

$$g(U(p), V(p)) = g(U^\psi(\psi(p)), V^\psi(\psi(p))).$$

(*34.28*) *Connection; covariant derivative.* One of the many ways in which geometers like to think of a *connection* ∇ *on* Σ is as a special kind of map

$$(U, W) \mapsto \nabla_U W$$

which, given two vector fields U and W on Σ, produces a new vector field, $\nabla_U W$, the *covariant derivative* of W along U. *The properties which characterize a connection are that if $f, g \in C^\infty(\Sigma)$ and U, V, W are vector fields on Σ, then:*

(34.29(i)) $$\nabla_{fU+gV} W = f\nabla_U W + g\nabla_V W,$$

(34.29(ii)) $$\nabla_U(fW) = (Uf)W + f\nabla_U W.$$

Suppose that for the moment we fix a local-coordinate system and write

$$U = U^j D_j, \quad W = W^k D_k,$$

where we use the most convenient of the notations

$$D_k \text{ or } D(k) \text{ for } D_{x(k)} = \partial/\partial x^k.$$

Suppose also that (locally)

(34.30) $$\nabla_{D(j)}(D_k) = \Gamma^i_{jk} D_i.$$

Then things begin to look familiar because the rules (34.29) yield:

(34.31) $$(\nabla_U W)^i = U(W^i) + \Gamma^i_{jk} U^j W^k.$$

Interpretation. Suppose that $t \mapsto X(t)$ is a smooth map from some interval (a, b) of

$\mathbb{R}$ into Σ forming an *integral curve* of the vector field U in the sense that

$$\dot{X}(t)=U(X(t)),$$

the velocity $\dot{X}(t)$ at time t being defined as at (34.16). An observer living in the *manifold with connection* (Σ, ∇) and moving according to X will measure the rate of change of $W(t)$ as the value of the covariant derivative $\nabla_U W$ at the point $X(t)$. *Without the special apparatus of a connection, no concept of such a rate of change exists,* because at different times, $W(t)$ belongs to different vector spaces $T_{X(t)}\Sigma$ amongst which no comparison is possible. (One must remember that in many classical contexts, one is making implicit use of the Euclidean connection.)

(*34.32*) *Transformation rule for Christoffel symbols.* Using (34.8) and (34.13), we find that the rules (34.29) forces the result

$$\Gamma^{y(q)}_{y(r),\,y(s)} = D^{x(j)}_{y(r)}\, D^{x(k)}_{y(s)}\, D^{y(q)}_{x(i)}\, \Gamma^{x(i)}_{x(j),\,x(k)} + D^{x(k)}_{y(r),\,y(s)}\, D^{y(q)}_{x(k)}, \tag{34.33}$$

where

$$D^{x(k)}_{y(r),\,y(s)} = \partial^2 x(k)/\partial y(r)\partial y(s).$$

The appearance of the final term in (34.33) shows that the *Christoffel symbols* $\Gamma^{\cdot}_{\cdot\cdot}$ which represent ∇ in local coordinates do not transform as the components of a tensor. This reflects the fact that if U, W and $\tilde{W}$ are vector fields, and $W(p)=\tilde{W}(p)$ at the point p, then

$$\nabla_U W \text{ and } \nabla_U \tilde{W} \text{ can differ at } p.$$

(*34.34*) *Torsion tensor.* Certain tensors—notably, the *torsion and curvature tensors*—are however naturally derived from ∇. Thus, consider the *torsion tensor*:

$$T(U, W) \equiv \nabla_U W - \nabla_W U - [U, W]. \tag{34.35}$$

We find that, in local coordinates,

$$T(U, W) = (T^i_{jk}\, U^j\, W^k)\, D_i,$$

where

$$T^i_{jk} = \Gamma^i_{jk} - \Gamma^i_{kj},$$

and T does transform as a tensor. The connection ∇ is called *torsion-free* if $T(U, W)=0$ for all vector fields U and W, equivalently, if

$$\Gamma^i_{jk} = \Gamma^i_{kj} \textit{ for all local-coordinate systems.} \tag{34.36}$$

(*34.37*) *Parallel displacement for* (Σ, ∇). If we stick to local-coordinate formulation, this is old-familiar stuff.

Suppose that $t \mapsto X(t)$ is a smooth curve in Σ and that, for each t, $W(t)$ is a tangent vector at $X(t)$. We assume that in some (then in any) local-coordinate system,

$$W(t)=W^i(t)D_i,$$

where the functions $W^i(\cdot)$ are smooth. (In the language of (34.76) below, $(X(\cdot), W(\cdot))$ is a smooth curve in the *tangent bundle* $T\Sigma$ of Σ.)

Then we say that (for the connection ∇) *W moves by parallel displacement along X* if

$$\text{(34.38)} \qquad \text{‘}\nabla_{\dot{X}} W=0\text{’}.$$

We get around the difficulty caused by the fact that W will most definitely not in general arise (as $W(t)=W(X(t))$) from a vector field by expressing (34.38) in local coordinates:

$$\text{(34.39)} \qquad \dot{W}^i+\Gamma^i_{jk}U^j W^k=0, \qquad U=\dot{X}.$$

On using transformation rules (34.13) and (34.33), we find that equation (34.39) corresponds to a coordinate-free concept.

(*34.40*) *Riemannian connection.* We continue on very familiar ground.

Let (Σ, g) be a Riemannian manifold, so that g is a Riemannian metric on Σ. Then a connection ∇ on Σ is called *Riemannian* if

(34.41(i)) ∇ *is torsion-free,*
(34.41(ii)) *parallel displacement under ∇ preserves inner products.*

(*34.42*) THEOREM (Fundamental theorem of Riemannian geometry). *Let (Σ, g) be a smooth d-dimensional Riemannian manifold. Then there exists precisely one Riemannian connection ∇ on (Σ, g). Moreover, ∇ satisfies*

$$\text{(34.43)} \qquad Ug(V, W)=g(\nabla_U V, W)+g(V, \nabla_U W)$$

for all vector fields U, V, W.

Proof. We proved the uniqueness of ∇ by local-coordinate methods at (32.16), and the argument given there applies without modification to the present situation.

You should check that the elegant—and very important—rule (34.43) corresponds exactly to (32.18(i)). There are sophisticated ways of looking at things which allow one to move directly to (34.43) without going into coordinates. See §§ I.4–I.9 of Helgason [1]. □

(*34.44*) *Geodesics.* What we said at (32.26) completely covers the notion of a geodesic for our present context. A curve X in Σ is a *geodesic* if and only if

$$\text{‘}\nabla_U U=0\text{’}, \quad U=\dot{X},$$

to which we can give a local-coordinate expression:

(34.45) $$\ddot{x}^i + \Gamma^{x(i)}_{x(j),\, x(k)} \dot{x}^j \dot{x}^k = 0,$$

if $(x^k(\cdot)\colon 1 \leqslant k \leqslant d)$ are local coordinates for $X(\cdot)$.

....

That completes our 'intrinsifying' operation on concepts for which we had earlier presented extrinsic versions. We now take the theory further, but begin with an example.

(*34.46*) *Important example: hyperbolic plane; geodesic polar coordinates.* Consider

$$\Sigma = \{(x, y)\colon x \in \mathbb{R},\ y > 0\}.$$

We therefore write (x, y) rather than (x^1, x^2) for this particular choice of local (in fact, global) coordinates for Σ. We equip Σ with the Lobachevski–Poincaré metric described in terms of (x, y) coordinates as follows:

(34.47) $$(ds)^2 = y^{-2}\{(dx)^2 + (dy)^2\}, \quad g_{..} = y^{-2} I.$$

Let us expand a little on (34.47) to be sure that everything is clear. Suppose that we have a smooth curve:

$$X\colon [0, 1] \to \Sigma$$
$$t \to X(t) \text{ with coordinates } x(t), y(t).$$

Then, arc-length s along the curve is defined by

$$s(t) = \int_0^t \|\dot{X}(u)\|_g \, du,$$

where $\dot{X}(t)$ denotes the tangent vector (see (34.16))

$$\dot{X}(t) = \dot{x}(t) D_x + \dot{y}(t) D_y$$

to the curve at $X(t)$. Next,

$$\dot{s}(t)^2 = \|\dot{X}(t)\|_g^2 = g(\dot{X}(t), \dot{X}(t)) = (\dot{x}(t)\, \dot{y}(t)) g_{..} (\dot{x}(t)\, \dot{y}(t))^T,$$

where (at $X(t)$)

$$g_{..} = \begin{pmatrix} g(D_x, D_x) & g(D_x, D_y) \\ g(D_y, D_x) & g(D_x, D_x) \end{pmatrix} = y(t)^{-2} I,$$

checking out with

$$\dot{s}(t)^2 = y(t)^{-2}\{\dot{x}(t)^2 + \dot{y}(t)^2\}.$$

From formula (32.16), we have

$$\Gamma^x_{\cdot\cdot} = y^{-1}\begin{pmatrix} 0 & -1 \\ -1 & 0 \end{pmatrix}, \quad \Gamma^y_{\cdot\cdot} = y^{-1}\begin{pmatrix} 1 & 0 \\ 0 & -1 \end{pmatrix}. \tag{34.48}$$

Here is the calculation for one of the entries in (34.48):

$$\Gamma^x_{xy} = \tfrac{1}{2}g^{xx}(D_x g_{xy} + D_y g_{xx} - D_x g_{xy}) + \tfrac{1}{2}g^{xy}(\ldots),$$

and since $g^{xy} = 0 = g_{xy}$, $g^{xx} = y^2$, and $g_{xx} = y^{-2}$, we have

$$\Gamma^x_{xy} = \tfrac{1}{2}y^2 D_y(y^{-2}) = -y^{-1}.$$

We consider geodesics with

$$(x(0), y(0)) = (0, 1) \quad \text{and} \quad (\dot{x}(0), \dot{y}(0)) = (\cos\alpha, \sin\alpha).$$

Because of (34.48), the geodesic equations (34.45) take the form:

$$\ddot{x} - 2y^{-1}\dot{x}\dot{y} = 0, \tag{34.49}$$

$$\ddot{y} + y^{-1}(\dot{x}^2 - \dot{y}^2) = 0. \tag{34.50}$$

We find from (34.49) that

$$\dot{x} = y^2\cos\alpha, \tag{34.51}$$

and substitution of this result into (34.50) gives

$$d(\dot{y}/y) + (\dot{x}\cos\alpha)dt = 0,$$

so that

$$\dot{y}y^{-1} + x\cos\alpha = \sin\alpha. \tag{34.52}$$

Next, (34.51) and (34.52) give

$$y\dot{y} + x\dot{x} = \dot{x}\tan\alpha,$$

whence

$$(x - \tan\alpha)^2 + y^2 = \sec^2\alpha.$$

Thus, *the tracks of geodesics through* (0, 1) *appear to those who think of* Σ *as the conventional half-plane as semicircles with centres on the x-axis*. (If $\alpha = \pi/2$, then the track follows the line $x = 0$.)

It is an easy exercise to complete the solution to find that

$$y(t) = \{\cosh t - \sinh t \sin\alpha\}^{-1} > 0,$$

$$x(t) = y(t)\sinh t\cos\alpha.$$

Now $(x(t), y(t))$ is the point at distance t (in the L-P metric) along the geodesic passing through (0, 1) at an angle α. (Think about the fact that 'at angle α' is correct for the L-P metric.) The pair (r, θ), which we now write instead of (t, α), gives local *geodesic polar coordinates* for $\Sigma\backslash\{(0, 1)\}$:

(34.53(i)) $$x(r, \theta) = y(r, \theta)\sinh r \cos\theta,$$

(34.53(ii)) $$y(r, \theta) = \{\cosh r - \sinh r \sin\theta\}^{-1}.$$

On working out the partial derivatives, we find that

(34.54) $$(ds)^2 = (dr)^2 + (\sinh r)^2(d\theta)^2,$$

a formula which will make a striking reappearance in § 36.

What we have done above is to illustrate how equations (34.53) can be derived by bare-hands calculation using the Γ symbols. By using some of the remarkable special features of the hyperbolic plane (Σ, g), formulae (34.53) can be proved very quickly. See § 36.

(*34.55*) *Exponential map on* $T_p\Sigma$; *normal coordinates.* We return to the study of a general smooth d-dimensional Riemannian manifold Σ with Riemannian connection ∇. Normal coordinates play quite a big part in certain aspects of the theory. Here, we use them to provide further motivation for the definition of the Laplace–Beltrami operator, and hope that by so doing we help consolidate your intuition about the subject generally. You will see that normal coordinates are a natural extension of the geodesic polar coordinates which we used in studying the hyperbolic plane.

Let p be a fixed point of our manifold, and let $v \in T_p\Sigma$. We can find a unique geodesic $X(\cdot)$ (defined up to explosion time) with

(34.56) $$X(0) = p, \quad \dot{X}(0) = v.$$

For v close to 0 in $T_p\Sigma$, $\{X(t): t \leqslant 1\}$ will be defined, and we then define the exponential map (on a neighbourhood of 0 in $T_p\Sigma$) via

(34.57) $$\mathrm{Exp}(v) = X(1).$$

Take some local-coordinate system of x's near p with associated functions g_{ij} and Γ^i_{jk}. Let $(e_1, e_2, \ldots, e_d)$ be an orthonormal basis for $T_p\Sigma$, so that

(34.58) $$e^j_m g_{jk} e^k_r = \delta_{mr}, \quad g^{jk} = \sum_m e^j_m e^k_m.$$

Consider the map

$$z \mapsto \mathrm{Exp}(v), \quad v \equiv z^m e_m,$$

of a neighbourhood of 0 in $\mathbb{R}^d$ to Σ. For small h, we shall have, utilizing the geodesic property of $X(\cdot)$,

(34.59) $$\begin{aligned} X^i(h) = \mathrm{Exp}(hv)^i &= X^i(0) + h\dot{X}^i(0) + \tfrac{1}{2}h^2\ddot{X}^i(0) + o(h^2) \\ &= X^i(0) + hv^i - \tfrac{1}{2}h^2\Gamma^i_{jk}v^j v^k + o(h^2). \end{aligned}$$

Thus, at the point p,

$$\partial x^i/\partial z^m = e^i_m, \tag{34.60}$$

so that the Jacobian D^x_z is non-singular. By the inverse-function theorem, we can therefore conclude that

$$z \mapsto \operatorname{Exp}(z^m e_m) \tag{34.61}$$

is a diffeomorphism of some neighbourhood of 0 *in* $\mathbb{R}^d$ *onto a neighbourhood of* p *in* Σ. In other words, *the values* $(z^1, z^2, \ldots, z^d)$ *form a set of local coordinates for a neighbourhood of* p *in* Σ. However, these are very special coordinates in view of (34.61): they are called *normal coordinates*, and *give as good a 'flattening' of* Σ *onto* $T_p\Sigma$ *as is possible. This 'flattening' is reflected in the fact that*

$$\Gamma^{z(i)}_{z(j),z(k)} = 0 \quad \text{at } p. \tag{34.62}$$

Equation (34.62) follows because $t \mapsto tz$ is a geodesic for every z.

Because taking normal coordinates at p corresponds as closely as possible to taking Cartesian coordinates for $\mathbb{R}^d$, it is very plausible that *one can obtain the correct versions of concepts for* Σ *by translating classical concepts for* $\mathbb{R}^d$ *via the use of normal coordinates.* Let us see how this works for two important concepts: *volume element* and the *Laplace–Beltrami operator.*

(*34.63*) *Volume element.* For obvious reasons we shall momentarily write n rather than d for the dimension of Σ. If $(z^1, z^2, \ldots, z^n)$ are normal coordinates near p, it is reasonable to regard

$$dz^1\, dz^2 \ldots dz^n$$

as the 'element of volume' at p. From (34.58), (34.60) and the Jacobian formula, we have

$$\begin{aligned} dx^1\, dx^2 \ldots dx^n &= \det(D^x_z)\, dz^1\, dz^2 \ldots dz^n \\ &= \det(g)^{-1/2}\, dz^1\, dz^2 \ldots dz^n, \end{aligned}$$

where g refers to the x-coordinate system. Thus, *the volume element in an arbitrary coordinate system should be*

$$\det(g)^{1/2}\, dx^1\, dx^2 \ldots dx^n. \tag{34.64}$$

Expression (34.64) is coordinate-free, and is of course correct.

(*34.65*) *Laplace–Beltrami operator.* For $f \in C^\infty(\Sigma)$, it seems very natural to seek to define the Laplace–Beltrami operator by

$$(\Delta f)(p) = \left(\sum \partial^2/(\partial z^m)^2 f\right)(p), \tag{34.66}$$

where $(z^1, z^2, \ldots, z^d)$ (restoring d!) are normal coordinates at p. We have to prove that such a definition does not depend on the choice of normal coordinates, and that the resulting definition is indeed appropriate. Now,

$$D_{z(m),z(m)}f = D^{x(i)}_{z(m),z(m)} D_{x(i)}f + D^{x(j)}_{z(m)} D^{x(k)}_{z(m)} D_{x(j),x(k)}f,$$

whence, in view of (34.60), idea (34.66) leads to the formula

(34.67) $$\Delta = g^{jk}(D_{jk} - \Gamma^{i}_{jk} D_i),$$

which, as we have seen, has other good reasons to qualify as the Laplace–Beltrami operator. For any coordinate system normal at p, we have $g_{..} = I$ and $\Gamma^{\cdot}_{..} = 0$, so the 'well-defined' property holds. The 'div grad' and other formulations of Δ—all of course in agreement—may be found in the textbooks.

(*34.68*) *Structural equations.* Cartan's structural equations come fully into their own in the theory of Lie groups, but have importance for general manifolds too.

Suppose that vector fields $W_1, W_2, \ldots, W_d$ are orthonormal at every point of some region C of Σ. We can find such fields in a neighbourhood of any point p by picking any vector fields orthonormal at p itself (for example, $\partial/\partial z^k$ $(1 \leqslant k \leqslant d)$, where the z's are normal coordinates at p), and then using Gram–Schmidt orthogonalization near p.

We wish to define the connection ∇ within C relative to the 'basis' of W's by *finding functions* K^{i}_{jk} *on* C *such that*

(34.69) $$\nabla_j W_k \equiv \nabla_{W(j)} W_k = K^{i}_{jk} W_i.$$

We calculate the functions $K^{\cdot}_{..}$ in terms of functions c^{i}_{jk} described by the Lie-bracket formulae:

(34.70) $$[W_j, W_k] = c^{i}_{jk} W_i.$$

The idea is to use the fundamental property (34.43) of the Riemannian connection via the equation

(34.71) $$0 = W_i g(W_j, W_k) = g(\nabla_i W_j, W_k) + g(W_j, \nabla_i W_k).$$

Since the Riemannian connection is torsion-free, we have, from (34.35),

$$\nabla_i W_k - \nabla_k W_i = [W_i, W_k].$$

Thus, (34.71) may be rewritten:

(34.72) $$g(\nabla_i W_j, W_k) + g(\nabla_k W_i, W_j) = g(W_j, [W_k, W_i]),$$

or

(34.73(i)) $$K^{k}_{ij} + K^{j}_{ki} = c^{j}_{ki}.$$

By cyclic permutation,

(34.73(ii)) $$K^{i}_{jk} + K^{k}_{ij} = c^{k}_{ij},$$

(34.73(iii)) $$K^{j}_{ki} + K^{i}_{jk} = c^{i}_{jk}.$$

Add the second and third of these equations, and subtract the first from the sum, to obtain:

$$2K^i_{jk} = c^i_{jk} + c^k_{ij} + c^j_{ik}. \tag{34.74}$$

Obviously, the calculation just performed is analogous to that used in the proof of Theorem 32.15.

(*34.75*) *A formula for the Laplace–Beltrami operator.* We continue to assume that $W_1, W_2, \ldots, W_d$ are orthonormal vector fields in some region C. We now prove that, on C,

$$\Delta = \sum_m (W_m)^2 - \left(\sum_m K^r_{mm}\right) W_r. \tag{34.76}$$

Proof of (*34.76*). We have:

$$\nabla_{W(j)} W_k = K^i_{jk} W_i, \quad \nabla_{D(j)} D_k = \Gamma^i_{jk} D_i,$$

the second equation referring to a local-coordinate system. Hence, since $W_j = W^q_j D_q$, etc., we have

$$W^q_j \nabla_{D(q)}(W^r_k D_r) = K^i_{jk} W^s_i D_s,$$

so that

$$W^q_j (D_q W^r_k) D_r + W^q_j W^r_k \Gamma^s_{qr} D_s = K^i_{jk} W^s_i D_s. \tag{34.77}$$

However, we know that

$$\sum_m W^q_m W^r_m = g^{qr}, \quad \text{and} \quad \Delta = g^{qr}(D_{qr} - \Gamma^s_{qr} D_s).$$

Moreover,

$$\sum_m (W_m)^2 = \sum_m W^q_m W^r_m D_{qr} + \sum_m W^q_m (D_q W^r_m) D_r.$$

So, equation (34.76) follows immediately on putting $j = k = m$ in equation (34.77) and summing over m. □

(*34.78*) *Orientation. The main point of the next subsections is to study* $\mathrm{BM}^{\mathrm{hor}}$ ($O(\Sigma)$) *and the marvellous way in which it can be used to construct an arbitrary 'non-singular' diffusion on a manifold by constructing a Riemannian metric from the diffusion operator.*

(*34.79*) *Tangent bundle,* $T\Sigma$; *horizontal vector fields on* $T\Sigma$. The *tangent bundle* $T\Sigma$ of Σ is obtained as the (*disjoint*) union

$$T\Sigma = \bigcup_p T_p\Sigma,$$

a point of $T\Sigma$ being a *pair* (p, v), where v is a tangent vector at p. We can construct a chart for $T\Sigma$ from a chart (G, φ) for Σ by using the $2d$-tuple

$$(x^1, x^2, \ldots, x^d, v^1, v^2, \ldots, v^d)$$

as local coordinates for (p, v), where

$$x = \varphi(p), \quad v = v^i \partial/\partial x^i \quad \text{at } p.$$

A curve $(X(t), V(t))$ in $T\Sigma$ with local coordinates $(x^i(t), v^i(t))$ will have tangent vector at $(X(t), V(t))$ represented by

$$\dot{x}^i(t) D_{x(i)} + \dot{v}^i(t) D_{v(i)}.$$

Now, if (X, V) moves *horizontally* on $T\Sigma$, that is, V moves by parallel displacement along X, then

$$\dot{v}^i(t) + \Gamma^i_{jk} \dot{x}^j(t) v^k(t) = 0.$$

This leads to the idea that a tangent vector

(34.80(i)) $$u^i D_{x(i)} + \lambda^i D_{v(i)} \text{ to } T\Sigma \text{ at } (p, v)$$

is called *horizontal* if

(34.80(ii)) $$\lambda^i + \Gamma^i_{jk} u^j v^k = 0.$$

The vector

(34.81) $$u^i D_{x(i)} - \Gamma^i_{jk} u^j v^k D_{v(i)}$$

is called the *horizontal lift* to $(p, v) \in T\Sigma$ of the vector $u^i D_{x(i)}$ in $T_p\Sigma$.

These ideas provide a second way of looking at a connection: as something which for each point (p, v) in $T\Sigma$ yields a direct-sum decomposition of the $2d$-dimensional tangent space $T_{(p,v)}T\Sigma$ into d-dimensional horizontal and vertical subspaces at p, and which does this in a way which varies 'smoothly' with (p, v).

(*34.82*) *Horizontal fields on $M(d, \Sigma)$*. The bundle $T\Sigma$ is easily studied because it is a *vector bundle*: the '*fibre over p*', $T_p\Sigma$, is a vector space. The problem with the orthonormal frame bundle $O(\Sigma)$ is that the fibre over p is the set of orthonormal frames at p, which is a 'round' object considerably more difficult to study. We shall use the natural imbedding of $O(\Sigma)$ in a certain vector bundle $M(d, \Sigma)$. We define

$$M_p(d) \equiv T_p\Sigma \times T_p\Sigma \times \ldots \times T_p\Sigma \quad (d\text{-fold product}),$$

and

$$M(d, \Sigma) \equiv \bigcup_p M_p(d).$$

Thus, a typical point of $M(d, \Sigma)$ is of the form

$$(p, v_1, v_2, \ldots, v_d),$$

where $p \in \Sigma$ and $v_1, v_2, \ldots, v_d$ are tangent vectors at p. You will immediately understand that a tangent vector

$$u^i D_{x(i)} + \sum_i \lambda^i_m \partial/\partial v^i_m \text{ to } M(d, \Sigma) \text{ at } (p, v_1, v_2, \ldots, v_d)$$

is called *horizontal* if

$$\lambda^i_m + \Gamma^i_{jk} u^j v^k_m = 0, \tag{34.83}$$

and that

$$u^i D_{x(i)} - \sum_i \Gamma^i_{jk} u^j v^k_m \partial/\partial v^i_m \tag{34.84}$$

is the *horizontal lift* to $TM(d, \Sigma)$ at $(p, v_1, v_2, \ldots, v_d)$ of the tangent vector $u^i D_{x(i)}$ to Σ at p.

The notation $M(d, \Sigma)$, suggested by writing $M(d)$ for the set of $d \times d$ matrices, is not standard.

We define the *canonical projection* $\pi: M(d, \Sigma) \to \Sigma$ via

$$\pi(p, v_1, v_2, \ldots, v_d) = p.$$

We further define *canonical vector fields* $H_1, H_2, \ldots, H_d$ on $M(d, \Sigma)$ as follows: the value of H_m at $(p, v_1, v_2, \ldots, v_d)$ is the horizontal lift to $TM(d, \Sigma)$ at $(p, v_1, v_2, \ldots, v_d)$ of the vector v_m at p.

Note that the fields H_m on $M(d, \Sigma)$ are defined in *coordinate-free* fashion.

(*34.85*) *Horizontal Brownian motion with drift on* $\mathrm{O}(\Sigma)$. Let V be a vector field on Σ and let H_V be its horizontal lift to $M(d, \Sigma)$. We wish to construct a diffusion on Σ which is Brownian motion with drift V. This can be described in a global coordinate-free way as follows.

(*34.86*) FUNDAMENTAL THEOREM (Eells, Elworthy). *Let B be a Brownian motion on $\mathbb{R}^d$. Let $X(0) \in \Sigma$, and let*

$$E(0) = (e_1(0), e_2(0), \ldots, e_d(0))$$

be an orthonormal frame for $T_{X(0)}\Sigma$. Then there exists a unique process

$$(X(t), E(t)) = (X(t), e_1(t), e_2(t), \ldots, e_d(t))$$

in $M(d, \Sigma)$, with initial value $(X(0), E(0))$ and defined up to 'explosion time', such that, simultaneously for all f in $C^\infty(M(d, \Sigma))$,

$$\partial f(X(t), E(t)) = (H_m f)(X(t), E(t)) \partial B^m + (H_V f)(X(t), E(t)) \partial t. \tag{34.87}$$

For every t, $E(t)$ is an orthonormal frame at $X(t)$, so that (X, E) lives in $\mathrm{O}(\Sigma)$. The process X is a diffusion on Σ with generator $\frac{1}{2}\Delta + V$.

Sketch of proof. In local coordinates, equation (34.87) takes the form:

(34.88(i)) $$\partial X^i = e^i_m \partial B^m + v^i(X)\partial t,$$

(34.88(ii)) $$\partial e^i_r = -\Gamma^i_{jk}(\partial X^j)e^k_r.$$

Compare equations (33.22). We are well aware that equation (34.88(ii)) signifies (as it must do) that E moves by stochastic parallel displacement along the path X, so that E will remain orthonormal.

Equations (34.88) are nice and familiar. So, for given $(X(0), E(0))$, the coordinate-free equation (34.87) has a pathwise unique solution $(X_\cdot, E_\cdot)$ up until the time of exit of X from a chart around $X(0)$. We say unto you: believe therefore that we can define a pathwise unique solution up to explosion time of X from Σ. Ye of little faith (but good taste) can be given proper instruction at the feet of Elworthy [1].

So (by the usual arguments, you understand) the process $(X_\cdot, E_\cdot)$ is unique in law and is Markovian. Let us see why X itself is Markovian. Let K be a fixed orthogonal matrix. If

$$\tilde{e}_n(t) \equiv K^m_n e_m(t), \quad \text{so that} \quad e_m = \sum_n K^m_n \tilde{e}_n,$$

then

$$\partial X^i = \tilde{e}^i_n \partial \tilde{B}^n + v^i(X)\partial t, \quad \text{where} \quad \tilde{B}^n \equiv \sum_m K^m_n B^m,$$

$$\partial \tilde{e}^i_r = -\Gamma^i_{jk}(\partial X^j)\tilde{e}^k_r.$$

Since $\tilde{B}$ is a BM($\mathbb{R}^d$), it follows that $(X_\cdot, KE_\cdot)$ is identical in law to the solution of (34.87) started from $(X(0), KE(0))$. *This 'invariance of the law of X under the action of $O(n)$ on the fibres of $O(\Sigma)$' shows that X is Markovian.* See § 10.6 of Dynkin [2] for how to make this intuitive idea precise.

As an *exercise*, you can prove that the *Bochner formula*:

(34.89) $$(\tfrac{1}{2}\Delta + V)h = (\tfrac{1}{2}\mathscr{A}^{\text{hor}} + H_V)(h \circ \pi), \quad \mathscr{A}^{\text{hor}} \equiv \sum_m H^2_m,$$

holds for any smooth function h on Σ. (Recall that π is the natural projection of $O(\Sigma)$ to Σ.) The operator $\mathscr{A}^{\text{hor}}$ is called *Bochner's horizontal Laplacian.* On taking $f = h \circ \pi$ in (34.87). we see that the law of X solves the local-martingale problem for $\tfrac{1}{2}\Delta + V$:

(34.90) $$h(X_t) - h(X_0) - \int_0^t (\tfrac{1}{2}\Delta + V)h(X_s)\, ds \textit{ is a local martingale.}$$

Thus, the generator of X extends $\tfrac{1}{2}\Delta + V$ on $C^\infty_K(\Sigma)$.

We can show, using (34.87), that *the solution of the local-martingale problem associated with (34.90) is unique* by 'lifting to $O(\Sigma)$'. For let X (or, more properly, its law) solve the local-martingale problem for Σ associated with (34.90), and define a frame motion via (34.88(ii)). Then the process $(X_\cdot, E_\cdot)$ (or, more properly, its law) solves the martingale problem for $O(\Sigma)$ associated with the operator $\frac{1}{2}\mathscr{A}^{\text{hor}}+H_V$. We know this solution to be unique, and because of the invariance of the law of the X-component of $(X_\cdot, E_\cdot)$ under the action of $O(n)$ on the fibres of $O(\Sigma)$, we obtain the required uniqueness assertion for (34.90). □

(*34.91*) *Why go up to the frame bundle?* We have seen that $O(\Sigma)$ carries global canonical vector fields H_m such that Bochner's horizontal Laplacian $\mathscr{A}^{\text{hor}}$ may be written $\sum_m H_m^2$.

If it were possible to find vector fields $W_0, W_1, \ldots, W_d$ on Σ such that

(34.92) $$\Delta=\sum_m W_m^2+W_0,$$

then, of course, we would be able to construct a $BM(\Sigma)$ process directly. We know from (34.76) that we can achieve (34.92) locally. But, to be able to do so globally, we would have to be able to find vector fields $W_0, W_1, \ldots, W_d$ on Σ none of which ever vanishes. However, for example, by the Hairy Ball Theorem, any (smooth) vector field on the sphere S^2 must vanish somewhere.

By contrast, as we shall see in the next section, S^3 is a Lie group and is therefore 'parallelizable'. For general Σ, the bundle $O(\Sigma)$ is giving us a nice structure with some of the good features of a Lie group.

In this regard, it is interesting to recall that we saw after (33.20) that the orthonormal frame bundle for 'bad' S^2 is the Lie group $O(3)$ (which, as we shall see in § 35, is closely related to S^3). The study of the process $BM^{\text{hor}}(O(S^2))$ is taken up again at the end of § 38.

We have only given the first part of the answer to the question at (34.91). To gain an idea of how $BM^{\text{hor}}(O(\Sigma))$ is *used*, see Malliavin [3], Elworthy [1], and Ikeda and Watanabe [1].

(*34.93*) *The Riemannian structure induced by a non-singular diffusion.* Let Σ be a smooth d-dimensional manifold, and let $\mathscr{G}$ be a smooth second-order elliptic operator on $C^\infty(\Sigma)$. Thus, in local coordinates,

(34.94) $$\mathscr{G}=\tfrac{1}{2}a^{ij}(x)D_{x(i)}D_{x(j)}+b^i(x)D_{x(i)},$$

where, for each x, the $d\times d$ matrix $a(x)$ is symmetric and strictly positive-definite. Check that

(34.95) (a^{ij}) *transforms as a contravariant* 2-*tensor.*

Thus, if we set

(34.96) $$g_{\cdot\cdot}(x)=(a_{\cdot\cdot}(x))^{-1},$$

then $(g_{\cdot\cdot})$ *is a covariant* 2-*tensor*, that is,

(34.97) $g_{\cdot\cdot}$ *defines a Riemannian metric on* Σ.

If Δ denotes the Laplace–Beltrami operator associated with this metric, then

$$\mathscr{G}=\tfrac{1}{2}\Delta+V,$$

where V is a vector field, and we are in the context of Theorem 34.86.

(*34.98*) **Exercise.** Check that the $b^{\cdot}(x)$ in (34.94) does not transform as a vector field. To be able to talk about 'the first-order part' of a second-order operator, you need a connection—see Meyer [9, 10].

(*34.99*) *Orientation.* We finish this huge section by explaining how *the Nash imbedding theorem allows us to regard the general smooth d-dimensional manifold as a submanifold of some* $\mathbb{R}^q$, whereupon the 'extrinsic' techniques of §§ 31 and 32 are no less general than the 'intrinsic' techniques explained in this section.

(*34.100*) DEFINITION (imbedding, submanifold, isometric imbedding). *Let M and N be two smooth manifolds (of possibly different dimension). A smooth map Φ: $M\to N$ is called an* imbedding *if Φ is one–one, and the derivative of Φ is one–one everywhere on M. In this case, we say that $\Phi(M)$ is a* submanifold *of N. If additonally M and N are Riemannian manifolds with metric tensors g and h respectively, then Φ is called an* isometric imbedding *if for all $x\in M$, and $U, V\in T_xM$,*

$$g_x(U, V)=h_{\Phi(x)}(U^{\Phi}, V^{\Phi}).$$

(*34.101*) **Exercise.** Check that if Σ is a regular submanifold of $\mathbb{R}^q$, then it is a submanifold of $\mathbb{R}^q$ with Φ as the natural injection of Σ into $\mathbb{R}^q$.

(*34.102*) *Important examples* ('irrational string-wrapping'). Let c be an irrational number, and consider a map

$$\Phi\colon \mathbb{R}\to T^2\subseteq\mathbb{R}^3,$$

where T^2 is the 2-torus (imbedded in $\mathbb{R}^3$ as explained in a moment). The point $\Phi(t)$ will have angular coordinates (t, ct). Formally, let $a>1$, and set

$$\Phi(t)=((a+\cos t)\cos ct, (a+\cos t)\sin ct, \sin t).$$

Because c is irrational, Φ is one–one, $\Phi(\mathbb{R})$ is dense in T^2, and Φ is not a homeomorphism of $\mathbb{R}$ for the relative topology of $\Phi(\mathbb{R})$. However, Φ is an imbedding, so that $\Phi(\mathbb{R})$ is a submanifold both of T^2 and of $\mathbb{R}^3$.

It is rather bizarre that, for this example, we can even so construct a smooth function P mapping $\mathbb{R}^3$ to the space of 3×3 matrices such that, for $x=\Phi(t)$, $P(x)$ is orthogonal projection on to the space spanned by $\dot{\Phi}(t)$. Thus, we can construct Brownian motion on $\Phi(\mathbb{R})$ directly by the

$$\partial X=P(X)\partial B$$

construction.

To get an example in which P cannot be extended smoothly off $\Phi(\mathbb{R})$, replace the torus $T^2 = S^1 \times S^1$ by the product of the circle S^1 and a 'figure of eight'. Now play the same game with irrational winding. □

With the warning provided by these examples in mind, we quote a very important theorem of Riemannian geometry, due to Nash in its first form, and improved by Gromov to the form stated here.

(*34.103*) THEOREM (Nash, Gromov). *Every d-dimensional smooth Riemannian manifold can be isometrically imbedded in* $\mathbb{R}^q$, *where* $q = \frac{1}{2}d(d+1) + 3d + 5$.

The theorem says more than this; see Gromov and Rohlin [1] for a detailed survey of this and various other imbedding results.

(*34.104*) *Use of Theorem 34.103 to construct* $\mathrm{BM}(\Sigma)$, $\mathrm{BM}^{\mathrm{hor}}(\mathrm{O}(\Sigma))$, *etc.* First, we consider an important special case.

(*a*) *The special case when* Σ *is compact.* In this case, the imbedding Φ, as a continuous one–one map on a compact Hausdorff space, is a homeomorphism from Σ to $A \equiv \Phi(\Sigma)$. Let $y \in \Sigma$, and let $x \equiv \Phi(y)$. Then $T_x A = (T_y \Sigma)^{\Phi}$ is a d-dimensional subspace of $\mathbb{R}^q$. Let $P(x)$ denote the $q \times q$ matrix representing orthogonal projection onto (the subspace through 0 parallel to) $T_x A$. We explain below that *the map* $x \mapsto P(x)$ *on* A *has a smooth extension* (also denoted by P) to $\mathbb{R}^q$. Thus we can construct a Brownian motion X on $A = \Phi(\Sigma)$ via

$$\partial X = P(X)\partial B,$$

and then $Y \equiv \Phi^{-1}X$ defines a $\mathrm{BM}(\Sigma)$. See (31.15).

(*b*) *The general case.* It is part of our definition (34.6) that Σ is σ-compact. One can show (see Warner [1]) that there exist a sequence (G_n) of open subsets of Σ and a sequence (K_n) of compact subsets of Σ such that

$$G_n \subseteq K_n^{\cdot} \subseteq G_{n+1} \quad (\forall n), \quad \bigcup K_n = \Sigma.$$

For $y \in \Sigma$ and $x \equiv \Phi(y)$, let $P(x)$ again denote the $q \times q$ matrix describing orthogonal projection on to $(T_y \Sigma)^{\Phi}$. Now, Φ is a homeomorphism from K_n to $A_n \equiv \Phi(K_n)$. By the extension result described below, we can find a bounded smooth extension P_n to $\mathbb{R}^q$ of the map $x \mapsto P(x)$ on A_n. Then we can construct a process X_n, living on $\Phi(G_n)$ until the first exit time τ_n from $\Phi(G_n)$, and satisfying

$$\partial X_n = P_n(X_n)\partial B = P(X_n)\partial B.$$

If $X_n(0) = X_{n+1}(0) \in \Phi(G_n)$, then Corollary 11.10 shows that $X_{n+1} = X_n$ up until time τ_n. Thus, the X_n processes fit compatibly to define a process

$$\{X(t) : t < \tau \equiv \sup \tau_n\}$$

with $X(t) = X_n(t)$ for $t < \tau_n$.

Then $\Phi^{-1}(X)$ is $\mathrm{BM}(\Sigma)$ defined up to its explosion time τ. □

Of course, the imbedding theorem also allows us to construct $\mathrm{BM}^{\mathrm{hor}}(O(\Sigma))$ from (33.2).

(*34.105*) *Existence of the smooth extension P on $\mathbb{R}^q$.* The existence of a smooth extension of the map $x \mapsto P(x)$ on $\Phi(\Sigma)$ in case (*a*) (or on $\Phi(K_n)$ in general case (*b*)) can be deduced from Whitney's extension theorem (Whitney [1], or Appendix A of Abraham and Robbins [1]).

(*34.106*) *References.* Two classics on differential geometry are Helgason [1] and Kobayashi and Nomizu [1]. Easier going are Bishop and Crittenden [1] and Warner [1].

35. Brownian motions on Lie groups. We begin this section by explaining without differential geometry some basic facts about Brownian motions on groups of matrices. Brownian motion on SO(3), the group of rotations of $\mathbb{R}^3$, is studied in some detail. We then look at what various ideas from differential geometry have to say for Lie groups and their Brownian motions.

Let $\mathbb{G}$ be a multiplicative group of $n \times n$ matrices, closed in $GL(n, \mathbb{R})$ carrying the relative topology of $\mathbb{R}^{n^2}$ and with no isolated points. The Lie algebra, $A(\mathbb{G})$, of $\mathbb{G}$ is the set of $n \times n$ matrices A such that

$$e^{tA} \in \mathbb{G}, \quad \forall t \in \mathbb{R}.$$

Because

$$\exp(t(A_1 + A_2)) = \lim \{\exp(tA_1/k)\exp(tA_2/k)\}^k,$$

$A(\mathbb{G})$ is clearly a vector space. It is also true that $A(\mathbb{G})$ is a *Lie algebra* in that it is stable under the *Lie-bracket product*:

$$[A_1, A_2] \equiv A_1A_2 - A_2A_1.$$

Indeed, if

$$C(t, k) \equiv \exp(tA_1/k)\exp(tA_2/k)\exp(-tA_1/k)\exp(-tA_2/k),$$

then, for large k,

$$C(t, k) = I + t^2(A_1A_2 - A_2A_1)/k^2 + O(k^{-3}),$$

so that it is highly plausible (and true) that

$$\exp(t^2[A_1, A_2]) = \lim C(t, k)^{k^2} \in \mathbb{G}.$$

(*35.1*) *Examples.* (i) Let $\mathbb{G}$ be $GL(n, \mathbb{R})$, the General Linear group of non-singular Real $n \times n$ matrices. Then $A(\mathbb{G})$ is the set of all real matrices.

(ii) Let $\mathbb{G}$ be $SL(n, \mathbb{R})$, the Special Linear group of Real $n \times n$ matrices which are 'special' in that they have determinant 1. Then, because

$$\det(e^{tA}) = e^{\mathrm{trace}(tA)},$$

$A(\mathbb{G})$ consists of matrices of zero trace.

(iii) Let $\mathbb{G}$ be SO(n), the set of Special (determinant 1) Orthogonal $n \times n$ matrices. Then A($\mathbb{G}$) is the set of skew-symmetric $n \times n$ matrices. For, if A is skew-symmetric, then (with T signifying transpose)

$$e^{tA}(e^{tA})^T = e^{tA}e^{-tA} = I,$$

so that e^{tA} is orthogonal. Conversely, if $H_t \equiv e^{tA}$ is orthogonal for every t, then

$$0 = (d/dt)(H_t^T H_t)|_{t=0} = A^T + A.$$

(35.2) DEFINITION (left Brownian motion on $\mathbb{G}$). *A process G with values in $\mathbb{G}$ is called a* left(-invariant) Brownian motion *on G if*

(i) *G is continuous,*

(ii) *for each $s \geqslant 0$, the process $\{G_s^{-1}G_{s+t}: t \geqslant 0\}$ is independent of the process $\{G_r: r \leqslant s\}$, and*

(iii) *for each $s \geqslant 0$, the processes*

$$\{G_s^{-1}G_{s+t}: t \geqslant 0\} \text{ and } \{G_t: t \geqslant 0\}$$

are identical in law.

Here are the analytic facts about left Brownian motions on $\mathbb{G}$.

It can be proved (see Yosida [2], Hunt [2]) that *each left Brownian motion on $\mathbb{G}$ is an FD diffusion on $\mathbb{G}$*. Its transition function is specified by the action of its generator $\mathscr{L}$ on C_K^∞ functions. The operator $\mathscr{L}$ can be written in Stratonovich form

$$\mathscr{L} = \tfrac{1}{2}\sum_{q=1}^{d} U_q^2 + V, \tag{35.3}$$

where U_q $(1 \leqslant q \leqslant d)$ and V are 'left-invariant vector fields' on $\mathbb{G}$. You can surely guess the definition of 'left-invariant vector field' given at (35.26) below. What matters is that there is a one–one correspondence between left-invariant vector fields U and elements A of the Lie algebra A($\mathbb{G}$) of $\mathbb{G}$ described as follows. Given A in A($\mathbb{G}$), the associated left-invariant vector field U is the first-order operator on smooth functions f on $\mathbb{G}$ defined by the equation:

$$Uf(G) = (d/dt)f(Ge^{tA})|_{t=0} = \sum_i\sum_j (GA)_{ij}\partial f/\partial G_{ij}. \tag{35.4}$$

Let us go straight on to the probability which corresponds to the above analysis.

Suppose that $\mathscr{L}$ is at (35.3), and let A_q $(1 \leqslant q \leqslant d)$ and C be the elements of A($\mathbb{G}$) which correspond to the vector fields U_q $(1 \leqslant q \leqslant d)$ and V respectively. Let $(\beta^1, \beta^2, \ldots, \beta^d)$ be a standard d-dimensional Brownian motion, and let

$$A(t) = \beta^q(t)A_q + tC. \tag{35.5}$$

Then, $\{A(t)\}$ *is a Brownian motion (continuous process with time-homogeneous independent additive increments*) on $\mathbb{A}(\mathbb{G})$. Now let G be a process taking values in the set $M(n)$ of all $n\times n$ matrices which solves the Stratonovich SDE:

(35.6) $$\partial G = G\partial A,$$

where $G(0)\in\mathbb{G}$. Then G *takes values in* $\mathbb{G}$ *and is a left Brownian motion on* $\mathbb{G}$ *with generator* $\mathscr{L}$. We say that G is obtained by *product-integral injection* of the Brownian motion $\{A(t)\}$ on $\mathbb{A}(\mathbb{G})$.

(*35.7*) *Examples.* (i) Calculations with cofactors confirm that if (35.6) holds, then

$$\partial(\det G) = (\det G)(\operatorname{trace}\partial A).$$

Thus, if $\mathbb{G}=\mathrm{SL}(n,\mathbb{R})$ and $G(0)\in\mathbb{G}$, we have $G(t)\in\mathbb{G}$, $\forall t$.

(ii) Let $\mathbb{G}=\mathrm{SO}(n)$, and let G solve (35.6) with $G(0)\in\mathbb{G}$. Then, since $A^T=-A$,

$$\partial G^T = (\partial A^T)G^T = (-\partial A)G^T,$$

and

$$\partial(GG^T) = G(\partial A)G^T + G(-\partial A)G^T = 0,$$

so that $GG^T=I$, $\forall t$. □

In the two examples just given, we have shown by direct verification that

(35.8) *if* A *is* $\mathbb{A}(\mathbb{G})$*-valued, and* $G(0)\in\mathbb{G}$, *then the solution of equation* (*35.6*) *is* $\mathbb{G}$*-valued.*

One can prove this in general (see §4.8 of McKean [1]) by showing that, with probability 1,

(35.9) $$G(t) = \lim G(0)\prod_i \exp(\Delta A_n^i(t)),$$

where

$$t_n^i \equiv t\wedge(i2^{-n}),\quad \Delta A_n^i(t)\equiv A(t_n^i)-A(t_n^{i-1}).$$

It now follows from Itô theory that G is a diffusion on $\mathbb{G}$; and that its generator is $\mathscr{L}$ follows from (30.18). For any fixed element K of $\mathbb{G}$, it follows from (35.6) that

$$\partial(K^{-1}G) = (K^{-1}G)\partial A,$$

so that $K^{-1}G$ is again a diffusion with generator $\mathscr{L}$. It is now immediate that G is a left Brownian motion on $\mathbb{G}$.

Note that, since

$$\partial(G^{-1}) = -G^{-1}(\partial G)G^{-1},$$

we have

(35.10) $$\partial(G^{-1}) = -(\partial A)G^{-1},$$

in agreement with the directly verifiable fact that G^{-1} *is a right Brownian motion on* $\mathbb{G}$.

(*35.11*) *Brownian motion on* SO(3). We have seen at (35.1(iii)) that the Lie algebra of SO(3) consists of the set SKEW(3) of skew-symmetric 3 × 3 matrices. Now, an element A of SKEW(3) has the form

$$A = A(x) = \begin{pmatrix} 0 & -x_3 & x_2 \\ x_3 & 0 & -x_1 \\ -x_2 & x_1 & 0 \end{pmatrix}$$

where $x = (x_1, x_2, x_3)^T \in \mathbb{R}^3$. Thus, for $w \in \mathbb{R}^3$,

$$A(x)w = x \times w,$$

where $\times$ is the vector product. You can check that

$$[A(x), A(y)] = A(x \times y),$$

so that the Lie bracket can be regarded as the vector product.

Now let A denote the process $A(B)$, where B is the standard Brownian motion on $\mathbb{R}^3$. Then A is *canonical Brownian motion on* SKEW(3). Note that

$$(dA)(dA) = -2I\,dt. \tag{35.12}$$

Define *canonical Brownian motion* H *on* SO(3) as the solution H of the Stratonovich SDE

$$\partial H = H\partial A, \quad H(0) = I. \tag{35.13}$$

Now,

$$\partial H = (\partial \tilde{A})H, \quad \partial \tilde{A} \equiv H(\partial A)H^{-1}, \quad A(0) = 0.$$

Note that, since H is orthogonal and A skew,

$$\partial \tilde{A}^T = H(-\partial A)H^{-1} = -\partial \tilde{A},$$

so that $\tilde{A}$ takes values in SKEW(3). Moreover,

$$\begin{aligned} d\tilde{A} &= H(dA)H^{-1} + \tfrac{1}{2}(dH)(dA)H^{-1} + \tfrac{1}{2}H(dA)d(H^{-1}) \\ &= H(dA)H^{-1} + \tfrac{1}{2}H(dA)(dA)H^{-1} + \tfrac{1}{2}H(dA)(-dA)H^{-1} \\ &= H(dA)H^{-1}, \end{aligned}$$

so that $\tilde{A}$ is a local martingale. We have

$$d\tilde{A}_{ij} = \sum_p \sum_q H_{ip}(dA_{pq})H_{jq},$$

and, on using the orthogonality of H and the fact that A is BM(SKEW(3)), we find that

$$d\tilde{A}_{ij} d\tilde{A}_{kl} = (I_{ik}I_{jl} - I_{il}I_{jk})\,dt.$$

Lévy's theorem now shows that

(35.14) *$\tilde{A}$ is also a canonical* BM(SKEW(3)), *confirming that H is a right (as well as a left) Brownian motion on* SO(3).

Note. Theorem 35.70 below tells which Lie groups support bi-invariant Brownian motions.

(*35.15*) *Study of the rotational structure of* BM(SO(3)). Let U be a unit vector in $\mathbb{R}^3$, so that $U \in S^2$, and let $\theta \in \mathbb{R}$. Then, elementary geometry shows that the element $H(\theta, U)$ of SO(3) which represents a right-handed rotation through an angle θ about the axis U is given by

$$H(\theta, U) = I + (\sin\theta)A(U) + (1-\cos\theta)A(U)^2.$$

Since

$$(d/d\theta)H(\theta, U) = A(U)H(\theta, U),$$

we have

$$H(\theta, U) = \exp[\theta A(U)].$$

It is well known that every element H of SO(3) is of the form $H = H(\theta, U)$ for some θ and U. (Indeed, $+1$ must be an eigenvalue of H, and U is a corresponding eigenvector.)

The following lemma, the proof of which is left as a (rather complicated) exercise, shows how the evolution of a semimartingale of rotations can be described.

(*35.16*) LEMMA. *Let U be a semimartingale in $\mathbb{R}^3$ such that $U_t \in S^2$ for all t. Let θ be a semimartingale in $\mathbb{R}$. Then*

$$H(\theta, U)^{-1}\partial H(\theta, U) = A(\partial X).$$

where

(35.17) $$\partial X = (\partial\theta)U + (\sin\theta)\partial U + (1-\cos\theta)(\partial U \times U).$$

We can now describe BM(SO(3)) as a motion of rotations. We remark that we encounter a familiar problem in that the identity matrix is a singularity of our coordinate system. So, on a first reading, assume that $0 < \theta(0) < 2\pi$. (That 0 and 2π serve as entrance boundaries for the θ motion will be clear after our discussion of 1-dimensional diffusions in Part 7.)

(*35.18*) THEOREM. *Let θ be a diffusion on* $(0, 2\pi)$ *satisfying an SDE*

$$d\theta = db + \tfrac{1}{2}\cot\tfrac{1}{2}\theta\, dt,$$

where b is a BM($\mathbb{R}$). *Let*

$$\varphi_t \equiv \int_0^t (4\sin^2\tfrac{1}{2}\theta)^{-1}\, ds.$$

Let R be a BM(S^2) *independent of θ, and let*

$$U_t \equiv R(\varphi_t).$$

Then $H \equiv H(\theta, U)$ is a BM(SO(3)).

We recommend that you prove this theorem as an exercise by showing that, if H is as described in the theorem, then the process X defined by (35.17) is a BM($\mathbb{R}^3$). Though this is an exercise in stochastic calculus good for the soul, it does not really explain Theorem (35.18).

The explanation lies in the fact, fundamental in physics, that SO(3) *can be regarded as the unit sphere S^3 in $\mathbb{R}^4$ with antipodal points identified*. We now recall how this is done.

(*35.19*) *Use of quaternions.* Hamilton introduced a quaternion as a formal expression

$$Q = a + x,$$

where a, the 'real' part of Q, is a real number, and x, the 'imaginary' part of Q, is a vector in $\mathbb{R}^3$. For $a, b \in \mathbb{R}$ and $x, y \in \mathbb{R}^3$, define

$$(a+x)+(b+y)=(a+b)+(x+y),$$
$$(a+x)(b+y)=(ab-x\cdot y)+(ay+bx+x\times y),$$

where $x \cdot y$ is the scalar product, and $x \times y$ the vector product. Obviously, a quaternion Q can be thought of as a point (a, x) in $\mathbb{R}^4$. We define (for $Q = a + x$)

$$\|Q\|^2 = |a|^2 + \|x\|^2,$$

so that $\|Q\|$ is the standard norm of Q in $\mathbb{R}^4$. The norm is multiplicative in that $\|Q_1 Q_2\| = \|Q_1\|\,\|Q_2\|$. If $Q \neq 0$, let

$$Q^{-1} \equiv (a - x)/\|Q\|^2.$$

Then $QQ^{-1} = Q^{-1}Q = 1$, where $1 = 1 + 0$. The quaternions form a skew field. (The last statement includes the fact that quaternionic multiplication is associative.)

Let $\theta \in \mathbb{R}$ and $U \in S^2 \subset \mathbb{R}^3$. Think of a vector v in $\mathbb{R}^3$ as a pure imaginary quaternion $0 + v$. Then (you check!)

$$H(\theta, U)v = Q(\tfrac{1}{2}\theta, U)\,v\,Q(\tfrac{1}{2}\theta, U)^{-1}, \tag{35.20}$$

where $Q(\frac{1}{2}\theta, U)$ is the unit quaternion (element of S^3):

$$Q(\tfrac{1}{2}\theta, U) = (\cos\tfrac{1}{2}\theta) + (\sin\tfrac{1}{2}\theta)U.$$

Two elements Q_1 and Q_2 of S^3 give rise to the same rotation H if and only if $Q_1 = Q_2$, or $Q_1 = -Q_2$. This explains the sense in which the (simply connected) space S^3 is the double cover of SO(3).

Now, S^3 is a group under quaternionic multiplication. The Lie algebra of S^3 is $\mathbb{R}^3$ thought of as the space of pure imaginary quaternions, and with twice the vector product as Lie bracket. We can define a Brownian motion Q on S^3 by solving

$$\partial Q = Q\partial B, \quad Q(0) \in S^3, \tag{35.21}$$

where B is a $\mathrm{BM}(\mathbb{R}^3)$. We find that Q is a two-sided Brownian motion on S^3.

What explains Theorem 35.18 is the following fact.

(*35.22*) LEMMA. *Suppose that Q is a* $\mathrm{BM}(S^3)$ *run at rate* $\frac{1}{4}$ *and expressed in 'right invariant' form:*

$$\partial Q = \tfrac{1}{2}(\partial B)Q,$$

where B is some $\mathrm{BM}(\mathbb{R}^3)$. *For each v in* $\mathbb{R}^3$, *define* $Hv = QvQ^{-1}$. *Then:*

$$\partial H = A(\partial B)H,$$

so that H is a $\mathrm{BM}(\mathrm{SO}(3))$.

Proof. Since multiplication of pure imaginary quaternions is just the vector product, we have

$$\begin{aligned}\partial(Hv) &= (\partial Q)vQ^{-1} + Qv\partial(Q^{-1})\\ &= \tfrac{1}{2}\partial B \times Hv - (Hv) \times \tfrac{1}{2}\partial B = (\partial B) \times Hv,\end{aligned}$$

for all v in $\mathbb{R}^3$. □

Because of this lemma, Theorem (35.18) is just the skew-product representation for $\mathrm{BM}(S^3)$, the result one dimension up from that in (31.33). Let us give this representation in its natural form without the 'rate $\frac{1}{4}$' etc. needed to tie it in with $\mathrm{BM}(\mathrm{SO}(3))$ and caused by the 'double' cover.

(*35.23*) *Skew-product representation for* $\mathrm{BM}(S^3)$. Let

$$Q = (\cos\alpha) + (\sin\alpha)U \quad (0 < \alpha < \pi, \ U \in S^2)$$

be $\mathrm{BM}(S^3)$ written in quaternionic form and satisfying

$$\partial Q = Q\partial B$$

where B is a $\mathrm{BM}(\mathbb{R}^3)$. Then,

$$\begin{aligned}&(-\sin\alpha)\partial\alpha + (\sin\alpha)\partial U + (\cos\alpha)(\partial\alpha)U\\ &\quad = -(\sin\alpha)U\cdot\partial B + (\cos\alpha)\partial B + (\sin\alpha)U \times \partial B,\end{aligned}$$

so that

$$\begin{aligned}\partial\alpha &= U\cdot\partial B,\\ \partial U &= U \times \partial B + (\cot\alpha)\{\partial B - (U\cdot\partial B)U\},\\ dU\cdot dB &= (\cot\alpha)(3-1)dt = 2(\cot\alpha)dt,\end{aligned}$$

and

$$d\alpha = U\cdot dB + (\cot\alpha)dt = d\beta + (\cot\alpha)dt,$$

where β is a $\mathrm{BM}(\mathbb{R})$ process. Show that

$$d\alpha\, dU_i = 0,$$
$$dU_i\, dU_j = (\sin\alpha)^{-2}(I_{ij} - U_i U_j)\,dt,$$
$$dU = U \times dB + (\cot\alpha)\{dB - (U \cdot dB)U\} - U(\sin\alpha)^{-2}\,dt.$$

Thus, by the martingale characterization of $\mathrm{BM}(S^2)$,

$$U = R\left(\int_0^t (\sin\alpha_s)^{-2}\,ds\right),$$

where R is a $\mathrm{BM}(S^2)$ independent of α.

For the skew-product representation of $\mathrm{BM}(S^d)$, achieved by Markovian rather than SDE methods, see §7.15 of Itô and McKean [1].

(*35.24*) *Orientation* (for you, not for the manifold!). The matrix groups which we have been studying have a natural manifold structure as submanifolds of $\mathrm{GL}(n, \mathbb{R})$. We now make some brief comments on general Lie groups and their geometry, explaining how matrix groups fit into the picture.

We study how the 'manifold' Brownian motion associated with a left-invariant metric on a Lie group is obtained by product-integral injection and how parallel transport of frames then looks relative to the absolute parallelism associated with left-invariance.

It is worth emphasizing that *one has to be rather careful about the relationship between the geometry and the algebra.* Only if the group supports a bi-invariant metric is there a really tidy match-up between Riemannian theory and the algebra. Conditions under which a matrix Lie group does admit a bi-invariant metric are given at Theorem 35.70.

(*35.25*) *Lie groups.* A Lie group $\mathbb{G}$ is a smooth d-dimensional manifold with a group structure (which we write in multiplicative notation) such that

(i) $(G, H) \mapsto GH$ *is smooth,*
(ii) $G \mapsto G^{-1}$ *is a diffeomorphism.*

To formulate (i), we need to know how to define local coordinates on a product manifold $\Sigma_1 \times \Sigma_2$. This is obvious: use product charts:

$$(p_1, p_2) \mapsto (\varphi_1(p_1), \varphi_2(p_2)).$$

(*35.26*) *Left-invariant vector fields* (*LIVFs*). A vector field U on $\mathbb{G}$ is called *left-invariant* if for every K in $\mathbb{G}$,

$$U^{\lambda(K)} = U,$$

where $\lambda(K)$ is the left shift $G \mapsto KG$. Thus, for every G and K,

$$U(KG)f = U(G)(f \circ \lambda(K)).$$

Here, of course, f is defined and smooth near KG, $U(KG)$ is the tangent vector at KG defined by the field U, etc.

Note that *a LIVF U is determined by its value $U(I)$ at the identity element I of $\mathbb{G}$.*

It is clear how right-invariant vector fields (RIVFs) are defined. The natural isomorphism between LIVFs and RIVFs is explained at (35.41) below.

(*35.27*) *One-parameter subgroups*: exp(v) (*with small* 'e'). Let v be a tangent vector at I for our Lie group $\mathbb{G}$. Let L_v be the associated LIVF. Consider the integral curve for L_v starting from I:

(35.28) $$\dot{q}(t) = L_v(q(t)), \quad q(0) = I.$$

That such a curve exists and is unique follows from the theory of differential equations. It is quickly verified that for any fixed s, the function

$$\tilde{q}(t) \equiv q(s)^{-1} q(t+s)$$

also defines a solution of (35.28). Hence $\tilde{q} = q$ and

(35.29) $$q(t+s) = q(s)\, q(t).$$

Thus, q is a homomorphism from $\mathbb{R}$ into $\mathbb{G}$: we say that q defines a one-parameter subgroup.

We see that the tangent space $T_I\mathbb{G}$ to $\mathbb{G}$ at I is the set of derivatives at I of one-parameter subgroups.

The function q defined above is written:

(35.30) $$q(t) = \exp(tv) \quad \text{or} \quad q(t) = e^{tv}.$$

Note the small 'e' on exp in contrast to the Exp map on $T_I\mathbb{G}$ which may be defined when $\mathbb{G}$ has extra structure (metric, or just connection). *Keep the two types of exponential quite separate in your mind.*

(*35.31*) *Note on matrix groups.* We know (Hille–Yosida theorem!) that any matrix solution of (35.29) is of the form

(35.32) $$e^{tA} = I + tA + (tA)^2/2! + \ldots$$

for some matrix A. For a matrix group $\mathbb{G}$, the tangent space $T_I\mathbb{G}$ at the identity can therefore be regarded as the set $\mathrm{A}(\mathbb{G})$ of matrices A such that

$$e^{tA} \in \mathbb{G}, \quad \forall t,$$

with A identified with the tangent vector to the curve $t \mapsto e^{tA}$ at I. If we regard our matrix group $\mathbb{G}$ as a submanifold of $\mathbb{R}^N$, where $N = n^2$, then, as already observed at

(35.4), A can be thought of as the vector

(35.33) $$\sum_i \sum_j A_{ij}\,\partial/\partial G_{ij}, \quad \text{tangent to } \mathbb{G} \text{ at } I.$$

(*35.34*) *Coordinatization via the exponential map.* For our Lie group $\mathbb{G}$, the map

(35.35) $$\exp\colon T_I\mathbb{G} \to \mathbb{G}$$

is a diffeomorphism of a neighbourhood of 0 in the vector space $T_I\mathbb{G}$ onto a neighbourhood of the identity in $\mathbb{G}$. For a given basis $(e^1, e^2, \ldots, e^d)$ of $T_I\mathbb{G}$, the map (35.35) therefore yields a coordinatization of a neighbourhood of I in $\mathbb{G}$. (The situation is very similar to that involving normal coordinates, and the proof is analogous to that given for that situation in (34.55). But remember: you have been warned not to confuse exp with Exp.)

(*35.36*) *Lie algebra.* If U and V are LIVFs, then $[U, V]$ is another LIVF.

(*35.37*) **Exercise.** Prove the statement just made by showing that if λ is any smooth map from one smooth manifold Σ_1 to another Σ_2, and U and V are vector fields on Σ_1, then

$$[U^\lambda, V^\lambda] = [U, V]^\lambda.$$

Hint. For $f \in C^\infty(\Sigma_2)$, $U^\lambda V^\lambda f = UV(f\circ\lambda)$. In coordinates, this is just the chain rule of elementary partial differentiation. □

The Lie algebra of $\mathbb{G}$ *is the set of LIVFs on* $\mathbb{G}$ *with the obvious vector-space structure over* $\mathbb{R}$ *and the bracket product.*

(*35.38*) *The matrix case.* Suppose that L_A and L_B are LIVFs on a matrix group $\mathbb{G}$ corresponding to the elements A and B (respectively) of $A(\mathbb{G})$. Thus,

$$L_A = \sum_i \sum_j (GA)_{ij}\,\partial/\partial G_{ij}, \quad L_B = \sum_r \sum_s (GB)_{rs}\,\partial/\partial G_{rs},$$

and, since

$$(\partial/\partial G_{ij})(GB)_{rs} = \delta_{ir} B_{js},$$

we have

(35.39) $$[L_A, L_B] = L_{[A, B]}.$$

Now consider what happens for the RIVFs R_A and R_B associated with A and B. We have

$$R_A = \sum_i \sum_j (AG)_{ij}\,\partial/\partial G_{ij}, \quad R_B = \sum_r \sum_s (BG)_{rs}\,\partial/\partial G_{rs},$$

and, since

$$(\partial/\partial G_{ij})(BG)_{rs} = \delta_{js} B_{ri},$$

we have

(35.40) $$[R_A, R_B] = -R_{[A, B]}.$$

(*35.41*) *Switching from LIVFs to RIVFs.* To understand the contrast between (35.39) and (35.40) in the general situation, consider the inverse map $\psi: \mathbb{G} \to \mathbb{G}$,

(35.42) $$\psi(G) \equiv G^{-1}.$$

Suppose that $v \in T_I\mathbb{G}$, that L_v is the LIVF which takes the value v at I, and that R_v is the RIVF which takes the value v at I. If $q(t) = e^{tv}$, then

$$(L_v f)(G) = (d/dt) f(Gq(t))|_{t=0},$$

and

$$(L_v^{\psi} f)(G^{-1}) = (d/dt) f(q(-t)G^{-1})|_{t=0} = -(R_v f)(G^{-1}).$$

Thus,

(35.43) $$L_v^{\psi} = -R_v.$$

Since, by Exercise (35.37), $[L_u, L_v]^{\psi} = [L_u^{\psi}, L_v^{\psi}]$, we see that

(35.44) $$[L_u, L_v] = L_w \textit{ if and only if } [R_u, R_v] = -R_w.$$

(*35.45*) *Product-integral injection for general Lie groups.* Let $(e_1, e_2, \ldots, e_d)$ be a basis for $T_I\mathbb{G}$ for our Lie group $\mathbb{G}$, and let $L_1, L_2, \ldots, L_d$ be the LIVFs which agree at I with $e_1, e_2, \ldots, e_d$ respectively. Let $v \in T_I\mathbb{G}$, and, with B being a Brownian motion on $\mathbb{R}^d$, let

$$Z = B^m e_m + tv$$

be a (bi-invariant) Brownian motion on the additive Lie group $T_I\mathbb{G}$. We obtain a left invariant Brownian motion X by product-integral injection of Z by what is formally the relation

$$\dot{X} = L_{\dot{Z}}(X),$$

that is, the tangent vector $\dot{X}$ to $\mathbb{G}$ at $X(t)$ is the value at $X(t)$ of the LIVF which is $\dot{Z}$ at the identity. We make rigorous sense of this in a now-familiar way: for $f \in C^{\infty}(\mathbb{G})$,

(35.46) $$\partial f(X(t)) = (L_m f)(X(t))\,\partial B^m + (L_v f)(X(t))\,\partial t,$$

where L_v is the LIVF which takes the value v at I. We are going to take the existence and uniqueness of solution of (35.46) for given $X(0)$ for granted: we have done enough of that sort of thing.

(*35.47*) *Left-invariant metrics.* It should be obvious how to construct the general left-invariant metric on the Lie group $\mathbb{G}$: pick a basis $(e_1, e_2, \ldots, e_d)$ for $T_I\mathbb{G}$, let $L_1, L_2, \ldots, L_d$ be the associated LIVFs, and insist that g makes $L_1, L_2, \ldots, L_d$ everywhere orthonormal.

(*35.48*) *Example: affine group.* Consider the group:

(35.49) $$\mathbb{G} = \left\{ \begin{pmatrix} y & x \\ 0 & 1 \end{pmatrix} : y > 0, x \in \mathbb{R} \right\}$$

of 'orientation-preserving' affine transformations of $\mathbb{R}$. We first look at left-invariant entities, then at right-invariant entities, and then at how they interrelate.

We take a basis (e_1, e_2) for $T_I\mathbb{G}$, equivalently a basis (A_1, A_2) for $\mathrm{A}(\mathbb{G})$:

$$A_1=\begin{pmatrix}0 & 1\\ 0 & 0\end{pmatrix}\leftrightarrow e_1=\partial/\partial x,\quad A_2=\begin{pmatrix}1 & 0\\ 0 & 0\end{pmatrix}\leftrightarrow e_2=\partial/\partial y.$$

and take the left-invariant Riemannian structure which makes these an orthonormal basis for $T_I\mathbb{G}$. At

$$G=\begin{pmatrix}y & x\\ 0 & 1\end{pmatrix}\quad\text{in } \mathbb{G},$$

we shall have

$$GA_1=\begin{pmatrix}0 & y\\ 0 & 0\end{pmatrix}\leftrightarrow L_1=y\partial/\partial x,\quad GA_2=\begin{pmatrix}y & 0\\ 0 & 0\end{pmatrix}\leftrightarrow L_2=y\partial/\partial y.$$

Clearly,

$$(ds)^2=y^{-2}\{(dx)^2+(dy)^2\}$$

gives the corresponding left-invariant metric, and $(\mathbb{G}, g)$ is the hyperbolic plane. The Brownian motion on the Riemannian manifold $(\mathbb{G}, g)$ has generator

(35.50) $$\tfrac{1}{2}y^2(D_{xx}+D_{yy})=\tfrac{1}{2}(L_1^2+L_2^2-L_2),$$

and so, by Theorem 30.14 on 'Stratonovich' generators, is described by product-integral injection

(35.51) $$\partial\begin{pmatrix}y & x\\ 0 & 1\end{pmatrix}=\begin{pmatrix}y & x\\ 0 & 1\end{pmatrix}\partial\begin{pmatrix}b-\frac{1}{2}t & a\\ 0 & 0\end{pmatrix}$$

where (a, b) is a $\mathrm{BM}(\mathbb{R}^2)$.

The matrix exponential ($=\exp$) map

$$(u, v)\mapsto\exp\left\{t\begin{pmatrix}v & u\\ 0 & 0\end{pmatrix}\right\}=\begin{pmatrix}e^{vt} & uv^{-1}(e^{vt}-1)\\ 0 & 1\end{pmatrix}$$

is completely different from the map

$$(u, v)\mapsto\mathrm{Exp}(uD_x+vD_y)\quad\text{at } I$$

which was calculated for the hyperbolic plane at (34.53).

We shall use (X, Y) rather than (x, y) in dealing with right-invariant entities. The RIVFs R_1 and R_2 agreeing with D_X and D_Y at I are respectively

$$R_1=D_X,\quad R_2=XD_X+YD_Y,$$

and the right-invariant metric which makes these RIVFs everywhere orthonormal is given by

$$(dS)^2=(dX)^2-2XY^{-1}(dX)(dY)+Y^{-2}(1+X^2)(dY)^2.$$

The Brownian motion on the manifold $\mathbb{G}$ equipped with this metric has generator

$$\tfrac{1}{2}(R_1^2+R_2^2+R_2),$$

and so is obtained via

$$\partial\begin{pmatrix} Y & X \\ 0 & 1 \end{pmatrix} = \partial\begin{pmatrix} B+\frac{1}{2}t & A \\ 0 & 0 \end{pmatrix}\begin{pmatrix} Y & X \\ 0 & 1 \end{pmatrix}$$

where (A, B) is a $\mathrm{BM}(\mathbb{R}^2)$.

Under the correspondence

$$\psi: G \to G^{-1}, \qquad \psi\begin{pmatrix} y & x \\ 0 & 1 \end{pmatrix} = \begin{pmatrix} Y & X \\ 0 & 1 \end{pmatrix},$$

we have

$$L_1^\psi = -R_1, \quad L_2^\psi = -R_2, \quad a \mapsto -A, \quad b \mapsto -B,$$

and it all makes sense because

$$\partial(G^{-1}) = -G^{-1}(\partial G)G^{-1}.$$

(*35.52*) *Structural constants for Lie groups.* Let $L_1, L_2, \ldots, L_d$ be a basis for LIVFs. Then, since $[L_j, L_k]$ is left-invariant, there must exist *constants* c_{jk}^i such that

(35.53) $$[L_j, L_k] = c_{jk}^i L_i.$$

It follows that if we choose the left-invariant metric g on $\mathbb{G}$ which makes $L_1, L_2, \ldots, L_d$ everywhere orthonormal, then

(35.54) $$\nabla_{L(j)} L_k = K_{jk}^i L_i,$$

where the *constants* K_{jk}^i are given by (34.74):

(35.55) $$2K_{jk}^i = c_{jk}^i + c_{ij}^k + c_{ik}^j.$$

Moreover, it follows from (34.76) and (35.46) that the Brownian motion X on the manifold $(\mathbb{G}, g)$ may be defined by product-integral injection via

(35.56) $$\partial f(X(t)) = (L_q f)(X(t))\,\partial B^q + v^q (L_q f)(X(t))\,\partial t,$$

where

(35.57) $$v^q = -\tfrac{1}{2}\sum_m K_{mm}^q.$$

(*35.58*) **Exercise.** Calculate the constants K for the affine group and verify that (35.56) agrees with (35.51).

(*35.59*) *Rotation of frames of* $\mathrm{BM}^{\mathrm{hor}}(\mathrm{O}(\Sigma))$ *for a LI metric.* We continue to work with the assumptions of subsection (35.52), so that equation (35.56) defines the Brownian motion X on $\mathbb{G}$ for the left-invariant (LI) metric g.

Now this process can also be obtained via the Eells–Elworthy stochastic development of Theorem 34.86. Then, the process X carries with it a g-orthonormal frame

$$E(t)=(e_1(t), e_2(t), \ldots, e_d(t))$$

moving by parallel displacement along X. To understand something of the relationship between the two types of construction, we ask:

how does the frame $E(t)$ move relative to the absolute parallelism induced by left-invariance?

Let us put this question in a more specific form. We can write

(35.60) $$e_m(t)=H^r_m(t)\,L_r(X(t)),$$

where $H(t)$ is an orthogonal $d\times d$ matrix. *How does the process H evolve?*

It is harmless and useful to pretend that B^q and X are smooth functions of t, and to write

$$\dot{B}^q=\varepsilon^q, \quad \dot{X}=U.$$

Then (35.56) may be written formally:

$$\dot{X}=(\varepsilon^q+v^q)\,L_q.$$

Because $e_m(t)$ moves by parallel displacement along X, we have

(35.61) $$0=\nabla_U e_m=\dot{H}^r_m\,L_r(X(t))+H^s_m(t)\,(\varepsilon^q+v^q)\,\nabla_{L(q)}\,L_s.$$

Using (35.54), we find that

$$\dot{H}^r_m+H^s_m\,K^r_{qs}(\varepsilon^q+v^q)=0.$$

Thus, restoring rigorous formulation, we have shown that in (35.60), *the matrix H performs the left-invariant Brownian motion on $\mathrm{O}(d)$ given by*:

(35.62) $$\partial H=H\,\partial A,$$

where A is the $\mathrm{SKEW}(d)$-valued additive Brownian motion

(35.63) $$A^r_s(t)=-K^r_{qs}(B^q(t)+tv^q).$$

Note that

$$K^r_{qs}+K^s_{qr}=0$$

(whence A is skew) follows from

$$L_q g(L_r, L_s)=0.$$

(*35.64*) *Bi-invariant metrics.* As has already been stated, the study of a Lie group as a Riemannian manifold becomes considerably more tidy when the metric is bi-invariant (that is, both left-invariant and right-invariant). For example, *if g is a bi-invariant metric for a Lie group $\mathbb{G}$, then the geodesics of $(\mathbb{G}, g)$ through I are exactly* *the one-parameter subgroups, and* exp *and* Exp *agree.* This is a consequence of the simple formula:

(35.65) $$\nabla_U V = \tfrac{1}{2}[U, V] \quad \textit{for LIVFs } U \textit{ and } V$$

for the 'bi-invariant' case. This formula can be derived by arguments similar to those used in our study of structure constants at (35.52). See, for example, Cheeger and Ebin [1].

Even when there is no bi-invariant metric, we can always define a unique '*bi-invariant connection*' which satisfies (35.65) for LIVFs. It is the demand for a *metric* which can sometimes cause trouble. See Helgason [1] or Poor [1].

(35.66) Bi-invariant Brownian motions. We now explain that:

(35.67) *a Lie group* $\mathbb{G}$ *supports a non-singular bi-invariant Brownian motion* X *if and only if* $\mathbb{G}$ *supports a bi-invariant metric* g.

Proof. This is fairly obvious. On the one hand, if $\mathbb{G}$ supports a bi-invariant metric g, then the Brownian motion X on the manifold $(\mathbb{G}, g)$ is a bi-invariant Brownian motion on the group $\mathbb{G}$.

On the other hand, if the group $\mathbb{G}$ supports a non-singular Brownian motion X, then the metric g induced by the generator of X in the manner described at (34.90) is bi-invariant. □

Result (35.67) makes it of probabilistic interest to know which Lie groups support bi-invariant metrics. The complete solution to this problem is known. We describe this solution without proof for the case of matrix groups. In this way, we can avoid setting up the Adjoint representation in the general case. (Not that this is anything but easy, but we do not wish this section to beat the record established by § 34!)

(35.68) Which matrix groups support bi-invariant metrics? Let $\mathbb{G}$ be a matrix Lie group, and let $\mathrm{A}(\mathbb{G})$ be its Lie algebra in the sense that $A \in \mathrm{A}(\mathbb{G})$ if and only if

$$e^{tA} \in \mathbb{G}, \quad \forall t.$$

Now since, by elementary matrix theory,

$$\exp(tGAG^{-1}) = G\exp(tA)G^{-1},$$

we find that, for $G \in \mathbb{G}$,

$$\mathrm{Ad}(G)\colon \mathrm{A}(\mathbb{G}) \to \mathrm{A}(\mathbb{G}),$$

where

$$\mathrm{Ad}(G)A \equiv GAG^{-1}.$$

Since $\mathrm{Ad}(GH) = \mathrm{Ad}(G)\,\mathrm{Ad}(H)$, we see that *the map*

(35.69) $$G \mapsto \mathrm{Ad}(G)$$

is a homomorphism from G into the group GL(A($\mathbb{G}$)) *of invertible linear transformations of the vector space* A($\mathbb{G}$). *This map is called the Adjoint representation of* $\mathbb{G}$.

Clearly, the 'size' of the range Ad($\mathbb{G}$) is a measure of the 'degree of non-Abelianness' of $\mathbb{G}$. The next result is therefore satisfying.

(*35.70*) THEOREM. $\mathbb{G}$ *supports a bi-invariant metric if and only if* Ad($\mathbb{G}$) *is a relatively compact subset of* GL(A($\mathbb{G}$)).

For proof, see Cheeger and Ebin [1] or Poor [1].

(*35.71*) *Examples.* (i) If $\mathbb{G}$ is compact, then, since Ad is continuous, Ad($\mathbb{G}$) is compact, and $\mathbb{G}$ supports a bi-invariant metric. We saw at (35.14) that the compact group SO(3) supports a bi-invariant Brownian motion.

(ii) If $\mathbb{G} = GL(n, \mathbb{R})$, where $n \geqslant 2$, and Z is an element of $\mathbb{G}$ such that $\mathrm{Ad}(Z) = I$, then

$$ZAZ^{-1} = A$$

for every $n \times n$ matrix A, so that Z is the identity. In other words, Ad is an *isomorphism* of $\mathbb{G}$, so that Ad($\mathbb{G}$) is not compact, and $\mathbb{G}$ does not support a bi-invariant metric.

(iii) If $\mathbb{G}$ is the group of all non-zero quaternions, and $A \in \mathrm{A}(\mathbb{G}) = \mathbb{R}^4$, then

$$GAG^{-1} \text{ has the same norm as } A,$$

so that Ad($\mathbb{G}$) is a subgroup of the compact group SO(4). Thus, as we already know, $\mathbb{G}$ does support a bi-invariant metric.

(*35.72*) *References.* See Baxendale [4] for work on Brownian motions in the diffeomorphism group, and references mentioned there.

36. Dynkin's Brownian motion of ellipses; hyperbolic space interpretation; etc. We first give a bare-hands approach to the study of Brownian motions of ellipses of unit area. We then explain some of the underlying geometry and in particular how the set of ellipses of unit area may be represented in terms of the hyperbolic plane of (34.46).

There is a one–one correspondence between the set $\mathscr{E}$ of *ellipses E of unit area* and the set $\mathscr{P}$ of *positive-definite symmetric* 2×2 *matrices Y of determinant* 1 described by

(36.1) $$E = \{z \in \mathbb{R}^2 : z^T Y z = 1\}.$$

Dynkin ([3]) proved some striking results on Brownian motions of ellipses and ellipsoids. He postulated that a BM($\mathscr{E}$) process $\{E_t\}$, equivalently, a BM($\mathscr{P}$) process $\{Y_t\}$, should have the following invariance property:

(36.2) *the transition probabilities of* $\{Y_t\}$ *should be invariant under maps* $\{Y_t\} \mapsto \{K^T Y_t K\}$, *where K denotes a fixed element of* SL(2, ℝ).

Now there is an obvious way to construct a process Y satisfying (36.2), namely, let $\{G_t\}$ be a right-invariant Brownian motion on SL(2, ℝ), and set

(36.3) $$Y_t = G_t^T G_t.$$

We want

(36.4) $$\partial G = (\partial B) G,$$

where B takes values in the Lie algebra of SL(2, ℝ). Thus we want Trace $(B) = 0$, and the 'canonical' way to do this proves to be to take

(36.5) $$B_t = \beta_t - \tfrac{1}{2}(\text{Trace } \beta_t) I,$$

where β is a 2×2-matrix-valued process such that $(\beta_{11}, \beta_{12}, \beta_{21}, \beta_{22})$ is a BM($\mathbb{R}^4$). Suppose that Y is defined via (36.3–5), with $G(0) \in$ SL(2, ℝ) of course.

One can study many of the interesting features of the behaviour of Y in a bare-hands way, at the cost of doing a number of calculations. You can find these carried through for the n-dimensional situation in Norris, Rogers and Williams [1]. In the present context, you can practise your Itô calculus to establish that

(36.6) $$dG = (dB)G + \tfrac{1}{4} G\,dt, \qquad dY = G^T(dB + dB^T)G + 2Y\,dt,$$

and

(36.7) $$dY_{ij}\,dY_{kl} = 2Y_{ij}Y_{kl}\,dt + 2a_{ij,kl}\,dt,$$

where $a_{ij,kl}$ is the matrix element in the table

	11	12	21	22
11	0	0	0	−2
12	0	1	1	0
21	0	1	1	0
22	−2	0	0	0

Note that the Markovian character of the Y process is clear from (36.6), (36.7) and the local-martingale-problem results.

(36.8) *The eigenvalue motion for Y.* Write the eigenvalues of Y as e^u and e^{-u}. Then

(36.9) $$\text{Trace}(Y) = Y_{11} + Y_{22} = e^u + e^{-u}.$$

Thus

(36.10) $$(e^u - e^{-u})\partial u = \partial Y_{11} + \partial Y_{22},$$

and, on squaring (36.10) and using (36.7), we obtain

$$(\partial u)^2 = 2\partial t, \quad [u]_t = 2t.$$

We can therefore write

(36.11) $$u = 2^{1/2} b + F \quad (b \text{ a BM}(\mathbb{R}),\ F \text{ of finite variation}).$$

But now, on using Itô's formula to give the finite-variation parts of (36.9), and using (36.6), we find that

(36.12) $$du = 2^{1/2}\,db + (\coth u)dt.$$

From this, we can conclude (see below for proof) that

(36.13) $$u_t \neq 0 \quad (t > 0),$$

so we can assume that $u_t > 0$ for $t > 0$. Since $\coth u$ is approximately 1 for large u, it is 'obvious' from (36.12) that

(36.14) $$t^{-1} u_t \to 1 \quad \text{as } t \to \infty \quad \text{(a.s.)}.$$

Thus, very roughly speaking, for large t, the semi-axes of our ellipse have lengths about e^t and e^{-t}.

A proof of (36.14) by Khasminskii's method is given in § 37. Here is a direct proof.

Proofs of (36.13) and (36.14). Suppose that u solves (36.12) with $u_0 > 0$. Let

$$\mathscr{A} \equiv D_u^2 + (\coth u) D_u$$

be the generator of u, and let $f(u) \equiv \log \sinh u$. Since $\mathscr{A} f = 1$, we see that

$$M_t \equiv \log \sinh u_t - t$$

defines a local martingale M. Clearly, M cannot explode to $+\infty$, so, by (IV.34.13), neither can it explode to $-\infty$. Hence u never hits 0.

Next, since $b(t) = o(t)$ and (because we have taken $u(\bullet) > 0$) $\coth u > 1$, it is clear from (36.12) that

$$\liminf u_t / t \geqslant 1, \quad \text{a.s.}$$

But now, since $u_t \to \infty$, a.s., so that $\coth u_t \to 1$, a.s., we can further deduce from (36.12) that (36.14) holds. □

Remark. We see that if $\lambda_t^{(1)}$ and $\lambda_t^{(2)}$ are the eigenvalues of $G_t^T G_t$, then, in an obvious sense,

$$\lim t^{-1} \log\{\lambda_t^{(1)}, \lambda_t^{(2)}\} = \{-1, 1\}.$$

In the n-dimensional case (Dynkin [3], Orihara [1], Norris, Rogers and Williams [1]),

$$\lim t^{-1} \log\{\lambda_t^{(1)}, \lambda_t^{(2)}, \ldots, \lambda_t^{(n)}\} = (-n+1, -n+3, \ldots, n-3, n-1). \quad □$$

(*36.15*) *The eigenvector motion.* We can write

(36.16) $$Y=\begin{pmatrix}\cos\gamma & -\sin\gamma\\ \sin\gamma & \cos\gamma\end{pmatrix}\begin{pmatrix}e^{u} & 0\\ 0 & e^{-u}\end{pmatrix}\begin{pmatrix}\cos\gamma & \sin\gamma\\ -\sin\gamma & \cos\gamma\end{pmatrix},$$

where γ specifies the direction of the minor axis of the ellipse. On applying Itô's formula to such equations as

$$Y_{11}=e^{u}\cos^2\gamma+e^{-u}\sin^2\gamma,$$

we find that

$$du\,d\gamma=0,\quad \text{and}$$

(36.17) *γ is a local martingale such that $d[\gamma]=(\frac{1}{2}\operatorname{cosech}^2 u)dt$.*

Hence,

(36.18) $$\gamma_t=W([\gamma]_t)=W\left(\int_0^t \tfrac{1}{2}\operatorname{cosech}^2 u_s ds\right),$$

where W is a BM($\mathbb{R}$) independent of u. Since u grows linearly in the sense of (36.13), $[\gamma]_\infty$ exists finitely, so that

(36.19) $$\lim \gamma_t=W([\gamma]_\infty) \textit{ exists.}$$

(*36.20*) *The generator $\mathscr{G}$ of Y in (u,γ) and (a,γ) coordinates.* Because of (36.12) and (36.17), the generator of the Y motion in (u,γ) coordinates is

(36.21) $$\mathscr{G}=D_u^2+(\coth u)D_u+(4\sinh^2 u)^{-1}D_\gamma^2,$$

where $D_u=\partial/\partial u$, etc. Thus,

(36.22) $$(2\sinh^2 u)\,\mathscr{G}=\tfrac{1}{2}(D_a^2+D_\gamma^2)$$

where

$$a=\tfrac{1}{2}\log\tanh\tfrac{1}{2}u.$$

Thus,

(36.23) *the (a,γ) motion is just* BM($\mathbb{R}^2$) *run with a random clock,*

and from this fact and the fact that γ_∞ is the γ-value when a reaches 0, we see that, for any fixed initial non-circular ellipse,

(36.24) *γ_∞, the total winding performed by the minor axis, has a Cauchy distribution.*

(*36.25*) *The $X=GG^T$ motion.* Again let B be as at (36.5), and consider the flow

(36.26) $$z_0\mapsto z_t \quad \text{on} \quad \mathbb{R}^2$$

associated with the equation

(36.27) $$\partial z=(\partial B)z.$$

We find that $z_t = G_t z_0$, where

$$\partial G = (\partial B) G,$$

exactly as at (36.4), where $G_0 = I$. Since $z_0 = G_t^{-1} z_t$, we see that the flow (36.26) maps the unit circle into the ellipse

$$z^T \tilde{X}_t z = 1$$

at time t, where $\tilde{X}_t = (G_t^{-1})^T G_t^{-1}$. Now $(G^{-1})^T$ is identical in law to G, so the $\tilde{X}$ motion on $\mathbb{R}^2$ is identical in law to the process X with

$$X_t = G_t G_t^T. \tag{36.28}$$

The X process arguably has as much right to be called a 'Brownian motion' of ellipses as the Y process. However, although X has the good 'flow' property, it is *not* a Brownian motion on a Riemannian manifold, as can be shown via (34.76).

Now X clearly has the same eigenvalues as Y. However, the kind of analysis given for Y establishes that

(36.29) *the γ_X process for X does not tend to a limit, but, as $t \to \infty$, it behaves ever more like a* BM($\mathbb{R}$) *process $\tilde{W}$; indeed, we have a skew-product representation*

$$\gamma_X(t) = \tilde{W}\left(\int_0^t \{1 - \operatorname{sech}^2 u\}\, du\right).$$

See Norris, Rogers and Williams [1] for an elementary account of the n-dimensional analogue.

Part—but only part—of the discrepancy between the behaviour of X and Y can be explained by the fact that Y is driven only by the symmetric part of B; see equation (36.6).

(*36.30*) *The geometry.* Having pointed out the existence of the X process and its different behaviour, we now return to a study of the Y process, now in terms of Brownian motion on the hyperbolic plane. We note that the generator $\mathscr{G}$ of Y can be written $\mathscr{G} = \frac{1}{2}\Delta$, where Δ is the Laplace–Beltrami operator associated with the metric tensor

$$(ds)^2 = \tfrac{1}{2}(du)^2 + (2\sinh^2 u)(d\gamma)^2, \tag{36.31}$$

which is nearly (but intriguingly different from) the metric (34.54) for the hyperbolic plane in geodesic polar coordinates from (0, 1). So, what is going on?

We can write a positive-definite symmetric 2×2 matrix of unit determinant as

$$Y = \begin{pmatrix} \zeta + \xi & \eta \\ \eta & \zeta - \xi \end{pmatrix}$$

where (ξ, η, ζ) belongs to the *hyperboloid sheet in* $\mathbb{R}^3$ described by

(36.32) $$\zeta^2 - \xi^2 - \eta^2 = 1, \quad \zeta > 1.$$

The obvious parametrization

(36.33) $$\xi = \sinh r \cos\theta, \quad \eta = \sinh r \sin\theta, \quad \zeta = \cosh r$$

corresponds to (36.16) with

(36.34) $$r = u, \quad \theta = 2\gamma.$$

If we substitute (36.33) into the *pseudo-Riemannian metric*

(36.35) $$(ds)^2 = (d\zeta)^2 - (d\xi)^2 - (d\eta)^2$$

suggested by (36.32), we obtain

(36.36) $$\begin{aligned} -(ds)^2 &= (dr)^2 + (\sinh^2 r)(d\theta)^2 \\ &= 2\{\tfrac{1}{2}(du)^2 + (2\sinh^2 u)(d\gamma)^2\}. \end{aligned}$$

Thus, modulo a trivial sign change in the metric, *the hyperbolic plane can be regarded as the hyperboloid sheet (36.32) in the space* $\mathbb{R}^3$ *equipped with the pseudo-Riemannian metric (36.35).* Moreover, we have proved the following result.

(36.37) THEOREM. *If* (r_t, θ_t) *represents Brownian motion on the hyperbolic plane in geodesic polar coordinates from* (0, 1), *then*:

$$(u_t, \gamma_t) \equiv (r_{2t}, \tfrac{1}{2}\theta_{2t})$$

gives Dynkin's Brownian motion of ellipses in (u, γ) *coordinates.*

(Of course, such geometry was very well known to Dynkin who used it most effectively. But the bare-hands approach is not without other applications.)

The invariance property (36.2) of Y implies that:

(36.38) *the metric at (36.36) is invariant under transformations*

$$Y \mapsto K^T Y K, \quad K \in \mathrm{SL}(2, \mathbb{R}),$$

of $\mathscr{P}$.

If we use the (x, y) coordinatization of the hyperbolic plane (so that $y > 0$), then we have, for the hyperbolic-plane metric,

$$(ds)^2 = y^{-2}\{(dx)^2 + (dy)^2\} = (dr)^2 + (\sinh^2 r)(d\theta)^2.$$

One of the key properties of the hyperbolic plane is that:

(36.39) *this metric is invariant under the action of* $\mathrm{SL}(2, \mathbb{R})$ *qua the group of linear fractional transformations*

$$z = x + iy \mapsto K(z) = (az + b)/(cz + d),$$

$$K = \begin{pmatrix} a & b \\ c & d \end{pmatrix} \in \mathrm{SL}(2, \mathbb{R}).$$

Proof of (36.39). If

$$Z = K(x+iy) = X + iY,$$

then, using $ad - bc = 1$, we find that

$$Y = y|cz+d|^{-2},$$

and (using deterministic complex-variable calculus) we obtain

$$dZ = (cz+d)^{-2}dz, \quad Y^{-2}dZd\bar{Z} = y^{-2}dzd\bar{z},$$

as required. □

(*36.40*) *Application.* The geodesic from (0, 1) in the y direction is easily shown to be given by

(36.41) $$(x, y)(t) = (0, e^t).$$

Take a fixed α and define

$$K = \begin{pmatrix} \cos\alpha & -\sin\alpha \\ \sin\alpha & \cos\alpha \end{pmatrix}.$$

Then, because K (acting as a fractional linear transformation) preserves the Riemannian structure of the hyperbolic plane, the image under K of the geodesic (36.41) is another geodesic. But

$$K((x, y)(t)) = (X, Y)(t),$$

where

$$X(t) = -Y(t)\sinh t \sin 2\alpha,$$
$$Y(t) = (\cosh t - \sinh t \cos 2\alpha)^{-1},$$

so that $(X, Y)(\bullet)$ is the geodesic through (0, 1) starting from (0, 1) at unit speed in a direction $\frac{1}{2}\pi + 2\alpha$. This greatly simplifies the calculation leading to (34.53).

(*36.42*) **Exercise** (recommended!). Suppose that G is our right-invariant Brownian motion on SL(2, $\mathbb{R}$), so that

$$\partial G = (\partial B)G,$$

where B is as at (36.5). Now, define

$$z_t = G_t(i),$$

so that G_t operates on $i = (0, 1)$ as a fractional linear transformation. Show that z_t is a diffusion process on the hyperbolic plane, and find the generator $\mathscr{L}$ of $\{z_t\}$. The reason why z does not behave as one would hope is explained in Pauwels and Rogers [1].

(*36.43*) *References.* There are several ways of looking at the work in this section.

(*a*) *Products of random matrices.* See Bougerol and Lacroix [1], and the papers referred to therein.

More generally, for Lyapunov exponents, and for the multiplicative Oseledec theorem, see Carverhill [1, 2, 3], and papers in Arnold and Wihstutz [1].

(*b*) *Symmetric-space approach.* Consider the map

$$\mathrm{SL}(2, \mathbb{R}) \to \mathscr{P}$$

defined by

$$G \mapsto G^T G$$

via which we introduced the Y process. For $G, K \in \mathrm{SL}(2, \mathbb{R})$, we have

$$G^T G = K^T K$$

if and only if G belongs to the right coset $\mathrm{SO}(2)K$ of $\mathrm{SO}(2)$ in $\mathrm{SL}(2, \mathbb{R})$. In this way, we think of $\mathscr{P}$ as the quotient space

$$\mathrm{SL}(2, \mathbb{R})/\mathrm{SO}(2)$$

of right cosets, thereby regarding $\mathscr{P}$ as a symmetric space. See Helgason [1].

In an analogous way, we can regard the hyperbolic plane as the symmetric space $\mathrm{SL}(2, \mathbb{R})/\mathrm{SO}(2)$ derived from considering the map

$$\mathrm{SL}(2, \mathbb{R}) \to \text{Hyperbolic plane}$$

$$G \mapsto G(i).$$

See §X.3 of Helgason [1].

Now see Dynkin [3], Orihara [1], Malliavin and Malliavin [1], and Baxendale [4].

(*c*) *Manifolds of negative curvature.* The hyperbolic plane is the simplest example of a manifold of constant negative curvature.

The Martin-boundary approach of Dynkin amounts to the fact that the Martin boundary can be identified as the sphere at infinity. This concept is formulated in terms of geodesic polar coordinates from a point. For the state-of-the-art results on generalizations to manifolds of (non-constant) negative curvature, see Ancona [1] and Kifer [1].

Other aspects of Brownian motions on manifolds of negative curvature have been the object of intense study. Cheeger and Ebin [1] is a nice account of some of the underlying differential geometry. See Gray and Pinsky [1], and W. S. Kendall [2, 3, 4].

37. Khasminskii's method for studying stability; random vibrations. In §5, we studied a 'random vibration' problem:

(37.1(i)) $$dx = y\,dt,$$

(37.1(ii)) $$dy = -2c\omega y\,dt - \omega^2 x\,dt + \sigma x\,dB.$$

(We have written (x, y) for (U, V).) As mentioned in § 5, Khasminskii has given us the technique for studying linear stochastic systems of this type. *The trick is to use polar coordinates.* This will work in any dimension, but, to keep things simple, let us first look at 2-dimensional problems.

(*37.2*) *Semimartingales in polar coordinates.* Let (x, y) be a semimartingale in $\mathbb{R}^2$. Write

$$x = r\cos\theta, \quad y = r\sin\theta,$$

so that

$$\theta = \arctan(y/x), \quad \log(r) = \tfrac{1}{2}\log(x^2+y^2).$$

Itô's formula gives

(37.3) $$d\theta = r^{-2}\{-y\,dx + x\,dy\} - r^{-4}\{xy\,d([y]-[x]) + (x^2-y^2)\,dx\,dy\},$$

and

(37.4) $$d(\log r) = r^{-2}(x\,dx + y\,dy) + \tfrac{1}{2}r^{-4}\{(x^2-y^2)\,d([y]-[x]) - 4xy\,dx\,dy\}.$$

(*37.5*) *Khasminskii's idea.* Let $Z = (x, y)^T$, the superscript T signifying transpose. Suppose that

(37.6) $$dZ = \sigma(Z)\,dB + b(Z)\,dt, \quad (B \text{ a } \mathrm{BM}(\mathbb{R}^d))$$

where $\sigma(\cdot)$ is a *linear* map from $\mathbb{R}^2$ to $L(\mathbb{R}^d, \mathbb{R}^2)$ and b is a *linear* map from $\mathbb{R}^2$ to $\mathbb{R}^2$. We see from (37.3) that

(37.7) $\{\theta_t : t \geq 0\}$ *satisfies an autonomous SDE and hence is Markovian.*

Moreover, equation (37.4) shows that if $r_0 \neq 0$, then

(37.8) $$\log(r_t) - \log(r_0) = M_t + \int_0^t f(\theta_s)\,ds,$$

where M is a local martingale null at 0 such that $[M]_t \leq Ct$, $\forall t$, for some constant C. Thus, by Theorem IV.34.11, for some Brownian motion β,

(37.9) $$t^{-1}M_t = t^{-1}\beta([M]_t) \to 0.$$

By the ergodic theorem (see § 53), we shall have, almost surely,

(37.10) $$t^{-1}\log(r_t) = t^{-1}\int_0^t f(\theta_s)\,ds \to \lambda \equiv \int_0^{2\pi} f(\theta)\rho(\theta)\,d\theta,$$

where $\rho(\cdot)$ is the invariant probability density for $\{\theta_t : t \geq 0\}$. The constant λ is the *principal Lyapunov exponent* for the equation (37.6). If $\lambda < 0$, then the motion is (*pathwise*) *asymptotically exponentially stable.*

The same idea works not only in n dimensions, but also with more general noise.

(*37.11*) *Example: Lyapunov exponent for Brownian motion of ellipses.* Consider the equation (36.27) in new notation

$$\partial(x\ y)^T = (\partial B)(x\ y)^T,$$

where B is as at (36.5):

$$\begin{pmatrix} B_{11} & B_{12} \\ B_{21} & B_{22} \end{pmatrix} = \begin{pmatrix} \frac{1}{2}(\beta_{11}-\beta_{22}) & \beta_{12} \\ \beta_{21} & \frac{1}{2}(\beta_{22}-\beta_{11}) \end{pmatrix}$$

where $(\beta_{11}, \beta_{12}, \beta_{21}, \beta_{22})$ is a BM($\mathbb{R}^4$). Then:

$$(x_t\ y_t)^T = G_t(x_0\ y_0)^T,$$

where $\{G_t\}$ is our right-invariant Brownian motion on SL(2, $\mathbb{R}$) at (36.4).

We have

$$d(x\ y)^T = (dB)(x\ y)^T + \tfrac{1}{4}(x\ y)^T dt,$$
$$d[x] = (y^2 + \tfrac{1}{2}x^2)dt, \quad d[x, y] = -\tfrac{1}{2}xydt,$$
$$d[y] = (x^2 + \tfrac{1}{2}y^2)dt.$$

Substitution in (37.4) yields

$$d(\log r) = 2^{-1/2}dw + \tfrac{1}{2}dt,$$

for some BM($\mathbb{R}$) process w. Hence, if e^u and e^{-u} denote the eigenvalues of $G^T G$, then

$$u(t) \sim t, \quad \text{a.s.},$$

as was shown at (36.14).

Observe that

$$\mathbf{E}\{r_t^p\} = \mathbf{E}\exp(2^{-1/2}pw_t + \tfrac{1}{2}pt) = \exp(tp(p+2)/4),$$

so that we have found the *pth moment Lyapunov exponent*:

$$\mu_p = \lim_{t\to\infty} t^{-1}\log\mathbf{E}\{r_t^p\} = p(p+2)/4.$$

(*37.12*) *Effect of noise on stability: example.* One of the interesting questions in this area is how a stochastic ('noise') term can affect the stability properties of a deterministic system. We give an example suggested by work of Baxendale.

Consider the 3-dimensional process X where

$$\partial X = \begin{pmatrix} -1 & 0 & 0 \\ 0 & -1 & 0 \\ 0 & 0 & \mu \end{pmatrix} X\partial t + \sigma X \times \partial B, \tag{37.13}$$

where $\mu > 0$, $\sigma \geqslant 0$, B is a BM($\mathbb{R}^3$), and $\times$ denotes the vector product. The noise adds a rotation to the deterministic situation.

Since $\mu>0$, the deterministic ($\sigma=0$) system is unstable.
Write

$$X=(x\ y\ z)^T,\ B=(\alpha\ \beta\ \gamma)^T,$$

$$u=r^2=x^2+y^2+z^2,\quad v=z/r.$$

Then,

$$\partial u=2X^T\partial X=2\{-r^2+(\mu+1)z^2\}\partial t,$$

so that u and r are FV processes. Moreover,

$$\partial(\log r)=r^{-1}\partial r=\{-1+(\mu+1)v^2\}\partial t. \tag{37.14}$$

It is intuitively obvious that when σ is very large ('$\sigma=\infty$'), then $X_t/\|X_t\|$ is uniform on the sphere, so that (Archimedes' theorem) v_t is uniform on $(-1, 1)$, and, from (37.14),

$$\lim t^{-1}\log r_t=(\mu-2)/3.$$

Thus, if $\mu<2$, sufficient noise will stabilize the system.

To work out what happens when $0<\sigma<\infty$, we seek to utilize the obvious fact that v is Markovian. We have

$$dz=(\mu-\sigma^2)z\,dt+\sigma x\,d\beta-\sigma y\,d\alpha,$$

$$(dz)^2=\sigma^2r^2(1-v^2)\,dt.$$

Since r is of finite variation,

$$dv=r^{-1}dz-r^{-2}z\,dr,$$

and we find that v is Markovian on $[-1, 1]$ with generator

$$\mathscr{A}=\tfrac{1}{2}\sigma^2(1-v^2)D_v^2+\{(\mu+1)(1-v^2)-\sigma^2\}vD_v$$

and invariant probability density ρ given by

$$\rho=\text{const.}\exp\{(\mu+1)v^2/\sigma^2\}.$$

Hence, by the ergodic theorem and (37.14),

$$t^{-1}\log r_t\to\lambda=\int_{-1}^{1}\{-1+(\mu+1)v^2\}\rho(v)\,dv. \qquad \square$$

(*37.15*) *References*. All that we have done is to draw attention to a most interesting area of considerable practical use in engineering. We have done nothing to indicate its depth. The recent appearance of the proceedings of the important Bremen conference (Arnold and Wihstutz [1]) makes further comment from us superflous. For some nice examples fully worked out, see Baxendale [1, 2].

For the definitive theorem on Lyapunov exponents and much more, see Carverhill [1, 2, 3].

For large-deviation results concerning

$$\lim t^{-1} \log P\{r_t > c\},$$

etc., see Stroock [4]. A wonderful paper on this area by Baxendale and Stroock will appear soon.

Kozin and Prodmorou [1] gives the solution to the random vibration problem (37.1). This problem shares with some other phase–space problems the amusing feature that the θ-diffusion has shunts.

38. Hörmander's theorem; Malliavin calculus; stochastic pullback; curvature. The fundamental theorem on *the existence of smooth transition densities for diffusion processes* is due to Hörmander, and some theorem it is!

Our first aim in this section is to describe, in language which avoids the jargon of Schwartz distributions and hypoellipticity, the theorem's consequences for diffusion theory, restricting attention initially to the case when the underlying manifold is $\mathbb{R}^n$.

Lie brackets will play a central role, and we explain their significance in terms of *pullbacks.* The stochastic version of pullbacks, one example of the *Itô–Watanabe–Bismut–Kunita formula*, now features large in papers on filtering and control; so do not dismiss stochastic pullbacks as abstract gobbledegook!

The Malliavin calculus forms a towering synthesis of many of the ideas which we study in this book. Our purpose is not to make a detailed study of this calculus but to draw your attention to things which might help you when you move on to study what the great people in that field have achieved.

We begin with what is the standard simplest example which illustrates Hörmander's theorem. But do not complain! You have already seen a much more interesting motivating example: $\text{BM}^{\text{hor}}(\text{O}(S^2))$—see just before (33.21).

(*38.1*) *Example.* Let B be a $\text{BM}(\mathbb{R})$, and let X be a process in $\mathbb{R}^2$ such that

$$\partial X^1 = \partial B, \quad \partial X^2 = X^1 \partial t. \tag{38.2}$$

Even though X is driven by a 1-dimensional Brownian motion, so that X is a singular diffusion, $X(t)$ will have smooth density for each $t > 0$. If, for example, $(X^1, X^2)(0) = (0, 0)$, then X is a Gaussian process, and for $t > 0$, $(X^1, X^2)(t)$ has non-singular covariance matrix

$$\begin{pmatrix} t & t^2/2 \\ t^2/2 & t^3/3 \end{pmatrix}.$$

We can write (38.2) in the form

$$\partial f(X) = (Uf)(X)\partial B + (Wf)(X)\partial t,$$

where U and W are the vector fields

$$U = D_1, \quad W = X^1 D_2$$

on $\mathbb{R}^2$. Then $[U, W] = D_2$, so that the tangent space to $\mathbb{R}^2$ at a point x is already spanned by the vectors $U(x)$ and $[U, W](x)$. This fact certainly allows us to apply Hörmander's theorem.

(*38.3*) *Example.* Let B be a BM($\mathbb{R}^3$), and let X be a process in $\mathbb{R}^3 \setminus \{0\}$ satisfying

$$\partial X = X \times \partial B, \tag{38.4}$$

where $\times$ is the vector product, and with $X(0)$ a fixed point. We know that X stays on the surface of a sphere, so that X will not have smooth density in $\mathbb{R}^3$. Write (38.4) as

$$\partial f(X) = (U_q f)(X) \partial B^q,$$

where

$$U_1 = X^3 D_2 - X^2 D_3, \quad U_2 = X^1 D_3 - X^3 D_1, \quad U_3 = X^2 D_1 - X^1 D_2.$$

As every physicist knows, $[U_1, U_2] = +U_3$, etc. Thus, if $\mathscr{H}$ is the Lie algebra generated by U_1, U_2 and U_3, then, for every H in $\mathscr{H}$, the vector $H(x)$ is perpendicular to the vector x, and so tangential to the sphere of radius $\|x\|$. So, the vectors $H(x)$, where $H \in \mathscr{H}$, do not span $T_x\mathbb{R}^3$, and, rightly, Hörmander's theorem will not apply to this example. However, in this case, $\mathbb{R}^3 \setminus \{0\}$ is 'foliated' into spheres centre 0, and we can use Hörmander's theorem to show that X has smooth density relative to surface-area measure on the sphere on which it starts.

(*38.5*) *Our basic SDE. Let σ^i_q $(1 \leqslant i \leqslant n,\ 1 \leqslant q \leqslant m)$ and c^i $(1 \leqslant i \leqslant n)$ be smooth functions on $\mathbb{R}^n$ with globally bounded first derivatives.* Let B be a BM($\mathbb{R}^m$). Then the equation

$$\partial X = \sigma(X) \partial B + c(X) \partial t \tag{38.6}$$

has a strong solution; and indeed, as will prove important later, the Diffeomorphism Theorem 13.8 applies. We can write (38.6) in the form

$$\partial f(X) = (U_q f)(X) \partial B^q + (Wf)(X) \partial t, \tag{38.7}$$

where U_q and W are the vector fields:

$$U_q = \sigma^i_q(x) D_i, \quad W = c^i(x) D_i.$$

We know from (30.20) that equation (38.7) has Itô form:

$$df(X) = (U_q f)(X) dB^q + (\mathscr{L} f)(X) dt, \tag{38.8}$$

where

$$\mathscr{L} = \tfrac{1}{2}\Sigma(U_q)^2 + W = \tfrac{1}{2}a^{ij}(x) D_i D_j + b^i(x) D_i,$$

where $a = \sigma\sigma^T$ and where b is the true-drift vector field:

$$b^i = c^i + \tfrac{1}{2}\sum_q U_q(\sigma_q^i).$$

Suppose now that $h \in C_K^\infty((0, \infty) \times \mathbb{R}^n)$; that is, h is smooth on $(0, \infty) \times \mathbb{R}^n$ and h has compact support. Then Itô's formula gives the familiar extension of (38.8):

$$dh(t, X_t) = (U_q h)(t, X_t)dB^q + \left(\frac{\partial}{\partial t} + \mathscr{L}\right)h(t, X_t)dt,$$

so that

$$M^h(t) \equiv h(t, X_t) - \int_0^t \left(\frac{\partial}{\partial s} + \mathscr{L}\right)h(s, X_s)ds \tag{38.9}$$

defines a martingale M^h. Note that the compact-support property of h implies that M^h is null at 0, and that, for some t_0, $M^h(t) = M^h(t_0)$ for all $t \geqslant t_0$.

We know from Lemma 13.6 and result (22.5) that X is Markovian with FD transition function which we denote by $\{P(t)\}$. Thus, for fixed x in $\mathbb{R}^n$, we have, on taking $\mathbf{E}^x$ expectations in (38.9) and making the notational switch $(s \to t)$,

$$\int_{t=0}^\infty dt \int_y P(t, x, dy)\left(\frac{\partial}{\partial t} + \mathscr{L}_y\right)h(t, y) = 0. \tag{38.10}$$

Now, for $g \in C_K^\infty((0, \infty) \times \mathbb{R}^n)$, integration by parts shows that:

$$\int_{t=0}^\infty \int_y g(t, y)\left(\frac{\partial}{\partial t} + \mathscr{L}_y\right)h(t, y)\,dtdy$$

$$= \int_{t=0}^\infty \int_y h(t, y)\left(-\frac{\partial}{\partial t} + \mathscr{L}_y^*\right)g(t, y)\,dtdy,$$

where $\mathscr{L}_y^*$ is the adjoint to $\mathscr{L}_y$:

$$\mathscr{L}_y^* g = \tfrac{1}{2}D_{y(i),\,y(j)}\{a^{ij}(y)g\} - D_{y(i)}\{b^i(y)g\}. \tag{38.11}$$

The fact that (38.10) is true for all h in $C_K^\infty((0, \infty) \times \mathbb{R}^n)$ says that the density $p(t, x, y)$ relative to $dtdy$ of the measure $dtP(t, x, dy)$ on $(0, \infty) \times \mathbb{R}^n$, which certainly exists as a Schwartz distribution, satisfies ('in the sense of distributions') the equation

$$\frac{\partial}{\partial t}p(\cdot, x, \cdot) = \mathscr{L}_y^* p(\cdot, x, \cdot). \tag{38.12}$$

Hörmander's theorem is concerned with deciding when Schwartz distribution solutions of (38.12) are in fact good old-fashioned smooth functions.

(38.13) Formulating the basic hypothesis. We now introduce Hörmander's key hypothesis, which requires us to define a certain Lie algebra $\mathscr{H}$ (at the same time correcting a silly slip in the introduction to Williams [13]).

Define

$$\mathscr{A}_0 = \mathrm{Lie}(U_1, U_2, \ldots, U_m),$$

$$\mathscr{A}_k = \mathrm{Lie}([U, W]: U \in A_{k-1}\}, \quad k \geqslant 1.$$

Here, of course, Lie(F), where F is a family of vector fields, is the Lie algebra generated by F: that is, the smallest vector space over $\mathbb{R}$ of vector fields on $\mathbb{R}^n$ which contains all elements of F and which is stable under the bracket $[\cdot, \cdot]$ operation.

Now define

(38.14) $$\mathscr{H} = \mathrm{Lie}(\mathscr{A}_0, \mathscr{A}_1, \mathscr{A}_2, \ldots).$$

Note that W *need not belong to* $\mathscr{H}$.

Here, then, is the basic assumption:

(*38.15*) HYPOTHESIS (Hörmander's hypothesis). *$\mathscr{H}$ is full in the sense that, for every x in $\mathbb{R}^n$, the vectors $\{H(x)$: $H \in \mathscr{H}\}$ span (and therefore make up the whole of) $T_x\mathbb{R}^n$.*

Now we can state the theorem.

(*38.16*) THEOREM (Hörmander's theorem). *Under Hypothesis 38.15, the FD diffusion X has a transition density $p(t, x, y)$:*

$$(P_t f)(x) = \int_y p(t, x, y) f(y)\, dy,$$

where $p(\cdot, \cdot, \cdot)$ is a smooth function on $(0, \infty) \times \mathbb{R}^n \times \mathbb{R}^n$. *Moreover, the function p satisfies Kolmogorov's forward equation:*

(38.17) $$\frac{\partial}{\partial t} p(\cdot, x, \cdot) = \mathscr{L}_y^* p(\cdot, x, \cdot) \text{ for each } x,$$

and Kolmogorov's backward equation:

(38.18) $$\frac{\partial}{\partial t} p(\cdot, \cdot, y) = \mathscr{L}_x p(\cdot, \cdot, y) \text{ for each } y.$$

For each x, the function $p(\cdot, x, \cdot)$ is the minimal positive solution of (38.17) such that, for $f \in C_0(\mathbb{R}^n)$,

$$\lim_{t \downarrow 0} \int_y p(t, x, y) f(y)\, dy = f(x).$$

(*38.19*) *Comments.* The final statement that the probabilistic recipe gives the fundamental (minimal positive) solution is covered by § 3.5 of McKean [1]. The shortest proofs of the remaining (Hörmander) parts are purely analytical. See

Hörmander [1], and, for a good recent exposition with other references, Chaleyat-Maurel and el Karoui [2].

(*38.20*) *Remark.* The reason that we must not build W into $\mathscr{H}$ by definition (though it might get in by accident!) can be seen by considering

$$(X^1(t), X^2(t)) = (B(t), t),$$

which certainly does not have a smooth density in $\mathbb{R}^2$.

(*38.21*) *Very important special case: Weyl's Lemma.* If the operator $\mathscr{L}$ is strictly elliptic in that, for each x, the matrix $a(x)$ is non-singular, then, at each x, the vectors $U_1(x), U_2(x), \ldots, U_m(x)$ already span $T_x\mathbb{R}^n$, so that the hypothesis (38.15) is clearly satisfied, and a smooth transition density exists. This special case of Hörmander's theorem, which covers very many of the applications one meets in practice, had earlier been obtained by Hermann Weyl. See § 3.5 of McKean [1].

(*38.22*) *Malliavin calculus.* In papers which have profoundly influenced the whole subject, Malliavin ([1, 2]) conceived of the idea of giving a purely probabilistic proof of Hörmander's theorem via a new calculus of variations, now appropriately called the Malliavin calculus. This calculus (in a variety of forms) has been pursued with brilliance and dedication by many of the world's best probabilists, and has become a technique of formidable power.

As Richard Durrett did write: 'As David Williams might say, to visit some of the most beautiful Mayan ruins in Mexico, one must hike for several hours through the jungle.' Well, let us try to provide you with a periscope, so that as you brave your way through the technical side of the Malliavin calculus (which, if not exactly a jungle, is certainly luxuriant enough to make your travel invigorating), you can peep over a thicket or two.

First, there is the matter of Lie brackets.

(*38.23*) *Pullback interpretation of Lie brackets.* We have already seen several uses of the *method of moving frames*. For example, the covariant derivative $\nabla_U W$ for a Riemannian manifold is the rate of change of W reported to us by an observer moving along an integral curve of U and carrying with him an orthonormal frame which moves by parallel displacement.

The *Lie derivative* $[U, M]$ of a vector field M (please bear with the notation!) along U:

$$[U, M] \equiv UM - MU,$$

may be interpreted in analogous fashion, but the moving frame is derived from the flow of diffeomorphisms associated with the vector field U. The Lie derivative does not require any Riemannian or connection structure, and so may be defined on any smooth manifold. In consequence, 'metric' concepts for frames (such as that of being 'orthonormal') now play no part.

We study the Lie derivative as it applies to vector fields on $\mathbb{R}^n$. (The extension to general manifolds is trivial.) So, suppose that U is a vector field on $\mathbb{R}^n$, say of at most linear growth. Associated with U is the flow of diffeomorphisms

$$(t, x) \mapsto X(t, x)$$

of the differential equation

$$\dot{X} = U(X),$$

so that

$$(d/dt)X(t, x) = U(X(t, x)), \quad X(0, x) = x. \tag{38.24}$$

Let $Y(t, x)$ be the Jacobian matrix with (i, k)th entry

$$Y^i_k(t, x) = \partial X^i(t, x)/\partial x^k.$$

Then it is known from the theory of ordinary differential equations (compare (13.12)) that the equation obtained by formal differentiation of (38.24) relative to x^k is valid:

$$(d/dt)\,Y(t, x) = U'(X(t, x))\,Y(t, x), \tag{38.25}$$

where

$$U'(z)^i_j = \partial U^i(z)/\partial z^j.$$

Write $\varphi(t)$ for the diffeomorphism:

$$\varphi(t)(x) = X(t, x)$$

and set

$$\rho(t) \equiv \varphi(t)^{-1}.$$

Our observer moving along $X(\bullet, x)$ chooses at time 0 the frame

$$(e_1, e_2, \ldots, e_n), \text{ where } e_k = D_k \text{ at } x.$$

His frame at time t is that induced by the flow in the sense that

$$e_k(t) = e_k^{\varphi(t)} = Y^i_k(t, x)D_i.$$

Let M be another vector field on $\mathbb{R}^n$. The observer will regard $M(t) \equiv M(X(t, x))$ as having coordinates $F^k(t, x)$, where

$$M^i(X(t, x)) = Y^i_k(t, x)F^k(t, x). \tag{38.26}$$

Putting it another way,

$$F^k(t, x)e_k = M(X(t, x))^{\rho(t)}.$$

We say that $F^k(t, x)e_k$ is the *pullback* into $T_x\mathbb{R}^n$ (induced by the flow) of the vector $M(X(t, x))$ at $X(t, x)$. *The pullback operators $\{\rho(t)\}$ are serving a rôle very similar to that of a connection, pulling back the different tangent spaces at points of the curve $X(\bullet, x)$ into a common vector space $T_x\mathbb{R}^n$ within which we can differentiate.*

We concentrate on a fixed x which can therefore be suppressed from our notation.

Now, our observer will report to us that the rate of change of $M(X(t))$ at time t is

$$\dot{F}^k(t)e_k(t) = (\dot{F}^k(t)e_k)^{\varphi(t)}.$$

The crucial thing (proved below) is that

(38.27) $$\dot{F}^k(t)e_k(t) = [U, M](X(t)).$$

Note that *equation (38.27) can be written in pullback (and coordinate-free) notation*:

(38.28) $$(d/dt)(\{M(X(t))\}^{\rho(t)}) = \{[U, M](X(t))\}^{\rho(t)}.$$

In coordinates, equation (38.28) corresponds to the matrix equation:

(38.29) $$(d/dt)(ZM) = Z[U, M],$$

where $Z(t) \equiv Y(t)^{-1}$.

Statements (38.27), (38.28) and (38.29) are clearly equivalent. With a view to the stochastic generalization, we prove (38.29) as follows.

Proof of (38.29). Since $Z = Y^{-1}$,

$$\dot{Z} = -Z\dot{Y}Z = -ZU',$$

from (38.25). Also, by definition of $\dot{X}$ as a vector,

$$(d/dt)M^i(X(t)) = \dot{X}(M^i)(X(t)) = U(M^i)(X(t)),$$

or $\dot{M} = U(M)$ for short. Hence,

$$(d/dt)(ZM) = -ZU'M + ZU(M).$$

But

$$(U'M)^i = (D_j U^i)M^j = M(U^i).$$

Result (38.29) follows. □

(38.30) Stochastic pullback. Equations (38.27), (38.28) and (38.29) have important stochastic analogues.

By the Diffeomorphism Theorem 13.8, our basic SDE

(38.31) $$\partial X = \sigma(X)\partial B + c(X)\partial t = U_q(X)\partial B^q + W(X)\partial t$$

generates a flow of diffeomorphisms $\{\varphi(t)\}$:

$$\varphi(t)(x) = X(t, x), \quad \text{etc.}$$

Moreover, if we set

$$Y^i_j(t, x) = \partial X^i(t, x)/\partial x^j,$$

then, by Theorem 13.12,

(38.32) $$\partial Y = U'_q(X)\, Y \partial B^q + W'(X)\, Y \partial t,$$

where

$$(U'_q)^i_j = D_j U^i_q = D_j \sigma^i_q, \quad (W')^i_j = D_j W^i = D_j c^i.$$

Set

$$Z(t, x) = Y(t, x)^{-1}.$$

Since $\partial Z = -Z(\partial Y)Z$, we obtain from (38.32):

(38.33) $$\partial Z = -ZU'_q(X)\partial B^q - ZW'(X)\partial t.$$

If M is a vector field on $\mathbb{R}^n$, then we can derive the stochastic analogue of equation (38.29), which is *one of the main Itô–Watanabe–Bismut–Kunita formulae for describing the evolution of tensor fields along diffusion curves*:

(38.34) $$\partial(ZM) = Z\{[U_q, M]\partial B^q + [W, M]\partial t\}.$$

Exercise. Prove (38.34) using (38.33) and the Stratonovich formula:

$$\partial M^i(X(t)) = (D_j M^i)\partial X^j.$$

It is very easy.

(*38.35*) *Malliavin–Bismut 'perturbation vector fields' along X.* Malliavin's *calculus of variations* takes the basic SDE (38.31), and *perturbs the driving Brownian motion* B. Since X is a strong solution and therefore a function of B, we can consider *'differentiating X along the perturbation' to produce a vector field M along X, the evolution of which can be studied by the pullback technique.* How this helps lead to a proof of Hörmander's theorem, we shall indicate shortly.

Even amongst all the virtuoso displays which have been achieved with the Malliavin calculus, Bismut's contribution, both to theory and to applications, has been particularly dazzling. The first of Bismut's many innovations was to explain that one can utilize with full effectiveness the simplest kind of perturbation:

(38.36) $$B \to B^\varepsilon, \quad B^{\varepsilon q}(t) = B^q + \varepsilon^r \int_0^t H^q_r(s)ds,$$

where $(\varepsilon^r) \in \mathbb{R}^n$ and $H^q(1 \leqslant q \leqslant m, 1 \leqslant r \leqslant n)$ is a previsible matrix-valued process.

(*Note. This is very much 'Up-periscope!' time,* so that we are trying to peep beyond the technicalities which would really start to hem us in at this stage if we were trying to be fully rigorous.)

Let X^ε be the output from our strong-solution machine for our SDE (38.31) when the input is B^ε rather than B:

(38.37) $$\partial X^\varepsilon = U_q(X^\varepsilon)(\partial B^q + \varepsilon^r H^q_r \partial t) + W(X^\varepsilon)\partial t.$$

There is a point to be checked: we must be sure that this input does not jam the machine. We return to this later.

You have correctly guessed what comes next: differentiate (38.37) with respect to ε, and set $\varepsilon=0$, writing

(38.38) $$M_r^i = \partial X^i/\partial \varepsilon^r, \text{ evaluated when } \varepsilon=0.$$

(Ever onward! . . .) We find that

(38.39) $$\partial M = U_q'(X)M\partial B^q + W'(X)M\partial t + U_q(X)H^q\partial t.$$

Since each column of M is a vector at $X(t)$, it is natural to investigate an analogue of (38.34). Now, from (38.33) and (38.39), we find that

(38.40) $$\partial(ZM) = ZU_q(X)H^q\partial t,$$

so that ZM is an FV process.

For reasons which we shall indicate shortly, the intuitive idea is to choose

(38.41) $$H_r^q = \{ZU_q\}^r,$$

whereupon we obtain the Malliavin covariance matrix:

(38.42) $$(ZM)(t) = \int_0^t \sum_q \pi_q(s)\,ds,$$

where $\pi_q(t)$ is $\|(ZU_q)(t)\|^2$ times the orthogonal projection onto the direction of $(ZU_q)(t)$. Interest centres on the question of the *invertibility of* $(ZM)(t)$, and we see that in particular this requires that, by time t, *the vectors* ZU_q *must (between them) have pointed in directions which span* $\mathbb{R}^n$. Since equation (38.34) implies that

(38.43) $$\partial(ZU_q) = Z[U_p, U_q](X(t))\partial B^p + Z[W, U_q](X(t))\partial t,$$

the role of the Lie brackets and the appropriateness of making a hypothesis such as Hörmander's become immediately apparent.

But though this explains that there is sensible geometry going on, it says nothing as yet about the analysis. *How does all this help prove the existence of smooth densities?* To answer that means bringing on to the stage some other characters; and, as in all the best stories, the most important of these is an old friend whom we have often met before.

We should, however, pause to notice that, because of (38.41), and with $[\cdot, \cdot]$ denoting quadratic covariation rather than Lie brackets for the only time in this section, we have

(38.44) $$(ZM)_j^i(t) = [R_i, R_j](t),$$

where

(38.45) $$R_i(t) = \int_0^t \sum_q H_i^q\,dB^q.$$

(*38.46*) *Malliavin–Bismut integration by parts.* After the clues at (38.36) and (38.45), it is clear that what is going to save the day is our old friend the Cameron–Martin–Girsanov theorem, and that it is going to do so in exactly the same way as it yielded the explicit martingale-representation result in §IV.41.

Let us restrict attention to the time-interval [0, 1], and try to see through our periscope how we might finally be able to reach a point where we can prove that $X(1)$ has smooth density. We ignore boundedness conditions and other encumbrances to the progress of the honest pilgrim.

Let $\mathbf{P}$ be the Wiener-measure law of $B=\{B(t): t \leqslant 1\}$, and define

$$d\mathbf{P}^\varepsilon/d\mathbf{P}=\exp\left\{-\varepsilon^r R_r(1)-\tfrac{1}{2}\sum_q \int_0^1 (\varepsilon^r H_q^r)^2 dt\right\} \quad \text{on } \mathscr{F}_1. \tag{38.47}$$

Then, by the Cameron–Martin–Girsanov theorem (IV.38),

(38.48) B^ε *defines a Brownian motion relative to* $\mathbf{P}^\varepsilon$.

So, the strong-solution machine will process (38.37) properly, and produce X^ε as the same function of B^ε as X is of B. We therefore have, for nice functions f on $\mathbb{R}^n$,

$$\mathbf{E}[f(X^\varepsilon(1))\exp\{\ldots\}]=\mathbf{E}f(X(1)), \tag{38.49}$$

where $\exp\{\ldots\}$ is the factor on the right-hand side of (38.47). Differentiate (38.49) with respect of ε^j, put $\varepsilon=0$, and rearrange to obtain

$$\mathbf{E}\{(D_k f)(X(1))M_j^k(1)\}=\mathbf{E}\{f(X(1))R_j(1)\}, \tag{38.50}$$

the fundamental Malliavin–Bismut integration-by-parts formula.

Having reached this stage, it is time to recall a Fourier-theory lemma which represents the last step in the argument proper. We can then see more clearly how the steps which we have indicated thus far need modification for a final version.

(*38.51*) *A Fourier-theory lemma.* Let $C_b^r(\mathbb{R}^n)$ consist of functions f on $\mathbb{R}^n$ such that if α is a multi-index

$$\alpha=(\alpha(1), \alpha(2), \ldots, \alpha(n))$$

with

$$|\alpha| \equiv \alpha(1)+\alpha(2)+\ \ldots\ +\alpha(n) \leqslant r,$$

then

$$D^\alpha f \equiv D_1^{\alpha(1)} D_2^{\alpha(2)} \ldots D_n^{\alpha(n)} f$$

exists and is bounded and continuous on $\mathbb{R}^n$. For any function f on $\mathbb{R}^n$, we define $\|f\|_\infty \equiv \sup|f(x)|$, as usual.

The required result is the following.

(*38.52*) LEMMA. *Suppose that* $m \in \mathbb{N}$, *and that* ξ *is a random variable with values in* $\mathbb{R}^n$ *such that, for some constant* K_m *and all* f *in* $C_b^{m+n+1}(\mathbb{R}^n)$,

$$|\mathbf{E}(D^\alpha f)(\xi)| \leqslant K_m \|f\|_\infty \tag{38.53}$$

whenever $|\alpha| \leqslant m+n+1$. *Then* ξ *has a density function* p *belonging to* $C_b^m(\mathbb{R}^n)$.

Proof. Let μ be the law of ξ and let φ denote its characteristic function:

$$\varphi(\theta) = \mathbf{E}\exp(i\theta\cdot\xi) \quad (\theta\in\mathbb{R}^n).$$

On taking $f(x) = \exp(i\theta\cdot x)$ in (38.53), we find that

$$|\theta|^\alpha|\varphi(\theta)| \leqslant C_m, \quad \text{if } |\alpha| \leqslant m+n+1,$$

where we use the notation

$$\gamma^\alpha = \gamma_1^{\alpha(1)}\gamma_2^{\alpha(2)}\ldots\gamma_n^{\alpha(n)}, \quad \gamma\in\mathbb{R}^n.$$

Hence, if β is a multi-index with $|\beta| \leqslant m$, then

$$|\theta|^\beta|\varphi(\theta)| \leqslant C_m\prod_k|\theta_k|^{-1-1/n},$$

and μ has density p with

$$(D^\beta p)(y) = (2\pi)^{-n}(-i)^{|\beta|}\int\exp(-i\theta\cdot y)\theta^\beta\varphi(\theta)\,d\theta,$$

the integral converging absolutely and uniformly, etc.. □

(*38.54*) *Putting the pieces together.* The right-hand side of (38.50) is bounded by $K\|f\|_\infty$, where $K = \mathbf{E}|R_j(1)|$. So, if one were to be extremely naïve, one could hope to show that $M(1)$ is sufficiently nicely invertible to conclude directly from (38.50) that

(38.55) $$|\mathbf{E}(D_jf)(X(1))| \leqslant \text{constant}.\|f\|_\infty,$$

the first step towards using Lemma 38.52 to prove smoothness of density of $X(1)$. This naïve idea does at least serve to emphasize the need to choose the perturbation (38.36) so as to give $M(1)$ the best chance of being invertible with $M(1)^{-1}$ satisfying good boundedness conditions. The obvious choice is the one described at (38.41), and we assume that made. That done, it is essential to prove that $M(1)^{-1}$ does indeed exist, and to obtain good L^p bounds on its elements. In addition to utilizing the Lie algebra structure, this step requires martingale and semimartingale inequalities which are deep extensions of the 'exponential' inequality of §IV.37, the Burkholder–Davis–Gundy inequalities of §IV.42, etc.

It is ridiculously naïve to suppose that we can just 'erase' the factor $M(1)$ from (38.50). We must allow from the beginning for the fact that this factor will appear. The way to do this is to look at the system (X, Z, M) as a single Markovian entity described by equations (38.31), (38.33) and (38.39), and to develop the whole approach with (X, Z, M) replacing X. We hope that by obtaining an analogue of (38.50) for a matrix-valued function of type

$$F(X, Z, M)(1) = f(X(1))M(1)^{-1},$$

we will be able to proceed properly to the conclusion (38.55). We further need to arrange our argument so that we can inductively obtain results (38.53) (with $\xi = X(1)$) for all values of m.

Achieving this means assuming initially that first derivatives of σ and b are globally bounded, and that all higher derivatives are of polynomial growth. However, a fundamental localization lemma of Kusuoka and Stroock [2] allows one to drop these restrictions later. (Of course, the C^∞ property of σ and b must be maintained.)

You can see that, all in all, the Malliavin-calculus proof of Hörmander's theorem is going to be quite a hike. But Hörmander's theorem is not by any means the only benefit. See if the References subsection (30.60) can persuade you to get your boots on.

(*38.56*) *Enter curvature.* As we have said, in many applications, $\mathscr{L}$ is strictly elliptic, and Weyl's lemma will do just as well as Hörmander's theorem. Hörmander's theorem is therefore of special importance only for singular diffusions. *When do they arise?*

One obvious situation, generalizing that of Example 38.1, is in connection with phase-space descriptions. Thus, for example, for a generalized Ornstein–Uhlenbeck velocity process, we have

$$\partial X = Y \partial t,$$

$$\partial Y = \sigma(Y)\partial B + c(Y)\partial t,$$

where σ and c are smooth, and σ is never zero. Then Hörmander's theorem applies, and, for $t > 0$, $(X, Y)(t)$ has smooth density in $\mathbb{R}^2$.

Another interesting application was mentioned in remarks following (33.20). We saw that $\mathrm{BM}^{\mathrm{hor}}(\mathrm{O}(S^2))$ is essentially the same as the 'singular' left-invariant Brownian motion on SO(3) of type

$$\partial G = G \begin{pmatrix} 0 & -\partial\alpha & -\partial\beta \\ \partial\alpha & 0 & 0 \\ \partial\beta & 0 & 0 \end{pmatrix},$$

where (α, β) is a $\mathrm{BM}(\mathbb{R}^2)$. This may be written in obvious notation:

$$\partial f(G) = (U_2 f)(G)\partial\beta + (U_3 f)(G)\partial\alpha,$$

and since (in equally obvious notation) $U_1 = [U_2, U_3]$, Hörmander's condition is satisfied on SO(3) in that

$$\{U_2(G), U_3(G), [U_2, U_3](G)\}$$

spans $T_G(\mathrm{SO}(3))$ for every G in SO(3). One can now deduce that, for $t > 0$, $G(t)$ has smooth density relative to the Haar measure on SO(3) (which is got in the obvious way from the uniform measure on the sphere S^3).

What is happening is that *the Lie brackets of horizontal vector fields on* $O(S^2)$ *are providing vertical components, and this is what curvature is all about.*

To do a little more to introduce curvature, suppose that U and W are vector fields on a smooth d-dimensional Riemannian manifold Σ. Let H_U and H_W be the horizontal lifts of U and W to $T\Sigma$. Thus, for example, at the point of $T\Sigma$ with coordinates $(x^i, v^i D_{x(i)})$, we have

(38.57) $$H_U = U^i(x) D_{x(i)} - \Gamma^i_{jk}(x) U^j(x) v^k D_{v(i)}.$$

The horizontal component of $[H_U, H_W]$ is clearly $H_{[U,W]}$. A simple calculation based on (38.57) shows that

(38.58) the vertical component of $[H_U, H_W]$ is

$$[H_U, H_W] - H_{[U,W]} = -R^i_{jkl} v^j U^k W^l D_{v(i)},$$

where R is the *curvature tensor*:

(38.59) $$R^i_{jkl} = D_k \Gamma^i_{lj} - D_l \Gamma^i_{kj} + \Gamma^i_{kp} \Gamma^p_{lj} - \Gamma^i_{lp} \Gamma^p_{kj}.$$

Trivial exercise. Check that this agrees with the more usual formulation of the curvature tensor: if U, V, W are vector fields on Σ, then

$$\begin{aligned} R(U, W)V &\equiv \nabla_U \nabla_W V - \nabla_W \nabla_U V - \nabla_{[U,W]} V \\ &= R^i_{jkl} V^j U^k W^l D_{x(i)}. \end{aligned}$$

With the arrival on stage of the curvature tensor, our overture to stochastic differential geometry is at an end; and that is as it should be. We hope that our overture has served its purpose in introducing themes and harmonic patterns (if not harmonic forms) which will persuade you to appreciate the work proper performed under more expert direction. Recommended performances include:

(*38.60*) *References.* For the Lie-bracket formulae for the evolution of vector fields and for important applications, see Kunita [4].

With regard to the Malliavin calculus:

(*a*) *Bismut approach.* For good routes into this area, see Bichteler and Fonken [1] and Zakai [1].

Norris [1] gives a full systematic account of the 'Bismut' method of proving Hörmander's theorem.

Bismut's creativity, already praised in the text, is indicated by the sample Bismut [2, 3, 4, 5] of his work. Look at the title of Bismut [5], and see Hsu [2].

(*b*) *Processes with jumps.* See Bass and Cranston [1], Bichteler and Jacod [1], and Bismut [3].

(*c*) *Connection with the theory of large deviations.* See Bismut [4].

(*d*) *Other approaches.* Though we have outlined only the Bismut approach, *we stress that there are other approaches to the Malliavin calculus of very great power and of very great intrinsic interest.* Here is a selection of a few of the main papers:

Malliavin [1, 2], Stroock [1, 2], Kusuoka and Stroock [1, 2], Sections V.7–8 of Ikeda and Watanabe [1], Ikeda and Watanabe [2].

The Taniguchi symposium, Itô [7], contains other important papers.

A naïve introduction, explaining how to represent Malliavin's process by the Brownian sheet, is contained in Williams [13].

(*e*) *Quasi-everywhere properties of Brownian motion.* The Malliavin calculus in Malliavin's own formulation automatically creates a fascinating concept of properties of Brownian motion which are valid 'quasi-everywhere'. This concept is formulated in terms of the infinite-dimensional capacity on the space of paths which is associated with Malliavin's process. Fukushima [2] and Lyons [3] are amongst papers which must tempt you into that field.

(*38.61*) *Further references to stochastic differential geometry in general.* As we reach the end of this Part on stochastic differential geometry, hosts of papers not mentioned in the text spring to mind, including such major achievements as Schwartz [1]. Elworthy [2] collects many interesting papers.

Lyons [2] is a most striking example of how probability can be used to solve an important problem on potential theory on Riemannian manifolds: what happens to the family of bounded/positive harmonic functions under change of metric to an equivalent metric? The technique used is fascinating and powerful.

D. G. Kendall [2] asks the question: suppose you have n points moving independently in $\mathbb{R}^d$; how does the 'shape' of the n-tuple evolve (which forces him to create a theory of shape in [3])?

We have included in the bibliography at the end of the book several other papers all of which are well worth reading.

6. ONE-DIMENSIONAL SDEs

39. A local-time criterion for pathwise uniqueness. In dimension greater than one, essentially the only way to prove pathwise uniqueness and the existence of solutions for an SDE is the Itô result (Theorem 11.2) for Lipschitz coefficients. But in one dimension, the results can be sharpened considerably. The historic paper of Yamada and Watanabe [1] gave the best Hölder conditions on the coefficients of the SDE in one dimension, and Nakao [1] subsequently showed how to handle pathwise uniqueness for uniformly positive σ which were of bounded variation on each compact. Perkins [1] realized that the use of semimartingale local time provided a unified framework and simplified proofs for 1-dimensional SDEs, and using local time in a quite different way, Le Gall [1] provided short and simple proofs of the classic Yamada–Watanabe results, and an improvement of the Nakao result. We shall follow Le Gall's elegant approach, which also yields easily a powerful comparison result of Yamada [1]. All of these papers are well worth reading; the simplicity with which such useful results can be derived is truly delightful.

We consider pathwise uniqueness for the 1-dimensional SDE

$$dX_t = \sigma(X_t)dB_t + b(X_t)dt, \quad X_0 = x. \tag{39.1}$$

So suppose X and X' are two solutions of (39.1). Then

$$\begin{aligned}
X_t \vee X'_t &= X_t + (X'_t - X_t)^+ \\
&= x + \int_0^t \sigma(X_s)\,dB_s + \int_0^t b(X_s)\,ds + \int_0^t I_{\{X'_s - X_s > 0\}} d(X' - X)_s \\
&\quad + \tfrac{1}{2} l_t^0(X' - X), \quad \text{(on using Tanaka's formula IV.43.6)} \\
&= x + \int_0^t [\sigma(X_s) + (\sigma(X'_s) - \sigma(X_s)) I_{\{X'_s - X_s > 0\}}] dB_s \\
&\quad + \int_0^t [b(X_s) + (b(X'_s) - b(X_s)) I_{\{X'_s - X_s > 0\}}] ds + \tfrac{1}{2} l_t^0(X' - X) \\
&= x + \int_0^t \sigma(X_s \vee X'_s) dB_s + \int_0^t b(X_s \vee X'_s) ds + \tfrac{1}{2} l_t^0(X' - X).
\end{aligned}$$

Thus we have the following result.

(*39.2*) PROPOSITION. *If X and X' are solutions of the SDE* (39.1) *and if $l_\bullet^0(X' - X) = 0$, then $X \vee X'$ also solves the SDE.*

If in addition uniqueness in law holds for (39.1), *then pathwise uniqueness holds.*

Proof. The first assertion is immediate. Uniqueness in law implies that the two solutions X and $X \vee X'$ have the same law, therefore for each t, $X_t \geqslant X'_t$ a.s. Symmetrically, $X'_t \geqslant X_t$ a.s., and continuity of the paths forces pathwise uniqueness. □

To establish pathwise uniqueness, then, we try to prove that $l_\bullet^0(X' - X) = 0$ whenever X and X' are two solutions. The following trivial result helps us to do this.

(*39.3*) PROPOSITION. *If $Y \equiv X' - X$, and ρ: $\mathbb{R}^+ \to \mathbb{R}^+$ is increasing, $\int_{0+} \rho(u)^{-1} du = \infty$, then $l_\bullet^0(Y) = 0$ if*

$$\int_0^t \rho(Y_s)^{-1} I_{\{Y_s > 0\}} d[Y]_s < \infty \quad \text{a.s.} \tag{39.4}$$

Proof. By the occupation density formula,

$$\int_0^t \rho(Y_s)^{-1} I_{\{Y_s > 0\}} d[Y]_s = \int_0^\infty \rho(a)^{-1} l_t^a(Y)\,da,$$

and by the right continuity in a of $l_t^a(Y)$, if $l_t^0(Y)>0$, then the right side is infinite. □

40. The Yamada–Watanabe pathwise-uniqueness theorem. Here is the classic criterion of Yamada–Watanabe for pathwise uniqueness of a 1-dimensional SDE.

(*40.1*) THEOREM (Yamada–Watanabe). *Suppose that σ and b are measurable, and satisfy the conditions:*

(40.2(i)) *there exists increasing $\rho\colon \mathbb{R}^+ \to \mathbb{R}^+$ such that*

$$\int_{0+} \rho(u)^{-1} du = \infty,$$

and for all $x,y \in \mathbb{R}$,

$$(\sigma(x)-\sigma(y))^2 \leqslant \rho(|x-y|);$$

(40.2(ii)) *b is Lipschitz.*

Then pathwise uniqueness holds for (39.1).

Proof (Le Gall). Since

$$\int_0^t \rho(Y_s)^{-1} I_{\{Y_s>0\}} d[Y]_s = \int_0^t \rho(|X_s - X_s'|)^{-1}(\sigma(X_s)-\sigma(X_s'))^2 I_{\{Y_s>0\}} ds$$
$$\leqslant t,$$

by (40.2(i)), condition (39.4) holds and $l_\bullet^0(Y)=0$. Thus

$$|X_t - X_t'| = \int_0^t \operatorname{sgn}(X_s - X_s')\{\sigma(X_s)-\sigma(X_s')\}\, dB_s$$
$$+ \int_0^t \operatorname{sgn}(X_s - X_s')\{b(X_s)-b(X_s')\}\, ds,$$

whence

$$\mathbf{E}|X_t - X_t'| \leqslant K\mathbf{E}\int_0^t |X_s - X_s'|\, ds,$$

K being the Lipschitz constant of b, and this implies that $\mathbf{E}|X_t - X_t'|=0$, by Gronwall's Lemma 11.11. □

Remarks. (i) If σ satisfies a Hölder condition of order α:

$$|\sigma(x)-\sigma(y)| \leqslant C|x-y|^\alpha \quad (x,\ y \in \mathbb{R})$$

for some $C<\infty$, $\alpha>0$, then provided $\alpha \geqslant \frac{1}{2}$, condition (40.2(i)) holds. Thus, in particular, pathwise uniqueness holds for the Girsanov SDE (26.3) provided $\alpha \geqslant \frac{1}{2}$.

The Girsanov SDE shows that the Yamada–Watanabe criterion is best possible; it shows that pathwise uniqueness can fail if we only demand that σ satisfies a Hölder condition of order $\alpha<\frac{1}{2}$.

(ii) As another application of Theorem 40.1, we deduce the promised pathwise uniqueness of the Bessel SDE (IV.35.3), whose coefficients clearly satisfy the Yamada–Watanabe conditions (40.2).

(iii) Theorem (40.1) holds also for time-dependent coefficients; only the notation becomes more complicated.

41. The Nakao pathwise-uniqueness theorem. Here is (the extension of) Nakao's pathwise-uniqueness result.

(*41.1*) THEOREM (Nakao, Le Gall). *Suppose that* σ *and* b *are bounded measurable functions, and that there exist* $\varepsilon>0$ *and bounded increasing* $f\colon \mathbb{R}\to\mathbb{R}$ *such that*:

(41.2(i)) $$\sigma(x)\geqslant\varepsilon \text{ for all } x;$$

(41.2(ii)) $$(\sigma(x)-\sigma(y))^2\leqslant|f(x)-f(y)| \text{ for all } x,\ y.$$

Then pathwise uniqueness holds for (39.1).

Proof (Le Gall). We saw (see Lemma 28.7) that uniqueness in law holds if $b=0$; in general use Theorem 27.1 to take care of b, so by Proposition 39.2 it is sufficient to prove

$$\int_0^t \rho(Y_s)^{-1}I_{\{Y_s>0\}}d[Y]_s<\infty \quad \text{a.s.},$$

where $Y=X'-X$ is the difference of two solutions X and X', and ρ is a suitable increasing function such that $\int_{0+}\rho(u)^{-1}du=\infty$. We take $\rho(u)=u$, and consider for $\delta>0$,

$$\mathbf{E}\left[\int_0^t I_{\{Y_s>\delta\}}Y_s^{-1}(\sigma(X_s')-\sigma(X_s))^2ds\right]$$
$$\leqslant\mathbf{E}\left[\int_0^t I_{\{Y_s>\delta\}}Y_s^{-1}\{f(X_s')-f(X_s)\}\,ds\right].$$

We can choose C^∞ increasing functions f_n bounded by the same constant as f such that $f_n(x)\to f(x)$ except possibly at points of discontinuity of f. Then

(41.3) $$\mathbf{E}\left[\int_0^t I_{\{Y_s>\delta\}}Y_s^{-1}\{f_n(X_s')-f_n(X_s)\}\,ds\right]$$
$$=\mathbf{E}\left[\int_0^t I_{\{Y_s>\delta\}}\int_0^1 f_n'(X_s+u(X_s'-X_s))\,du\,ds\right]$$
$$\leqslant\int_0^1 du\,\mathbf{E}\left[\int_0^t f_n'(X_s+u(X_s'-X_s))\,ds\right].$$

If $Z_s^u \equiv X_s + u(X_s' - X_s)$, then we have an expression

$$Z_t^u = x + \int_0^t \sigma_s^u dB_s + \int_0^t b_s^u ds$$

where for some K, $|\sigma_s^u| + |b_s^u| + |\sigma_s^u|^{-1} \leqslant K$ for all $s \geqslant 0$, $u \in [0, 1]$. Using Tanaka's formula, it is not hard to show that for some $C = C_t$, for all $u \in [0, 1]$,

$$\sup_a \mathbf{E}\, l_t^a(Z^u) \leqslant C,$$

and so we have bounded (41.3) above by

$$\int_0^1 du\, \mathbf{E} \int f_n'(a) l_t^a(Z_u) \varepsilon^{-2}\, da \leqslant C\varepsilon^{-2} \int f_n'(a)\, da$$

$$\leqslant C\varepsilon^{-2} . 2 \sup_x |f(x)|.$$

Now let $n \to \infty$ in (41.3) to conclude

$$\mathbf{E}\left[\int_0^t I_{\{Y_s > \delta\}} Y_s^{-1} \{f(X_s') - f(X_s)\}\, ds\right] \leqslant 2C\varepsilon^{-2} \sup_x |f(x)|.$$

(The points of discontinuity of f can safely be ignored, because they form a countable set at worst and X and X' spend Lebesgue almost no time in that set, by the density of occupation formula and (41.2(i)).) Finally, let $\delta \downarrow 0$ to deduce that

$$\int_0^t I_{\{Y_s > 0\}} Y_s^{-1} d[Y]_s < \infty \quad \text{a.s.},$$

since its expected value is finite. □

(*41.4*) *Remarks.* (i) If σ and σ^{-1} are bounded, and if σ is of bounded variation on compacts, then clearly (41.2(i)–(ii)) are satisfied, and pathwise uniqueness holds. This is the result proved by Nakao [1].

(ii) The condition $0 < \varepsilon \leqslant \sigma(x) \leqslant K < \infty$ is not of itself sufficient to guarantee pathwise uniqueness; Barlow [3] constructs for each $\alpha \in (0, \frac{1}{2})$ a σ which satisfies this condition and is Hölder of order α, but for which pathwise uniqueness fails.

(iii) Again, the same proof works for time-dependent coefficients.

42. Solution of a variance control problem. We return to the example of § 1, where we attempt to maximize our choice of control u (which is to be a process previsible with respect to the natural filtration of X^u, and bounded between positive finite constants δ and K, with $\delta \leqslant K$) the 'payoff'

$$\mathbf{E} \int_0^\infty e^{-\alpha s} I_{[-a, a]}(X_s^u)\, ds,$$

where X^u is the controlled process

$$X_t^u = x_0 + \int_0^t u_s \, dB_s,$$

B is a Brownian motion, and x_0 is a given starting point. It was conjectured that the process

$$X_t = x_0 + \int_0^t u(X_s)\, dB_s \tag{42.1}$$

was optimal, where $u(x) = \delta$ for $|x| \leqslant a$; $u(x) = K$ for $|x| > a$. There exists a weak solution because of the 'change of time scale' method, and that is all we really need. However, it is interesting to note that by Theorem 41.1, pathwise uniqueness holds for (42.1), so that by Theorem 17.1, the SDE is exact. This deals with all questions of existence and uniqueness of solutions of (42.1), and it remains only to prove that X is optimal, which is another application of the martingale optimality principle (15.1).

Let

$$F(x) \equiv \mathbf{E}^x\left[\int_0^\infty e^{-\alpha s} I_{[-a,a]}(X_s)\, ds\right] \quad (x \in \mathbb{R}).$$

Since the process X is strong Markov (Theorem 21.1), if we define $\tau \equiv \inf\{t: |X_t| = a\}$, then, for $|x| < a$,

$$\begin{aligned} F(x) &= \mathbf{E}^x\left[\int_0^\tau e^{-\alpha s} ds + e^{-\alpha\tau}\int_0^\infty e^{-\alpha s} I_{[-a,a]}(X_{s+\tau})\, ds\right] \\ &= \alpha^{-1}(1 - \mathbf{E}^x e^{-\alpha\tau}) + \mathbf{E}^x e^{-\alpha\tau} F(a), \end{aligned}$$

since F is symmetric. Hence, for $|x| < a$,

$$F(x) = \alpha^{-1} - (\alpha^{-1} - F(a)) \frac{\cosh(\theta x/\delta)}{\cosh(\theta a/\delta)} \tag{42.2}$$

where $\theta^2 = 2\alpha$, and likewise for $|x| > a$,

$$F(x) = e^{-\theta|x-a|/K} F(a). \tag{42.3}$$

Thus F is continuous, and will have a continuous derivative at a if and only if

$$F(a) = \alpha^{-1} \frac{K \tanh(\theta a/\delta)}{\delta + K \tanh(\theta a/\delta)}. \tag{42.4}$$

So if we define F by (42.2), (42.3) and (42.4), then F has the properties:

$$\begin{cases} \tfrac{1}{2}K^2 F'' = \alpha F \text{ in } |x| > a; \\ \tfrac{1}{2}\delta^2 F'' = \alpha F - 1 \text{ in } |x| < a; \\ F \text{ is piecewise } C^2 \text{ with continuous derivative;} \\ F'' > 0 \text{ in } |x| > a,\ F'' < 0 \text{ in } |x| < a. \end{cases} \tag{42.5}$$

Consider now the process

$$Y_t^u \equiv \int_0^t e^{-\alpha s} I_{[-a,a]}(X_s^u)\,ds + e^{-\alpha t} F(X_t^u),$$

where u is an arbitrary control. By Itô's formula,

$$e^{\alpha t} dY_t^u = F'(X_t^u) dX_t^u + \{\tfrac{1}{2} u_t^2 F''(X_t^u) - \alpha F(X_t^u) + I_{[-a,a]}(X_t^u)\}\,dt.$$

The first term is a local martingale, the second is a decreasing finite-variation process, which is zero if $X^u = X$. Thus Y^u is a bounded supermartingale, and so

$$F(x_0) = Y_0 \geqslant \mathbf{E}^{x_0} Y_\infty^u = \mathbf{E}^{x_0}\left\{\int_0^\infty e^{-\alpha s} I_{[-a,a]}(X_s^u)\,ds\right\}$$

with equality if $X^u = X$. Thus X is optimal.

43. A comparison theorem. Finally, we consider a comparison theorem for 1-dimensional SDEs; there are many similar results (see, for example, Yamada [1], Yamada and Ogura [1]), but we give one which is due to Ikeda and Watanabe [1], proved by the local-time techniques of Le Gall. Again, we state the result for time-homogeneous coefficients, though it remains true for time-dependent ones.

(*43.1*) THEOREM (Ikeda–Watanabe). *Suppose that, for* $i = 1, 2$,

$$X_t^i = X_0^i + \int_0^t \sigma(X_s^i)\,dB_s + \int_0^t \beta_s^i\,ds, \tag{43.2}$$

and that there exist $b_i \colon \mathbb{R} \to \mathbb{R}$, $b_1(x) \geqslant b_2(x)$, *such that*

$$\beta_s^1 \geqslant b_1(X_s^1), \quad b_2(X_s^2) \geqslant \beta_s^2.$$

Suppose also that:

(43.3(i)) σ *satisfies the Yamada–Watanabe condition* (40.2(i));
(43.3(ii)) $X_0^1 \geqslant X_0^2$ a.s.;
(43.3(iii)) *one of* b_1, b_2 *is Lipschitz.*

Then $X_t^1 \geqslant X_t^2$ *for all* t a.s..

Proof. Exactly as in the proof of Theorem 40.1, if $Y \equiv X^2 - X^1$, then $l_\cdot^0(Y) = 0$. Thus,

$$(X_t^2 - X_t^1)^+ = \int_0^t I_{\{X_s^2 - X_s^1 > 0\}}\,(\sigma(X_s^2) - \sigma(X_s^1))\,dB_s$$

$$+ \int_0^t I_{\{X_s^2 - X_s^1 > 0\}}\,(\beta_s^2 - \beta_s^1)\,ds$$

implying, if b_1, say, is Lipschitz with constant K,

$$\begin{aligned} 0 \leqslant \mathbf{E}(X_t^2 - X_t^1)^+ &\leqslant \mathbf{E}\int_0^t I_{\{X_s^2 - X_s^1 > 0\}}(\beta_s^2 - \beta_s^1)\,ds \\ &\leqslant \mathbf{E}\int_0^t I_{\{X_s^2 - X_s^1 > 0\}}\{b_2(X_s^2) - b_1(X_s^1)\}\,ds \\ &\leqslant \mathbf{E}\int_0^t I_{\{X_s^2 - X_s^1 > 0\}}\{b_1(X_s^2) - b_1(X_s^1)\}\,ds \\ &\leqslant K\mathbf{E}\int_0^t (X_s^2 - X_s^1)^+\,ds. \end{aligned}$$

Hence $\mathbf{E}(X_t^2 - X_t^1)^+ = 0$ for all t. □

Ikeda and Watanabe [1] use this result to establish criteria for explosion of multidimensional diffusions by comparison with a 1-dimensional diffusion where the problem is easy. This is an entirely typical application of comparison theorems; see Yamada [1] for other nice examples.

Remark. Look ahead to Remark (48.2(i)).

7. ONE-DIMENSIONAL DIFFUSIONS

44. Orientation. A 1-dimensional diffusion will now be taken to mean a continuous strong Markov process with values in some interval $I \subseteq \mathbb{R}$ (see (45.1) for a precise definition).

If $\sigma: \mathbb{R} \to (0, \infty)$ and $b: \mathbb{R} \to \mathbb{R}$ are Lipschitz functions, then the solution of the SDE

(44.1) $$dX_t = \sigma(X_t)\,dW_t + b(X_t)\,dt,$$

where W is a $\mathrm{BM}^0(\mathbb{R})$, is a 1-dimensional diffusion. We are now going to concentrate on the idea of constructing a diffusion from a Brownian motion B by time transformation, rather than via an SDE. This is the reason for the notational switch from B to W in our SDE. It will be to B that the crucial application of the stochastic calculus will be made.

Recall our treatment of equation (44.1) in § 28. Let $s: \mathbb{R} \to \mathbb{R}$ be the continuous strictly increasing C^2 function defined by

(44.2) $$s(0) = 0, \quad s'(x) = \exp\left[-\int_0^x 2b(t)\sigma(t)^{-2}\,dt\right].$$

If $Y_t \equiv s(X_t)$, then, by Itô's formula,

$$dY_t = g(Y_t)\,dW_t,$$

where $g \equiv (s'\sigma)\circ s^{-1}$. We saw in § 28 how to obtain a weak solution to the SDE

$dY = g(Y)\,dW$ by time change: if B is a Brownian motion, and we define the strictly increasing PCHAF

(44.3) $$A_t \equiv \int_0^t g(B_u)^{-2}\,du,$$

and let γ be the continuous inverse to A, then

$$Y_t \equiv B(\gamma_t)$$

is a (weak) solution of the SDE $dY = g(Y)\,dW$. (In other words, $W = \int g(Y)^{-1}\,dY$ is a $\mathrm{BM}^0(\mathbb{R})$.)

Thus a weak solution of the SDE (44.1) can be obtained from a Brownian motion by two operations:

(44.4(i)) time change B by the inverse of the PCHAF A;
(44.4(ii)) apply a continuous strictly increasing function (s^{-1}) to the time change of B.

Our main aim is to prove that *every 1-dimensional diffusion which is 'regular' in the sense explained in the next section may be obtained from a Brownian motion by the two operations just described.*

This characterization of 1-dimensional diffusions was achieved to its fullest extent by Itô and McKean [1], following on from the essential work of Feller, Dynkin, Ray and Volkonskii. (See Itô and McKean's book for the history.) Freedman [1] and Breiman [1] have fine accounts, which we shall follow in places.

However, *we make essential use of the generalized Itô–Tanaka formula (IV.45.1) in what we hope is the kind of (probabilistic rather than analytic) treatment which is requested on the last page of Breiman* [1]. Essentially the same approach is used by Meléard [1].

There are more general diffusions than the solution X of an SDE of type (44.1), as you know from the Feller–McKean example (III.23). But because any diffusion on $\mathbb{R}$ may be approximated by such an X, the example (44.1) is a good one to have in mind. Of course, if the interval I has boundaries (as in the case of reflecting Brownian motion) we shall have to deal with these separately; and, of course, we shall have to make some proper definitions before we go any further!

45. Regular diffusions. Fix an interval $I \subseteq \mathbb{R}$, and let $\Omega \equiv C(\mathbb{R}^+, I)$ be the canonical space of continuous I-valued paths, with the canonical process $X_t(\omega) \equiv \omega(t)$ $(\omega \in \Omega)$, and filtration $\mathscr{F}_t^\circ \equiv \sigma(\{X_s\colon s \leqslant t\})$, $\mathscr{F}^\circ \equiv \sigma(\{X_s\colon s \geqslant 0\})$, and shift operators θ_t defined as usual by $(\theta_t\omega)(s) \equiv \omega(t+s)$.

(*45.1*) DEFINITION (canonical diffusion on I). *A* canonical diffusion *on I is a family $\{\mathbf{P}^x\colon x \in I\}$ of probability measures on* $(\Omega, \mathscr{F}^\circ)$ *such that*

(45.1(i)) $x \mapsto \mathbf{P}^x(A)$ *is measurable for all* $A \in \mathscr{F}^\circ$;
(45.1(ii)) $\mathbf{P}^x(X_0 = x) = 1$ *for all* $x \in I$;
(45.1(iii)) *for each* $\{\mathscr{F}^\circ_{t+}\}$*-stopping time* T, *and each* $x \in I$, $\mathbf{P}^{X(T)}$ *is a regular conditional* $\mathbf{P}^x$*-distribution of* $\theta_T \omega$ *on* $\{T < \infty\}$.

Important remarks. (i) We have chosen to define a canonical diffusion because of its simplicity. However, we know from Volume 1 that we cannot possibly restrict attention entirely to canonical processes. Diffusion theory gets its strength from non-canonical constructions of diffusions: via SDEs, or, as we are now going to see, by the time-substitution method, etc. So the proper setting is a sextuple

$$(\Omega, \mathscr{F}, \mathscr{F}_t, X_t, \theta_t, \mathbf{P}^x).$$

Of course, we can transfer some of the structure of a non-canonical diffusion to yield a canonical diffusion with the same '$\mathbf{P}^x$ laws' on the canonical space.

We do not really want to get involved with abstract sextuples. So, we ask you to assume for the moment that we are given a diffusion X in canonical form. Enough of the 'sextuple' formulation was indicated in § III.18 for you to get the idea of how the time-transformation method could be properly formulated, and you can turn to Blumenthal and Getoor [1] for the details. You may well be content on a first reading to see that the laws of the process constructed by the time-transformation method satisfy the requirements of Definition 45.1.

(ii) We demand the strong Markov property for $\{\mathscr{F}^\circ_{t+}\}$-stopping times in order to preclude examples like the 'hold-drift' example III.35.8.

(iii) Do notice that our diffusions are honest; if you want to know what happens when killing is allowed, look in Itô and McKean [1], or see Meléard [1] for an account using methods closer to those used here.

(*45.2*) DEFINITION (regular diffusion). *A diffusion is called* regular *if for all* $x \in \text{int}(I)$, $y \in I$,

$$\mathbf{P}^x(H_y < \infty) > 0,$$

where $H_y \equiv \inf\{t > 0: X_t = y\}$. *In particular, y can be a finite boundary point.*

The concept of regularity for diffusions is very like the concept of irreducibility for Markov chains: the behaviour of regular diffusions (as with irreducible Markov chains) is much more orderly, and it is possible to decompose the general diffusion into 'regular pieces' (see Itô and McKean [1]), so we lose little generality, and gain much simplification, by making the following assumption.

(*45.3*) ASSUMPTION. *All 1-dimensional diffusions considered from now on are regular.*

The programme now is as follows. Firstly we find the analogue of the strictly increasing functions s at (44.2), called the scale function of the diffusion. In general, s is not C^2; one can say only that s is continuous. Next, writing the PCHAF A (44.3) as

$$A_t \equiv \int_0^t g(B_s)^{-2}\,ds = \int m(da)\, l_t^a,$$

where $m(da) = g(a)^{-2}\,da$, our aim is to find the analogue of m (called the speed measure). In general, m will not have a density, although $0 < m(a, b) < \infty$ for all $a < b \in \mathrm{int}(I)$. We then show how to obtain the arbitrary 1-dimensional diffusion (possibly on a non-open interval) from Brownian motion by time-change, and finally we examine analytical properties of diffusions, the ergodic theorem and the 'coupling' method.

46. The scale function s. Before constructing the scale function s, we need a few preliminaries.

Recall that X is a regular diffusion. Hence, if $x \in \mathrm{int}(I)$, there is always a $y > x$ for which condition (46.2) below holds. (It is only if x is a boundary point that the condition has (extra) force.)

Lemma 46.1 says in particular that for $x \in \mathrm{int}(I)$, the diffusion enters (x, ∞) immediately with $\mathbf{P}^x$-probability 1; of course, you can argue similarly that the diffusion must also enter $(-\infty, x)$ immediately with $\mathbf{P}^x$-probability 1.

(*46.1*) LEMMA. (i) *Suppose that $x \in I$, and that for some $y \in I$, $y > x$,*

$$\mathbf{P}^x(H_y < \infty) > 0. \tag{46.2}$$

Then

$$\mathbf{P}^x(\exists \varepsilon > 0 \text{ such that } X_t \leqslant x,\ \forall t \leqslant \varepsilon) = 0. \tag{46.3}$$

(ii) *For $a, b \in I$, $a < b$, there exists $\delta > 0$ and $v < \infty$ such that, for all $x \in [a, b]$,*

$$\mathbf{P}^x(H \leqslant v) \geqslant \delta, \tag{46.4}$$

where $H \equiv H_a \wedge H_b$. Moreover, for all $p > 0$, and all $x \in [a, b]$

$$\mathbf{E}^x(H^p) < \infty. \tag{46.5}$$

In particular, H is $\mathbf{P}^x$-a.s. finite for all $x \in [a, b]$.

Proof. (i) Let $T \equiv \inf\{u > 0: X_u > x\}$. Then T is an $\{\mathcal{F}^\circ_{t+}\}$-stopping time, and by the Blumental 0–1 law (III.9), $\mathbf{P}^x(T > 0) = 0$ or 1. For the purpose of argument by contradiction, suppose that this probability equals 1. Define

$$\begin{aligned}\Lambda &= \{\exists \varepsilon > 0 \text{ such that } X_t \leqslant x \text{ for all } t \leqslant \varepsilon\}\\ &= \{T > 0\}.\end{aligned}$$

Then

$$\begin{aligned}0 &= \mathbf{P}^x(T<\infty, \exists\varepsilon>0 \text{ such that } X_{T+t}\leqslant x \text{ for all } 0\leqslant t\leqslant\varepsilon)\\ &= \mathbf{E}^x(I_\Lambda\circ\theta_T;\ T<\infty)\\ &= \mathbf{E}^x(\mathbf{P}^{X(T)}(\Lambda);\ T<\infty) \quad \text{by (45.1(iii))};\\ &= \mathbf{P}^x(T<\infty),\end{aligned}$$

since we suppose that $\mathbf{P}^x(\Lambda)=1$, and since $X_T=x$ on $\{T<\infty\}$. But $H_y>T$ and we suppose at (46.2) that H_y is finite with positive probability, a contradiction. Therefore $\mathbf{P}^x(T>0)=\mathbf{P}^x(\Lambda)=0$.

(ii) Fix $c\in(a,b)$. Since the diffusion is regular and c is in $\text{int}(I)$, there exist $\delta>0$ and $v<\infty$ such that

$$\mathbf{P}^c(H_a\leqslant v)\geqslant\delta, \quad \mathbf{P}^c(H_b\leqslant v)\geqslant\delta.$$

For any $x\in(a,c)$, we use the strong Markov property at H_x:

$$\begin{aligned}\mathbf{P}^c(H_a\leqslant v) &\leqslant \mathbf{P}^c(H_x<\infty, H_a\circ\theta_{H_x}\leqslant v)\\ &\leqslant \mathbf{P}^x(H_a\leqslant v).\end{aligned}$$

Therefore $\mathbf{P}^x(H_a\leqslant v)\geqslant\mathbf{P}^x(H_a\leqslant v)\geqslant\delta$, and the case of $x\in(c,b)$ is similar. Finally, from (46.4) it follows that, for all $n\in\mathbb{N}$, for all $x\in[a,b]$,

$$\mathbf{P}^x(H>nv)\leqslant(1-\delta)^n,$$

by the Markov property at time v, and induction on n. Result (46.5) now follows. □

Now fix a and b in I with $a<b$, let J denote the interval $[a,b]$, and define

(46.6) $$s_J(x)\equiv\mathbf{P}^x(H_b<H_a), \quad x\in J.$$

Clearly $s_J(a)=0$, $s_J(b)=1$, and for $a<x<y<b$,

(46.7) $$s_J(x)=\mathbf{P}^x(H_y<H_a)\,s_J(y)$$

so that s_J is non-decreasing.

(46.8) LEMMA. *The function s_J is continuous and strictly increasing.*

Proof. For $a<x<y<b$,

$$\begin{aligned}\mathbf{P}^x(H_y<H_a) &= \mathbf{P}^x\Big(\sup_{t\leqslant H_a} X_t\geqslant y\Big)\\ &\uparrow \mathbf{P}^x\Big(\sup_{t\leqslant H_a} X_t> x\Big)\end{aligned}$$

as $y\downarrow x$. But $H_a>0$, $\mathbf{P}^x$-a.s., so, by (46.3),

$$\mathbf{P}^x(H_y<H_a)\uparrow 1 \quad \text{as } y\downarrow x,$$

and the right-continuity of s follows from (46.7). A symmetric argument shows that s_J is left continuous.

To prove that s_J increases strictly, we must show that $\mathbf{P}^x(H_y<H_a)<1$ for $a<x<y<b$. Suppose that this were false. Define the stopping times $T_0=0$, $S_{n+1}=\inf\{t>T_n: X_t=x\}$, $T_{n+1}=\inf\{t>S_{n+1}: X_t=a \text{ or } y\}$, $n\in\mathbb{Z}^+$. Then, since $y\in\operatorname{int}(I)$,

$$0<\mathbf{P}^y(H_a<\infty)\leqslant\mathbf{P}^y\,(\text{for some } n,\ T_n<\infty \text{ and } X(T_n)=a).$$

But by the strong Markov property at S_n,

$$\mathbf{P}^y(T_n<\infty,\ X(T_n)=a)=\mathbf{E}^y[\mathbf{P}^x(H_a<H_y): S_n<\infty]$$
$$=0,$$

since we are assuming $\mathbf{P}^x(H_y<H_a)=1$. Hence $\mathbf{P}^y(H_a<\infty)=0$, a contradiction. □

Notice that for each $x\in[a,b]$,

(46.9) $$s_J(X(t\wedge H_a\wedge H_b))=\mathbf{P}^x[H_b<H_a|\mathscr{F}^\circ_{t+}]$$

is a continuous $\mathbf{P}^x$*-martingale.*

(*46.10*) DEFINITION (scale function). *Let X be a (regular) diffusion on I with laws $\{\mathbf{P}^x: x\in I\}$. A* scale function *for X is a continuous strictly increasing function $s: I\to\mathbb{R}$ such that, for all $a<x<b\in I$,*

(46.11) $$\mathbf{P}^x(H_b<H_a)=\frac{s(x)-s(a)}{s(b)-s(a)}.$$

If $s(x)=x$ is a scale function for X, we say that X *is in natural scale.*

It is trivial to verify that s is uniquely determined to within increasing affine transformations. The main result of this section is the following.

(*46.12*) THEOREM. *A diffusion X on I has a scale function s, and $Y\equiv s(X)$ is a diffusion in natural scale on $s(I)$.*

Proof. If I is compact, take $s=s_I$ as defined by (46.6). Otherwise, take increasing compact intervals J_n with union I; if I contains one of its endpoints, choose each J_n to contain that point as well. If $J_n=[a_n,b_n]$, and if $s_n\equiv s_{J_n}$, define $s^n: J_n\to\mathbb{R}$ as follows: let $s^1\equiv s_1$, and for $n\geqslant1$,

$$s^{n+1}(x)\equiv s^n(b_n)\{s_{n+1}(x)-s_{n+1}(a_n)\}\{s_{n+1}(b_n)-s_{n+1}(a_n)\}^{-1}$$
$$+s^n(a_n)\quad(x\in J_{n+1}).$$

Thus

$$s^{n+1}(a_n)=s^n(a_n),\quad s^{n+1}(b_n)=s^n(b_n)\quad\forall n\geqslant1.$$

Now define s by the formula

$$s(x) \equiv s^n(x) \quad \text{for } x \in J_n .$$

To ensure that s is well-defined, and to confirm that it has the defining property (46.11), we shall prove that if $\alpha \leqslant a < x < b \leqslant \beta$ and if $\tilde{s}: [\alpha, \beta] \to \mathbb{R}$ has the property

$$\text{(46.13)} \qquad \mathbf{P}^x(H_\alpha < H_\beta) = \frac{\tilde{s}(x) - \tilde{s}(\alpha)}{\tilde{s}(\beta) - \tilde{s}(\alpha)} \quad (\alpha \leqslant x \leqslant \beta)$$

then for $a \leqslant x \leqslant b$,

$$\text{(46.14)} \qquad \mathbf{P}^x(H_b < H_a) = \frac{\tilde{s}(x) - \tilde{s}(a)}{\tilde{s}(b) - \tilde{s}(a)}.$$

Indeed, by the strong Markov property at $H_a \wedge H_b$, if $x \in (a, b)$,

$$\mathbf{P}^x(H_\beta < H_\alpha) = \mathbf{P}^x(H_b < H_a)\mathbf{P}^b(H_\beta < H_\alpha) + \mathbf{P}^x(H_a < H_b)\mathbf{P}^a(H_\beta < H_\alpha).$$

Using (46.13), an easy rearrangement yields (46.14). Hence the restriction of s^{n+1} to J_n is a scale function on J_n, and so must be an increasing affine transformation of s^n. But s^n and s^{n+1} agree at the two ends of J_n, so that s^n must indeed be the restriction of s^{n+1} to J_n. It is trivial to check that Y is a regular diffusion in natural scale. □

(*46.15*) COROLLARY. *Let X be a diffusion on I and let $T \equiv \inf\{t \geqslant 0: X_t \notin \operatorname{int}(I)\}$. Then for each $x \in I$, $Y^T \equiv s(X^T)$ is a continuous* $\mathbf{P}^x$*-local martingale.*

Proof. If both endpoints are in I, then I is a compact interval and the result follows on taking $J = I$ at (46.9).

Next, suppose that exactly one endpoint is in I; without loss of generality, suppose that the upper endpoint b_I of I is not in I. Then if (b_n) is a sequence of points in I such that $b_n > x$ and $b_n \uparrow\uparrow b_I$, then $H(b_n) \uparrow \infty$ $\mathbf{P}^x$-a.s., for otherwise we would have $X(\lim H(b_n)) = b_I \notin I$. Since the lower endpoint a_I is in I, $T = H(a_I)$, and Y^T is a $\mathbf{P}^x$-local martingale reduced strongly by the stopping times $H(b_n)$.

Finally, if neither endpoint is in I, then I is an open interval; $I = (a_I, b_I)$. Taking $a_n \downarrow\downarrow a_I$, $b_n \uparrow\uparrow b_I$, then $s(X)$ is a $\mathbf{P}^x$-local martingale, reduced strongly by $H(a_n) \wedge H(b_n)$. □

If we can prove results about diffusions in natural scale, we can apply them to $Y = s(X)$ and deduce the corresponding results for X; thus we lose no generality by making the following assumption.

(*46.16*) ASSUMPTION. *All diffusions considered from now on will be in natural scale.*

47. The speed measure m; time substitution. The main aim of this section is to exhibit a (regular) diffusion in natural scale as a time-change of Brownian motion.

A number of different cases must be considered, depending on the nature of the endpoints of the interval I. The final result, however, has a pleasing unity; here is what it says.

(*47.1*) THEOREM. *Let* $\{\mathbf{P}^x: x \in I\}$ *be a (regular) diffusion in natural scale on the interval* I. *Then there exists a measure* m *on* I *such that, for each* $y \in I$, *there exists on some enrichment of* $(\Omega, \mathscr{F}^\circ, \mathbf{P}^y)$ *a Brownian motion* B *started at* y *such that the canonical process* X, *a diffusion with law* $\mathbf{P}^y$, *can be expressed as a time change of* B:

$$X_t \equiv B(\gamma_t), \tag{47.2}$$

where $\gamma.$ *is the right-continuous inverse to the PCHAF*

$$A_t \equiv \int_I m(dz) l_t^z, \tag{47.3}$$

and where $\{l_t^z: z \in \mathbb{R}, t \geqslant 0\}$ *is (a jointly continuous version of) the local time process of* B.

It is clear that once one knows the measure m, one can use it to construct a diffusion with law $\mathbf{P}^y$ from a Brownian motion, by way of the prescription (47.2)–(47.3); in particular, the measure m determines the laws $\{\mathbf{P}^y: y \in I\}$ of the diffusion. Such a measure is obviously important enough to deserve a name.

(*47.4*) DEFINITION (speed measure) *The measure* m *appearing in the statement of Theorem 47.1 is called the* speed measure *of the diffusion* $\{\mathbf{P}^x: x \in I\}$. *The speed measure also has the property*:

$$\text{for any } a < b \in \operatorname{int}(I), \quad 0 < m([a, b]) < \infty. \tag{47.5}$$

Remarks. (i) Notice that in regions where m is large, the diffusion moves slowly; thus, if the name 'speed' measure were not so well established, we would be tempted to call m the 'sloth' measure!

(ii) *The converse to Theorem 47.1 also holds*; given an interval I and a measure m on I satisfying (47.5), then if B is a Brownian motion started at $y \in I$, with local time process $\{l_t^z: z \in \mathbb{R}, t \geqslant 0\}$, and we define the PCHAF A by

$$A_t \equiv \int_I m(dz) l_t^z,$$

then $X_t \equiv B(\gamma_t)$ is a regular diffusion in natural scale on I. Here, of course, γ is the right-continuous inverse to A. It is easiest to see this if $I = \mathbb{R}$, for then A is finite-valued, continuous, and *strictly* increasing (because of (47.5)), $A_\infty = \infty$ a.s., and so the time-change γ is also finite-valued, continuous, and strictly increasing. The discussion of § III.21 shows that X is a strong Markov process on $\mathbb{R}$, whose paths are continuous (since B and γ are continuous), and so X is a diffusion. Properties of Brownian motion imply that X is regular and in natural scale.

The cases where $I \neq \mathbb{R}$ are essentially the same, but require care over the boundaries. Think about this again once you have read up to § 51 on boundary points.

The different cases of Theorem 47.1 depend on the different types of endpoint, which we classify as follows.

(*47.6*) DEFINITION (inaccessible, absorbing, reflecting endpoints). *The endpoint c of the interval I is called* inaccessible *if $c \notin I$. If $c \in I$, then c is called* absorbing *if $\mathbf{P}^c(H_y < \infty) = 0$ for all $y \in I \backslash \{c\}$, and c is called* reflecting *if $\mathbf{P}^c(H_y < \infty) > 0$ for some $y \in I \backslash \{c\}$.*

(*47.7*) *Proof of Theorem 47.1 when both endpoints are inaccessible.* The first case of Theorem 47.1 to be considered is that in which both endpoints are inaccessible, and therefore I is open.

For each interval $J \subset I$, define the function h_J by

$$h_J(x) \equiv \mathbf{E}^x(H_J) \quad (x \in J), \tag{47.8}$$

where

$$H_J \equiv \inf\{t \colon X_t \notin J\}. \tag{47.9}$$

If $J = [a, b]$, then H_J is almost surely equal to $H_a \wedge H_b$, since both of a and b are in $\operatorname{int}(I) = I$; and by Lemma 46.1(ii), h_J is finite-valued. The key observation is that if K is another interval, $J \subseteq K \subset I$, with endpoints $\alpha < \beta$ in I, then using the strong Markov property at $H_a \wedge H_b$, for $x \in J$,

$$\begin{aligned} h_K(x) &= \mathbf{E}^x(H_a \wedge H_b) + \frac{x-a}{b-a}\mathbf{E}^b(H_K) + \frac{b-x}{b-a}\mathbf{E}^a(H_K) \\ &= h_J(x) + \frac{x-a}{b-a} h_K(b) + \frac{b-x}{b-a} h_K(a), \end{aligned} \tag{47.10}$$

from which it follows immediately that

(47.11(i)) *h_K is strictly concave in K;*
(47.11(ii)) *$-h_J'' = -h_K''$ as measures in* $\operatorname{int}(J)$.

Hence *it is possible to define consistently a measure m on I by specifying that on any compact interval $J \subset I$,*

$$m(dx) \equiv -\tfrac{1}{2} h_J''(dx) \quad (x \in \operatorname{int}(J)). \tag{47.12}$$

Moreover, for $a < b$ in $\operatorname{int}(I)$, strict concavity of h_J implies that $m([a, b]) > 0$, and, since $[a, b]$ can be included in some open interval K whose closure is compact, it must be that $m([a, b]) < \infty$, since this is the increase in the gradient of $-\tfrac{1}{2}h_K$ over the interval J. Hence the speed measure m satisfies (47.5).

Now by Corollary 46.14, the diffusion X is a continuous local martingale, so by Theorem IV.34.11, on some enrichment of $(\Omega, \mathscr{F}^\circ, \mathbf{P}^y)$ there exists a Brownian motion B such that

$$X_t = B([X]_t).$$

It remains only to identify the increasing process $[X]$ as the inverse γ to the PCHAF A defined at (47.3).

To do this, take an interval $J = [a, b]$ containing y, and notice that

$$M_t \equiv h_J(X_t^H) + (t \wedge H) \equiv \mathbf{E}^y(H \mid \mathscr{F}_{t+}^\circ) \tag{47.13}$$

is a uniformly integrable $\{\mathscr{F}_{t+}^\circ\}$-martingale, where $H \equiv H_J = H_a \wedge H_b$. Thus, if τ is inverse to $[X]$,

$$\tilde{M}_t \equiv M(\tau_t) \equiv h_J(X^H(\tau_t)) + (\tau_t \wedge H) \tag{47.14}$$

is a uniformly integrable $\{\mathscr{F}^\circ(\tau_t +)\}$-martingale. From the continuity of $[X]$, it follows easily that

$$\tau_t < s \Leftrightarrow t < [X]_s \tag{47.15}$$

for any $s, t > 0$. Again from continuity of $[X]$, we deduce that

$$X^H(\tau_t) \equiv X(\tau_t \wedge H) = B([X]_{\tau(t) \wedge H}) = B(t \wedge \tilde{H}),$$

where $\tilde{H} \equiv [X]_H$. Thus

$$\tilde{M}_t = h_J(B(t \wedge \tilde{H})) + (\tau_t \wedge H). \tag{47.16}$$

Now *since h_J is concave, we can use the generalized Itô–Tanaka formula (IV.45.2) on the expression (47.16) for $\tilde{M}$*; using (47.12), we obtain

$$\tilde{M}_t = \tilde{M}_0 + \int_0^{t \wedge \tilde{H}} D_- h_J(B_s)\, dB_s - \int_I m(dz)\, l^z(t \wedge \tilde{H}) + (\tau_t \wedge H).$$

But the left-hand side is a martingale, so *the finite-variation term on the right must be zero*; therefore

$$\tau_t \wedge H = A(t \wedge \tilde{H}), \tag{47.17}$$

where A is the PCHAF of B:

$$A_t \equiv \int_I m(dz) l_t^z.$$

Now from (47.15) and (47.17) we conclude that on $[0, \tilde{H})$, τ is continuous and equal to A. By taking the interval J larger and larger, the first-exit time H_J increases to ∞ (since I is open) and hence τ is continuous, finite-valued, and equal to A on $[0, \zeta)$, where $\zeta \equiv [X]_\infty$. But τ is strictly increasing, since it is the inverse to the continuous increasing process $[X]$; thus $A = \tau$: $[0, \zeta) \to [0, \infty)$ is a homeomorphism with inverse $[X]$. Hence $[X] = \gamma$, and the theorem is proved in the case of open I.

(*47.18*) **Exercise.** Prove that for Brownian motion in the interval (a, b),

$$\mathbf{E}^x[H_a \wedge H_b] = (x-a)(b-x),$$

and deduce that for Brownian motion, speed measure is Lebesgue measure.

(*47.19*) *Time substitution for general I.* We are now going to prove Theorem 47.1 for the case where I has one accessible endpoint, and one inaccessible. Once we have done this, it will be clear that the same techniques can be used to establish Theorem 47.1 when both endpoints are accessible. The argument used is similar in general outline to that given for open I, but we separate the cases because the differences are sufficiently significant to require independent treatment.

We now suppose, without loss of generality, that the lower endpoint is accessible and is equal to 0. The upper endpoint b_I, which can be finite or infinite, is inaccessible, so that $I = [0, b_I)$. Two cases arise, according as 0 is absorbing, or reflecting.

(*47.20*) *The case where* 0 *is absorbing.* The easier case is that where 0 is absorbing. In this case, if we take an open interval J containing y, $J = (a, b) \subseteq I$, with $b < b_I$, and define h_J by (47.8)–(47.9) as before, then $H_J = H_a \wedge H_b$, h_J is finite-valued and (47.10) remains valid for open $K \supseteq J$ with $\sup K < b_I$. As before, we define m in J by $m(dx) = -\frac{1}{2}h_J''(dx)$, and m extends consistently to the whole of $\operatorname{int}(I)$. Since $X = X^T$ ($T \equiv \inf\{t\colon X_t \notin \operatorname{int}(I)\}$), X is a continuous local martingale (Corollary 46.15), and so is a time-change of some Brownian motion B on a suitable enrichment of $(\Omega, \mathscr{F}^\circ, \mathbf{P}^y)$. By exactly the same argument as before, if τ denotes the inverse to $[X]$, then

$$\tau = A\colon [0, \zeta) \to [0, H_0)$$

is a homeomorphism, where A is the PCHAF of B defined by

$$A_t \equiv \int_I m(dz) l_t^z,$$

and so $[X]$ is the inverse of A, at least on $[0, H_0)$. To deal with $[H_0, \infty)$, we set $m(\{0\}) = \infty$, so that $A(\zeta) = \infty$, and $\gamma(t) = \zeta$ for all $t \geqslant H_0$; thus $\gamma_t = [X]_t$ for all $t \geqslant 0$.

(*47.21*) *The case where 0 is reflecting.* Both the treatment of h_J and the time-changing of X require a different approach in the case where 0 is reflecting.

Taking $J = [0, b)$, where $y < b < b_I$, and defining h_J by (47.8)–(47.9) as before, we have that $H_J = H_b$, and, by a simple modification of the proof of Lemma 46.1(ii), h_J is again finite-valued. Noticing that for $0 \leqslant x < z < b$,

$$h_J(x) = \mathbf{E}^x[H_z] + h_J(z) > h_J(z),$$

we see that h_J is strictly decreasing in $[0, b)$, and so, if we define $h_J(x) = h_J(0)$ for all $x < 0$, then

(47.22) $\quad h_J$ *is concave in* $(-\infty, b]$, *and constant in* $(-\infty, 0]$.

Moreover, if $K=[0, c)$, with $b<c<b_I$, then for $x \in J$,

$$h_K(x)=h_J(x)+\mathbf{E}^b[H_c];$$

hence we can consistently define the measure m on $(-\infty, b_I)$ by

$$m(dx)=-\tfrac{1}{2}h_J''(dx)$$

for $x \in J$. Since h_J is constant in $(-\infty, 0)$, the measure m is concentrated in $[0, b_I)$, *and may give positive mass to the endpoint* 0.

For the time-changing of the diffusion X, we shall first show that

(47.23(i)) $X_\bullet^{H(b)}$ *is a (local) submartingale, hence a semimartingale*;
(47.23(ii)) $Z_\bullet \equiv X_\bullet - X_0 - L_\bullet$ *is a continuous local martingale,*

where $2L$ is the local time at 0 of X. From this, we shall deduce Theorem 47.1.

For simplicity, we shall abbreviate

$$H \equiv H_J = H_b.$$

(*47.24*) *Proof of* (*47.23*(*i*)). If $\rho_t \equiv \inf\{u>t\colon X_u^H=0\}$, then for $s<t$,

$$X_s^H=\mathbf{E}^x[X^H(t \wedge \rho_s)|\mathscr{F}_{s+}^\circ] \leqslant \mathbf{E}^x[X^H(t)|\mathscr{F}_{s+}^\circ]$$

since $\{X(t \wedge \rho_s \wedge H)\colon t \geqslant s\}$ is a $\mathbf{P}^x$-martingale (Corollary 46.15). Thus $X_\bullet^H$ is a submartingale. At this point, we need *the Meyer decomposition theorem* (*VI.32.3*) *to tell us that since X is a local submartingale, X is a semimartingale.*

(*47.25*) *Proof of* (*47.23*(*ii*)). Since X is a continuous semimartingale, it has a local time process. We shall denote its local time at 0 by $2L$, so that by Tanaka's formula (IV.43.6) applied to the nonnegative continuous semimartingale X,

$$X_t=X_t^+=X_0+Z_t+L_t,$$

where

$$Z_t=\int_0^t I_{\{X_s>0\}}\,dX_s.$$

We claim that Z is a local martingale. To prove this, we fix $\varepsilon \in (0, b)$, set $T_0=0$, and define for $n \geqslant 0$:

$$S_{n+1} \equiv \inf\{u>T_n\colon X_u \notin [0, \varepsilon)\}, \quad T_{n+1} \equiv \inf\{u>S_{n+1}\colon X_u=0\};$$

then

$$N_\bullet^{n,\varepsilon} \equiv \int_0^\bullet (S_n, T_n]_u\,dX_u \equiv X(T_n \wedge \bullet)-X(S_n \wedge \bullet)$$

is a $\mathbf{P}^x$-local martingale for each x (Corollary 46.15 again), and clearly the sum on

n of the $N^{n,\varepsilon}$ is another local martingale $N^{\varepsilon}_{\cdot}$. The quadratic variation process of N^{ε} is dominated by

$$[Z]_{\cdot} \equiv \int_0^{\cdot} I_{\{X_s>0\}}\, d[X]_s,$$

which can be made integrable by stopping at a suitable time, and then the (stopped L^2-bounded) martingales N^{ε} converge uniformly in L^2 to Z as $\varepsilon\downarrow 0$.

Now since

$$X_t = y + Z_t + L_t \geqslant 0,$$

and L grows only when X is zero, by (Skorokhod's) Lemma 6.2 we conclude that $L_t = 0 \vee (\sup\{-y-Z_s\colon s\leqslant t\})$. Moreover, on some enrichment of $(\Omega, \mathcal{F}^{\circ}, \mathbf{P}^y)$, there exists a Brownian motion W such that

$$Z_t = W([Z]_t)$$

(Theorem IV.34.11), and if

$$\tilde{L}_t = 0 \vee (\sup\{-y-W_s\colon s\leqslant t\}),$$

then according to Lévy's presentation of reflecting Brownian motion (6.5),

$$W_t^+ \equiv y + W_t + \tilde{L}_t$$

is a reflecting Brownian motion started at x. Further, since $\tilde{L}([Z]_t)$ is easily seen to be L_t, we have that

$$X_t = W^+([Z]_t).$$

Let $\{l_t^a\colon a\geqslant 0, t\geqslant 0\}$ be the local time process of W^+, let $\tau_t \equiv \inf\{u\colon [Z]_u > t\}$, and let $\tilde{H} \equiv [Z]_H$, where, as above, $H \equiv H_J$. By time-changing the UI $\{\mathcal{F}^{\circ}_{t+}\}$-martingale

$$M_t = h_J(X_t) + (t\wedge H) \equiv \mathbf{E}^y[H \mid \mathcal{F}^{\circ}_{t+}]$$

we obtain the UI $\{\mathcal{F}^{\circ}(\tau_t+)\}$-martingale

$$\begin{aligned}\tilde{M}_t \equiv M(\tau_t) &= h_J(X(\tau_t)) + (\tau_t\wedge H)\\ &= h_J(W^+(t\wedge\tilde{H})) + (\tau_t\wedge H).\end{aligned}$$

Now apply Itô's formula to $\tilde{M}$:

$$\tilde{M}_t = h(y) + \int_0^{t\wedge\tilde{H}} D_-h(W_s^+)\, dW_s^+ - \int_{[0,\infty)} m(da)\, l^a_{t\wedge\tilde{H}} + (\tau_t\wedge H). \tag{47.26}$$

Since $\tilde{M}$ is a martingale, the finite-variation part of the right-hand side of (47.26) must be zero, hence

$$\begin{aligned}0 &= \int_0^{t\wedge\tilde{H}} D_-h(0)\, d\tilde{L}_t - \int_{[0,\infty)} m(da)\, l^a_{t\wedge\tilde{H}} + (\tau_t\wedge H)\\ &= -\int_{[0,\infty)} m(da)\, l^a_{t\wedge\tilde{H}} + (\tau_t\wedge H).\end{aligned}$$

Hence immediately

$$\tau_t = A_t \equiv \int_{[0,\infty)} m(da) l_t^a \quad \text{for } t < \tilde{H}.$$

By taking J larger and larger as before, we obtain a representation of X as the time change of the reflecting Brownian motion W^+ by the inverse to the PCHAF A. Since W^+ can itself be represented as the time change of a Brownian motion $\tilde{W}$ by the PCHAF $\int_0^\bullet I_{[0,\infty)}(\tilde{W}_s)\, ds$, the theorem is proved in this case. □

The case of two accessible endpoints can now be deduced from the two cases we have treated in detail; we leave it to you to choose how much you want to do of this.

(*47.27*) *Speed measure and the generator.* We are deliberately underplaying the role of the infinitesimal generator in 1-dimensional diffusion theory, but the relationship between the speed measure and the generator is so transparent (and, in practice, the easiest way of obtaining the speed measure) that we feel we must say something about it. So let us suppose for simplicity that we have a diffusion in natural scale on ρ with a speed measure m satisfying

$$m(dx) = \rho(x)dx, \tag{47.28}$$

where $\rho(\bullet)$ is bounded and uniformly positive. Thus if B is a Brownian motion and

$$A_t \equiv \int m(dz) l_t^z \equiv \int_0^t \rho(B_s)\, ds$$

is the PCHAF by which we time-change, then the martingale

$$C_t^f = f(B_t) - f(B_0) - \int_0^t \tfrac{1}{2} f''(B_s)\, ds$$

time-changes to the martingale

$$\tilde{C}_t^f = f(X_t) - f(X_0) - \int_0^t \tfrac{1}{2}\rho(X_s)^{-1} f''(X_s)\, ds,$$

where $X_t \equiv B(\gamma_t)$, γ denoting the inverse to A. (See § 26 for more detail on time-changing). Thus the diffusion X is a weak solution to the SDE

$$dX_t = \rho(X_t)^{-1/2} dW_t \tag{47.29}$$

and X has generator

$$\mathscr{G} \equiv \frac{1}{2\rho(x)} \frac{d^2}{dx^2} \equiv \frac{1}{2} \frac{d^2}{dm dx}. \tag{47.30}$$

Hence if we have the generator of a 1-dimensional diffusion in natural scale, or the SDE which it satisfies, we can read off the speed measure from (47.28)–(47.30)!

This is true at least if ρ is of the nice form considered here, but if (say) ρ were allowed to vanish in places, there are potential difficulties involved in the time-changing. Nonetheless, the equivalence of the speed measure, SDE, and generator characterizations given here always provides a probable candidate for the speed measure, even when boundary points demand a more painstaking analysis.

(*47.31*) *Some notation.* The fact that a diffusion in natural scale can be obtained by time-changing a Brownian motion is astonishingly useful, as we shall see. *For the rest of this chapter, then, we shall where convenient assume that such a diffusion X is realized as*

$$X_t = B(\gamma_t),$$

where B is a Brownian motion with local time process $\{l_t^z: z\in\mathbb{R}, t\geqslant 0\}$ *and*

$$\gamma_t = \inf\left\{u:\ A_u \equiv \int m(dz) l_t^z > t\right\}.$$

The symbols B, A, l and γ will always be used with these meanings.

48. Example: the Bessel SDE. We now give a complete and thorough 'scale and speed' analysis of the SDE for the squared Bessel process:

$$dX_t = 2(X_t^+)^{1/2}\,dW_t + \alpha\,dt, \quad X_0 = 1, \tag{48.1}$$

where $\alpha\geqslant 0$. For integer $\alpha>0$, we encountered this SDE in § IV.35, where it arose as the SDE satisfied by the squared modulus of a BM($\mathbb{R}^\alpha$), but the SDE makes perfectly good sense for any $\alpha\in[0,\infty)$ (and, indeed, for any $\alpha\in\mathbb{R}$). Since the coefficients satisfy the conditions of the Yamada–Watanabe theorem (Theorem 40.1), *pathwise uniqueness holds for* (*48.1*), and therefore (Theorem 17.1) *uniqueness in law holds for* (*48.1*).

We can use Theorem 23.5 to prove that *there is a (weak) solution to* (*48.1*); or, at least, we could if the variance coefficient $\sigma(x)\equiv 2(x^+)^{1/2}$ were bounded. However, by replacing $\sigma(\cdot)$ by $\sigma_N(\cdot)\equiv\sigma(\cdot\wedge N)$, we can use Theorem 23.5 to prove the existence of a solution X^N to

$$dX_t^N = \sigma_N(X_t^N)\,dW_t + \alpha\,dt$$

which by Theorem 40.1 is pathwise unique, and this solution provides a solution to (48.1) up to $\inf\{t: X_t^N > N\}$; we can now paste the X^N to provide a solution X to (48.1). Since a weak solution exists, and pathwise uniqueness holds, by Theorem 17.1 *the SDE* (*48.1*) *is exact.* Hence, by Theorem 21.1, the solution is a strong Markov process, and therefore *a 1-dimensional diffusion.* It will be apparent that the diffusion is regular, and so it has a scale function and a speed measure, which we shall compute.

We are dealing in this example with two different ways of specifying a 1-dimensional diffusion; via the SDE (48.1), and via the scale function and the speed measure. If we have a 1-dimensional diffusion specified by an SDE, we can always use Itô's formula to work out the scale and speed, at least in int (I). However, since the scale and speed are frequently badly behaved at the boundary of I, Itô's formula may break down at the endpoints if these happen to be accessible. *Thus the SDE specification is not well adapted to investigating the boundary behaviour of a 1-dimensional diffusion*, which the scale and speed approach takes in its stride. The $\mathrm{BES}^2(\alpha)$ SDE for $0<\alpha<2$ is an example where the SDE approach runs into just this sort of difficulty.

(*48.2*) *Remarks.* (i) Since pathwise uniqueness holds for all the SDEs (48.1), we can use the stochastic comparison theorem (Theorem 43.1) to compare solutions for different α. In particular, we can compare any solution X^1 to (48.1) with the trivial solution $X^2_\cdot \equiv 0$ to the SDE

$$dX_t^2 = 2((X_t^2)^+)^{1/2}\,dW_t, \quad X_0^2 = 0,$$

which is the SDE (48.1) for $\alpha=0$; hence immediately

$$\mathbf{P}(X_t^1 \geqslant 0 \text{ for all } t\geqslant 0)=1.$$

Taking $\alpha=1$ in this example shows that the stochastic comparison theorem cannot be strict: even though $b^1(x)\geqslant b^2(x)+\varepsilon$ for all x, and $X_0^1 > X_0^2$, it does not follow that $X_t^1 > X_t^2$ for all t, since for $\alpha=1$, $X_t = W_t^2$ solves (48.1), and $X_t=0$ for arbitrarily large values of t.

(ii) Since uniqueness in law holds for the SDE (48.1), distributional properties of every solution can be deduced from the distributional properties of just one solution. We have just seen an example of this; since for $\alpha=1$ there is a solution of (48.1) such that X visits 0 for arbitrarily large values of t, we deduce the same property for every solution of (48.1).

(*48.3*) *The case where* $\alpha=2$. The case $\alpha=2$ is a critical case which must be dealt with separately. We shall prove the following:

(i) *The solution X is strictly positive:*

$$\mathbf{P}(X_t>0,\ \forall t>0)=1.$$

(ii) $\mathbf{P}\left(\sup_t X_t=\infty,\ \inf_t X_t=0\right)=1.$

(iii) *The scale fuction is*

$$s_2(x)=\log x.$$

(iv) *The speed measure of the diffusion $Y_\cdot \equiv s_2(X)$ in natural scale on $\mathbb{R}$ is*

$$m_2(dy)=e^y\,dy/4.$$

We saw in §IV.35 that there is a weak solution to (48.1) for $\alpha=2$, namely, the squared modulus of 2-dimensional Brownian motion, for which the properties (i) and (ii) are immediate. Since, therefore, any solution X lives in $(0, \infty)$, where log is C^∞, we can apply Itô's formula to $Y_\cdot \equiv \log(X_\cdot)$, yielding

(48.4) $$dY_t = 2X_t^{-\frac{1}{2}} dW_t = 2e^{-\frac{1}{2}Y_t} dW_t.$$

Thus log(•) is a scale function, since Y is a local martingale, and (iii) and (iv) follow as in §28.

(*48.5*) *The case where* $\alpha>2$. We shall show that for $\alpha>2$,

(i) *the solution* X *is strictly positive*:

$$\mathbf{P}(X_t>0,\ \forall t>0)=1.$$

(ii) $\mathbf{P}(X_t \to \infty \text{ as } t\to\infty)=1.$

(iii) *The scale function of* X *is*

$$s_\alpha(x) \equiv -x^{1-\alpha/2}.$$

(iv) *The speed measure of the diffusion* $Y_\cdot \equiv s_\alpha(X_\cdot)$ *in natural scale on* $(-\infty, 0)$ *is*

$$m_\alpha(dy) = [\,(\alpha-2)|y|^{c(\alpha)}\,]^{-2} I_{(-\infty,0)}(y)\,dy,$$

where $c(\alpha)\equiv(\alpha-1)/(\alpha-2)>0$.

The property (i) follows immediately by comparison with the solution to (48.1) for $\alpha=2$, which is strictly positive.

The form of the scale and speed given in (iii) and (iv) comes from the calculations of §28 applied to this example; since s_α is C^∞ in $(0, \infty)$ where X takes values, we can apply Itô's formula to $Y_\cdot \equiv s_\alpha(X_\cdot)$ to obtain

$$dY_t = (\alpha-2)|Y_t|^{c(\alpha)} dW_t, \quad Y_0 = -1,$$

from which (iii) and (iv) follow.

The fact that $s_\alpha(0)=-\infty$, and $s_\alpha(\infty)=0$ implies (ii).

(*48.6*) *The case where* $0<\alpha<2$. This case is the most interesting. We already know that the SDE (48.1) is exact, and that the solution is nonnegative; we shall show that

(i) 0 *is recurrent for* X:

$$\mathbf{P}(\exists n \text{ such that } X_t>0 \text{ for all } t>n)=0.$$

(ii) $\displaystyle\int_0^\infty I_{\{0\}}(X_t)\,dt=0.$

(iii) *The scale function of* X *is*

$$s_\alpha(x) \equiv x^{1-\alpha/2}.$$

(iv) *The speed measure of the diffusion* $Y \equiv s_\alpha(X_\cdot)$ *in natural scale on* $[0, \infty)$ *is*

$$m_\alpha(dy) = [(2-\alpha)y^{c(\alpha)}]^{-2} I_{(0,\infty)}(y)dy,$$

where $c(\alpha) = (\alpha-1)/(\alpha-2)$. *In particular,* m_α *puts no mass on the endpoint* 0.

As before, we compute the scale function to be

$$s_\alpha(x) = x^{1-\alpha/2} \quad (x>0),$$

so that $Y_\cdot \equiv s_\alpha(X_\cdot)$ satisfies

$$dY_t = (2-\alpha)\, Y_t^{c(\alpha)} dW_t, \tag{48.7}$$

at least up until the first hit on 0, *when the Itô's formula calculation on which* (48.7) *is based breaks down.* To obtain the complete picture of the diffusion, we are going to *construct* a solution to (48.1) by use of scale and speed. However, in turns out to be more convenient to change the scale first, and then change the time.

Thus we start with a reflecting Brownian motion U, $U_0 = 1$, with local time $2L$ at zero, to which we apply the convex function

$$f(x) = \begin{cases} 0 & \text{for } x<0, \\ x^p & \text{for } x \geqslant 0, \end{cases}$$

where $p \equiv 2/(2-\alpha) > 1$. Letting $V_t \equiv f(U_t)$, Itô's formula as extended via Lemma IV.45.9 yields

$$\begin{aligned} V_t - V_0 &\equiv f(U_t) - f(U_0) \\ &= \int_0^t f'(U_s)\,dU_s + \tfrac{1}{2}\int_0^t f''(U_s)\,ds. \end{aligned}$$

Let $\{l_t^y : y \in \mathbb{R}, t \geqslant 0\}$ be the local time process of U. Then $Z \equiv U - \frac{1}{2} l^0$ is a Brownian motion.

We now have

$$\begin{aligned} V_t - V_0 &= \int_0^t p U_s^{p-1}\, d(Z_s + \tfrac{1}{2} l_s^0) + \int_0^t \tfrac{1}{2} p(p-1) U_s^{p-2}\, ds \\ &= \int_0^t p V_s^{1-r}\, dZ_s + \int_0^t \tfrac{1}{2} p(p-1) V_s^{1-2r}\, ds, \end{aligned}$$

where $r \equiv p^{-1}$. Hence by Itô's formula, for any $h \in C_K^\infty(\mathbb{R})$,

$$C_t^h = h(V_t) - h(V_0) - \int_0^t \mathscr{L} h(V_s)\, ds \text{ is a local martingale,} \tag{48.8}$$

where

$$\mathscr{L}h(x) \equiv \tfrac{1}{4}p^2 x^{1-2r} \mathscr{G}h(x),$$

and where $\mathscr{G}$ is the generator of the $\mathrm{BES}^2(\alpha)$ process:

$$\mathscr{G}h(x) = 2xh''(x) + \alpha h'(x).$$

If now we define

$$\rho(x) \equiv \tfrac{1}{4}p^2 x^{1-2r} I_{\{x>0\}} + I_{\{x \leqslant 0\}},$$

and let

$$\varphi_t \equiv \int_0^t \rho(V_s)\,ds,$$

then since

$$\varphi_t = \int_0^t \left[\tfrac{1}{4}p^2 U_s^{p-2} I_{\{U(s)>0\}} + I_{\{U(s)=0\}}\right] ds,$$

it is clear that φ increases strictly, and $\sup_t \varphi_t = \infty$. Moreover, since (almost surely) $\mathrm{Leb}(\{t\colon U_t = 0\}) = 0$, we see that

$$\begin{aligned}\varphi_t &= \int_0^t \tfrac{1}{4}p^2 U_s^{p-2} I_{\{U(s)>0\}}\,ds \\ &= \int_{(0,\infty)} \tfrac{1}{4}p^2 y^{p-2} l_t^y\,dy\end{aligned}$$

which is finite for every t with probability 1, since $p-2>-1$. Let $\tau_t \equiv \inf\{u\colon \varphi_u > t\}$ be the (strictly increasing, continuous, finite-valued) inverse to φ, and let $\tilde{V}_t \equiv V(\tau_t)$. *We claim that $\tilde{V}$ is a (weak) solution to (48.1).* Indeed, for each $T>0$,

$$\begin{aligned}\int_0^T I_{\{0\}}(\tilde{V}_s)\,ds &= \int_0^{\tau(T)} I_{\{0\}}(V_t)\rho(V_t)\,dt \\ &= \int_0^{\tau(T)} I_{\{0\}}(U_t)\,dt \qquad (48.9)\\ &= 0,\end{aligned}$$

since $\mathrm{Leb}(\{t\colon U_t = 0\}) = 0$ a.s. But by time-changing as in §26, (48.8) tells us that

$$\tilde{C}_t^h = h(\tilde{V}_t) - h(\tilde{V}_0) - \int_0^t \mathscr{G}h(\tilde{V}_s)\,ds \textit{ is a local martingale.} \qquad (48.10)$$

and $\tilde{V}$ is a weak solution of (48.1).

Property (ii) is established (recall that we have uniqueness in law). Property (i) follows from the facts that $\varphi\colon [0,\infty) \to [0,\infty)$ is a homeomorphism, and that 0 is recurrent for U. Finally, (iii) and (iv) are implicit in the construction of $\tilde{V}$.

(*48.11*) *Remark on boundary properties of* 0 *when* $0<\alpha<2$. We have

$$m_\alpha(dy)=K_\alpha y^{-2c(\alpha)}$$
$$-2c(\alpha)=2(\alpha-1)/(2-\alpha)\in(-1,\infty).$$

Hence $\int_{0+} y\,m_\alpha(dy)<\infty$, confirming that 0 is recurrent for Y by Theorem 51.2(ii) below. Moreover, $\int_{0+} m_\alpha(dy)<\infty$. Now see Theorem 51.2(iii), noticing that that result confirms that 0 must be absorbing when $\alpha=0$.

(*48.12*) *Mind-boggling example.* Let X and X' be independent $\mathrm{BES}^2(\alpha)$ processes, with $\alpha<1$, and $X_0=X'_0=1$, say. Then

$$\mathrm{Leb}(\{t\colon X_t=0\})=\mathrm{Leb}(\{t\colon X'_t=0\})=0 \quad \text{a.s.}$$

so that X and X' are almost always strictly positive. One would imagine that, as X and X' are independent, $X+X'$ would be positive for all time. Yet, by Theorem IV.35.6, $Y\equiv X+X'$ is a $\mathrm{BES}^2(2\alpha)$, so 0 is recurrent for Y!

49. Diffusion local time. The time-changed Brownian motion X of Theorem 47.1 is a continuous semimartingale, so it has a local-time process $\{L_t^a\colon a\in I,\ t\geqslant 0\}$ which turns out to be simply the time-change of the local time $\{l_t^a\colon a\in\mathbb{R},\ t\geqslant 0\}$ of B.

(*49.1*) THEOREM. *The local time process of* X *is*

$$\{L_t^a\colon a\in\mathrm{int}(I),\ t\geqslant 0\}=\{l^a(\gamma_t)\colon a\in\mathrm{int}(I),\ t\geqslant 0\},$$

and the occupation-measure formula says that, for any bounded measurable function f *supported in* $\mathrm{int}(I)$,

$$\int_0^t f(X_s)\,ds=\int m(da)f(a)L_t^a. \tag{49.2}$$

Remark. By definition $L_\cdot^a=0$ if $a\notin I$, so by right continuity in a (Theorem IV.44.2), $L_\cdot^a=0$ if a is an upper boundary point of I. For a lower boundary point a,

$$L_t^a=\lim_{x\downarrow a} L_t^x.$$

Proof. The proof requires a lemma from real analysis.

(*49.3*) LEMMA. *Suppose that* $\{A_t\colon t\geqslant 0\}$ *and* $\{C_t\colon t\geqslant 0\}$ *are right-continuous increasing functions, that* γ *is the right-continuous function inverse to* A, *and that* $A_0=C_0=\gamma_0=0$. *Then for any bounded measurable function* $h\colon\mathbb{R}^+\to\mathbb{R}$, *and for any* $t\geqslant 0$,

$$\int_{(0,t]} h(s)\,dC(\gamma_s)=\int_{(0,\gamma_t]} h(A_s)\,dC_s. \tag{49.4}$$

Proof of lemma. It is elementary to check (49.4) for h of the form $I_{(0,a]}$; the validity of (49.4) extends immediately to step functions h by linearity, and then to all h by the monotone-class Theorem II.4. □

Proof of theorem. We do only the case $I=(-\infty, 0]$: the others are similar but easier. Let M denote the martingale

$$M_t \equiv (B_t \wedge 0)+\tfrac{1}{2}l_t^0,$$

and let $N_t \equiv M(\gamma_t)$. The process N is continuous, because N could only jump when γ jumps; and a jump of γ is an interval in which B makes an excursion into $(0, \infty)$, throughout which M does not change. If $S_n = \inf\{t: |X_t| > n \text{ or } l^0(\gamma_t) > n\}$, then $\gamma(S_n)$ is an $\{\mathcal{F}_t\}$-stopping time reducing M to a bounded martingale, and so, by Proposition IV.30.10(iv), N is an $\{\mathcal{F}(\gamma_t)\}$-local martingale. We are therefore in a position to apply the time-change formula (IV.30.11) to Tanaka's formula; for $a \in \operatorname{int}(I)$,

$$\begin{aligned} L_t^a &\equiv |X_t - a| - |X_0 - a| - \int_0^t \operatorname{sgn}(X_s - a)dX_s \\ &= |X_t - a| - |X_0 - a| - \int_0^t \operatorname{sgn}(X_s - a)\, d(N_s - \tfrac{1}{2}l^0(\gamma_s)) \\ &= |B(\gamma_t) - a| - |B_0 - a| - \int_0^{\gamma_t} \operatorname{sgn}(X(A_s) - a)\, d(B_s \wedge 0) \end{aligned}$$

by Lemma (49.3) and IV.30.11. Now $X(A_s) = B(\gamma(A_s)) = B_s$, provided s is a point of increase of A; otherwise, in a time-interval where A is constant, B makes an excursion into $(0, \infty)$, so that for all s in such an interval, $X(A_s) = 0$, and

$$\operatorname{sgn}(-a) = \operatorname{sgn}(X(A_s) - a) = \operatorname{sgn}(B_s - a) = \operatorname{sgn}((B_s \wedge 0) - a) = 1.$$

(This would be false if we had allowed $a = 0$, since $\operatorname{sgn}(0) = -1$.) Hence

$$L_t^a = |B(\gamma_s) - a| - |B_0 - a| - \int_0^{\gamma_t} \operatorname{sgn}((B_s \wedge 0) - a)\, d(B_a \wedge 0)$$

which is the local time at $a \in (-\infty, 0)$ at time γ_t for $B_. \wedge 0$, which is clearly the same as the local time at a at time γ_t for $B_.$; the first assertion is proved.

To prove (49.2), use (49.4) with $C = A$, $h(s) = f(X_s)$. The only slight problem is if one of the boundaries is absorbing, in which case A jumps to ∞ when B reaches that boundary, and $A(\gamma_s) = s$ only until X reaches the boundary. The demand that f be supported in $\operatorname{int}(I)$ takes care of this, as you are invited to check; we shall assume that neither boundary is absorbing, so that $A(\gamma_s) = s$ for all s, and by (49.4),

$$\begin{aligned} \int_0^t f(X_s)\, ds &= \int_0^{\gamma_t} f(X(A_s))\, dA_s \\ &= \int_0^{\gamma_t} f(B_s)\, dA_s, \end{aligned}$$

the set of times for which $X(A_s)=B_s$ having zero A-measure. Hence

$$\int_0^t f(X_s)\,ds = \int m(da)\int_0^{\gamma_t} f(B_s)l^a(ds)$$

$$= \int m(da)\,f(a)l^a(\gamma_t),$$

by definition of A, and by the fact that l^a grows only when $B=a$. □

50. Analytical aspects. The central result of 1-dimensional diffusion theory (Theorem 48.1) was originally proved by the analytic tools of Markov process theory (generators, resolvents) rather than by the semimartingale/sample-path techniques used here. There are plenty of problems where the analytic properties are extremely useful, so we investigate these here; but do notice that the sample-path approach gives very neat proofs of all the main results in this section.

(*50.1*) PROPOSITION. *The semigroup* $\{P_t: t\geqslant 0\}$ *of the (regular) diffusion* $\{\mathbf{P}^x: x\in I\}$ *is Feller (that is,* $P_t: C_b(I)\to C_b(I)$ *for each* $t>0$).

Proof. Fix $f\in C_b(I)$, $t>0$, $x\in I$. Without loss of generality, suppose that $|f(x)|\leqslant 1$ for all $x\in I$. Since the paths are continuous, given $\varepsilon>0$ there is a $\delta>0$ such that, if $|t-s|<\delta$, then $|P_tf(x)-P_sf(x)|<\varepsilon/2$. For $y>x$, $\mathbf{P}^y(H_x\geqslant\delta)$ is clearly monotone increasing in y. We claim that, for some y_0,

$$\mathbf{P}^y(H_x\geqslant\delta)<\varepsilon/2 \quad\text{for}\quad y\in(x, y_0).$$

Indeed, if this did not happen, then, for $z<x$ we would have

$$\begin{aligned}\mathbf{P}^x(H_z\geqslant\delta)&\geqslant\mathbf{P}^x(H_y<H_z,\,H_z\geqslant\delta)\\&\geqslant\mathbf{P}^x(H_y<H_z)\mathbf{P}^y(H_z\geqslant\delta)\\&\geqslant\mathbf{P}^x(H_y<H_z)\mathbf{P}^y(H_x\geqslant\delta),\end{aligned}$$

and, letting $y\downarrow x$, we could deduce that $\mathbf{P}^x(H_z\geqslant\delta)\geqslant\varepsilon/2$ for all $z<x$. But this contradicts the fact that $H_z\downarrow H_x$ almost surely as $z\uparrow x$. Thus, for $x<y<y_0$,

$$\begin{aligned}|P_tf(x)-P_tf(y)|&\leqslant|\mathbf{E}^y\{f(X_t)-P_tf(x);\,H_x<\delta\}|\\&\quad+|\mathbf{E}^y\{f(X_t)-P_tf(x);\,H_x\geqslant\delta\}|\\&\leqslant|\mathbf{E}^y\{P_{t-H_x}f(x)-P_tf(x);\,H_x<\delta\}|+\varepsilon/2\\&\leqslant\varepsilon.\end{aligned}$$

You should have viewed with some suspicion the step which stated that

$$\mathbf{E}^y\{f(X_t);\,H_x<\delta\}=\mathbf{E}^y\{P_{t-H_x}f(x);\,H_x<\delta\}.$$

But consider the following immediate consequence of the strong Markov property at time H_x:

$$\sum_{i=0}^{2^n-1} \mathbf{E}^y\{f\circ X(H_x+t-i2^{-n}\delta);\ i2^{-n}\delta\leqslant H_x<(i+1)2^{-n}\delta\}$$
$$=\sum_{i=0}^{2^n-1} \mathbf{E}^y\{P(t-i2^{-n}\delta)f(x);\ i2^{-n}\delta\leqslant H_x<(i+1)2^{-n}\delta\};$$

and now let $n\uparrow\infty$.

Argue symmetrically for $y<x$. □

A diffusion need not be FD. Let X be a BES(3) process and let $R=1/X$. Then for $f\in C_0((0,\infty))$, as $x\to\infty$

$$\mathbf{E}[f(R_t)|R_0=x]\to\mathbf{E}[f(1/X)|X_0=0],$$

which need not be zero. The point ∞ is an entrance boundary for R; see § 51.

The resolvent of a 1-dimensional diffusion has a nice form: it has a continuous density with respect to the speed measure m which can be written explicitly in terms of (generalized—perhaps unbounded) 'eigenfunctions' of the generator. We give a probabilistic definition of these eigenfunctions, since we have not yet said what the generator is!

Let

$$T_x\equiv\inf\{t>0: B_t=x\},\quad H_x\equiv\inf\{t>0: X_t=x\}.$$

Then for $x, a\in\operatorname{int}(I)$, $A(T_a)=H_a$, $\mathbf{P}^x$-a.s. Now pick a reference point q in $\operatorname{int}(I)$ and define for each $\lambda>0$ the functions $\psi_\lambda^+:\operatorname{int}(I)\to(0,\infty)$ and $\psi_\lambda^-:\operatorname{int}(I)\to(0,\infty)$ by

$$\text{(50.2(i))}\qquad \psi_\lambda^+(x)\equiv\begin{cases}\mathbf{E}^x[\exp(-\lambda A(T_q))] & x\leqslant q,\quad x\in\operatorname{int}(I)\\ 1/\mathbf{E}^q[\exp(-\lambda A(T_x))] & x\geqslant q,\quad x\in\operatorname{int}(I);\end{cases}$$

$$\text{(50.2(ii))}\qquad \psi_\lambda^-(x)\equiv\begin{cases}\mathbf{E}^x[\exp(-\lambda A(T_q))] & x\geqslant q,\quad x\in\operatorname{int}(I)\\ 1/\mathbf{E}^q[\exp(-\lambda A(T_x))] & x\leqslant q,\quad x\in\operatorname{int}(I).\end{cases}$$

Notice that we could define $\psi_\lambda^\pm$ equally well in terms of the Laplace transforms of H_x; for example,

$$\psi_\lambda^+(x)=\mathbf{E}^x[\exp(-\lambda H_q)]\quad x\leqslant q,\quad x\in\operatorname{int}(I).$$

Changing the reference point q will only replace $\psi_\lambda^\pm$ by fixed multiples—see equation (50.5) below.

Here are some properties of $\psi_\lambda^\pm$.

(50.3) PROPOSITION. *The functions $\psi_\lambda^\pm$ are strictly convex, continuous, strictly monotone, and positive and finite throughout* $\operatorname{int}(I)$. *Moreover, they solve the differential equation*

$$\text{(50.4(i))}\qquad \tfrac{1}{2}\frac{d^2f}{dm\,dx}=\lambda f;$$

that is, for $x, y\in\operatorname{int}(I)$ with $x<y$,

$$\text{(50.4(ii))}\qquad (D_-f)(y)-(D_-f)(x)=\int_{[x,y)}2\lambda f\,dm.$$

Sketch of proof. We leave you to deduce the continuity assertions from the regularity property.

For any $a<b\in \operatorname{int}(I)$,

$$\psi_\lambda^+(a)=\psi_\lambda^+(b)\mathbf{E}^a[\exp(-\lambda A(T_b))], \tag{50.5}$$

so ψ_λ^+ is strictly increasing. For $a<x<b\in \operatorname{int}(I)$,

$$\begin{aligned}\psi_\lambda^+(x)/\psi_\lambda^+(b)&=\mathbf{E}^x[\exp(-\lambda A(T_b))]\\&=\mathbf{E}^x[\exp(-\lambda A(T_b)); T_a<T_b]\\&\quad+\mathbf{E}^x[\exp(-\lambda A(T_b)); T_b<T_a]\\&<\mathbf{P}^x(T_a<T_b)\{\psi_\lambda^+(a)/\psi_\lambda^+(b)\}+\mathbf{P}^x(T_b<T_a)\end{aligned}$$

whence the strict convexity of ψ_λ^+ follows. Only (50.4) remains to be proved.

Suppose that X starts at $a<b$. Then equation (50.5) shows that

$$\exp[-\lambda A(t\wedge T_b)]\psi_\lambda^+\circ B(t\wedge T_b)=\mathbf{E}^a[\exp(-\lambda A(T_b))|\mathscr{F}_t]$$

is a local martingale. On applying the Itô–Tanaka formula, equation (50.4) appears as the necessary and sufficient condition for the finite-variation part of this continuous semimartingale to vanish. The argument for ψ_λ^- is similar. □

(*50.6*) *Remark.* The process $\psi_\lambda^+(X_t)\exp(-\lambda t)$ is an $\{\mathscr{F}(\gamma_t)\}$-local martingale. (Why?) This fact is frequently useful.

(*50.7*) THEOREM. *The resolvent $\{R_\lambda: \lambda>0\}$ of the diffusion X has a continuous density $r_\lambda(\cdot,\cdot)$ with respect to m in* $\operatorname{int}(I)$:

$$R_\lambda f(x)=\int_I m(dy)r_\lambda(x,y)f(y)$$

for $x\in \operatorname{int}(I)$, and bounded measurable f supported in $\operatorname{int}(I)$. *Explicitly,*

$$\begin{aligned}r_\lambda(x,y)&=c_\lambda\psi_\lambda^+(x)\psi_\lambda^-(y) \qquad && x\leqslant y\in \operatorname{int}(I),\\&=c_\lambda\psi_\lambda^-(x)\psi_\lambda^+(y) && y\leqslant x\in \operatorname{int}(I),\end{aligned} \tag{50.8}$$

where

$$(c_\lambda)^{-1}\equiv\tfrac{1}{2}\{\psi_\lambda^-(x)D\psi_\lambda^+(x^-)-\psi_\lambda^+(x)D\psi_\lambda^-(x^+)\}, \tag{50.9}$$

and D denotes differentiation. The definition of the inverse Wronskian c_λ is independent of the choice of $x\in \operatorname{int}(I)$. The resolvent density can be expressed probabilistically; for $x,y\in \operatorname{int}(I)$,

$$\begin{aligned}r_\lambda(x,y)&=\mathbf{E}^x\int_0^\infty \exp\{-\lambda A_u\}l^y(du)\\&=\mathbf{E}^x\int_0^\infty e^{-\lambda t}L^y(dt).\end{aligned} \tag{50.10}$$

Remarks. The only reason for care over the endpoints is because (as you have gussed!) they might be absorbing. If they are not, the expressions (50.8), (50.9) and (50.10) are valid throughout I.

Proof.

$$\begin{aligned} R_\lambda f(x) &= \mathbf{E}^x \int_0^\infty e^{-\lambda t} f(X_t)\,dt \\ &= \mathbf{E}^x \int_0^\infty \lambda e^{-\lambda s}\left(\int_0^s f(X_t)\,dt\right) ds \\ &= \mathbf{E}^x \int_0^\infty \lambda e^{-\lambda s} \int L_s^y f(y) m(dy) \\ &= \int m(dy) f(y) \mathbf{E}^x \int_0^\infty e^{-\lambda s} L^y(ds), \end{aligned}$$

providing the second form of (50.10), having used (50.8). The first form follows by a time change. Applying the strong Markov property at T_y and using the fact that l^y only grows when B is at y, we deduce that, for $x > y$,

$$r_\lambda(x, y) = \mathbf{E}^x[\exp(-\lambda A(T_y)]r_\lambda(y, y) = \frac{\psi_\lambda^-(x)}{\psi_\lambda^-(y)} r_\lambda(y, y).$$

Hence (50.8) will hold provided we can prove the result in the case when $x = y$. We shall give an excursion proof for this case in § VI.54. The fact that the definition of c_λ is independent of $x \in \text{int}(I)$ follows from the differential equation (50.4) satisfied by $\psi_\lambda^\pm$. □

Traditionally, (50.8) is proved by verifying that if we define the operator R_λ by (50.8), then R_λ is indeed inverse to $\lambda - \mathscr{G}$, where $\mathscr{G}$ is the generator:

$$\mathscr{G} = \tfrac{1}{2} \frac{d^2}{dm\,dx}.$$

See Breiman [1] or Mandl [1] for this approach. This would not help us, because we have not identified the generator—but, then again, we did not need to! Of course, ultimately analysis will provide results on 1-dimensional diffusions which probability on its own cannot achieve.
One of the most important of such results is the following.

(*50.11*) THEOREM. *There exists a continuous map*

$$p: (0, \infty) \times \text{int}(I) \times \text{int}(I) \to (0, \infty)$$

such that, for all bounded measurable f supported in int(I),

$$\mathbf{E}^x[f(X_t)] = \int m(dy)p(t, x, y)f(y).$$

Proof. See §4.11 of Itô and McKean [1].

51. Classification of boundary points. Still using the notation of (47.31) and § 50, we are going to characterize (in terms of the speed measure m) the behaviour of a diffusion in natural scale on $I=[0, \infty)$ or $(0, \infty)$ at the boundaries 0 and ∞. We need to notice two things:

(51.1(i)) *For all* $x, y>0$, $\mathbf{E}^x l^y(T_0)=2(y \wedge x)$.

(51.1(ii)) *As* $x \to \infty$, $\psi_\lambda^-(x) \downarrow \psi_\lambda^-(\infty) \geqslant 0$, *and* $D_\pm \psi_\lambda^-(x) \uparrow 0$. *Hence*

$$\psi_\lambda^-(x) = \psi_\lambda^-(\infty) + \int_x^\infty dy \int_y^\infty m(dv).2\lambda\psi_\lambda^-(v).$$

The first is proved by applying the optional sampling theorem to Tanaka's formula IV.43.7, and the second follows from (50.4).

(*51.2*) THEOREM.

(i) $\mathbf{P}^x(H_0<\infty)=0$ *for all* $x>0$ *if and only if* $\int_{0+} xm(dx)=\infty$.
(ii) $\mathbf{P}^x(H_0<\infty)=1$ *for all* $x>0$ *if and only if* $\int_{0+} xm(dx)<\infty$.
(iii) *If* $\int_{0+} xm(dx)<\infty$, *then* 0 *must be absorbing if* $\int_{0+} m(dx)=\infty$.
(iv) $\lim_{x\to\infty} \psi_\lambda^-(x)>0$ *if and only if* $\int^\infty xm(dx)<\infty$.

Proof. (i), (ii). Note the following implications: $\mathbf{P}^x(H_0<\infty)=0$ if and only if $\lim_{\varepsilon\downarrow 0} \mathbf{E}^x(\exp(-\lambda H_\varepsilon))=0$ if and only if $\lim_{\varepsilon\downarrow 0} \psi_\lambda^-(\varepsilon)=\infty$.

Suppose firstly that $\mathbf{P}^x[H_0<\infty\}>0$ for some (then all) $x>0$, equivalently that $\psi_\lambda^-(0+)<\infty$. Then

$$\begin{aligned}\infty > \psi_\lambda^-(0+) &> \psi_\lambda^-(x)\\ &> \psi_\lambda^-(\infty) + \int_x^1 dy \int_y^1 m(dv).2\lambda\psi_\lambda^-(1)\\ &= \psi_\lambda^-(\infty) + \int_x^1 (v-x)m(dv).2\lambda\psi_\lambda^-(1);\end{aligned}$$

and letting $x \downarrow 0$ yields the condition $\int_{0+} vm(dv)<\infty$.

Next we prove that $\int_{0+} xm(dx)<\infty$ implies that $\mathbf{P}^x(H_0<\infty)=1$ for all $x>0$. Taking (49.2) with $t=H_0, f=I_{(0,1]}$, we obtain

$$\mathbf{E}^x\left[\int_0^{H_0} I_{(0,1]}(X_s)\,ds\right]=\mathbf{E}^x\left[\int_{(0,1]} m(da)L^a(H_0)\right]$$
$$=\mathbf{E}^x\left[\int_{(0,1]} m(da)l^a(T_0)\right]$$
$$=2\int_{(0,1]} m(da)(a\wedge x)$$

by (51.1(i)) and because $L^a(H_0)=l^a(\gamma(H_0))=l^a(T_0)$. Moreover, since T_0 is finite a.s., $L^a(H_0)=l^a(T_0)$ is zero for all large enough a. Hence

$$H_0=\int_{(0,1]} m(da)L^a(H_0)+\int_{(1,\infty)} m(da)L^a(H_0)<\infty,\qquad \text{a.s.}.$$

(iii) Suppose that $t>T_0$. Then $l_t^0>0$, and, by continuity of $l_t^{\cdot}$, for some $\varepsilon>0$,

$$\min\{l_t^x: 0\leqslant x\leqslant\varepsilon\}>0.$$

Thus, since our hypothesis for this part says that $m(0,\varepsilon)=\infty$ for every $\varepsilon>0$, we find that

$$A_t>\int_{(0,\varepsilon]} m(da)l_t^a\geqslant m(0,\varepsilon]\min\{l_t^x: 0\leqslant x\leqslant\varepsilon\}=\infty.$$

The additive functional A jumps to ∞ at T_0, so the time-changed process X must remain in 0 after $A(T_0-)=H_0$; that is, 0 is absorbing for X.

(iv) Suppose that $\psi_\lambda^-(x)\downarrow\varepsilon>0$ as $x\to\infty$. Then, for $x>0$,

$$\infty>\psi_\lambda^-(x)\geqslant\varepsilon+\int_x^\infty dy\int_y^\infty m(dv).2\lambda\varepsilon$$
$$=\varepsilon+2\lambda\varepsilon\int_x^\infty (v-x)m(dv)$$

whence $\int^\infty vm(dv)<\infty$.

Conversely, under the assumption that $\int ym(dy)<\infty$, we have (recall the natural-scale assumption) for $x>1$,

$$\mathbf{E}^x(H_1)=\int_1^\infty 2(y-1)\wedge(x-1)m(dy)$$
$$\uparrow\int_1^\infty 2(y-1)m(dy)<\infty\quad\text{as}\quad x\uparrow\infty.$$

But if it were true that $\psi_\lambda^-(x)\downarrow 0$ as $x\uparrow\infty$, then the $\mathbf{P}^x$-distribution of H_1 would converge to the point mass at ∞, contradicting the fact that the $\mathbf{P}^x$-expectations of H_1 remain bounded. □

(*51.3*) DEFINITION (accessible, inaccessible, entrance, exit boundary point). *If* $\mathbf{P}^x(H_0<\infty)=0$ *for all* $x>0$, *then* 0 *is an* inaccessible *boundary point. Otherwise,* 0 *is an* accessible *boundary point. If* 0 *is accessible, but* $\int_{0+} m(dx)=\infty$, *then* 0 *is called an* exit *boundary point. If* $\int^{\infty} x m(dx)<\infty$, *then* ∞ *is called an* entrance *boundary point.*

(*51.4*) *Remarks.* The point ∞ is an entrance boundary point for the reciprocal of BES(3) considered in Example 48.4.

By extension, we call a boundary point of a regular diffusion not in natural scale accessible, exit, or entrance if the corresponding boundary point of the diffusion transformed to natural scale is accessible, exit, or entrance.

(*51.5*) Persuade yourself that the converse to Theorem 47.1 holds in general; given an interval I and a measure m satisfying (47.5), there exists a regular diffusion on I in natural scale with speed measure m.

52. Khasminskii's test for explosion. It is important in practice to be able to decide when the solution of an SDE explodes (i.e. reaches infinity in finite time). There are well-established techniques for handling this, dating back in essence to Khasminskii [1]; we shall only discuss the case of a condition *sufficient to ensure non-explosion*, although similar methods yield analogous sufficient conditions to ensure explosion, as you will see if you consult Friedman [1], Ikeda and Watanabe [1], McKean [1], Stroock and Varadhan [1], or, indeed, Khasminskii [1]!

The essential idea is a comparison with a 1-dimensional diffusion, so we begin with a 1-dimensional version of the test.

(*52.1*) THEOREM. *Suppose that* X *solves the 1-dimensional SDE*

$$dX_t=\theta(X_t)\,dW_t+\mu(X_t)dt,$$

where θ *and* μ *are continuous in* $[1,\infty)$, *and* θ *is strictly positive in* $[1,\infty)$. *Defining functions* s *and* m *from* $[1,\infty)$ *to* $\mathbb{R}^+$ *by*

(52.2) $$s'(x)=\exp\left\{-2\int_1^x \mu(y)\theta(y)^{-2}dy\right\},\quad s(1)=0,$$

$$m'(x)=[\theta(x)^2 s'(x)]^{-1},\quad m(1)=0,$$

then a sufficient condition for non-explosion is

(52.3) $$\int_1^{\infty} s'(x)\,dx\int_1^x m'(y)\,dy=\infty.$$

In more detail, if $T_n\equiv\inf\{t:X_t>n\}$, *then condition* (52.3) *implies*

(52.4) $$\mathbf{P}(\sup_n T_n=\infty)=1.$$

Remark. In fact, (52.3) and (52.4) are equivalent.

Proof. The function s is nothing other than the scale function of the diffusion, and m is the speed measure; as one can readily check, the generator $\mathscr{G}$ can be expressed as

$$\mathscr{G} \equiv \tfrac{1}{2}\theta(x)^2 \frac{d^2}{dx^2} + \mu(x)\frac{d}{dx} = \tfrac{1}{2}\frac{d^2}{dm\, ds}.$$

The aim is to find some increasing function $\psi : \mathbb{R} \to \mathbb{R}^+$ such that $\mathscr{G}\psi \leqslant \psi$ and $\psi(x) \to \infty$ as $x \to \infty$. The point of this is that for such a ψ,

$Y_t \equiv \psi(X_t)e^{-t}$ *is a nonnegative supermartingale,*

as one can easily verify from Itô's formula. Hence, if $X_0 = x$, for $n > x$,

$$\begin{aligned} Y_0 = \psi(x) &\geqslant \mathbf{E}[\psi(X(T_n))e^{-T_n}] \\ &= \psi(n)\mathbf{E}[e^{-T_n}] \end{aligned}$$

and so

$$\mathbf{E}[e^{-T_n}] \leqslant \psi(x)/\psi(n) \to 0 \quad \text{as } n \to \infty,$$

implying (52.4).

The standard method for constructing such a function is as follows. Set $\psi(x) = 1$ for all $x < 1$, and for $x \geqslant 1$ define successively

(52.5)
$$\begin{aligned} \psi_0(x) &\equiv 1, \\ \psi_{n+1}(x) &= 2\int_1^x s(dy)\int_1^y m(dv)\psi_n(v). \end{aligned}$$

Each ψ_n is finite-valued, nonnegative, and increasing, and it is not too hard to prove by induction that, for all n, for all $x \geqslant 1$,

(52.6)
$$\psi_n(x) \leqslant \psi_1(x)^n/n!.$$

Then we can define

(52.7)
$$\psi(x) \equiv \sum_{n \geqslant 0} \psi_n(x) \quad (x \geqslant 1)$$

and it is easy to check that ψ is well-defined, C^2, and $\mathscr{G}\psi = \psi$ in $[1, \infty]$. Because of the construction of ψ, it is clear from (52.6) that, for $x \geqslant 1$,

(52.8)
$$1 + \psi_1(x) \leqslant \psi(x) \leqslant \exp\{\psi_1(x)\}.$$

The condition (52.3) is simply the condition $\lim \psi_1(x) = \infty$, which implies that $\psi(x) \uparrow \infty$ as $x \uparrow \infty$, completing the proof. □

(52.9) **Exercise.** Modify the proof to show that (52.4) ⇒ (52.3). (*Hint:* ψ_1^+ defined at (50.2(i)) satisfies $\mathscr{G}\psi_1^+ = \psi_1^+$, and can be constructed recursively just as ψ was constructed in the proof.)

Now we consider the problem of explosion of diffusions in $\mathbb{R}^n$. Suppose that X solves the SDE

$$dX_t = \sigma(X_t)dB_t + b(X_t)dt, \tag{52.10}$$

where $b\colon \mathbb{R}^n \to \mathbb{R}^n$, $\sigma\colon \mathbb{R}^n \to L(\mathbb{R}^d, \mathbb{R}^n)$ are continuous, and define

$$\mu(r) \equiv \sup\{\text{trace}\, a(x) + 2x^T b(x)\colon x \in \mathbb{R}^n,\ |x|^2 = r\} \tag{52.11(i)}$$

$$\theta(r) \equiv \sup\{2(x^T a(x)x)^{1/2}\colon x \in \mathbb{R}^n,\ |x|^2 = r\}, \tag{52.11(ii)}$$

where, as usual, $a = \sigma\sigma^T$. The functions θ and μ are continuous.

(*52.12*) THEOREM (Khasminskii). *Suppose that* X *solves the diffusion* SDE (52.10), *and let* $T_n \equiv \inf\{t\colon |X_t| > n\}$. *Then assuming that* θ *is strictly positive in* $[1, \infty)$, *and defining* s *and* m *in terms of* μ *and* θ *by* (52.2), *the condition* (52.3) *implies that:*

$$\mathbf{P}\left(\sup_n T_n = \infty\right) = 1.$$

Remark. Positivity of θ in any $[\lambda, \infty)$ would, of course, do just as well.

Proof. Writing $R_t \equiv |X_t|^2$, we have by Itô's formula that

$$\begin{aligned} dR_t &= 2\sum_{i=1}^n X_t^i \sigma_q^i(X_t)dB_t^q + [2X_t^T b(X_t) + \text{trace}\, a(X_t)]dt \\ &= 2(X_t^T a(X_t)X_t)^{1/2} d\beta_t + [2X_t^T b(X_t) + \text{trace}\, a(X_t)]dt, \end{aligned}$$

where β is some Brownian motion. Now consider $\psi(R_t)e^{-t}$, where ψ is the function constructed in the proof of Theorem 52.1.

$$\begin{aligned} d(\psi(R_t)e^{-t}) = d(\text{local martingale}) + \{&\tfrac{1}{2}\cdot 4X_t^T a(X_t)X_t \psi''(R_t) \\ &+ (2X_t^T b(X_t) + \text{trace}\, a(X_t))\psi'(R_t) - \psi(R_t)\}e^{-t}dt \end{aligned}$$

and the finite-variation part of this is bounded above by

$$\{\tfrac{1}{2}\theta(R_t)^2\psi''(R_t) + \mu(R_t)\psi'(R_t) - \psi(R_t)\}e^{-t}dt, \tag{52.13}$$

by definition (52.11) of θ and μ, and by the fact that ψ is convex and increasing. But by the construction of ψ, the curly bracket in (52.13) is not positive, so that

$$\psi(R_t)e^{-t} \textit{ is a nonnegative supermartingale.}$$

The conclusion that $\mathbf{P}(\sup_n T_n = \infty) = 1$ follows immediately by the same argument as before. □

(*52.14*) *Remark.* The Khasminskii test is essentially a comparison with a 1-dimensional diffusion, so it would be nice to derive it from the stochastic comparison theorem (Theorem 43.1). The equality of the covariances of both processes which was needed there does not apply here, and some further work is required to cast the present problem in a form to which Theorem 43.1 can be applied; see Ikeda and Watanabe [1] for all the details.

53. An ergodic theorem for 1-dimensional diffusions. Let X be a (regular) diffusion on $\mathbb{R}$ in natural scale, with speed measure m. Thus X is recurrent: $\mathbf{P}^x(H_y < \infty) = 1$ for all x, y. The following ergodic theorem for X is well known; our proof follows §6.8 of Itô and McKean [1].

(*53.1*) THEOREM. *Suppose that f, g: $\mathbb{R} \to \mathbb{R}^+$ are measurable functions such that:*

$$\int (f(x)+g(x))m(dx) < \infty, \quad \int g(x)m(dx) \neq 0.$$

Then:

$$\frac{\int_0^t f(X_s)\,ds}{\int_0^t g(X_s)\,ds} \overset{\text{a.s.}}{\to} \frac{\int f(x)m(dx)}{\int g(x)m(dx)}$$

as $t \to \infty$.

Proof. Define the stopping times $S_0 = 0$,

$$T_n \equiv \inf\{t > S_n : X_t = 0\}, \quad S_{n+1} \equiv \inf\{t > T_n : X_t = 1\}$$

for $n \in \mathbb{Z}^+$. Recurrence implies that all the T_n are finite almost surely. If we now define

$$Y_n \equiv \int_{T_{n-1}}^{T_n} f(X_s)\,ds \quad (n \in \mathbb{N})$$

then, by the strong Markov property, the Y_n are independent with the same law. Suppose that we can show that

$$\mathbf{E}Y_1 = 2\int f(x)m(dx); \tag{53.2}$$

then the result follows, because, by the strong law of large numbers,

$$n^{-1}\int_0^{T_n} f(X_s)\,ds \overset{\text{a.s.}}{\to} \mathbf{E}Y_1$$

and for $t \in (T_n, T_{n+1}]$,

$$\sum_{i=1}^{n} Y_i \leqslant \int_{T_0}^{t} f(X_s)\,ds \leqslant \sum_{i=1}^{n+1} Y_i.$$

To prove (53.2), we split the integral at S_1 and calculate

$$\mathbf{E}\left[\int_{T_0}^{S_1} f(X_s)\,ds\right], \quad \mathbf{E}\left[\int_{S_1}^{T_1} f(X_s)\,ds\right].$$

With the notation of (48.6), if X were given in the form $B(\gamma_t)$, then, by the occupation measure formula (49.2),

$$\int_{S_1}^{T_1} f(X_s)\,ds = \int m(da) f(a) \{ l^a(\gamma_{T_1}) - l^a(\gamma_{S_1}) \}. \tag{53.3}$$

Now $B(\gamma_{S_1}) = 1$, and γ_{T_1} is by definition $\inf\{u > \gamma_{S_1} : B_u = 0\}$. Thus the process $\{l^a(\gamma_{T_1}) - l^a(\gamma_{S_1}) : a \geqslant 0\}$ has the same law as $\{\tilde{l}^a(H) : a \geqslant 0\}$, where $\{\tilde{l}^a(t) : a \in \mathbb{R}, t \geqslant 0\}$ is the local time process of some Brownian motion $\tilde{B}$ started at 1, and H is the first time that $\tilde{B}$ hits 0. But

$$\mathbf{E}[\tilde{l}^a(H)] = 2(a \wedge 1), \tag{53.4}$$

as we saw in (51.1). Equation (53.2) now follows from (53.3), (53.4), and the similar result for the integral from T_0 to S_1. □

As a special case of Theorem 53.1, if the speed measure had finite mass (so that we may renormalize m to be a probability measure π), then

$$t^{-1} \int_0^t f(X_s)\,ds \overset{\text{a.s.}}{\to} \int f(x) \pi(dx) \tag{53.5}$$

as $t \to \infty$.

(*53.6*) **Exercise.** Extend Theorem 53.1 to a recurrent diffusion in an interval other than $\mathbb{R}$.

54. Coupling of 1-dimensional diffusions. The coupling method has been extensively exploited in the last ten years or so to provide distributional limit theorems and related rates of convergence; we refer you to Griffeath [1], Lindvall [1] and the references therein for some indication of the scope of the method.

The basic situation is the following. On some common probability space we have two processes $\{X_t : t \geqslant 0\}$ and $\{X'_t : t \geqslant 0\}$ with values in a Polish space (say), S. The processes X and X' have the property that

$$\mathbf{P}(X_t = X'_t \text{ for all } t \geqslant T) = 1 \tag{54.1}$$

where T is the *coupling time*

$$T \equiv \inf\{t : X_t \equiv X'_t\}.$$

If μ_t (respectively, μ'_t) denotes the law of X_t (respectively, X'_t), we deduce the fundamental *coupling inequality*: for all $t \geqslant 0$,

$$\| \mu_t - \mu'_t \| \leqslant 2\mathbf{P}(T > t), \tag{54.2}$$

where $\| \cdot \|$ denotes the total variation norm of a measure.

(*54.3*) *Proof of* (*54.2*). Let $\nu \equiv \mu_t - \mu'_t$. Since $\nu(S)=0$, we have

$$\begin{aligned}\|\nu\| &\equiv \sup\left\{\left|\int f\,d\nu\right| : f \text{ measurable, } |f| \leqslant 1\right\}\\ &= \sup\{\nu(A)-\nu(A^c) : A \subseteq S \text{ measurable}\}\\ &= 2\sup\{|\nu(A)| : A \subseteq S \text{ measurable}\}.\end{aligned}$$

Now

$$\begin{aligned}\mu_t(A) &= \mathbf{P}(X_t \in A)\\ &= \mathbf{P}(X_t \in A,\ t \leqslant T) + \mathbf{P}(X_t \in A,\ T > t)\\ &= \mathbf{P}(X'_t \in A,\ t \leqslant T) + \mathbf{P}(X_t \in A,\ T > t)\end{aligned}$$

by (54.1). Hence

$$\begin{aligned}|\nu(A)| &= |\mathbf{P}(X_t \in A,\ T > t) - \mathbf{P}(X'_t \in A,\ T > t)|\\ &\leqslant \mathbf{P}(T > t).\end{aligned}$$

□

Here is a typical application of the coupling inequality to strong Markov processes. Suppose that X and Y are strong Markov processes (both defined on the same probability space) with the same transition function $\{P_t\}$ but with different initial laws λ and μ, and let X' be defined by

$$X'_t = \begin{cases} Y_t & \text{for } t \leqslant T\\ X_t & \text{for } t \geqslant T,\end{cases}$$

where $T \equiv \inf\{t : X_t = Y_t\}$. By the strong Markov property, X' is a Markov process with initial law μ and transition function $\{P_t\}$. Hence by the coupling inequality,

$$\|\lambda P_t(\cdot) - \mu P_t(\cdot)\| \leqslant 2\mathbf{P}(T > t).$$

Of particular interest is the case where the initial law λ is an *invariant* law for $\{P_t\}$: $\lambda P_t = \lambda$ for all $t \geqslant 0$. If $\mu = \delta_x$, then the coupling inequality says

$$\|\lambda - P_t(x, \cdot)\| \leqslant 2\mathbf{P}(T > t)$$

so if the coupling is *successful* (that is, $\mathbf{P}(T < \infty) = 1$), then the coupling inequality yields the ergodic result

(54.4) $$\|\lambda - P_t(x, \cdot)\| \to 0 \quad \text{as } t \to \infty.$$

This sort of limit result occurs widely in Markov process theory, and the coupling method will establish it *provided we can ensure that coupling is successful.* The choice of suitably dependent X and Y to guarantee successful coupling is often quite an intricate business, but for 1-dimensional diffusions, we have the following easy result.

(54.5) THEOREM. *Let $\{P_t\}$ be the transition function of a (regular) diffusion in natural scale on $\mathbb{R}$ with speed measure m of finite mass. Then for each x:*

$$\lim_{t\to\infty} \|\pi - P_t(x, \cdot)\| = 0,$$

where
$$\pi(dx) \equiv m(\mathbb{R})^{-1} m(dx).$$

Proof. Firstly, it is easy to see that π is an invariant measure for $\{P_t\}$ by a direct calculation of $\int m(dx)\lambda r_\lambda(x, y) = 1$ from the formula (50.8) for the resolvent density. Thus if we set up on some probability space independent diffusions X and Y, where the law of X_0 is π and the law of Y_0 is δ_x, all that is necessary is to prove that $\mathbf{P}(T < \infty) = 1$.

But X and Y are continuous local martingales, and, since X is recurrent, it follows from Corollary IV.34.13 that $[X]_\infty = \infty$ a.s., and likewise $[Y]_\infty = \infty$ a.s. But, as X and Y are independent,

$$[X - Y]_t = [X]_t + [Y]_t$$

and so $[X - Y]_\infty = \infty$. By Corollary IV.34.13 again, applied to the continuous local martingale $X - Y$,

$$\limsup_{t\to\infty} (X_t - Y_t) = \infty, \quad \liminf_{t\to\infty} (X_t - Y_t) = -\infty \quad \mathbf{P}\text{-a.s.}$$

so $\mathbf{P}$-a.s. $X_t - Y_t = 0$ for some t. □

CHAPTER VI

The General Theory

1. ORIENTATION

1. Preparatory remarks. In Chapter IV, we saw how to construct the stochastic integral $\int H\,dM$ for locally bounded previsible H and for local martingales M whose paths were either continuous or of finite variation. The aim as we begin this chapter is to extend the definition of the stochastic integral to all local martingales M, and we expect that the first step will be to extend to $M \in \mathcal{M}_0^2$, the space of L^2-bounded martingales. If it were true that each $M \in \mathcal{M}_0^2$ could be expressed as

(1.1) $$M = X + V,$$

where $X \in \mathrm{c}\mathcal{M}_0^2$ and $V \in \mathrm{IV}\mathcal{M}_0^2$, then the task would be complete; by Theorem IV.30.4, the representation (1.1) is unique, and we could define for locally bounded previsible H

$$\int H\,dM = \int H\,dX + \int H\,dV,$$

both integrals on the right having been defined in Chapter IV.

However, it is *not* true that every $M \in \mathcal{M}_0^2$ has a representation of the form (1.1); in § 2, we shall give an example of a martingale M in the orthogonal complement $\mathrm{d}\mathcal{M}_0^2$ of $\mathrm{c}\mathcal{M}_0^2$ (see § IV.24) *whose paths are of unbounded variation on every interval.* Nonetheless, it *is* true that

(1.2) $\mathrm{IV}\mathcal{M}_0^2$ *is a dense subspace of* $\mathrm{d}\mathcal{M}_0^2$

as we shall prove in Theorem 36.3, so that it is 'nearly' enough to consider only martingales of the form (1.1). Now we know from Theorem IV.18.4 that if $M \in \mathrm{IV}\mathcal{M}_0^2$ then

(1.3) $$M_t^2 - M_0^2 - \sum_{s \leq t} \Delta M_s^2 = 2\int_{(0,t]} M_{s-}\,dM_s,$$

and the right-hand side is a finite-variation local martingale, by Theorem IV.13.6. It is easy to deduce that it is in fact a UI martingale.

Thus for $M \in \mathrm{IV}\mathcal{M}_0^2$, we have constructed the process $[M]$ whose existence is the subject of Theorem IV.26:

$$[M]_t = \sum_{s \leq t} \Delta M_s^2.$$

Because $\mathrm{IV}\mathcal{M}_0^2$ is dense in $\mathrm{d}\mathcal{M}_0^2$, this leads us to suspect that if $M \in \mathcal{M}_0^2$ is written in its canonical decomposition $M = X + V$, where $X \in \mathrm{c}\mathcal{M}_0^2$, $V \in \mathrm{d}\mathcal{M}_0^2$, then

$$(1.4) \qquad M_t^2 - [X]_t - \sum_{s \leq t} \Delta V_s^2$$

$$= X_t^2 - [X]_t + 2X_t V_t + V_t^2 - \sum_{s \leq t} \Delta V_s^2 \text{ is a UI martingale,}$$

where the process $[X]$ was constructed in Theorem IV.30.1. *This turns out to be true, proving the key result Theorem IV.26 on which the construction of the stochastic integral depends.*

Thus, once we have established (1.2), the construction of stochastic integrals with respect to any local martingale is achieved. However, *the justification of (1.2) leads us into a profound study of the general theory of processes, with consequences of far greater interest and importance than the construction of stochastic integrals.* Let us indicate the route to be followed.

(*1.5*) *Compensators.* A central result is the following. (Recall that an increasing process is implicitly assumed to be an adapted R-process.)

(*1.6*) THEOREM. *Let A be an integrable increasing process such that $A_0 = 0$. Then there exists a unique previsible increasing process A^{p} such that $A_0 = 0$ and*

$$M \equiv A - A^{\mathrm{p}}$$

is a martingale.

The process A^{p} is called the *compensator* or *dual previsible projection* of A. It is trivial that A^{p} is also integrable.

Both existence and uniqueness parts of this result are extremely important, and profound. The uniqueness statement is a consequence of the following theorem.

(*1.7*) THEOREM. *Suppose that $M \in \mathrm{IV}\mathcal{M}_0$. If M is previsible, then $M_t = 0$ for all t.*

A special case of this appeared as Theorem IV.30.4; for if M is a continuous local martingale, then it is certainly previsible.

It is the existence statement of Theorem 1.6 which is the really profound result, however. To understand how we shall approach it, let us first consider the discrete-time analogue. If $(A_n)_{n \geq 0}$ is an integrable increasing process adapted to the filtration $(\mathscr{F}_n)_{n \geqslant 0}$, then the construction of the compensator is easy;

$$A_0^p \equiv 0,$$

$$\Delta A_n^p \equiv A_n^p - A_{n-1}^p = \mathbb{E}(\Delta A_n | \mathscr{F}_{n-1}) \equiv \mathbb{E}(A_n - A_{n-1} | \mathscr{F}_{n-1}), \quad n \geqslant 1.$$

Intuitively, then, we expect that A^p should be defined in the continuous-time setting by the formula

$$dA_t^p = \mathbb{E}(dA_t | \mathscr{F}_{t-}),$$

by analogy with the discrete-time construction. Of course, no direct meaning can be assigned to this statement, but the essence of it can be expressed in another way. In discrete time again, if $(C_n)_{n \geqslant 1}$ is a bounded measurable process, then

$$\mathbb{E}\left[\sum_{n \geqslant 1} C_n \Delta A_n^p \right] = \mathbb{E}\left[\sum_{n \geqslant 1} C_n \mathbb{E}(\Delta A_n | \mathscr{F}_{n-1}) \right]$$

$$= \mathbb{E}\left[\sum_{n \geqslant 1} \mathbb{E}(C_n | \mathscr{F}_{n-1}) \Delta A_n \right]$$

$$= \mathbb{E}\left[\sum_{n \geqslant 1} {}^p C_n \Delta A_n \right],$$

where ${}^p C_n \equiv \mathbb{E}(C_n | \mathscr{F}_{n-1})$. The process ${}^p C$ is previsible, and is easily seen to be characterized by the fact that

$$\mathbb{E}[{}^p C_T \colon T < \infty] = \mathbb{E}[C_T \colon T < \infty]$$

for all stopping times T which are previsible in the sense that

$$\{T \leqslant n\} \in \mathscr{F}_{n-1} \quad (n \geqslant 1).$$

Now this characterization of A^p can be carried across to the continuous-time setting: we define A^p by the property that, for all bounded measurable processes C,

(1.8) $$\mathbb{E}\left[\int_{(0, \infty)} C_s \, dA_s^p \right] = \mathbb{E}\left[\int_{(0, \infty)} {}^p C_s \, dA_s \right],$$

where ${}^p C$ is the unique, previsible process such that

(1.9) $$\mathbb{E}[{}^p C_T \colon T < \infty] = \mathbb{E}[C_T \colon T < \infty]$$

for all 'previsible' times T.

So you can see that to carry out this programme, we must answer the following questions:

(1.10(i)) *What is a previsible time?*
(1.10(ii)) *Does there exist a process ${}^p C$ satisfying (1.9)?*
(1.10(iii)) *If so, is ${}^p C$ unique?*
(1.10(iv)) *Does (1.8) determine a unique increasing process A^p?*
(1.10(v)) *If so, is the process A^p previsible?*

Briefly, the answers are as follows. A *previsible* time is defined to be a stopping time T such that $T>0$ a.s. and such that the process $I_{\{t\geq T\}}$ is previsible. There does exist a previsible process ${}^{p}C$ (called the *previsible projection* of C) satisfying (1.9), and this process is unique to within evanescence. This uniqueness assertion is a very important use of the *section theorems*; the existence is proved by a (more elementary) monotone-class argument, working from the case of $C = I_F(\omega)I_{(a,b]}(t)$. For the last two questions, it is not hard to prove that

(1.11) $$C \mapsto \mu(C) \equiv \mathbf{E}\left[\int_{(0,\infty]} {}^{p}C_s \, dA_s\right]$$

is a finite measure on bounded measurable processes C, and that moreover, if C is an evanescent process, then $\mu(C)=0$. These conditions imply that there exists an integrable increasing process A^{p} such that

$$\mu(C) = \mathbf{E}\left[\int C_s \, dA^{p}_s\right];$$

the key result that

(1.12) $$A^{p} \text{ is previsible} \Leftrightarrow \mu(C) = \mu({}^{p}C)$$

is the deepest part of the whole construction. From the definition (1.11) of μ, and the easy fact that ${}^{p}({}^{p}C) = {}^{p}C$, it follows then that the answers to (1.10(ii)–(v)) are all 'yes', and so Theorem 1.6 will hold.

The preceding construction of A^{p} does not in fact require that A should be adapted, and we shall see important cases where A is not adapted. For non-adapted processes, *optional* and *dual optional* projections take on considerable importance. The optional σ-algebra on $[0,\infty)\times\Omega$ is generated by the adapted R-processes in the same way as the previsible σ-algebra on $(0,\infty)\times\Omega$ is generated by the adapted L-processes. The *optional projection* ${}^{o}C$ of a bounded (measurable) process C is an optional process such that

$$\mathbf{E}[{}^{o}C_T \colon T<\infty] = \mathbf{E}[C_T \colon T<\infty]$$

for all optional times (i.e. stopping times) T. It is easy to deduce that, for finite-valued stopping times T,

(1.13) $${}^{o}C_T = \mathbf{E}(C_T \mid \mathscr{F}_T) \quad \text{a.s.}$$

For a raw (not necessarily adapted) integrable increasing process A, we construct the dual optional projection A^{o} of A by the definition

$$\mathbf{E}\left[\int_{(0,\infty)} C_s \, dA^{o}_s\right] = \mathbf{E}\left[\int_{(0,\infty)} {}^{o}C_s \, dA_s\right].$$

Although this apparently parallels the construction of A^{p}, the two are not independent; the 'optional' version of (1.12) is needed to establish (1.12).

However, the optional projection is not merely an auxiliary concept introduced to prove Theorem 1.6; in the *general filtering* problem, an observer sees a process $(Y_t)_{t\geq 0}$, which is a noisy observation of (some functional of) an underlying process $(X_t)_{t\geq 0}$ of interest, and must then estimate $f(X_t)$ at time t on the basis of what he has so far observed. If $(\mathscr{Y}_t)_{t\geq 0}$ is the (usual augmentation of the) filtration generated by Y, then the estimate formed will simply be $\mathbf{E}[f(X_t)|\mathscr{Y}_t]$, the *$(\mathscr{Y}_t)$-optional projection of $f(X_t)$*. The whole matter of filtering is of the greatest practical importance; we discuss several problems in filtering (§§ 8–11).

We stress that the most important projections are the optional projection ^{o}C and the dual previsible projection A^{p}.

The way of thinking which we have just outlined is the exact preparation necessary for *the* central result of this chapter: the *Meyer decomposition theorem*. Suppose that we have a submartingale Z; to avoid technicalities, some of which have real substance, let us suppose for now that Z is bounded. Since Z is a submartingale, *Z increases on average.* We might therefore still hope to find a compensator Z^{p} (a temporary notation this, and not a standard one) which is a previsible increasing process such that

$$Z - Z^{p} \text{ is a martingale } (M, \text{say}).$$

The Meyer decomposition theorem says that we can indeed do this. Its proof uses (1.12) in a crucial way, but since Meyer's result is clearly much more subtle than Theorem 1.6, a lot more is needed. See Part 6.

One important consequence of the decomposition

$$Z = M + Z^{p}$$

is that our (bounded) submartingale Z is a semimartingale.

Remark on (joint) measurability. Though we are assuming that all processes with which we deal are $\mathscr{B}[0, \infty) \times \mathscr{F}$ measurable, we do—as you have noticed—sometimes stress the fact.

2. Lévy processes. A Lévy process is defined to be an R-process with stationary independent increments. Lévy's analysis of the paths of these processes has been a strong motivating force for many of the key concepts of the modern theory. It is essential for a proper appreciation of semimartingales to understand how Lévy processes are built from Poisson measures and Brownian motion; and, especially, to understand how a Lévy process without Brownian component (such as the symmetric Cauchy process) can fail to be of finite variation.

Kai Lai Chung will say (inimitably) that the only proper way for probabilists to proceed is to suppose given a Lévy process X and to derive any analysis associated with the transition function of X from sample path properties. We have much sympathy with this view. In Meyer [8], you will find this approach carried out in exemplary fashion, and you should read that excellent paper.

However, there is another path, albeit less straight and narrow. One can determine analytically what form the transition function has to take, and then build a process X with this transition function. This is a much easier route (which makes us feel still more guilty about choosing it), and Breiman [1] is our splendid guide.

For now, we consider only real-valued Lévy processes. We can think of a Lévy process X as being a Markov process on $\mathbb{R}$ with transition function $\{P_t\}$ satisfying

$$P_t(x, x+\Gamma)=P_t(0, \Gamma) \quad (t>0, x\in\mathbb{R}, \Gamma\in\mathscr{B}(\mathbb{R})).$$

It is elementary to use the dominated convergence theorem to show that $\{P_t\}$ is an FD transition function. Hence we can assume that X is a strong Markov R-process.

(*2.1*) *Examples.*

(*a*) X has increasing paths (and then is a subordinator) if and only if

$$\mathbf{E}^0\exp(i\theta X_t)=\exp[t\psi(\theta)],$$

where ψ may be written

$$\psi(\theta)=ic\theta+\int_{(0,\infty)}(e^{i\theta b}-1)\nu(db)$$

for some $c\geqslant 0$ and some measure ν on $(0, \infty)$ satisfying

$$\int_{(0,\infty)}(b\wedge 1)\nu(db)<\infty.$$

See §II.64, where the significance of this representation in terms of Poisson integrals is explained: we have

$$X_t=ct+\int_{(0,\infty)}bQ((0, t]\times db),$$

where Q is a Poisson measure on $(0, \infty)\times(0, \infty)$ with expectation measure

$$\mathbf{E}Q(dt, db)=dt\cdot\nu(db).$$

(*b*) X has continuous paths if and only if X is of the form:

$$X_t=ct+\sigma B_t+X_0,$$

where $c\in\mathbb{R}$, $\sigma\in[0, \infty)$ and B is a Brownian motion independent of X_0. See Theorem I.28.12 and §III.14.

(*c*) If X is a Lévy process starting at 0, and τ is a subordinator starting at 0 and independent of X, then $X\circ\tau$ is a Lévy process starting at 0.

Exercise: Prove this.

(*d*) An important special case of (*c*), already discussed in I.28.19 and elsewhere, produces the Cauchy process, for which

$$P_t(x, dy) = \pi^{-1} t[t^2 + (y-x)^2]^{-1} dy,$$

or, equivalently,

$$\mathbf{E}^0 \exp(i\theta X_t) = \exp[-t|\theta|].$$

Now recall the classical Lévy–Khinchine theorem.

(2.2) THEOREM (Lévy, Khinchine). *Let X be a Lévy process started at* 0. *Then*:

(2.3) $$\mathbf{E}\exp(i\theta X_t) = \exp[t\psi(\theta)],$$

where $\psi(\theta)$ *has the representation*:

(2.4) $$\psi(\theta) = ic\theta - \tfrac{1}{2}\sigma^2\theta^2 + \int_{\mathbb{R}\setminus\{0\}} \{e^{i\theta b} - 1 - i\theta b/(1+b^2)\}\nu(db),$$

where $c \in \mathbb{R}$, $\sigma \geqslant 0$ *and* ν *is a measure on* $\mathbb{R}\setminus\{0\}$ *such that*

(2.5) $$\int_{\mathbb{R}\setminus\{0\}} (b^2 \wedge 1)\nu(db) < \infty.$$

See 9.5 of Breiman [1] for a well-motivated proof.

We shall write $\psi(\theta) = \psi_X(\theta)$ or $\psi(X, \theta)$ when we wish to emphasize that ψ is associated with the process X.

For probabilistic understanding, it is best to rewrite (2.4) as follows:

(2.6) $$\psi(\theta) = ic_1\theta - \tfrac{1}{2}\sigma^2\theta^2 + \int_{|b|\geqslant 1} \{e^{i\theta b} - 1\}\nu(db) + \int_{0<|b|<1} \{e^{i\theta b} - 1 - i\theta b\}\nu(db),$$

where

$$c_1 = c - \int_{|b|\geqslant 1} b\nu(db)/(1+b^2) - \int_{0<|b|<1} \{b/(1+b^2) - b\}\nu(db),$$

both integrals converging absolutely.

Now let $\psi(\theta)$ be as at (2.6). We shall construct a Lévy process X such that (2.3) holds.

On a suitable probability triple $(\Omega, \mathcal{F}, \mathbf{P})$, set up two independent random elements:

(i) a Brownian motion B,

(ii) a Poisson random measure Q on $(0, \infty) \times (\mathbb{R}\setminus\{0\})$ with expectation measure

$$\mathbf{E}Q(dt, db) = dt \cdot \nu(db).$$

It is clear from (2.6) that we are going to set

$$X_t = c_1 t + \sigma B_t + Y_t + Z_t, \tag{2.7}$$

where (see Example 2.1(*a*))

$$Y_t = \int_{|b| \geq 1} bQ((0, t] \times db)$$

and where Z must be a Lévy process independent of B and Y satisfying

$$\psi_Z(\theta) = \int_{|b| < 1} (e^{i\theta b} - 1 - i\theta b)\nu(db). \tag{2.8}$$

The problem is therefore reduced to that of constructing Z.

For $0 < \varepsilon < \delta$, define

$$Z_{\varepsilon,\delta}(t) \equiv \int_{\varepsilon \leq |b| < \delta} b\tilde{Q}((0, t] \times db) \tag{2.9}$$

$$\equiv \int_{\varepsilon \leq |b| < \delta} bQ((0, t] \times db) - t \int_{\varepsilon \leq |b| < \delta} b\nu(db),$$

where $\tilde{Q}$ is the compensated Poisson measure

$$\tilde{Q} \equiv Q - \mathbf{E}Q. \tag{2.10}$$

The independence properties of Q (see § II.37) guarantee that

(2.11) $Z_{\varepsilon,\delta}$ *has independent increments.*

Moreover, Riemann-sum approximation allows us to transfer the intuitive ideas:

$$\mathbf{E}\tilde{Q}(dt, db) = 0, \quad \text{var } \tilde{Q}(dt, db) = dt \cdot \nu(db)$$

into precise statements:

(2.12) $\mathbf{E}Z_{\varepsilon,\delta}(t) = 0$, so that, by (2.11), $Z_{\varepsilon,\delta}$ *is a martingale,*

and

$$\mathbf{E}\{Z_{\varepsilon,\delta}(t)^2\} = t \int_{\varepsilon \leq |b| < \delta} b^2 \nu(db). \tag{2.13}$$

Moreover, the calculation in § I.28 shows that

$$\psi(Z_{\varepsilon,\delta}, \theta) = \int_{\varepsilon \leq |b| < \delta} (e^{i\theta b} - 1 - i\theta b)\nu(db). \tag{2.14}$$

Doob's L^2 inequality now gives the estimate:

$$\mathbf{E}\{\sup_{s\leqslant t} Z_{\varepsilon,\delta}(s)^2\}\leqslant 4t\int_{\varepsilon\leqslant|b|<\delta} b^2\nu(db). \tag{2.15}$$

For $0<\varepsilon<\delta$, $Z_{\varepsilon,1}$ splits as the sum

$$Z_{\varepsilon,1}=Z_{\delta,1}+Z_{\varepsilon,\delta}$$

of two (independent Lévy) processes. Hence, because of (2.15) and the integrability condition (2.5), we see that if $\varepsilon(0)=1$, and $\varepsilon(n)\downarrow 0$ rapidly, then, *with probability* 1,

$$Z(t)=\lim_{n\to\infty} Z_{\varepsilon(n),1}(t) \tag{2.16}$$

exists uniformly on compact intervals. It is clear that Z inherits the Lévy-process property from that of the $Z_{\varepsilon(n),1}$ processes. Moreover, it follows from (2.14) and (2.16) that Z does indeed satisfy (2.8). Finally, we note that since the convergence at (2.16) takes place in L^2, the process Z inherits the martingale property from the $Z_{\varepsilon(n),1}$ processes. (We take for $\mathscr{F}_t$ the smallest σ-algebra with respect to which each B_s with $s\leqslant t$ is measurable, and each variable $Q(H\times\Gamma)$ is measurable, where $H\in\mathscr{B}((0,t])$, and $\Gamma\in\mathscr{B}(\mathbb{R}\backslash\{0\})$.)

In terms of the theory of stochastic integrals relative to random measures,

$$Z_t=\int_{0<|b|<1} b\tilde{Q}((0,t]\times db). \tag{2.17}$$

The construction of our Lévy process X is now complete.

(*2.18*) *Important remarks.* (i) Note that

$$X=M+A,$$

where

M is the martingale $\sigma B+Z$,
A is the finite variation process $A_t=c_1 t+Y_t$.

Hence, *X is a semimartingale.*

(ii) Let us suppose that all processes are stopped at time 1. Then the martingale $Z\equiv\{Z(t\wedge 1)\colon t\geqslant 0\}$ is bounded in L^2, and is the L^2-limit of the martingales $Z_{\varepsilon,1}\equiv\{Z_{\varepsilon,1}(t\wedge 1)\colon t\geqslant 0\}$. Now by the construction (2.9) of $Z_{\varepsilon,1}$, each of the martingales $Z_{\varepsilon,1}$ is of integrable variation; we shall see (at (36.4)) that each $Z_{\varepsilon,1}$ therefore belongs to $\mathrm{d}\mathscr{M}_0^2$, the orthogonal complement of $\mathrm{c}\mathscr{M}_0^2$ (see §IV.24 for definitions and properties). Thus Z, which is the L^2-limit of the $Z_{\varepsilon,1}$, must also belong to $\mathrm{d}\mathscr{M}_0^2$, which is a closed subspace of $\mathscr{M}_0^2$. Thus Z is certainly not a continuous martingale, *but neither is it of finite variation in general.* Indeed,

(2.19)
$$\mathbf{E}\Big[\sum_{s\leqslant 1}|\Delta Z_s|\Big]=\int_{|b|\leqslant 1}|b|\nu(db),$$

and it is perfectly possible for this to be infinite, even though ν satisfies the integrability condition (2.5). If the integral (2.19) diverges, then for $\lambda>0$,

$$\mathbf{E}\Big[\exp\Big\{-\lambda\sum_{s\leqslant 1}|\Delta Z_s|\Big\}\Big]=\exp\Big\{-\int_{|b|\leqslant 1}(1-e^{-\lambda|b|})\nu(db)\Big\}=0,$$

and so with probability 1,

$$\sum_{s\leqslant 1}|\Delta Z_s|=\infty.$$

In such a case, then, Z is an L^2-bounded martingale which is neither continuous nor of finite variation, *so the theory developed in Chapter IV will not allow us to define* $\int H\,dZ$.

2. DEBUT AND SECTION THEOREMS

3. Progressive processes. Recall from § II.73 that a process X with parameter-set $[0,\infty)$ is called *progressive* if, for each t, the map

$$X\colon [0,t]\times\Omega\to\mathbb{R}$$
$$(s,\omega)\mapsto X(s,\omega)$$

is $\mathscr{B}[0,t]\times\mathscr{F}_t$-measurable. This is of course a much stronger requirement than the $\mathscr{B}[0,\infty)\times\mathscr{F}$-measurability of the map X from $[0,\infty)\times\Omega$ to $\mathbb{R}$ which we are now assuming of X. In particular, the Fubini Lemma II.11.4 shows that

(3.1) *a progressive process is adapted.*

Lemma II.73.11 provides the following stronger result.

(3.2) LEMMA. *If X is a progressive process and T is a stopping time relative to $\{\mathscr{F}_t\}$, then X_T is $\mathscr{F}_T$-measurable.*

Peep ahead to § 17 for a correct appreciation of this result.

Now recall the important Lemma II.73.10, the proof of which is recalled for convenience.

(3.3) LEMMA. *An adapted R-process is progressive.*

Proof. Fix $t\geqslant 0$. For $n\in\mathbb{N}$, define for $s<t$,

$$Y^{(n)}(s,\omega)\equiv X([k+1]2^{-n}t,\omega),\quad k2^{-n}\leqslant s<(k+1)2^{-n}t;$$

and put $Y^{(n)}(t,\omega)\equiv X(t,\omega)$. Then $Y^{(n)}$ is trivially $\mathscr{B}[0,t]\times\mathscr{F}_t$-measurable, and $X=\lim Y^{(n)}$ on $[0,t]\times\Omega$. □

A subset F of $[0, \infty) \times \Omega$ is called *progressive* if its indicator I_F is a progressive process. The progressive sets form a σ-algebra, and a process is progressive if and only if it is measurable relative to this 'progressive σ-algebra'. You can easily prove this by monotone-class arguments.

Volume 1 contained many applications of the Début Theorem II.76. We recall it here, because we shall need to make heavy use of it. For proof, see §IV.50 of Dellacherie and Meyer [1].

(*3.4*) THEOREM (Choquet, Doob, Hunt). *Let F be a progressive subset of $[0, \infty) \times \Omega$. Then the début D_F of F, defined as follows:*

$$D_F(\omega) \equiv \inf\{t \geqslant 0 \colon (t, \omega) \in F\}$$

is a stopping time.

Remember that the Début theorem—unlike the Section Theorem 5.1 below—is true for all progressive F.

(*3.5*) *Example.* Let B be the canonical process $\text{CBM}_0(\mathbb{R})$, and let $\{\mathscr{F}_t\}$ be the usual augmentation of the natural filtration of B. For each ω, the set

$$\{s \colon B(s, \omega) \neq 0\}$$

is the disjoint union of open intervals, the excursion intervals of ω from 0. Consider the set

$$F \equiv \{(s, \omega) \colon s \text{ is the left-hand endpoint of an excursion interval of } \omega\}.$$

This set F is a famous example (due to Dellacherie and Meyer) of a set which is progressive but not optional.

Proof that F is progressive. For each time s, define

$$D_s \equiv \inf\{u > s \colon B(u) = 0\}.$$

Then, by elementary arguments or by the Début theorem,

$$\{D_s \leqslant v\} \in \mathscr{F}_v \subseteq \mathscr{F},$$

so that each D_s is $\mathscr{F}$-measurable. Since the map $s \mapsto D_s(\omega)$ is right-continuous, the map $(s, \omega) \mapsto D_s(\omega)$ is $\mathscr{B}[0, \infty) \times \mathscr{F}$-measurable. If we put

$$F_+ \equiv \{(s, \omega) \colon D(s, \omega) = s\},$$

then F_+ is $\mathscr{B}[0, \infty) \times \mathscr{F}$-measurable. A trivial modification of our argument shows that

$$F_+ \cap \{[0, t) \times \Omega\} \in \mathscr{B}[0, t) \times \mathscr{F}_t.$$

Hence,

$$F_+ \cap \{[0, t] \times \Omega\} = F_+ \cap \{[0, t) \times \Omega\} \cup (\{t\} \times \{D_t = t\}) \in \mathscr{B}[0, t) \times \mathscr{F}_t.$$

(Recall that since D_t is a stopping time, $\{D_t = t\} \in \mathscr{F}_t$.) We have just shown that the set F_+ is progressive. But

$$F = F_+^c \cap \{(s, \omega): B(s, \omega) = 0\}.$$

Hence F is progressive. □

4. Optional processes, $\mathcal{O}$; optional times. For the purposes of the general theory, progressive processes are not sufficiently well-behaved; we have to work with a sub-σ-algebra of the progressive σ-algebra, known as the optional σ-algebra.

(*4.1*) DEFINITION (optional σ-algebra $\mathcal{O}$; optional process). *The* optional σ-algebra $\mathcal{O}$ *of subsets of* $[0, \infty) \times \Omega$ *is defined to be the smallest σ-algebra on* $[0, \infty) \times \Omega$ *such that every adapted R-process is $\mathcal{O}$-measurable;*

$$\mathcal{O} \equiv \sigma(\{\text{adapted R-processes}\}).$$

A process X with parameter set $[0, \infty)$ *is called* optional *if X is $\mathcal{O}$-measurable as a map from* $[0, \infty) \times \Omega$ *to* $\mathbb{R}$.

By Lemma 3.3,

(4.2) *every optional process is progressive.*

The converse is not true; as already remarked, the set F in § 3 is progressive but not optional.

(*4.3*) DEFINITION (optional time). *A random time T (that is, an $\mathscr{F}_\infty$-measurable map from Ω to $[0, \infty]$) is called* optional *if* $[T, \infty) \in \mathcal{O}$, *where*

$$[T, \infty) \equiv \{(t, \omega): T(\omega) \leqslant t < \infty\}$$

as usual.

(*4.4*) LEMMA. *T is an optional time if and only if T is a stopping time.*

Proof. The process $I_{[T, \infty)}$ is an R-process, and so is optional if and only if it is adapted, that is, if and only if

$$\{\omega: I_{[T, \infty)}(t, \omega) = 1\} = \{\omega: T(\omega) \leqslant t\} \in \mathscr{F}_t. \qquad \square$$

The following extension of Lemma 4.4 and the fact that 'optional time' has long been synonymous with 'stopping time' in the literature explain why $\mathcal{O}$ is called the optional σ-algebra.

(*4.5*) THEOREM (Meyer). *Define*

$$\mathscr{T} \equiv \sigma(\{[T, \infty): T \text{ a stopping time}\}).$$

Then $\mathscr{T} = \mathcal{O}$.

We need the following lemma, of a now familiar type.

(4.6) LEMMA. *If S and T are stopping times with $S \leqslant T$, and if $Z \in \mathrm{m}\mathscr{F}_S$, then the process $Z[S, T)$ is $\mathscr{T}$-measurable.*

Proof. We can approximate Z uniformly by sums of terms of the form cI_Λ, where $\Lambda \in \mathscr{F}_S$. But

$$I_\Lambda[S, T) = [S_\Lambda, T_\Lambda) = [S_\Lambda, \infty)\backslash[T_\Lambda, \infty),$$

where, as usual, S_Λ denotes the random time

$$S_\Lambda \equiv \begin{cases} S \text{ on } \Lambda \\ +\infty \text{ on } \Omega\backslash\Lambda. \end{cases}$$

We know, and it is trivial to prove, that S_Λ and T_Λ are stopping times. □

Proof of Theorem 4.5. Let X be an adapted R-process. Fix $\varepsilon > 0$. Because of the above lemma, we need only show that X may be approximated to within ε by a process Y which is a sum of processes of the form $Z[S, T)$. Define

$$T_0 \equiv 0, \quad Z_0 \equiv X(T_0).$$

Now define inductively for $n \geqslant 0$,

$$T_{n+1} \equiv \inf\{t > T_n: |X_t - Z_n| > \varepsilon\}, \quad Z_{n+1} = X(T_{n+1}) \text{ if } T_{n+1} < \infty.$$

Since X has left limits, $T_n \uparrow \infty$. It is elementary to prove that each T_k is a stopping time, and, by Lemma 3.2, Z_n is $\mathscr{F}(T_n)$-measurable. Set

$$Y \equiv \sum_{n \geqslant 0} Z_n[T_n, T_{n+1}).$$

Since Y approximates X uniformly within ε, the proof is finished. □

Let us combine Theorem 4.5 with Lemma IV.6.9 into the following result.

(4.7) THEOREM. *Recall that $\mathscr{O}$ is a σ-algebra on $[0, \infty) \times \Omega$, and that $\mathscr{P}$ is a σ-algebra on $(0, \infty) \times \Omega$. We have*

$$\mathscr{O} = \sigma(\{[T, \infty): T \text{ a stopping time}\})$$

$$\mathscr{P} = \sigma(\{(T, \infty): T \text{ a stopping time}\}).$$

Every set in $\mathscr{P}$, when considered as a subset of $[0, \infty) \times \Omega$, is an element of $\mathscr{O}$. We therefore have:

(4.8) PREVISIBLE $\Rightarrow$ OPTIONAL $\Rightarrow$ PROGRESSIVE.

Proof. All that remains is to prove that if T is a stopping time, then $(T, \infty) \in \mathscr{O}$. But $T_n \equiv T + n^{-1}$ is a stopping time, and

$$(T, \infty) = \bigcup_n [T_n, \infty).$$ □

5. The 'optional' section theorem. We can now state the 'correct' version of the section theorem of which a 'first draft' was given as Theorem II.76. This theorem is deeper than, and at least as important as, the Début theorem.

First, note that by the Début theorem, if F is a progressive subset of $[0, \infty) \times \Omega$, then

$$\{\omega: (t, \omega) \in F \text{ for some } t\} = \{\omega: D_F(\omega) < \infty\} \in \mathscr{F}.$$

(*5.1*) THEOREM (Meyer's 'optional' section theorem). *Let F be an optional subset of $[0, \infty) \times \Omega$, and suppose that $\varepsilon > 0$. Then there exists a stopping time T such that*

(i) *$(T(\omega), \omega) \in F$ whenever $T(\omega) < \infty$;*

(ii) $\mathbf{P}[T < \infty] \geqslant \mathbf{P}[D_F < \infty] - \varepsilon$.

Note that (i) can be expressed more neatly: $[T] \subseteq F$, where

$$[T] = [T, T] = \{(t, \omega): t = T(\omega)\}.$$

For proof of Theorem 5.1, see IV.84 of Dellacherie and Meyer [1].

Important uses of the section theorems (both 'optional' and 'previsible') were made in §III.41 and §III.50, but the most common and important use of the section theorems is in establishing indistinguishability of two processes by examining behaviour at stopping times. We shall soon see how useful such results can be.

(*5.2*) LEMMA. *Let X and Y be two optional processes such that*

$$\mathbf{P}[X_T = Y_T] = 1 \tag{5.3}$$

for every finite stopping time T. Then X and Y are indistinguishable:

$$\mathbf{P}[X_t = Y_t, \forall t] = 1.$$

Proof. The set $F \equiv \{(t, \omega): X_t(\omega) \neq Y_t(\omega)\} \subseteq [0, \infty) \times \Omega$ is optional. Hence, if F is not evanescent, then, by the section theorem, there exists a stopping time U with $[U] \subseteq F$ and $\mathbf{P}[U < \infty] > 0$. By taking $T = U \wedge c$ for some suitably large constant c, we obtain an obvious contradiction with (5.2). Hence, F is evanescent. □

(*5.4*) LEMMA. *Let X and Y be two bounded optional processes such that*

$$\mathbf{E}[X_T; T < \infty] = \mathbf{E}[Y_T; T < \infty]$$

for every stopping time T (including those taking infinite values). Then X and Y are indistinguishable.

Proof. First take $F \equiv \{(t, \omega): X_t(\omega) > Y_t(\omega)\}$. Apply the same argument as for Lemma 5.2. □

Note that though we can restrict our attention to bounded stopping times in Lemma 5.2, we cannot do so in Lemma 5.4. To see this, let X be any bounded R-martingale, and let $Y \equiv 0$.

(5.5) *Example*: a progressive set which is not optional. Let F be the progressive set described in at 3.5. Recall that F is the set of left-hand endpoints of excursions of $\text{CBM}_0(\mathbb{R})$ away from 0. We can use the Section Theorem 5.1 to prove that S is not optional. For if T is a stopping time with graph contained in F, then $B_T=0$, and by the strong Markov property and the instantaneous-return property now familiar to us, B must return to 0 immediately after time T. This contradicts the definition of F. (Our wording has been a little imprecise, but it is obvious how to tighten it to prove that if $[T]\subseteq F$, then $\mathbf{P}[T<\infty]=0$, while $\mathbf{P}[D_F<\infty]=1$.)

6. Warning (not to be skipped). In Example 5.5, the process $Y=1-I_{\{0\}}\circ B$ is optional (indeed previsible). The example therefore illustrates that:

(6.1) *if Y is optional (or even previsible), the process Z with*

$$Z_t \equiv \liminf_{u \downarrow\downarrow t} Y_u$$

need not be optional.

On the other hand, it is true generally (under the usual conditions which we are now assuming) that:

(6.2) *If Y is progressive, then the process Z with*

$$Z_t \equiv \liminf_{u \downarrow\downarrow t} Y_u$$

is progressive, and the process V with

$$V_t \equiv \liminf_{u \to t} Y_u$$

is optional (yes, optional!). See Theorem IV.90 of Dellacherie and Meyer [1].

The possible pitfalls for one's intuition which are so strikingly illustrated by the contrast between (6.1) and (6.2) mean that applications of the début and section theorems are often not as straightforward as they look. So at some stage (which we leave to your own choosing) you must read Dellacherie [1] and Dellacherie and Meyer [1] for proper understanding. For example, your first attempt to prove the following very important result could well come undone over points such as (6.1) and (6.2). Consult Theorem IV.28 of Dellacherie [1] to see how deep this result really is.

(6.3) THEOREM. (Meyer, Mertens, Rao). *Let X be an optional process. Suppose that whenever (T_n) is a uniformly bounded sequence of stopping times such that $T_n \downarrow T$, we have*

(6.4) $$\mathbf{P}[X(T_n)\to X(T)]=1.$$

Then, a.s., X is right-continuous.

ADDENDUM. *Under the extra condition that X is bounded, the theorem remains true if (6.4) is weakened to the form:*

$$\lim \mathbf{E}[X(T_n)] = \mathbf{E}[X(T)].$$

3. OPTIONAL PROJECTIONS AND FILTERING

7. Optional projection ^{o}X of X. We have become so used to dealing with adapted processes that, though it is obvious, it is worth mentioning that the following theorem is important only when X is not adapted to the $\{\mathcal{F}_t\}$ filtration. Though we have a convention that all processes are measurable, we stress the measurability of X in the statement of the theorem.

(*7.1*) THEOREM, DEFINITION (optional projection). *Let X be a bounded (measurable) process. Then there exists an optional process ^{o}X, unique to within evanescence, such that for every stopping time T (including those taking infinite values)*

(7.2) $$^{o}X_T I_{\{T<\infty\}} = \mathbf{E}[X_T I_{\{T<\infty\}} | \mathcal{F}_T].$$

The process ^{o}X is called the optional projection *of X.*

(*7.3*) *Remarks.* (i) Theorem 7.1 extends in obvious fashion to the case when X is nonnegative and not necessarily bounded.

(ii) Let T be a stopping time. Then immediately from (7.2),

(7.4) $$\mathbf{E}[^{o}X_T;\ T<\infty] = \mathbf{E}[X_T;\ T<\infty].$$

By applying (7.4) to T_Λ, where $\Lambda \in \mathcal{F}_T$, we deduce that (7.4) implies (7.2); either may be taken as the definition of ^{o}X.

(iii) The example after Lemma 5.4 shows that if X is bounded, and if Y is a bounded optional process such that $\mathbf{E}[X_T] = \mathbf{E}[Y_T]$ for all *bounded* stopping times T, then Y need not be ^{o}X.

(*7.5*) *Notation.* As in Dellacherie and Meyer [1], the following notations are used (the last three will be defined before long):

Optional projection: ^{o}X,
Previsible projection: ^{p}X,
Dual optional projection: X^{o},
Dual previsible projection: X^{p}.

The dual projections are defined only for certain FV processes X.

(*7.6*) *Proof of Theorem 7.1.* The uniqueness part of the theorem is immediate from Lemma 5.4, so we turn our attention to the existence result.

Let $\mathscr{H}$ be the class of bounded processes X (considered as functions on $[0,\infty)\times\Omega$) for which an optional projection satisfying (7.2) exists. It is immediate that $\mathscr{H}$ satisfies the assumptions of monotone-class theorem II.4.

Let $\mathscr{C}$ be the class of subsets of $[0,\infty)\times\Omega$ of the form

$$C=[a,b)\times F, \text{ where } a<b, \text{ and } F\in\mathscr{F}. \tag{7.7}$$

Then $\mathscr{C}$ is a π-system, and $\sigma(\mathscr{C})=\mathscr{B}[0,\infty)\times\mathscr{F}$. For C as at (7.7), choose Z to be an R-version of the martingale $t\mapsto \mathbf{E}[I_F|\mathscr{F}_t]$. Then the R-process

$$^{\circ}I_C(t,\omega)\equiv I_{[a,b)}(t)Z_t(\omega)$$

serves as an optional projection of the process I_C. Use of the monotone-class theorem finishes the proof. □

(*7.8*) LEMMA. *If X is right-continuous, then so is $^{\circ}X$.*

Proof. Apply Theorem 6.3. □

In fact, it is known that

$$\text{if } X \text{ is an R-process, then so is } {}^{\circ}X. \tag{7.9}$$

See Theorem VI.47 of Dellacherie and Meyer [1].

We therefore have the following result which will allow us to calculate the optional projection in most cases of interest.

(*7.10*) THEOREM. *Let X be a bounded R-process. Then a process Y is indistinguishable from $^{\circ}X$ if and only if Y is an adapted R-process such that, for each t,*

$$Y_t=\mathbf{E}[X_t|\mathscr{F}_t], \quad \text{a.s.} \tag{7.11}$$

Proof. Only the 'if' part needs proof. If Y satisfies (7.11) for each t, then, for each t, $\mathbf{P}[{}^{\circ}X_t=Y_t]=1$. Since Y is an R-process by assumption, and $^{\circ}X$ is an R-process by (7.9), it now follows that Y and $^{\circ}X$ are indistinguishable. □

(*7.12*) **Important exercise.** We shall return to the material of this exercise later. Let $(\Omega,\mathscr{F},\{\mathscr{F}_t\},\mathbf{P},B_t)$ relate to $\mathrm{CBM}_0(\mathbb{R})$, so that, for example, $\mathbf{P}$ is Wiener measure and $\{\mathscr{F}_t\}$ has been augmented as usual. Figure VI.1 illustrates the random times which we now introduce.

Let

$$\tau\equiv\inf\{t: B_t=1\}; \quad \sigma\equiv\sup\{t<\tau: B_t=0\}.$$

There exists a unique ρ in $(0,\sigma)$ such that

$$B_\rho=\sup\{B_t: t<\sigma\}.$$

Put

$$\alpha\equiv\sup\{t<\rho: B_t=0\}.$$

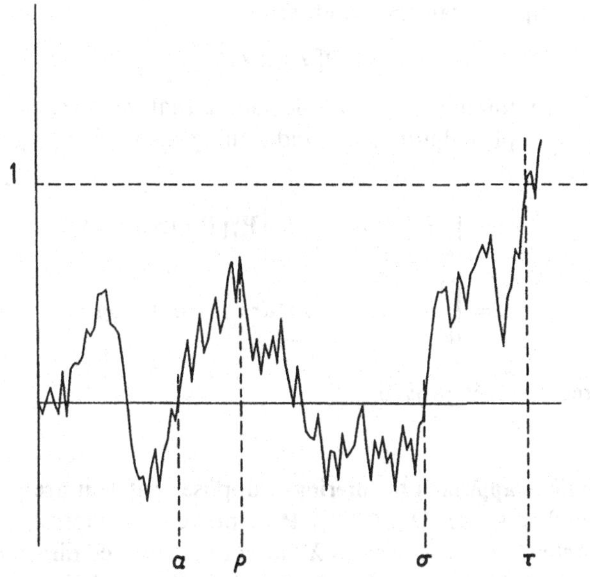

Figure VI.1

Finally, put

$$M_t \equiv \sup\{B_s : s < t\}.$$

Prove that (with $x^+ \equiv \max(x, 0)$ as usual):

$$^{o}I_{[\sigma,\infty)}(t) = B^+_{t\wedge\tau}; \quad ^{o}I_{[\rho,\infty)}(t) = M_{t\wedge\tau};$$

$$^{o}I_{[\alpha,\infty)}(t) = M_{t\wedge\tau} - B^+_{t\wedge\tau} \log M_{t\wedge\tau}.$$

Hint. Each of these three results follows quickly from the strong Markov property together with the following obvious fact for $\mathrm{CBM}_0(\mathbb{R})$: for $a<x<b$ (and with H_a denoting the hitting time of a),

$$\mathbf{P}^x[H_b < H_a] = (x-a)/(b-a). \tag{7.13}$$

Solution. The result for σ is trivial. The result for ρ is easy: just consider at time t what happens after B next reaches the value $M(t)$.

Now for the problem concerning α. Imagine that we have watched the Brownian path up to time t. We know the values $m = M_t$, and $b = B_t$. Let us suppose that $m < 1$, so that $\tau > t$.

If $b \leqslant 0$, we have

$$\mathbf{P}[\alpha \leqslant t \mid \mathcal{F}_t] = \mathbf{P}[\rho \leqslant t \mid \mathcal{F}_t] = m.$$

Let us now suppose that $b>0$, and write

$$y=\mathbf{P}[\alpha>t|\mathscr{F}_t],$$

(the 'greater than' is intended!). The whole point is that we can calculate y in terms of certain absolute probabilities, all conditioning then being forgotten. Indeed,

$$\begin{aligned} y &= \int_b^1 \mathbf{P}^b\left[\max_{t<H_0} B_t \in dr\right] \mathbf{P}^0[B(\rho)>m \vee r] \\ &= \int_b^1 br^{-2}(1-m\vee r)\,dr = 1-m+b\log m, \end{aligned}$$

and the desired result is proved.

8. The innovations approach to filtering. Suppose that you are given on some filtered probability space $(\Omega, \mathscr{F}, \{\mathscr{F}_t\}, \mathbf{P})$ a process of interest, $\{X_t: 0\leqslant t\leqslant 1\}$, called the *signal process*. The process X cannot be observed directly; all that you can see is the *observation process* $\{Y_t: 0\leqslant t\leqslant 1\}$ (Both X and Y are $\{\mathscr{F}_t\}$-adapted). The problem of filtering is concerned with finding $\mathbf{E}[f(X_t)|\mathscr{Y}_t]$, where $\{\mathscr{Y}_t\}$ is the usual augmentation of $\{\mathscr{Y}_t^\circ\}$, the filtration generated by the observation process Y. In the language we are using, *filtering is concerned with finding the* $\{\mathscr{Y}_t\}$-*optional projection of* $f(X_t)$. We shall consider only processes on $[0, 1]$, though this restriction is not essential, and, for these sections on filtering, we shall follow a well-established and convenient convention on notation.

(*8.1*) CONVENTION. *The* $\{\mathscr{Y}_t\}$-*optional projection of a process H will be denoted by* $\hat{H}$.

The principal results in this area appeared in a paper of Fujisaki, Kallianpur and Kunita [1], building on the notion of the *innovations process* introduced by Kailath [1]. We shall follow the treatment of Fujisaki, Kallianpur and Kunita, who consider the situation where the observation process takes the form

(8.2) $$Y_t=\int_0^t h_s\,ds+W_t,$$

where W is an $\{\mathscr{F}_t\}$-Brownian motion in $\mathbb{R}^n$, $W_0=0$, and h is an $\{\mathscr{F}_t\}$-adapted R-process with values in $\mathbb{R}^n$ such that

(8.3) $$\mathbf{E}\int_0^1 |h_s|^2\,ds<\infty.$$

Then the main result of (non-linear) filtering theory is summarized in the following theorem.

(*8.4*) THEOREM (Fujisaki, Kallianpur and Kunita). (i) *The process*

$$N_t \equiv Y_t - \int_0^t \hat{h}_s ds \tag{8.5}$$

is a $\{\mathscr{Y}_t\}$*-Brownian motion in* $\mathbb{R}^n$.

(ii) *If* Z *is an* L^2*-bounded* $\{\mathscr{Y}_t\}$*-martingale,* $Z_0 = 0$*, then there exists a* $\{\mathscr{Y}_t\}$*-previsible process* $C = (C^1, \ldots, C^n)$ *such that:*

$$\mathbf{E}\left[\int_0^1 \sum_{i=1}^n (C_s^i)^2 \, ds\right] < \infty,$$

and such that:

$$Z_t = \int_0^t (C_s, dN_s) \equiv \int_0^t \sum_{i=1}^n C_s^i dN_s^i. \tag{8.6}$$

(*8.7*) DEFINITION (innovations process) *The* $\{\mathscr{Y}_t\}$*-Brownian motion N defined by* (*8.5*) *is called the* innovations process.

(*8.8*) *Remarks.* (i) The integrability conditions imposed are somewhat stronger than necessary; we aim for transparency rather than maximality!

(ii) If it were true that (the usual augmentation of) the filtration generated by N were equal to $\{\mathscr{Y}_t\}$, then the second part of the theorem would be a restatement of the Brownian stochastic integral representation Theorem IV.36.1. However, no good sufficient conditions are known in general to ensure this, and the Tsirel'son example shows what could go wrong. So there really is something to be proved here.

(iii) The monotone-class argument given in the proof of Theorem 19.6 below makes it clear that if g is a $\{\mathscr{Y}_t\}$ optional process, then there exists a $\{\mathscr{Y}_t\}$ previsible process $\tilde{g}$ such that, almost surely, $g(t) = \tilde{g}(t)$ except for countably many t. Since N is continuous, and $[N]$ is continuous, it is good sense to allow $g \in L^2(N)$ and to define $g \bullet N \equiv \tilde{g} \bullet N$.

Proof of Theorem 8.4. (i) Let T be a $\{\mathscr{Y}_t\}$-stopping time, $T \leqslant 1$. Then since $N_1^* \equiv \sup_{t \leqslant 1} |N_t|$ is dominated by the square-integrable random variable $W_1^* + \int_0^1 \{|h_s| + |\hat{h}_s|\} \, ds$, N_T is integrable and

$$\begin{aligned}\mathbf{E} N_T &= \mathbf{E}\left[W_T + \int_0^T (h_s - \hat{h}_s) \, ds\right] \\ &= \int_0^1 \mathbf{E}[h_s - \hat{h}_s; s < T] \, ds = 0,\end{aligned}$$

since $\{s<T\}\in\mathscr{Y}_s$, and $\hat{h}_s=\mathbf{E}[h_s|\mathscr{Y}_s]$. Thus, by Lemma II.53.4, N is a UI martingale (and even an L^2-bounded martingale). The quadratic variation process of N satisfies

$$[N^i, N^j]_t=[W^i, W^j]_t=t\delta_{ij}$$

and so (Theorem IV.34.1) N is a Brownian motion.

(ii) The basic idea is very simple; we change measure to make Y into a Brownian motion, and then use the integral representation result Theorem IV.36.1. However, real applications of the Cameron–Martin–Girsanov change of measure are seldom entirely straightforward, and a little care is needed. We make use of the following (very plausible) result.

(*8.9*) LEMMA. *Suppose that S is a* $\{\mathscr{Y}_t^\circ\}$*-stopping time. Then*

$$\mathscr{Y}_S^\circ=\sigma(\{Y_{t\wedge S}\colon 0\leqslant t\leqslant 1\}).$$

Proof. See (IV.100) of Dellacherie and Meyer [1].

Now fix $K\in\mathbb{N}$, and let

$$T\equiv\inf\left\{t\colon \left|\int_0^t(\hat{h}_s, dN_s)\right|+\int_0^t|\hat{h}_s|^2ds=K\right\}\wedge 1.$$

If now

$$\rho_t\equiv\exp\left[-\int_0^t(\hat{h}_s, dN_s)-\tfrac{1}{2}\int_0^t|\hat{h}_s|^2ds\right],$$

then $\rho_{T\wedge t}$ is a bounded and uniformly positive martingale. We can define a probability measure $\mathbf{Q}$ on $(\Omega,\mathscr{Y}_1)$ by $d\mathbf{Q}/d\mathbf{P}=\rho_T$, and $\mathbf{P}$ and $\mathbf{Q}$ are equivalent. (*Exercise.* Use the fact that N is a *continuous* $\{\mathscr{Y}_t\}$ adapted process to show that T is a.s. equal to a $\{\mathscr{Y}_t^\circ\}$ stopping time S, and that ρ_T is a.s. equal to a $\{\mathscr{Y}_S^\circ\}$-measurable variable. (*Hint.* See (II.74.2) and contrast (II.75.3).) We ignore the 'a.s.' qualifications below.) An application of Theorem IV.38.4 shows that, under $\mathbf{Q}$,

$$\tilde{Y}_t\equiv N_t+\int_0^{t\wedge T}\hat{h}_s\,ds \text{ is a Brownian motion.}$$

We shall prove that if $Z\in L^2(\mathscr{Y}_T^\circ,\mathbf{P})$, $\mathbf{E}(Z|\mathscr{Y}_0)=0$, then Z has a representation

$$Z=\int_0^1(C_s, dN_s);$$

by letting $K\uparrow\infty$, we obtain (8.6). (*Exercise.* Check the details of this for yourself.)

If we let $\tilde{\mathscr{Y}}_t^\circ\equiv\sigma(\{\tilde{Y}_s\colon 0\leqslant s\leqslant t\})$, then by Lemma 8.9, $\tilde{\mathscr{Y}}_T^\circ=\mathscr{Y}_T^\circ$, so that $Z\in L^2(\tilde{\mathscr{Y}}_T^\circ,\mathbf{P})=L^2(\tilde{\mathscr{Y}}_T^\circ,\mathbf{Q})\subseteq L^2(\tilde{\mathscr{Y}}_1,\mathbf{Q})$. Also, $\rho_T^{-1}\in L^\infty(\tilde{\mathscr{Y}}_T,\mathbf{Q})$, so that, applying the Brownian motion integral representation Theorem IV.36.1 to $\tilde{Y}$, we obtain

$$Z\rho_T^{-1} = \int_0^1 (H_s, d\tilde{Y}_s)$$

for some $\{\tilde{\mathscr{Y}}_t\}$-previsible square-integrable process H. Since $Z\rho_T^{-1}$ is $\tilde{\mathscr{Y}}_T$ measurable, the process H is null on $(T, 1]$ and hence is equal to a $\{\mathscr{Y}_t\}$-previsible process. Thus if

$$M_t \equiv \mathbf{E}^Q[Z\rho_T^{-1}|\tilde{\mathscr{Y}}_t] = \int_0^{T\wedge t} (H_s, d\tilde{Y}_s),$$

then applying Itô's formula to the $\mathbf{P}$-martingale $M_t\rho_{t\wedge T}$ yields

$$M_t\rho_{t\wedge T} = \int_0^{t\wedge T} \rho_s(H_s - M_s\hat{h}_s, dN_s),$$

and letting $t\uparrow 1$ gives

$$Z = \int_0^1 \rho_s(H_s - M_s\hat{h}_s, dN_s),$$

as required. □

(*8.10*) COROLLARY. *Every $\{\mathscr{Y}_t\}$-local martingale null at* 0 *is continuous.*

Proof. Let $\{M_t: 0 \leqslant t \leqslant 1\}$ be a $\{\mathscr{Y}_t\}$-local martingale null at 0. Since M is locally a UI martingale, it is sufficient to prove the result for a UI martingale. But then there exist $M_1^{(n)} \in L^2(\mathscr{Y}_1)$ such that $\mathbf{E}|M_1 - M_1^{(n)}| \leqslant 3^{-n}$ and the martingales $M_t^{(n)} \equiv \mathbf{E}[M_1^{(n)}|\mathscr{Y}_t]$ are continuous, by Theorem 8.4. By Doob's submartingale maximal inequality II.70.1 applied to the submartingale $|M - M^{(n)}|$,

$$\mathbf{P}((M - M^{(n)})_1^* > 2^{-n}) \leqslant 2^n \mathbf{E}|M_1 - M_1^{(n)}|$$

and the $M^{(n)}$ converge uniformly to M by Borel–Cantelli. □

If we assume some more structure, we can obtain a really useful result for the integrand in the stochastic integral. We shall prove a fundamental result of Fujisaki, Kallianpur and Kunita [1].

(*8.11*) THEOREM. (Fujisaki, Kallianpur and Kunita). *Suppose that:*

(8.12(i)) *X is a continuous adapted process with values in a Polish space S;*
(8.12(ii)) *W is a Brownian motion in $\mathbb{R}^n$;*

(8.12(iii)) *$f \in C(S)$ is such that $\sup_{t\leqslant 1} \mathbf{E}|f(X_t)|^2 < \infty$, and*

there exists an $\{\mathscr{F}_t\}$-optional process $(\mathscr{G}f_t)_{0\leqslant t\leqslant 1}$ such that:

(8.13) $$\mathbf{E}\int_0^1 |\mathscr{G}f_s|^2 ds < \infty$$

and such that:

$$M_t \equiv f(X_t) - f(X_0) - \int_0^t \mathscr{G} f_s \, ds \text{ is a martingale.} \tag{8.14}$$

Then M is bounded in L^2 and there exists an $\{\mathscr{F}_t\}$-optional process $\alpha \equiv (\alpha^1, \ldots, \alpha^n)$ such that:

$$[M, W^i]_t = \int_0^t \alpha^i \, ds. \tag{8.15}$$

Then, if $f_t \equiv f(X_t)$, we have the filtering equations:

$$\hat{f}_t = \hat{f}_0 + \int_0^t \widehat{\mathscr{G} f_s} \, ds + \int_0^t (\widehat{f_s h_s} - \hat{f}_s \hat{h}_s + \hat{\alpha}_s, dN_s). \tag{8.16}$$

Proof. Firstly, we notice that if C is any measurable process such that

$$\mathbf{E} \int_0^1 |C_s| \, ds < \infty,$$

and if $V_t \equiv \int_0^t C_s \, ds$, then

$$\hat{V}_t - \int_0^t \hat{C}_s \, ds \text{ is a } UI \ \{\mathscr{Y}_t\}\text{-martingale.} \tag{8.17}$$

The uniform integrability is easy, and if $T \leqslant 1$ is any $\{\mathscr{Y}_t\}$-stopping time, then

$$\begin{aligned} \mathbf{E} \hat{V}_T = \mathbf{E} V_T &= \mathbf{E} \int_0^1 I_{(0,T]}(s) C_s \, ds \\ &= \int_0^1 \mathbf{E}[C_s; s \leqslant T] \, ds \\ &= \int_0^1 \mathbf{E}[\hat{C}_s; s \leqslant T] \, ds \end{aligned}$$

since $\{s \leqslant T\} \in \mathscr{Y}_s$, and $\hat{C}$ is the $\{\mathscr{Y}_t\}$-optional projection of C. Rearrangement gives (8.17).

Because of the Kunita–Watanabe inequalities IV.28.5, the continuous finite-variation process $[M, W]$ has a density with respect to Lebesgue measure, which may be taken to be $\{\mathscr{F}_t\}$-optional, by the argument which led to (8.17). Thus the statement (8.15) is justified.

An application of (8.17) to the case where $C = \mathscr{G} f$ together with the assumed integrability of f and $\mathscr{G} f$ shows that

$$U_t \equiv \hat{f}_t - \hat{f}_0 - \int_0^t \widehat{\mathscr{G} f_s} \, ds \text{ is an } L^2\text{-bounded } \{\mathscr{Y}_t\}\text{-martingale}$$

which is null at 0, so, by Theorem 8.4, there is a representation

$$U_t = \int_0^t (\varphi_s, dN_s).$$

Now the proof is completed by calculating the $\{\mathscr{Y}_t\}$-optional projections of $f_t Y_t$ in two different ways. Firstly,

$$\begin{aligned} f_t Y_t &= \int_0^t f_s \, dY_s + \int_0^t Y_s \, df_s + \int_0^t \alpha_s \, ds \\ &= \int_0^t f_s (dW_s + h_s ds) + \int_0^t Y_s (dM_s + \mathscr{G} f_s ds) + \int_0^t \alpha_s \, ds \\ &= \int_0^t [f_s h_s + Y_s \mathscr{G} f_s + \alpha_s] \, ds + \{\mathscr{F}_t\}\text{-martingale.} \end{aligned}$$

Thus using (8.17)

$$\widehat{f_t Y_t} = \hat{f}_t Y_t = \int_0^t [\widehat{f_s h_s} + Y_s \widehat{\mathscr{G} f_s} + \hat{\alpha}_s] \, ds + \{\mathscr{Y}_t\}\text{-martingale.} \tag{8.18}$$

The other way,

$$\begin{aligned} \hat{f}_t Y_t &= \int_0^t \hat{f}_s \, dY_s + \int_0^t Y_s \, d\hat{f}_s + \int_0^t \varphi_s \, ds \\ &= \int_0^t [\hat{f}_s \hat{h}_s + Y_s \widehat{\mathscr{G} f_s} + \varphi_s] \, ds + \{\mathscr{Y}_t\}\text{-martingale.} \end{aligned} \tag{8.19}$$

By Corollary 8.10, the $\{\mathscr{Y}_t\}$-martingales are both continuous, so the difference of the finite-variation parts of (8.18) and (8.19) is a continuous martingale of finite variation and null at 0, which is therefore null (Theorem IV.30.4). The filtering equations (8.16) follow. □

9. The Kalman–Bucy filter. The Kalman–Bucy filter is a celebrated example in widespread use in real-world problems of filtering. The signal process X is assumed to satisfy an Ornstein–Uhlenbeck SDE

$$X_t = X_0 + B_t + \int_0^t a X_s \, ds$$

and the observation process Y is given by

$$Y_t = W_t + \int_0^t c X_s \, ds,$$

where B and W are independent Brownian motions on $\mathbb{R}$ independent of the Gaussian variable X_0. Since the bivariate process (X_t, Y_t) is Gaussian, the conditional law of X_t given $\mathscr{Y}_t$ will be Gaussian, and all we have to do is to identify the conditional mean and conditional variance. This is a problem to which Theorem 8.11 applies, where if f is C^2 and of polynomial growth,

$$\mathscr{G}f_t \equiv \tfrac{1}{2}f''(X_t) + aX_t f'(X_t).$$

Applying this to the two cases $f(x)=x$ and $f(x)=x^2$, and letting $v_t \equiv \widehat{X_t^2} - \hat{X}_t^2$ denote the conditional variance, we find that

$$\hat{X}_t = \mathbf{E}X_0 + c\int_0^t v_t\,dN_t + \int_0^t a\hat{X}_s\,ds \tag{9.1}$$

$$\widehat{X_t^2} = \mathbf{E}X_0^2 + c\int_0^t \{\widehat{X_s^3} - \hat{X}_s\widehat{X_s^2}\}\,dN_s + \int_0^t (1+2a\widehat{X_s^2})\,ds. \tag{9.2}$$

But if $Z \sim N(\mu, \sigma^2)$, it is elementary that $\mathbf{E}Z^3 = \mu(\mu^2+3\sigma^2)$, and hence

$$\begin{aligned}\widehat{X_s^3} - \hat{X}_s\widehat{X_s^2} &= \hat{X}_s\{\hat{X}_s^2 + 3v_s - \widehat{X_s^2}\}\\ &= 2v_s\hat{X}_s.\end{aligned}$$

Thus

$$\begin{aligned}dv_t &\equiv d(\widehat{X_t^2} - \hat{X}_t^2)\\ &= 2cv_t\hat{X}_t\,dN_t + (1+2a\widehat{X_t^2})dt - 2\hat{X}_t(cv_t\,dN_t + a\hat{X}_t\,dt) - c^2v_t^2\,dt\\ &= (1+2av_t - c^2v_t^2)\,dt,\end{aligned}$$

so that *the conditional variance v_t is the (deterministic) solution of the ordinary differential equation*

$$v_t = \mathrm{var}(X_0) + \int_0^t (1+2av_s - c^2v_s^2)\,ds, \tag{9.3}$$

and the conditional mean satisfies

$$\hat{X}_t = \mathbf{E}X_0 + c\int_0^t v_s\,dN_s + \int_0^t a\hat{X}_s\,ds. \tag{9.4}$$

The differential equation (9.3) has an explicit solution. If α and $-\beta$ are the roots of the quadratic $1+2ax-c^2x^2$, α and β both positive, and if $\lambda = c^2(\alpha+\beta)$, then

$$v(t) = \frac{\gamma\alpha e^{\lambda t} - \beta}{\gamma e^{\lambda t}+1}$$

where

$$\gamma = \frac{\sigma^2+\beta}{\alpha-\sigma^2},$$

and $\sigma^2 = \mathrm{var}(X_0)$. Note that $v(\infty) = \alpha$.

There is a vector analogue of the Kalman filter which you can derive for yourself as an exercise if you wish, but do notice that the (matrix) differential equation for the conditional variance has in general no closed-form solution.

10. The Bayesian approach to filtering; a change-detection filter. An alternative, and equally useful, approach to filtering is via the likelihood-ratio with respect to Wiener measure. We illustrate the main ideas with a simple example which has real-world applications. Let T be a random variable with values in $(0, \infty)$, with probability density function φ, and tail probability

$$g_t \equiv \mathbf{P}[T > t].$$

Define

$$A = I_{[T, \infty)}, \quad Y_t = \int_0^t A_s\, ds + B_t, \tag{10.1}$$

where B is a Brownian motion independent of A. For example, T might be the time of onset in a patient of some illness which must be detected. We want to find

$$\hat{A}_t \equiv \mathbf{E}[A_t | \mathscr{Y}_t] = \mathbf{E}[A_t | Y_s;\ s \leqslant t],$$

where $\{\mathscr{Y}_t\}$ is the usual augmentation of the natural filtration of Y.

The key idea, due to Kallianpur and Striebel, is to use Bayes' formula, working with likelihood-ratio relative to Wiener measure. Let $\mathbf{Q}_s$ be the law of Y given that $T = s$, that is, the law of the process $\{Y_s(u): u \geqslant 0\}$, where

$$Y_s(u) \equiv B(u) + (u - s)^+.$$

Let $\mathbf{P}$ be Wiener measure. Then, on $\mathscr{Y}_t$, for $t \geqslant s$,

$$d\mathbf{Q}_s/d\mathbf{P} = \exp\{Y_t - Y_s - \tfrac{1}{2}(t-s)\}.$$

Hence, Bayes' theorem gives

$$\hat{A}_t = H_t/(H_t + g_t), \tag{10.2}$$

where

$$\begin{aligned} H_t &= \int_0^t \varphi(s) \exp\{Y_t - Y_s - \tfrac{1}{2}(t-s)\}\, ds \\ &= \exp\{Y_t - \tfrac{1}{2}t\} \int_0^t \varphi(s) \exp\{-Y_s + \tfrac{1}{2}s\}\, ds. \end{aligned} \tag{10.3}$$

Itô's formula now yields the Zakai equation:

$$dH_t = H_t\, dY_t + \varphi(t)\, dt. \tag{10.4}$$

A further application of Itô's formula—to (10.2), using (10.4)—shows that

$$d\hat{A} = \hat{A}(1 - \hat{A})\, dN + (1 - \hat{A})\varphi_t\, dt/g_t, \tag{10.5}$$

where N is the innovations process

$$N_t = Y_t - \int_0^t \hat{A}_s \, ds.$$

We know that N is a Brownian motion relative to the $\{\mathscr{Y}_t\}$ filtration. In this case, one can show that (10.5) has a strong solution for $\hat{A}$ in terms of N, so that N generates the same augmented filtration as does Y.

In the terminology developed later, we can describe (10.5) as showing in particular that the process

$$\alpha_t = \int_0^t (1 - \hat{A}_s)\varphi_s \, ds / g_s \tag{10.6}$$

is the *dual previsible projection* or *compensator* of A relative to the filtration generated by Y. Intuitively, $\dot{\alpha}_t \, dt$ is our estimate of the probability that T will fall in the interval $(t-dt, t]$ given the behaviour of Y up to time $t-dt$. See (22.8).

(*10.7*) *Analysis via the innovations approach.* Let us suppose that the density φ is continuous. The process $X_t \equiv (t, I_{\{t \geq T\}})$ is a Markov process on $\mathbb{R}^+ \times \{0, 1\}$ with generator $\mathscr{G}$ given by

$$\mathscr{G}f(t, 0) = f'(t, 0) + \varphi_t[f(t, 1) - f(t, 0)]/g_t$$

$$\mathscr{G}f(t, 1) = f'(t, 1)$$

when $f(\cdot, i)$ is C^1, $i = 0, 1$. Thus we are in a position to apply Theorem 8.11 with

$$f(t, i) = I_{\{i=1\}}, \quad h_t = I_{\{X_t = i\}}.$$

The filtering equations (8.16) reduce to (10.5), as you can easily check. The case $\varphi_t = e^{-t}$ is particularly interesting in that we are estimating a function of a finite Markov chain given noisy observation of an additive functional of the chain. We shall have more to say about this in the next section.

(*10.8*) *Illustrative example.* If we think of the time T of change as the time of onset of some disease in a patient, the appearance of which will require the call-out of a doctor, then it is clear that a false alarm will be a costly and time-consuming nuisance, to be avoided if at all possible. One might think that the way to avoid false alarms is to improve the accuracy of observation, so that the noise B in (10.1) will be replaced by the noise εB. In fact, however small we make ε, provided it is still positive, *the probability of a false alarm will not be altered.* And this is true in far greater generality still.

(*10.9*) THEOREM. *Suppose that* $T: (\Omega, \mathscr{F}) \to [0, \infty]$ *is a random time, suppose that* $(\Omega, \{\mathscr{Y}_t\}, \mathbf{P})$ *satisfies the usual conditions,* $\mathscr{Y}_t \subseteq \mathscr{F}$ *for all* t, *and let*

$$A_t \equiv I_{\{t \geq T\}}.$$

If we write C for the $\{\mathscr{Y}_t\}$*-optional projection of A, if we fix* $\xi \in (0, 1)$*, and let*

$$\tau \equiv \inf\{t : C_t \geqslant \xi\},$$

then assuming that $\Delta C_t \leqslant 0$ *for all t,*

$$\mathbf{P}(T \leqslant \tau \mid \tau < \infty) = \xi. \tag{10.10}$$

Proof. By definition of the optional projection,

$$\mathbf{E}[C_\tau : \tau < \infty] = \mathbf{E}[A_\tau : \tau < \infty] = \mathbf{P}(T \leqslant \tau < \infty).$$

By the assumption that $\Delta C_t \leqslant 0$ (satisfied if C is continuous, as in the above example), $C_\tau = \xi$ on $\{\tau < \infty\}$, and (10.10) follows. □

(*10.11*) *Remarks.* This result says that if we call the doctor out when the observations show that with probability at least ξ the patient has the disease, then we are going to be in error with probability $1-\xi$, however accurate our observation. (Bayesian statisticians may see this as tautology.)

11. Robust filtering. Useful though the filtering equations (8.16) are, they do have limitations; one is that the integrands in the SDE for $\hat{f}$ involve estimates of other functions of the signal process, and only in special cases can we expect to be able to do anything about these; and another is that in order to form the estimate $\hat{f}$ in practice, we have to perform a stochastic integral numerically, which can be problematic. Under suitable conditions, one can reduce the problem to solving some system of differential equations *where the only appearance of randomness is in the coefficients.* This avoids the need to do a numerical stochastic integration. We illustrate the method for the case where the signal process X is a Markov chain with finite state-space I, and $\mathbf{Q}$-matrix Q.

For $x \in I$, let $e_x(t) \equiv I_{\{x\}}(X_t)$, let $h_t = h(X_t)$, and let $p_t(x) \equiv \widehat{e_x(t)}$. Then assuming that W is independent of X, the filtering equations (8.16) for $p_t(x)$ become

$$dp_t(x) = [h(x)p_t(x) - \hat{h}_t p_t(x)]\,dN_t + (Q^* p_t)(x)\,dt, \tag{11.1}$$

where Q^* is the adjoint of Q,

$$(Q^* f)(x) \equiv \sum_{y \in I} q_{yx} f_y.$$

By Itô's formula,

$$\begin{aligned} d(\log p_t(x)) &= p_t(x)^{-1} dp_t(x) - \tfrac{1}{2} p_t(x)^{-2} d[p_t(x)] \\ &= (h(x) - \hat{h}_t)\,dN_t + p_t(x)^{-1}(Q^* p_t)(x)\,dt \\ &\quad - \tfrac{1}{2}\{h(x) - \hat{h}_t\}^2\,dt \\ &= h(x)(dN_t + \hat{h}_t\,dt) + p_t(x)^{-1}(Q^* p_t)(x)\,dt \\ &\quad - \tfrac{1}{2}h(x)^2\,dt - \hat{h}_t\,dN_t - \tfrac{1}{2}\hat{h}_t^2\,dt. \end{aligned}$$

If now we define the process $\gamma_t(x)$ by

$$d(\log \gamma_t(x)) = h(x)(dN_t + \hat{h}_t dt) + p_t(x)^{-1}(Q^* p_t)(x)dt - \tfrac{1}{2}h(x)^2 dt,$$
$$\gamma_0(x) = p_0(x),$$

then it is clear that $p_t(x)/\gamma_t(x)$ is independent of x; indeed,

$$d(\log p_t(x)/\gamma_t(x)) = -\hat{h}_t dN_t - \tfrac{1}{2}\hat{h}_t^2 dt,$$

so that if we can find $\gamma_t(\bullet)$, we can obtain $p_t(x)$ simply by renormalizing. But

$$d(\log \gamma_t(x)) = h(x)dY_t + \gamma_t(x)^{-1}(Q^*\gamma_t)(x)dt - \tfrac{1}{2}h(x)^2 dt,$$

so that, if we write $\theta_t(x) \equiv \exp(-h(x)Y_t)\gamma_t(x)$, then

(11.2) $$d(\log \theta_t(x)) = [\exp(h(x)Y_t)\theta_t(x)]^{-1}(Q^* \exp(h(\bullet)Y_t)\theta_t)(x)dt - \tfrac{1}{2}h(x)^2 dt,$$

and $\theta_t(\bullet)$ can be obtained by integrating with respect to Lebesgue measure only.

If the underlying Markov process were a diffusion, say, then the formal analogue of (11.2) will still hold. But you see the technical problems; the analogue of Q^* is the adjoint $\mathscr{G}^*$ of the generator $\mathscr{G}$ of the diffusion, and so (11.2) becomes a *stochastic partial differential equation*. For more on this, see Davis [1], Kunita [1] and Pardoux [1]. The robust filtering equation (11.2) is widely used in filtering on Markov chains in such areas as automatic speech recognition.

(*11.3*) *References.* Because of the subject's interest and real-world applicability, we have sprinkled the bibliography rather liberally with important papers on filtering.

4. CHARACTERIZING PREVISIBLE TIMES

12. Previsible stopping times; PFA theorem. Recall the previsible σ-algebra $\mathscr{P}$ on $(0, \infty) \times \Omega$ from IV.6.

(*12.1*) DEFINITION (previsible (stopping) time). *A stopping time T is called* previsible *if*

(i) $T > 0$,
(ii) $[T, \infty) \in \mathscr{P}$.

The restriction that $T > 0$ is part of our convention about time 0.

Since $(T, \infty) \in \mathscr{P}$ for any stopping time T, condition (ii) is equivalent to
(ii)* $[T] \in \mathscr{P}$.

One of the main ways in which previsible stopping times arise is described in Theorem 12.3 below which hinges on the following lemma, in which D_F is the début of F.

(*12.2*) LEMMA. *Let F be a previsible subset of* $(0, \infty)\times\Omega$ *such that* $[D_F]\subseteq F$. *Then D_F is a previsible stopping time.*

Proof. If $[D_F]\subseteq F$, then $[D_F, \infty)=F\cup(D_F, \infty)\in\mathscr{P}$. □

The début of an optional set is always an optional time, but the début of a previsible set is not always a previsible time—why?

(*12.3*) THEOREM. *Let X be a previsible R-process on* $(0, \infty)\times\Omega$.

(i) *If $\varepsilon>0$ and $T\equiv\inf\{t: |X_t-X_{t-}|>\varepsilon\}$, then T is previsible.*

(ii) *Let*

$$J\equiv\{(t, \omega): t>0,\ X_t(\omega)\neq X_{t-}(\omega)\}$$

be the jump set of X. Then there exists a sequence (T_n) of previsible stopping times such that

$$J\subseteq\bigcup_n[T_n].$$

Obviously, the (T_n) sequence will not have any monotonicity property in general.

Proof of (i). Since X is previsible by assumption, and X_- is previsible because it is an adapted L-process, the set

$$F\equiv\{(t, \omega): |X_t(\omega)-X_{t-}(\omega)|>\varepsilon\}$$

is previsible. Moreover, $T=D_F$, and (why?) $[T]\subseteq F$. Now apply Lemma 12.2.

Proof of (ii). Let $U(1, m)$, $U(2, m)$, ... be the first, second, ... time that $|X_t-X_{t-}|>1/m$. Now arrange the double sequence $(U(k, m))$ as a single sequence (T_n). □

(*12.4*) DEFINITION (fair, announceable (stopping) times) *Let T be a stopping time with $T>0$.*

(i) *We call T* fair *if for every bounded R-martingale M,*

$$\mathbf{E}(M_T)=\mathbf{E}(M_{T-}).$$

(ii) *We call T* announceable *if there exists a sequence (T_n) of stopping times which announce T in the sense that $T_n\uparrow\uparrow T$. Recall that this means that:*

$$T_n(\omega)\leqslant T_{n+1}(\omega)<T(\omega)\quad\forall n,\quad T_n(\omega)\uparrow T(\omega).$$

Note. In the definition of 'fair',

$$M_\infty\equiv M_{\infty-}\equiv\lim_{t\to\infty} M_t.$$

(*12.5*) *Notes on terminology.*

(*a*) '*Fair*'. We have invented the adjective 'fair' and the mnemonic 'PFA' for Theorem 12.6. A bounded martingale M is often regarded as a player's fortune in a fair game. A stopping time T is fair if, on average, the player's fortune does not change at time T (irrespective of the fair game M he is playing).

(*b*) *'Previsible' and 'announceable'*. The original definition of previsible (or 'predictable') stopping time was the present definition of announceable stopping time. If T is announceable, then—with hindsight!—the observer can appreciate that 'instantaneously before T', he could have predicted (or announced) that T was about to occur. The French *prévisible* is apparently sometimes used in business contexts with this 'with hindsight' connotation, while the English 'predictable' is normally intended to refer to the future proper. But do read Note (*c*). Once you have grasped the 'foreseeable with hindsight' idea(!), you can appreciate why adapted L-processes should be regarded as previsible.

(*c*) It is helpful to add that the following strict-sense interpretations underlie the mathematical terminology: a *prediction* is a true statement about the future; a mere calculated guess is a *projection*.

(*12.6*) THEOREM (PFA theorem; Meyer). *The following statements about a stopping time T with $T>0$ are equivalent:*

(P) *T is previsible;*

(F) *T is fair;*

(A) *T is announceable.*

We shall give a complete proof of this result in § 16. We defer the proof because we believe that it is best for you to see first how the theorem may be used. Our proof that (F)⇒(A) is based on work of Chung and Walsh, and it allows us to be very explicit about the business of announceability, as follows.

(*12.7*) THEOREM (after Chung and Walsh [2]). *Let T be a stopping time with $T>0$. Let g be a strictly increasing continuous function from* $[0, \infty]$ *to* $[0, 1]$ *with* $g(0)=0$. *Let M be an R-modification of the martingale*

$$M_t = \mathbf{E}[g(T) \mid \mathcal{F}_t].$$

Set

$$V_t \equiv M_t - g(T\wedge t) = \mathbf{E}[g(T) - g(T\wedge t) \mid \mathcal{F}_t],$$

and

$$T_n \equiv \inf\{t \colon V_t < 1/n\}.$$

Then T is announceable if and only if (T_n) announces T.

The proof of Theorem 12.7 will be given in § 16 as part of the proof of the PFA theorem.

Shortly, we shall see how Theorem 12.6 applies to Markov processes.

13. Totally inaccessible and accessible stopping times. We now introduce two further classes of stopping times.

(*13.1*) DEFINITION (accessible, totally inaccessible (stopping) times). *Let T be a stopping time. Then T is called* totally inaccessible *if, for every previsible stopping time S,*

$$\mathbf{P}[T=S<\infty]=0;$$

T is called accessible *if for every totally inaccessible stopping time S,*

$$\mathbf{P}[T=S<\infty]=0.$$

Note. According to our conventions, time 0 is totally inaccessible, while in Dellacherie and Meyer [1], time 0 is previsible. Time ∞ is agreed by everyone to be both previsible and totally inaccessible.

Now, a previsible stopping time is clearly accessible. Our examples will show that, for certain filtrations, there exist accessible stopping times which are not previsible. The general relation between accessible and previsible stopping times is given by the following lemma.

(*13.2*) LEMMA. *T is accessible if and only if there exist previsible stopping times $S_1, S_2, S_3, \ldots$ such that*

$$[T]\subseteq\bigcup_n [S_n].$$

(*13.3*) *Remarks.* (i) According to Lemma 13.2 then, for each ω, there exists an $n=n(\omega)$ such that $T(\omega)=S_n(\omega)$. We shall say that the sequence (S_n) *catches T.* It should be emphasized that the sequence (S_n) will in general have no monotonicity property. Thus, for example, (S_n) might enumerate some dense subset of $(0, \infty)$.

(ii) In a number of important situations—and we characterize these exactly in § 19—all accessible stopping times are previsible.

(iii) The 'if' part of Lemma 13.2 is trivial. The 'only if' part should be seen as a corollary of the decomposition of a general stopping time which we now describe. We trust that you will not object to our using two-letter symbols AC and IN for subsets of Ω in the following lemma. The shorthand 'A for announceable' has already been used, so we use 'AC for accessible'.

(*13.4*) LEMMA. *Let T be a stopping time. Then there exists an essentially unique decomposition:*

$$\{T<\infty\}=\mathrm{AC}\cup\mathrm{IN}$$

of the set $\{T<\infty\}$ as the disjoint union of two sets AC *and* IN *such that T_{AC} is accessible and T_{IN} is totally inaccessible.*

Proof. For $\Lambda\in\mathscr{F}$, say that $\Lambda\in\mathscr{H}$ if there exists a countable family (S_n) of previsible stopping times such that:

$$\Lambda=\bigcup_n\{T=S_n<\infty\};$$

then, $\Lambda \in \mathscr{F}_T$ and $\Lambda \subset \{T<\infty\}$. Let $h \equiv \sup\{\mathbf{P}(\Lambda): \Lambda \in \mathscr{H}\}$. For $k \in \mathbb{N}$, choose Λ_k in $\mathscr{H}$ with $\mathbf{P}(\Lambda_k) > h - 1/k$. Set $\mathrm{AC} \equiv \bigcup_k \Lambda_k$. Then $\mathrm{AC} \in \mathscr{H}$, and the rest is obvious. □

Proof of the 'only if' part of Lemma 13.2. This is also now obvious, since if T is accessible, then $\{T<\infty\} = \mathrm{AC}$, and $\mathrm{AC} \in \mathscr{H}$. □

(*13.5*) *Martingale characterization of totally inaccessible stopping times.* The 'fair' property characterizes previsible stopping times in terms of the behaviour of martingales at those times. For an analogous characterization of totally inaccessible stopping times, see Theorem 21.10.

14. Some examples. These examples serve as hors d'oeuvres for Meyer's previsibility theorem (§ 15) which completely characterizes the stopping times of FD processes.

(*14.1*) *Example.* Let T be an exponentially distributed random variable of rate 1 on some complete triple $(\Omega, \mathscr{F}, \mathbf{P})$, so that $\mathbf{P}[T>t] = e^{-t}$, $t \geqslant 0$. Let $X \equiv I_{[T,\infty)}$, so that X is a Markov chain, with two states 0 and 1, run under its $\mathbf{P}^0$ law $\mathbf{P}$. Let $\mathscr{F}_t^\circ \equiv \sigma(\{X_s : s \leqslant t\})$, and let $\{\mathscr{F}_t\}$ be the usual augmentation of $\{\mathscr{F}_t^\circ\}$. Since X has FD transition function, we know that X is strong Markov relative to $\{\mathscr{F}_t\}$. Of course, T is a stopping time.

Let S be any stopping time satisfying $S<T$. By applying the strong Markov theorem at time S (lack of memory of exponential times), we see that $T–S$ is also exponentially distributed with rate 1. Hence $S=0$, a.s.; so that T is certainly not previsible since it cannot be announceable.

(*14.2*) **Exercise.** Argue that T is totally inaccessible. (*Hint.* If S is previsible and (S_n) announce S, prove firstly that $\mathbf{E}[\exp\{-\lambda(T-S_n)\}; S_n<T] = (1+\lambda)^{-1}\,\mathbf{P}(S_n<T)$.

(*14.3*) *Example.* We consider the exploding birth process with instantaneous return discussed in § III.26. When at point i of $\mathbb{N}$, our particle stays there for an exponentially distributed time of rate q_i, and then jumps to $i+1$. The jump rates are chosen so that $\sum q_i^{-1} < \infty$, so that (a.s.) explosion occurs during a finite time. At each of the successive explosion times $\eta_1, \eta_2, \eta_3, \ldots$, the particle jumps (or 'branches') from ∞ to a state in $\mathbb{N}$ in accordance with a given probability measure μ on $\mathbb{N}$. For definiteness, let us work with the P^1 law.

Assume for now that $0<\mu_1<1$. Let T be the time of the first jump from ∞ to 1. Then T must be one of the successive explosion times. In symbols, $[T] \subseteq \bigcup_n [\eta_n]$; in words, (η_n) catches T. Thus T is accessible, or perhaps previsible. It is intuitively obvious that T is *not* previsible, because we cannot tell immediately before an explosion time whether or not that particular explosion time will see the first jump from ∞ to 1. We can *prove* that T is not previsible by showing that T is not fair. If we write

$$M_t \equiv \mathbf{E}[T|\mathcal{F}_t],$$

then

$$M_t = \begin{cases} T & \text{on } \{t \geq T\}, \\ t + r(X_t) + c & \text{on } \{t < T\}, \end{cases}$$

where

$$r(j) \equiv \sum_{k \geq j} q_k^{-1}, \qquad c \equiv \mu_1^{-1} \sum_{j \geq 2} \mu(j) r(j).$$

Since $M(T-) - M(T) = c$, it is clear that T is not fair. (Your worries about the fact that M is not bounded will vanish when we come to the PFA Theorem. The point is that M is uniformly integrable—see (16.2).)

Incidentally, it is worth noting what happens to the martingale M at time $\eta = \eta_1$. We have

$$M_\eta = \begin{cases} \eta & \text{if } \eta = T, \\ \eta + r(X_\eta) + c & \text{if } \eta \neq T, \end{cases}$$

$$M_{\eta-} = \eta + c.$$

Hence,

$$M_\eta - M_{\eta-} = r(X_\eta) I_{\{X_\eta \neq 1\}} - c I_{\{X_\eta = 1\}}.$$

and it is easily verified—see § III.26.3—that $\mathbf{E}M(\eta) = \mathbf{E}M(\eta-)$. We have mentioned this just to clarify how T fails to be fair even though T is caught by (η_n) and every η_n is fair.

A final point to note in connection with this example is that if $\mu(1) = 1$, then $T = \eta$ (we continue to ignore 'a.s.' qualifications), so T is previsible. In this case, ∞ is a *degenerate* branch point from which the particle branches deterministically (to 1).

(*14.4*) *Example.* This example generalizes Example 14.1. Let T be a random variable with values in $[0, \infty]$ carried by some complete triple $(\Omega, \mathcal{F}, \mathbf{P})$. Let $X \equiv I_{[T,\infty)}$, let $\mathcal{F}_t^\circ \equiv \sigma(\{X_s\colon s \leq t\})$, and let $\{\mathcal{F}_t\}$ be the usual augmentation of $\{\mathcal{F}_t^\circ\}$. Let F be the distribution function of T.

You have probably already made the following correct guesses about this situation:

(i) T is previsible if and only if $\mathbf{P}[T = t_0] = 1$ for some $t_0 > 0$;

(ii) T is accessible if and only if the distribution of T is purely atomic with at least two atoms and with no atom at 0;

(iii) T is totally inaccessible if and only if F is continuous on $(0, \infty)$;

(iv) the totally inaccessible part of T is T_{IN}, where

$$\mathrm{IN} = \{\omega\colon F \text{ is continuous at } T(\omega)\}.$$

When you have read on to § 22, you will understand why this is correct.

Note that though X is not a Markov process, the process Z with $Z_t \equiv (X_t, t)$ is even a time-homogeneous Markov process. You can see that in the 'previsible' case (i), $Z(T-)$ is the degenerate branch point $(0, t_0)$ from which Z branches deterministically to $(1, t_0)$. In case (ii), $Z(T-)$ is one of the atoms of the distribution of T, and each such atom is a branch point. In case (iii), Z is an FD process for which $Z(T-) \neq Z(T)$ on $\{0 < T < \infty\}$.

(*14.5*) **Exercise.** Describe the Ray–Knight compactification for the 'accessible' case (ii).

15. Meyer's previsibility theorem for Markov processes. For Markov processes, things now really begin to take shape.

Firstly, we discuss the case of FD processes.

(*15.1*) THEOREM (Meyer). *Let X be an FD process. Fix an initial law μ. Let $\{\mathscr{F}_t^\mu\}$ be the usual $\mathbf{P}^\mu$ augmentation of $\{\mathscr{F}_t^\circ\}$. Then the following statements about an $\{\mathscr{F}_t^\mu\}$ stopping time T with $T > 0$ are equivalent:*

(i) *T is $\{\mathscr{F}_t^\mu\}$ previsible;*

(ii) *T is $\{\mathscr{F}_t^\mu\}$ accessible;*

(iii) *$X(T-) = X(T)$ (a.s. $\mathbf{P}^\mu$) on $\{T < \infty\}$.*

(*15.2*) COROLLARY. *A stopping time T is $\{\mathscr{F}_t^\mu\}$ totally inaccessible if and only if $X(T-) \neq X(T)$ on $\{0 < T < \infty\}$.*

(*15.3*) COROLLARY. *Let $X = (X_t, \Omega, \mathscr{F}, \mathscr{F}_t, \mathbf{P})$ be $\mathrm{CBM}_0(\mathbb{R})$, so that $\mathbf{P}$ is Wiener measure, etc. Then every strictly positive stopping time is previsible.*

Proof of Theorem 15.1. Let us agree that 'martingale', 'previsible', 'accessible', etc., refer to the set-up $(\Omega, \mathscr{F}^\mu, \{\mathscr{F}_t^\mu\}, \mathbf{P}^\mu)$ which satisfies the usual conditions. In particular, we shall write $\mathbf{P}$ for $\mathbf{P}^\mu$, $\mathbf{E}$ for $\mathbf{E}^\mu$, $\mathscr{F}_t$ for $\mathscr{F}_t^\mu$, etc.

If T is previsible (hence announceable), then Blumenthal's quasi-left-continuity Theorem III.11.1 shows that $X(T-) = X(T)$ on $\{T < \infty\}$. If T is accessible, then, since T is caught by a sequence of previsible times, we again have $X(T-) = X(T)$ on $\{T < \infty\}$. We have established that (ii) ⇒ (iii), and we know that (i) ⇒ (ii) as a matter of definition.

Assume Theorem 15.4 below for a moment. It follows from this result that if $X(T-) = X(T)$, then, for every bounded R-martingale M, we have $M(T-) = M(T)$, a.s.; thus T is fair, and hence T is previsible.

(*15.4*) THEOREM. *Let M be a UI R-martingale. Then, except on some* **P**-*null ω-set, the following statement holds:*

$$\forall t, \quad (X_{t-}(\omega) = X_t(\omega)) \Rightarrow (M_{t-}(\omega) = M_t(\omega));$$

in other words, a sample path of M can have a discontinuity only where the corresponding sample path of X has a discontinuity.

Proof of Theorem 15.4. Recall that an R-martingale M is UI if and only if it is of the form:

$$M_t = \mathbf{E}[\xi \mid \mathscr{F}_t] \tag{15.5}$$

for some $\xi \in L^1 \equiv L^1(\Omega, \mathscr{F}, \mathbf{P})$. Let U be the class of ξ in L^1 such that the R-modification of the martingale at (15.5) is (a.s., $\mathbf{P}$) continuous wherever X is continuous. We must prove that $U = L^1$. Again, we are guided by Chung and Walsh [2], but we can use the FD property (which they do not assume) to simplify things.

It is obvious that U is a linear subspace of L^1. Suppose that $M_t = \mathbf{E}[\xi \mid \mathscr{F}_t]$ and $N_t = \mathbf{E}[\eta \mid \mathscr{F}_t]$, with M and N in R-modifications. Then, by the submartingale inequality II.70.1,

$$\mathbf{P}[\sup |M_t - N_t| > \varepsilon] < \varepsilon^{-1}\mathbf{E}|\xi - \eta|.$$

It is now an easy exercise in the use of the first Borel–Cantelli lemma to show that U is closed in L^1.

We can show by direct calculation that U contains every element ξ of L^1 of the form:

$$\xi = f_1(X_{t_1}) f_2(X_{t_2}) \ldots f_n(X_{t_n}), \tag{15.6}$$

where $n \in \mathbb{N}$, $t_1 < t_2 < \ldots < t_n \in [0, \infty)$, and $f_1, f_2, \ldots, f_n \in C_b(E_\partial)$. Suppose, for example, that $0 < s < u$ and that $f, g \in C_b(E_\partial)$. Then, we can calculate an R-modification of M, where

$$M_t \equiv \mathbf{E}[f(X_s) g(X_u) \mid \mathscr{F}_t] \tag{15.7}$$

as follows. For $t \geqslant u$, we have

$$M_t = f(X_s) g(X_u).$$

For $s \leqslant t < u$, we have

$$M_t = f(X_s)\mathbf{E}[g(X_u) \mid \mathscr{F}_t] = f(X_s) P_{u-t} g(X_t).$$

For $0 \leqslant t < s$, we have

$$\begin{aligned} M_t &= \mathbf{E}[\mathbf{E}\{f(X_s) g(X_u) \mid \mathscr{F}_s\} \mid \mathscr{F}_t] \\ &= \mathbf{E}[f(X_s) P_{u-s} g(X_s) \mid \mathscr{F}_t] = \{P_{s-t}(f \cdot P_{u-s} g)\}(X_t). \end{aligned}$$

Now, by the FD property, the map $(v, x) \mapsto P_v g(x)$ is continuous; and it is easy to deduce that M is continuous whenever X is continuous.

Theorem 15.4 therefore follows if we assume Lemma 15.8. (This is the last such jump ahead in this proof!)

(*15.8*) LEMMA. *Let U be a closed linear subspace of L^1 which contains every element ξ of the form (15.6). Then $U = L^1$.*

Proof of Lemma 15.8. If $U \neq L^1$, then there exists a non-trivial element α in L^∞ such that $\mathbf{E}[\alpha\xi] = 0$ for all ξ in U. Now consider the class $\mathscr{H}$ of ξ in $\mathrm{b}\mathscr{F}^\circ$ such that $\mathbf{E}[\alpha\xi] = 0$. It is immediate that $\mathscr{H}$ satisfies the hypotheses of monotone-class Theorem II.3.2. Moreover, $\mathscr{H}$ contains the algebra $\mathscr{C}$ of linear combinations of elements of the form (15.6). Hence, since $\sigma(\mathscr{C}) = \mathscr{F}^\circ$, we have $\mathscr{H} = \mathrm{b}\mathscr{F}^\circ$, and α must be trivial. This contradiction establishes the lemma. □

(*15.9*) *The case of Ray processes.* For Ray processes, the situation is necessarily more complicated; for example, we know from Examples 14.2 and 14.3 that the equivalence of 'previsible' and 'accessible' no longer holds. The following result is the best one can hope for in the 'Ray' context.

(*15.10*) THEOREM (Meyer, Getoor). *Let X be an honest Ray process on a compact metric space F. Let F_{br} denote the set of branch points of F. Let T be an $\{\mathscr{F}_t^\mu\}$ stopping time.*

(i) *If $T > 0$ and $X(T-) = X(T)$, then T is previsible.*

(ii) *The totally inaccessible part of T is T_{IN}, where*

$$\mathrm{IN} = \{T=0\} \cup \{0 < T < \infty,\ X(T-) \neq X(T),\ X(T-) \notin F_{\mathrm{br}}\}.$$

A full proof of Theorem 15.10 is given in Chapter 7 of Getoor [1]. Meyer [1] has a marvellous proof of part (i) of the theorem.

16. Proof of the PFA theorem. We prove that

$$\mathrm{A} \Rightarrow \mathrm{P}, \quad \mathrm{A} \Rightarrow \mathrm{F}, \quad \mathrm{F} \Rightarrow \mathrm{A}, \quad \mathrm{P} \Rightarrow \mathrm{F}.$$

(*16.1*) *Proof that $A \Rightarrow P$.* Suppose that T is announced by a sequence (T_n) of stopping times. For each n, the process $I_{(T_n, \infty)}$ is an adapted L-process, and hence is previsible. But

$$I_{[T, \infty)} = \lim I_{(T_n, \infty)}.$$

Hence T is previsible.

(*16.2*) *Proof that $A \Rightarrow F$.* Let (T_n) announce T. For a bounded (or UI) martingale M, the optional-sampling theorem and martingale convergence theorem imply that:

$$M(T-) = \lim M(T_n) = \lim \mathbf{E}[M(T) \mid \mathscr{F}(T_n)] = \mathbf{E}[M(T) \mid \bigvee_n \mathscr{F}(T_n)].$$

Hence, $\mathbf{E}M(T-) = \mathbf{E}M(T)$, and T is fair.

(16.3) Proof that $F \Rightarrow A$. As already stated (Theorem 12.7), this proof is an adaptation of an illuminating argument in Chung and Walsh [2].

Let T be a fair stopping time. Let g be a continuous strictly increasing function from $[0, \infty]$ to $[0, 1]$ such that $g(0)=0$. Let

$$M_t \equiv \mathbf{E}[g(T)|\mathscr{F}_t],$$

taking an R-modification of the martingale M. Define

$$V_t \equiv M_t - g(T \wedge t) = \mathbf{E}[g(T) - g(T \wedge t)|\mathscr{F}_t].$$

The last equality holds because $T \wedge t$ is $\mathscr{F}_t$-measurable. It is now clear that V is a nonnegative supermartingale.

Set

$$S \equiv \inf\{t > 0: V_{t-} = 0 \text{ or } V_t = 0\},$$

so that, since $V_T = 0$, a.s., we have $S \leqslant T$, a.s. Now, by Theorem II.78.1, it is almost surely true that $V_t = 0$, $\forall t \geqslant S$. Thus $\mathbf{E}V_S = 0 = \mathbf{E}[g(T) - g(T \wedge S)]$, whence $S \geqslant T$ a.s., and so $S = T$ a.s.

Now set

$$T_n \equiv \inf\{t: V_t < 1/n\}.$$

Since $T = S$, a.s., we have $T_n \uparrow T$. Note that the definition of T_n implies that $V(T_n-) \geqslant 1/n$ if $T_n > 0$. We now use the fact that T is fair to evaluate:

$$\mathbf{E}V(T-) = \mathbf{E}M(T-) - \mathbf{E}g(T) = \mathbf{E}M(T) - \mathbf{E}g(T) = 0.$$

Hence, a.s., $V(T-) = 0$, and it is impossible to have $T_n = T$ (or else $1/n \leqslant V(T_n-) = V(T-) = 0$). We have shown that (T_n) announces T.

(16.4) Proof that $P \Rightarrow F$. If you wish to understand the subject properly, you need to understand the proof of the section theorems, and this proof that P $\Rightarrow$ F, much of which is based on §IV.76 of Dellacherie and Meyer [1], is a good introduction to the methods required. We did try for some time to find a quick proof that P $\Rightarrow$ F, but though many 'proofs' would immediately spring to the mind of anyone familiar with stochastic-integral theory, they all presuppose that P $\Rightarrow$ A (though it sometimes takes a little thought to spot exactly where!).

(16.5) Step 1. Let $\mathscr{J}$ be the set of finite unions of stochastic intervals of the form $[U, V)$, where U and V are announceable stopping times with $U \leqslant V$. Since A $\Rightarrow$ P, we have $\mathscr{J} \subseteq \mathscr{P}$. It is a simple exercise to show that $\mathscr{J}$ is an algebra. (Compare (IV.6.6), and use the definition of 'announceable'.)

Next, we have

$$\sigma(\mathscr{J}) = \mathscr{P}.$$

This is obvious because, by (IV.6.9),

$$\mathscr{P} = \sigma(\{(T, \infty): T \text{ a stopping time}\})$$

and

$$(T, \infty) = \bigcup_n [T + 1/n, \infty),$$

and, of course, $T + 1/n$ is announceable.

If J is a set in $\mathscr{J}$ and $D(J)$ is the debut of J, then it is obvious (from 'right-continuity') that $[D(J)] \subseteq J$, and it is easy to show that $D(J)$ is announceable.

(*16.6*) *Step 2.* An element K of $\mathscr{P}$ is said to belong to $\mathscr{J}_\delta$ if K is the intersection of countably many elements of $\mathscr{J}$. Let $K \in \mathscr{J}_\delta$. It is obvious (from 'right-continuity') that $[D(K)] \subseteq K$. *We now prove that $D(K)$ is announceable.*

Let $\mathscr{S}$ be the class of all announceable stopping times S with $S \leqslant D(K)$. Let

$$c \equiv \sup_{S \in \mathscr{S}} \mathbf{E}[k(S)], \quad \text{where} \quad k(S) \equiv S/(1+S).$$

Choose S_n in $\mathscr{S}$ with $\mathbf{E}[k(S_n)] > c - 1/n$. Let

$$W_n \equiv S_1 \vee S_2 \vee \ldots \vee S_n,$$

and let $W \equiv \uparrow \lim W_n$. You can easily prove that $W_n \in \mathscr{S}$, $\forall n$, and that:

(16.7(i)) $\quad W \in \mathscr{S}$ (that is, W is announceable, and $W \leqslant D(K)$),

(16.7(ii)) $\quad S \in \mathscr{S} \Rightarrow \mathbf{P}[W \geqslant S] = 1.$

The stopping time W is called the *essential supremum* of $\mathscr{S}$, and is unique modulo null sets. A similar idea was used in § II.75.

We now prove that $D(K) = W$, a.s., so that $D(K)$ is announceable. Since $K \in \mathscr{J}_\delta$, we can write K as

$$K = J_1 \cap J_2 \cap \ldots, \text{ where } J_n \in \mathscr{J}.$$

[Aside. If $J_n = B_n \times \Omega$, where $B_n = [1 - 1/n, 1) \cup [2, \infty)$, then $D(J_n) = 1 - 1/n$, while $D(K) = 2$. This explains the business of introducing W.] Now write

(16.8) $$K_n \equiv J_n \cap [W, \infty) \in \mathscr{J}.$$

Then, since $W \leqslant D(K)$, we have

(16.9) $$K = K_1 \cap K_2 \cap \ldots.$$

But we now have from (16.8) and (16.9)

$$W \leqslant D(K_n) \leqslant D(K),$$

so that, by (16.7(ii)),

$$D(K_n) = W, \quad \text{a.s.}$$

It is now obvious that $D(K) = W$, a.s., since $[W] = [D(K_n)] \subseteq K_n$ for each n, and so $[W] \subseteq K = \bigcap_n K_n$. So,

(16.10) $\quad$ *for $K \in \mathscr{J}_\delta$, $D(K)$ is announceable.*

(*16.11*) *Step 3.* Now (at last!) let T be a previsible stopping time. For $C \in \mathscr{P}$, define

$$\mu(C) \equiv \mathbf{P}(\{\omega: (T(\omega), \omega) \in C\}),$$

so that μ is a measure on $((0, \infty) \times \Omega, \mathscr{P})$ supported by $[T]$. (You should check this!) A set C in $\mathscr{P}$, with complement C' relative to $(0, \infty) \times \Omega$, will be said to belong to $\mathscr{C}$ if, for each $\varepsilon > 0$, there exist sets K_ε, K_ε^* in $\mathscr{J}_\delta$ with

$$K_\varepsilon \subseteq C, \quad \mu(C \setminus K_\varepsilon) < \varepsilon, \quad K_\varepsilon^* \subseteq C', \quad \mu(C' \setminus K_\varepsilon^*) < \varepsilon.$$

You can check that (in the language of Theorem II.2) $\mathscr{C}$ is a d-system containing $\mathscr{J}$, so that $\mathscr{C} = \mathscr{P}$. In particular since $[T] \in \mathscr{P}$, for each $\varepsilon > 0$, there exists K_ε in $\mathscr{J}_\delta$ such that

$$K_\varepsilon \subseteq [T], \quad \mu([T] \setminus K_\varepsilon) < \varepsilon.$$

Now, set

$$T_n \equiv D(K_1) \wedge D(K_{1/2}) \wedge \ldots \wedge D(K_{1/n}).$$

(*16.12*) *Step 4.* We now have the following situation:

(16.13(i)) each T_n is announceable;
(16.13(ii)) $T_n \downarrow T$;
(16.13(iii)) almost surely, for each ω, there exists $N(\omega)$ such that for $n \geqslant N(\omega)$, $T_n(\omega) = T(\omega)$.

Now let M be a bounded R-martingale. Since T_n is announceable (and hence fair),

$$\mathbf{E}M(T_n-) = \mathbf{E}M(T_n).$$

But we can now let $n \uparrow \infty$, using (16.13(ii)) on the right-hand side and (16.13(iii)) on the left-hand side, to conclude that

$$\mathbf{E}M(T-) = \mathbf{E}M(T),$$

and T is fair. (Cheers!) □

17. The σ-algebras $\mathscr{F}(\rho-)$, $\mathscr{F}(\rho)$, $\mathscr{F}(\rho+)$. Let ρ be a random time, that is, an $\mathscr{F}(\infty)$ measurable map $\rho: \Omega \to [0, \infty]$, where

$$\mathscr{F}(\infty) \equiv \bigvee_{t \geqslant 0} \mathscr{F}(t)$$

(*17.1*) DEFINITION. *If* $\rho > 0$, *define the* strict pre-ρ σ-algebra $\mathscr{F}(\rho-)$ *as follows:*

(17.2) $\mathscr{F}(\rho-) \equiv \sigma(\{X(\rho)$: *X is a previsible process on* $(0, \infty]\})$.

For any random time ρ, *define*

(17.3) $\mathscr{F}(\rho) \equiv \sigma(\{X(\rho)$: *X is an optional process on* $[0, \infty]\})$,

(17.4) $\mathscr{F}(\rho+) \equiv \sigma(\{X(\rho)\colon X \text{ is a progressive process on } [0,\infty)\})$.

Nota bene. In each case, $X(\infty)$ is allowed to be an arbitrary $\mathscr{F}_\infty$ measurable random variable.

The idea behind these definitions is that

$\mathscr{F}(\rho-)$ consists of events which depend on what happens strictly before time ρ;
$\mathscr{F}(\rho)$ consists of events which depend on what happens up to and including time ρ;
$\mathscr{F}(\rho+)$ consists of events which depend on what happens up to, including, and immediately after, time ρ.

You must agree that the definitions are supremely elegant. Now let us try to convince you that they are appropriate!

We have already made much use of the σ-algebra $\mathscr{F}(T)$, where T is a stopping time, but with a different definition of $\mathscr{F}(T)$.

(*17.5*) LEMMA. *Let T be a stopping time. Define*

$\mathscr{F}^{\text{old}}(T) \equiv \{\Lambda \in \mathscr{F}_\infty\colon \Lambda \cap \{T \leqslant t\} \in \mathscr{F}_t, \forall t\}$,

$\mathscr{F}^{\text{old}}(T+) \equiv \{\Lambda \in \mathscr{F}_\infty\colon \Lambda \cap \{T < t\} \in \mathscr{F}_t, \forall t\}$,

$\mathscr{F}^{\text{new}}(T) \equiv \sigma(\{X(T)\colon X \text{ is optional}\})$,

$\mathscr{F}^{\text{new}}(T+) \equiv \sigma(\{X(T)\colon X \text{ is progressive}\})$.

Then all four σ-algebras just defined are equal.

So, (17.3) and (17.4) tie in with our previous work on stopping times.

Proof. If X is progressive then $X(T)$ is $\mathscr{F}^{\text{old}}(T)$-measurable, by Lemma 3.2. Hence, $\mathscr{F}^{\text{new}}(T+) \subseteq \mathscr{F}^{\text{old}}(T)$. Because $\{\mathscr{F}_t\}$ is right-continuous, we know that $\mathscr{F}^{\text{old}}(T) = \mathscr{F}^{\text{old}}(T+)$. Since it is trivial that $\mathscr{F}^{\text{new}}(T) \subseteq \mathscr{F}^{\text{new}}(T+)$, all that remains is to prove that $\mathscr{F}^{\text{old}}(T) \subseteq \mathscr{F}^{\text{new}}(T)$.

Let $\Lambda \in \mathscr{F}^{\text{old}}(T)$. Then

$$I_\Lambda = X(T), \quad X = I_{[T_\Lambda,\infty)},$$

and X is an adapted R-process, so optional. Hence, $\Lambda \in \mathscr{F}^{\text{new}}(T)$. □

The 'old' and 'new' superscripts can be dropped!

Because of Theorem 4.7, we have, for any random time ρ,

(17.6) $\mathscr{F}(\rho-) = \sigma(\{\{T<\rho\}\colon T \text{ a stopping time}\})$, if $\rho > 0$,

(17.7) $\mathscr{F}(\rho) = \sigma(\{\{T \leqslant \rho\}\colon T \text{ a stopping time}\})$.

These are again elegant, and obviously right intuitively. But note the elusiveness of $\mathscr{F}(\rho+)$!

Originally, $\mathscr{F}(\rho-)$ was defined in a different way. We now build this into the following lemma which is useful in 'practice'.

(*17.8*) LEMMA. *Let ρ be a random time with $\rho>0$. Then:*

$$\mathscr{F}(\rho-)=\sigma(\{\Lambda\cap\{\rho>u\}: u\geqslant 0, \Lambda\in\mathscr{F}_u\}).$$

Proof. This is a trivial consequence of the fact that if $\Gamma=\Lambda\cap\{\rho>u\}$, then

$$I_\Gamma=X(\rho), \quad \text{where } X=I_{(u_\Lambda,\infty)}.$$

For, by Exercise IV.6.8, such processes X generate $\mathscr{P}$. □

(*17.9*) LEMMA. (i) *If S and T are stopping times with $S<T$, then: $\mathscr{F}(S)\subseteq\mathscr{F}(T-)$.*

(ii) *If T is a previsible stopping time announced by a sequence (T_n), then:*

$$\mathscr{F}(T-)=\bigvee_n \mathscr{F}(T_n).$$

(*17.10*) COROLLARY. *If M is a bounded (or UI) R-martingale, and T is a previsible stopping time, then:*

$$\mathbf{E}[M(T)|\mathscr{F}(T-)]=M(T-).$$

Proof of Lemma. (i) If $\Lambda\in\mathscr{F}_S$, then:

$$\begin{aligned}\Lambda&=\Lambda\cap\{S<T\}\\&=\bigcup_{q\in\mathbb{Q}}(\Lambda\cap\{S\leqslant q\}\cap\{q<T\})\\&\in\mathscr{F}(T-) \text{ by Lemma 17.8.}\end{aligned}$$

(ii) Use Lemma 17.8 as follows. Let $u\geqslant 0$ and $\Lambda\in\mathscr{F}_u$. Then:

$$\Lambda\cap\{T>u\}=\bigcup_n[\Lambda\cap\{T_n>u\}]\in\bigvee_n\mathscr{F}(T_n).$$

Thus $\mathscr{F}(T-)\subseteq\bigvee_n\mathscr{F}(T_n)$, and the other inclusion follows from the first part. □

Proof of corollary. This is now just a restatement of the result used at (16.2). □

The following result is very frequently used.

(*17.11*) LEMMA. *If T is a previsible stopping time and $\Lambda\in\mathscr{F}(T-)$, then T_Λ is previsible.*

Proof. There exists a previsible process X such that $I_\Lambda=X(T)$. Hence,

$$[T_\Lambda]=[T]\cap\{(t,\omega): X(t,\omega)=1\}\in\mathscr{P}.$$ □

We conclude this section with an example in which $\mathcal{F}(\rho+) \neq \mathcal{F}(\rho)$. Inevitably, this is based on the ideas previously used in §§ 3 and 5 to give an example of a progressive process which is not optional.

(*17.12*) *Example.* Let X be our old friend $\mathrm{CBM}_0(\mathbb{R})$, Wiener process, with the usual augmented filtration, etc. Let

$$\tau \equiv \inf\{t: |X_t| = 1\}, \quad \sigma \equiv \sup\{t < \tau: X_t = 0\}.$$

Let $\xi = X(\tau)$, so that $\xi = \mathrm{sgn}(X(\sigma+))$ in an obvious sense. By using (6.2), we can show that ξ is measurable relative to $\mathcal{F}(\sigma+)$. But ξ is not measurable relative to $\mathcal{F}(\sigma)$. Can you prove this? *Hint*: Find $\mathbf{E}(\xi | \mathcal{F}_\sigma)$.

18. Quasi-left-continuous filtrations. Our filtration $\{\mathcal{F}_t\}$ is said to be *quasi-left-continuous* (qlc) if $\mathcal{F}(T-) = \mathcal{F}(T)$ for every *previsible* stopping time T.

(*18.1*) THEOREM (Meyer). *The following statements are equivalent*:

(i) $\{\mathcal{F}_t\}$ *is qlc.*

(ii) *For every bounded (and then for every UI) R-martingale M and every previsible stopping time T, we have*:

$$M(T-) = M(T), \quad \text{a.s.}$$

(iii) *Every accessible stopping time is previsible.*

Proof that (*i*) $\Rightarrow$ (*ii*). Suppose that (i) holds. Let T be a previsible stopping time, and let M be a UI R-martingale. Then $M(T)$ is $\mathcal{F}(T)$ measurable, and so, since $\mathcal{F}(T) = \mathcal{F}(T-)$ by (i), $M(T)$ is $\mathcal{F}(T-)$ measurable. Hence, by (17.10), $M(T) = M(T-)$, a.s.

Proof that (*ii*) $\Rightarrow$ (*iii*). Suppose that (ii) holds (for bounded M). Let T be accessible, and let M be a bounded R-martingale. Then, since T is caught by a sequence (T_n) of previsible stopping times and $M(T_n) = M(T_n-)$ a.s., for each n, it is clear that $M(T) = M(T-)$ a.s. Hence T is fair; so, by the PFA theorem, T is previsible.

Proof that (*iii*) $\Rightarrow$ (*i*). Suppose that (iii) holds. Let T be previsible, and let $\Lambda \in \mathcal{F}(T)$. Let $S \equiv T_\Lambda$. Then S is caught by the sequence (T, ∞) consisting of two previsible times. Hence S is accessible. By (iii), S is previsible. Hence $I_{[S,\infty)}$ is previsible, and $I_\Lambda = I_{[S,\infty)}(T)$ is $\mathcal{F}(T-)$ measurable. □

The following result was announced as Theorem III.11.6.

(*18.2*) THEOREM (Meyer). *Let $(X, \Omega, \{\mathcal{F}_t\}, \{\mathbf{P}^\mu\})$ be an FD process. Let (T_n) be a sequence of $\{\mathcal{F}_t\}$ stopping times with $T_n \uparrow\uparrow T$. Then*:

$$\mathcal{F}(T) = \bigvee_n \mathcal{F}(T_n).$$

Proof. From Theorem 18.1 and Meyer's previsibility Theorem 15.1, it follows immediately that:

$$\mathscr{F}^{\mu}(T) = \bigvee_n \mathscr{F}^{\mu}(T_n), \quad \forall \mu.$$

We leave you to check that you can take intersections over μ. □

5. DUAL PREVISIBLE PROJECTIONS

19. The previsible section theorem; the previsible projection ${}^{p}X$ of X. The 'optional' section theorem (Theorem 5.1) has a full previsible analogue.

(*19.1*) THEOREM (Meyer). *Let F be a previsible subset of $(0, \infty) \times \Omega$, and suppose that $\varepsilon > 0$. Then there is a previsible stopping time T such that*:

(i) $[\![T]\!] \subseteq F$,

(ii) $\mathbf{P}\{T < \infty\} \geqslant \mathbf{P}\{D_F < \infty\} - \varepsilon$.

You will find a slightly stronger result as Theorem IV.85 of Dellacherie and Meyer [1], and can read the proof there.

We can deduce the next two results from Theorem 19.1 in exactly the same way as we deduced Lemmas 5.2 and 5.4 from Theorem 5.1.

(*19.2*) LEMMA. *Let X and Y be two previsible processes such that, for every finite previsible stopping time T,*

$$\mathbf{P}[X_T = Y_T] = 1.$$

Then X and Y are indistinguishable.

(*19.3*) LEMMA. *Let X and Y be two bounded previsible processes such that*:

$$\mathbf{E}[X_T; T < \infty] = \mathbf{E}[Y_T; T < \infty]$$

for every previsible stopping time T (including those taking infinite values). Then X and Y are indistinguishable.

Here is a very important application.

(*19.4*) THEOREM. *A previsible local martingale M is continuous.*

Note. Here, we understand that M is a local martingale with R-paths on $[0, \infty)$, the restriction to $(0, \infty)$ of which is previsible.

Proof. Suppose first that M is a previsible uniformly integrable martingale. Now let T be a finite previsible stopping time. By Corollary 17.10, we have

$$\mathbf{E}[M(T) | \mathscr{F}(T-)] = M(T-), \quad \text{a.s.}$$

But, since M is previsible, $M(T)$ is $\mathscr{F}(T-)$ measurable, so that, almost surely, $M(T)=M(T-)$. The process M is previsible by assumption, and $M_-=\{M(t-): t>0\}$ is previsible because it is an adapted L-process. By Lemma 19.2, $M=M_-$, and M is continuous.

Since, if M is a local martingale, then $M-M_0$ may be localized to a uniformly integrable martingale, all that remains is to prove the following result.

(*19.5*) LEMMA. *If X is a previsible process on* $(0, \infty)$, *and T is any stopping time, then the stopped process X^T is previsible.*

Proof. $X^T=XI_{(0,\,T]}+X_T(T, \infty)$. But X is previsible, $(0, T]$ is previsible, and X_T is $\mathscr{F}(T)$ measurable. The lemma follows. □

The 'previsible' counterpart of the optional projection Theorem 7.1 is the following.

(*19.6*) THEOREM, DEFINITION (previsible projection). *Let X be a bounded (measurable) process. Then there exists a unique previsible process ${}^{\mathrm{p}}X$ on* $(0, \infty)$ *such that for every previsible stopping time T,*

$$(19.7)\qquad {}^{\mathrm{p}}X_T I_{\{T<\infty\}}=\mathbf{E}[X_T I_{\{T<\infty\}}\,|\,\mathscr{F}(T-)].$$

The process ${}^{\mathrm{p}}X$ is called the previsible projection *of X.*

The proof will be given in a moment. As with ${}^{\circ}X$, ${}^{\mathrm{p}}X$ is of course unique only modulo evanescence.

It is immediate from the definition of ${}^{\mathrm{p}}X$ that:

$$(19.7)^*\qquad \mathbf{E}[{}^{\mathrm{p}}X_T;\, T<\infty]=\mathbf{E}[X_T;\, T<\infty].$$

Conversely, if we let T be a previsible stopping time, and let $\Lambda\in\mathscr{F}(T-)$, then by Lemma 17.11, T_Λ is a previsible stopping time, so by applying (19.7)* to T_Λ we deduce (19.7). Either of (19.7) or (19.7)* may therefore be taken as the definition of ${}^{\mathrm{p}}X$.

Proof of Theorem 19.6. The proof copies *mutatis mutandis* that of Theorem 7.1. The uniqueness of ${}^{\mathrm{p}}X$ is of course guaranteed by Lemma 19.2. If $C=(a,b]\times F$, where $0<a<b$ and $F\in\mathscr{F}$, we can take

$${}^{\mathrm{p}}I_C(t,\omega)=I_{(a,\,b]}(t)Z_t(\omega),$$

where $Z_t(\cdot)$ is the left-hand limit at t of the R-martingale $\mathbf{E}[I_F|\mathscr{F}_t]$. Use of the monotone-class theorem completes the proof. □

One last simple but useful result is the following.

(*19.8*) LEMMA. *Let X be a bounded process, and let H be a bounded previsible process. Then:*

$${}^{\mathrm{p}}(HX)=H.{}^{\mathrm{p}}X.$$

Proof. For a previsible stopping time T, $H_T I_{\{T<\infty\}}$ is $\mathscr{F}(T-)$ measurable. Hence,

$$\mathbf{E}[H_T X_T I_{\{T<\infty\}} | \mathscr{F}(T-)] = H_T {}^{\mathrm{p}}X_T I_{\{T<\infty\}}. \qquad \square$$

(*19.9*) **Exercise.** Prove the analogue of Lemma 19.8 for optional projections.

20. Doléans' characterization of FV processes. It is a good idea at this stage for you to re-read §7 of Chapter IV.

As explained there, an FV process is taken to be adapted unless it is described as 'raw'. The same goes for increasing processes. Thus, an increasing process is an adapted process the paths of which are non-decreasing R-processes on $[0, \infty)$; and, for a raw increasing process A, we drop the hypothesis that A is adapted.

We are going to concentrate on raw FV processes null at 0, that is, on raw FV_0 processes.

Again, as in §IV.7, for a raw FV_0 process A, we let V_A denote the total-variation process of A. Then A can be written as the difference of two raw increasing processes

$$A = \tfrac{1}{2}(V + A) - \tfrac{1}{2}(V - A).$$

And again, we call a raw FV_0 process A of integrable variation and say that A is a raw IV_0 process if

$$\mathbf{E}V_A(\infty) < \infty.$$

(*20.1*) *The signed measure μ_A induced by A.* Let A be a raw IV_0 process. For any bounded (measurable) process X, we define as before:

$$(X \bullet A)_\infty(\omega) \equiv \int_{(0,\infty)} X(t,\omega)\, dA(t,\omega). \tag{20.2}$$

Now put

$$\mu_A(X) \equiv \mathbf{E}[(X \bullet A)_\infty]. \tag{20.3}$$

This provides the 'functional' definition of a bounded signed measure μ_A on $((0, \infty) \times \Omega, \mathscr{B}(0, \infty) \times \mathscr{F})$. Obviously, μ_A is a (positive) measure if and only if A is a raw increasing process. Note that if $|\mu_A|$ denotes the total-variation measure associated with μ_A, then

$$|\mu_A| = \mu_{V_A}.$$

(*20.4*) DEFINITION (signed **P**-measure). *A bounded signed measure μ on $((0, \infty) \times \Omega, \mathscr{B}(0, \infty) \times \mathscr{F})$ is called a* signed **P**-measure *if*

$$\mu(H) = 0$$

for every evanescent subset H of $(0, \infty) \times \Omega$.

(*20.5*) THEOREM (Doléans). *A bounded signed measure μ on $((0, \infty) \times \Omega, \mathscr{B}(0, \infty) \times \mathscr{F})$ is of the form $\mu = \mu_A$ for some raw* IV_0 *process A if and only if μ is a signed* **P**-*measure. Moreover, A is then unique modulo evanescence.*

The real punch comes with the next result. Obviously, we call A previsible if the restriction of A to $(0, \infty) \times \Omega$ is previsible.

(*20.6*) THEOREM (Doléans). *Suppose that A is a raw* IV_0 *process and that* $\mu = \mu_A$.

(i) *Then A is optional if and only if μ commutes with the optional projection in the sense that:*

$$\mu(X) = \mu({}^{\mathrm{o}}X)$$

for every bounded (measurable) X.

(ii) *Moreover, A is previsible if and only if μ commutes with the previsible projection in the sense that:*

$$\mu(X) = \mu({}^{\mathrm{p}}X)$$

for every bounded X.

The proofs of Theorems 20.5 and 20.6 are interesting and illuminating. Even so, we defer them to §§ 26–27 to show how the theorems allow us to introduce dual previsible projections.

Part (ii) of Theorem 20.6 is a key step in the marvellous Doléans proof of the Meyer decomposition theorem.

21. Dual previsible projections, compensators. The intuitive significance of dual previsible projections should become clear from the examples in §§ 22–3; we show here how Theorems 20.5 and 20.6 allow us to *define* dual previsible projections.

Let A now denote a raw IV_0 process. Then A corresponds to a bounded signed **P**-measure μ on $(0, \infty) \times \Omega$. If X is a bounded process, set

$$\mu^{\mathrm{p}}(X) \equiv \mu({}^{\mathrm{p}}X).$$

Then μ^{p} is a signed **P**-measure (why?) on $(0, \infty) \times \Omega$ which commutes with the previsible projection, so that μ^{p} corresponds to a previsible IV_0 process A^{p}. Let us put this another way.

(*21.1*) THEOREM. *Let A be a raw* IV_0 *process. Then there exists a unique previsible* IV_0 *process A^{p} such that, for every bounded X,*

$$\mathbf{E}[(X \bullet A^{\mathrm{p}})_\infty] = \mathbf{E}[({}^{\mathrm{p}}X \bullet A)_\infty]. \tag{21.2}$$

(*21.3*) DEFINITION (dual previsible projection, compensator). *The process A^{p} is the* dual previsible projection *or* compensator *of A.*

Let us give an alternative description of A^{p} which is more useful in calculations.

(*21.4*) THEOREM. *Let A be a raw* IV_0 *process. Then A^{p} is the unique previsible* IV_0 *process such that ${}^{\mathrm{o}}A - A^{\mathrm{p}}$ is a martingale.*

One can interpret this theorem as a special case of the Meyer decomposition Theorem 29.7; if A is increasing, the Meyer decomposition of the 'class (D) submartingale' ${}^\circ A$ *takes the form*

$$ {}^\circ A = ({}^\circ A - A^p) + A^p. $$

Proof. Suppose that $0 \leqslant s < t$ and that $\Lambda \in \mathcal{F}_s$. Let X be the (previsible) indicator function of $(s_\Lambda, t_\Lambda]$. Then (21.2) shows that

$$ \mathbf{E}[A_t - A_s; \Lambda] = \mathbf{E}[A_t^p - A_s^p; \Lambda], $$

so that

$$ \mathbf{E}[A_t - A_s | \mathcal{F}_s] = \mathbf{E}[A_t^p - A_s^p | \mathcal{F}_s]. $$

But since $\mathbf{E}[A_t | \mathcal{F}_s] = \mathbf{E}[\mathbf{E}\{A_t | \mathcal{F}_t\} | \mathcal{F}_s] = \mathbf{E}[{}^\circ A_t | \mathcal{F}_s]$, we have

$$ \mathbf{E}[{}^\circ A_t - {}^\circ A_s | \mathcal{F}_s] = \mathbf{E}[A_t - A_s | \mathcal{F}_s], $$

so that ${}^\circ A - A^p$ is a martingale.

If α is another previsible IV_0 process such that ${}^\circ A - \alpha$ is a martingale, then $\alpha - A^p$ is a previsible martingale of finite variation. But, by (19.4), a previsible martingale is continuous, and, by (IV.30.4), a continuous martingale of finite variation is constant. Since both α and A^p are null at 0, $\alpha = A^p$. □

(*21.5*) **Exercise** (*Dellacherie's fomula*). Let A be a raw IV_0 process. Let S and T be two stopping times with $S \leqslant T$. Prove that, for every bounded X,

$$ \mathbf{E}\left[\int_{(S,T]} X_t \, dA_t \Big| \mathcal{F}_S\right] = \mathbf{E}\left[\int_{(S,T]} {}^p X_t \, dA_t \Big| \mathcal{F}_S\right]. $$

This formula is often used in the literature.

Solution. Let $\Lambda \in \mathcal{F}_S$, and let J be the (previsible) indicator of $(S_\Lambda, T_\Lambda]$. Then, by Lemma 19.8,

$$ {}^p(XJ) = ({}^p X) J. $$

(*21.6*) *Localization.* Theorem 21.4 makes clear the following extension of the concept of compensator.

(*21.7*) THEOREM, DEFINITION. *Let A be an (adapted, hence optional) element of* $\mathrm{IV}_{0,\mathrm{loc}}$. *Then there exists a unique previsible element A^p of* $\mathrm{IV}_{0,\mathrm{loc}}$ *such that $A - A^p$ is a local martingale. The process A^p is called the* dual previsible projection *or* compensator *of A.*

It is often not that easy to decide when a process is in $\mathrm{IV}_{0,\mathrm{loc}}$, so we have to be careful in applying this theorem. An interesting application is made in §23.

(*21.8*) *When is A^p continuous?* This question, like the analogous question which arises in connection with the Meyer decomposition, is important, particularly in the case when A is increasing.

(*21.9*) THEOREM. *Let A be an integrable raw increasing process. Then A^{p} is continuous if and only if A is* regular *in that*

$$\mathbf{E}(A_T)=\mathbf{E}(A_{T-})$$

for every finite previsible T.

Proof. The process A^{p} is previsible—what else? The process A^{p}_{-} is also previsible because it is an adapted L-process. Hence A^{p} is continuous (that is, $A^{\mathrm{p}}=A^{\mathrm{p}}_{-}$) if and only if $A^{\mathrm{p}}(T)=A^{\mathrm{p}}(T-)$ for every previsible T. Now let T be previsible. Let $X\equiv I_{[T]}$, so that X is previsible. Then

$$\begin{aligned}\mathbf{E}(A_T-A_{T-})&=\mathbf{E}(X\bullet A)_\infty=\mathbf{E}(X^{\mathrm{p}}\bullet A)_\infty\\&=\mathbf{E}(X\bullet A^{\mathrm{p}})=\mathbf{E}(A^{\mathrm{p}}_T-A^{\mathrm{p}}_{T-}).\end{aligned}$$

The result is now immediate. □

We now give the martingale characterization of totally inaccessible stopping times promised in § 13.

(*21.10*) THEOREM. (i) *Let T be a totally inaccessible stopping time, and let $A=I_{[T,\infty)}$. Then A^{p} is continuous. The process $A-A^{\mathrm{p}}$ is a UI martingale which is continuous except for a single jump of $+1$ at T (when $T<\infty$).*

(ii) *Let T be any stopping time. Suppose that there exists a UI martingale M which is continuous except for a single strictly positive jump at T (when $T<\infty$). Then T is totally inaccessible.*

Proof of (*i*). The fact that A^{p} is continuous follows from Theorem 21.9, and the fact that $A-A^{\mathrm{p}}$ is a martingale follows from Theorem 21.4.

Proof of (*ii*). This is immediate from the fact that a previsible time S is fair, so that there is no chance that S can equal T on $\{T<\infty\}$. □

(*21.11*) **Exercise.** We saw (Corollary 15.3) how Meyer's previsibility theorem implies that any strictly positive stopping time of Brownian motion must be previsible. Give a 'general theory' proof of this by using the fact that the martingale $A-A^{\mathrm{p}}$ must be continuous (Theorem IV.36.1).

22. Cumulative risk. We hope that the examples in this and later sections will establish that dual previsible projections are interesting and intuitive.

(*22.1*) *The discrete-parameter case.* Let $(\Omega,\mathcal{F},\{\mathcal{F}_n:n\geqslant 0\},\mathbf{P})$ be a set-up satisfying the 'usual conditions'. Let A be an integrable raw increasing process in the obvious sense. Then

$${}^{\mathrm{o}}A_k=\mathbf{E}[A_k|\mathcal{F}_k]\quad(\forall\, k)$$

and, since ${}^{\circ}A - A^{p}$ is a martingale,

$$A_k^p - A_{k-1}^p = \mathbf{E}[{}^{\circ}A_k - {}^{\circ}A_{k-1} | \mathscr{F}_{k-1}]$$
$$= \mathbf{E}[A_k - A_{k-1} | \mathscr{F}_{k-1}].$$

(*22.2*) *Cumulative risk in discrete time.* Let $(\Omega, \mathscr{F}, \mathbf{P})$ be a complete triple, and let $T: \Omega \to \{1, 2, 3, \dots\}$ be $\mathscr{F}$-measurable. Let $A \equiv I_{[T, \infty)}$ in the obvious sense, let $\mathscr{F}_n^{\circ} \equiv \sigma(\{A_m: m \leqslant n\})$, and throw in the null sets to produce $\mathscr{F}_n$. Then T is a stopping time, and A is adapted. Let $F_n \equiv \mathbf{P}[T \leqslant n]$. Then, from (22.1),

$$A_k^p - A_{k-1}^p \equiv \begin{cases} (F_k - F_{k-1})/(1 - F_{k-1}) & \text{if } k-1 < T, \\ 0 \quad \text{if } k-1 \geqslant T. \end{cases}$$

Note that $A_k - A_{k-1}$ is the conditional probability or '*risk*' that T occurs at time k given that T has not occurred before time k.

(*22.3*) *Cumulative risk in continuous time.* Let $(\Omega, \mathscr{F}, \mathbf{P})$ be a complete triple, and let $T: \Omega \to (0, \infty)$ be $\mathscr{F}$-measurable. Let $A \equiv I_{[T, \infty)}$, let $\mathscr{F}_t^{\circ} \equiv \sigma(\{A_s: s \leqslant t\})$, and let $\{\mathscr{F}_t\}$ be the usual augmentation of $\{\mathscr{F}_t^{\circ}\}$. Then T is a stopping time, and A is adapted. The obvious guess at A^p in the light of (22.2) is

$$A_t^p = h(T \wedge t), \quad h(u) \equiv \int_{(0, u]} dF(v)/[1 - F(v-)], \tag{22.4}$$

where $F(v) \equiv \mathbf{P}[T \leqslant v]$.

Regard A^p as defined via (22.4). Then A^p is a Borel function h of the continuous adapted (hence previsible) process $T \wedge t$. Hence, A^p is previsible. We must now show that $A - A^p$ is a martingale.

At time s, the only information available to the observer is the value of T if $T \leqslant s$. Thus, modulo null sets, $\mathscr{F}_s$ consists of sets $\{T \in B\}$, where $B \in \mathscr{B}[0, s]$, and their complements. You will very quickly convince yourself that we need only check that for $s < t$,

$$\mathbf{E}[A_t^p - A_s^p; T > s] = \mathbf{E}[A_t - A_s; T > s].$$

We evaluate the left-hand side in the obvious way:

$$\int_{u > s} dF(u) \int_{(s, t \wedge u]} dF(v)[1 - F(v-)]^{-1}$$
$$= \int_{(s, t]} dF(v)\,[1 - F(v-)]^{-1} \int_{[v, \infty)} dF(u)$$
$$= F(t) - F(s)$$

as required.

(*22.5*) **Exercise.** Confirm that the statements made in Example 14.4 are correct.

Assume now that F is continuous, which amounts to assuming that T is totally inaccessible. Then, from (22.4),

$$A^{\mathrm{p}} = -\log[1 - F(T \wedge t)].$$

In particular,

$$A^{\mathrm{p}}_{\infty} = A^{\mathrm{p}}_{T} = -\log[1 - F(T)],$$

and, since $F(T)$ is uniformly distributed on $[0, 1]$,

(22.6) *A^{p} has the exponential distribution of rate* 1.

Of course, the results just described will not surprise actuaries. But let us see what really underlies (22.6).

(*22.7*) *Cumulative risk for a general totally inaccessible T.* Let T be a finite totally inaccessible stopping time for some 'general' set-up $(\Omega, \mathcal{F}, \{\mathcal{F}_t\}, \mathbf{P})$ satisfying the usual conditions. The word 'general' here conveys the fact that we are no longer tied to the situation in which the only information available at time t is whether or not T has occurred.

Let $A \equiv I_{[T,\infty)}$. We know from Theorem 21.10 that A^{p} is continuous, and that $M \equiv A - A^{\mathrm{p}}$ is a martingale continuous except for a jump of $+1$ at T. Let $\lambda > 0$, and let

$$X_t \equiv \exp(\lambda M_t) \leqslant \mathrm{e}^{\lambda}.$$

Then because T is totally inaccessible, ${}^{\mathrm{p}}X = X_-$, and so

$$\mathbf{E}[\exp(-\lambda A^{\mathrm{p}}_T)] = \mathbf{E}[({}^{\mathrm{p}}X \cdot A)_\infty] = \mathbf{E}[(X \cdot A^{\mathrm{p}})_\infty]$$

$$= \mathbf{E}\int_0^\infty \exp(-\lambda A^{\mathrm{p}}_s)\, dA^{\mathrm{p}}_s.$$

Hence, $\mathbf{E}[\exp(-\lambda A^{\mathrm{p}}_T)] = (1+\lambda)^{-1}$, so that A^{p} again has the exponential distribution of rate 1.

Exercise. Prove the fact, used implicitly above, that $A^{\mathrm{p}}_T = A^{\mathrm{p}}_\infty$.

(*22.8*) *Risk estimation in the presence of noise.* Return to example (22.3), supposing now that T has a density φ. Let g_t denote the tail probability $g_t \equiv \mathbf{P}[T > t]$. Then, (22.4) may be written as

$$A^{\mathrm{p}}_t = \int_0^t (1 - A_s)\, \varphi_s / g_s\, ds.$$

The good sense of equation 10.6 in our change-detection filter is now apparent.

23. Some Brownian motion examples. Please re-read Exercise 7.12, the notation of which we now use. Thus, B is $\mathrm{CBM}_0(\mathbb{R})$, M is the maximum-to-date of B (so

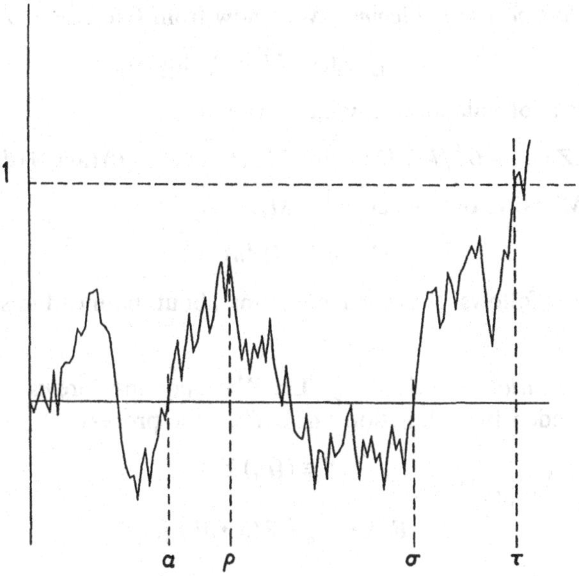

Figure VI.2

that M is not a martingale!), τ is the hitting time of 1, σ is the time of the last visit to 0 before τ, ρ is the time of the maximum of B over $[0, \sigma]$, and α is the time of the last visit to 0 before time ρ. Figure VI.2 repeats Figure VI.1 to illustrate the situation.

Let A be the raw increasing process $I_{[\rho,\infty)}$. We know from Exercise 7.12 that ${}^{\circ}A(t) = M(t \wedge \tau)$. Since $M(t \wedge \tau)$ is previsible and increasing, it is immediate from Theorem 21.4 that $A^{\mathrm{p}}(t) = M(t \wedge \tau)$.

Now, let A be the raw increasing process $I_{[\sigma,\infty)}$. Then ${}^{\circ}A(t) = B^{+}(t \wedge \tau)$. Thus, A^{p} is the unique previsible increasing process such that $B^{+}(t \wedge \tau) - A^{\mathrm{p}}(t)$ is a martingale. Tanaka's formula IV.43.6 makes it clear that $A^{\mathrm{p}}(t) = \frac{1}{2}l(t \wedge \tau)$, where l is the (semimartingale) local time at 0 of B. For notational convenience, we introduce

$$L_t \equiv \tfrac{1}{2} l_t,$$

the local time at 0 of B in the Itô–McKean normalization.

We have proved the first two parts of the following lemma.

(*23.1*) LEMMA.

$$I^{\mathrm{p}}_{[\rho,\infty)} = M(t \wedge \tau), \quad I^{\mathrm{p}}_{[\sigma,\infty)} = L(t \wedge \tau),$$

$$I^{\mathrm{p}}_{[\alpha,\infty)} = -\int_0^{t \wedge \tau} \log(M_s)\, dL_s.$$

Proof of the last part of the lemma. We know from Exercise 7.12 that for $t \leqslant \tau$,

$$Z \equiv {}^{o}I_{[a,\infty)}(t) = M_t - B_t^+ \log M_t.$$

Hence, by Itô's formula and Tanaka's formula,

$$dZ = (1 - B^+/M)\,dM - (\log M)\,dL - I_{(0,\infty)}(B)\,(\log M)\,dB.$$

But, since dM grows only when $B^+ = M$,

$$(1 - B^+/M)\,dM = 0.$$

The result now follows. If you have qualms about 'boundedness', worry them away. □

(*23.2*) *An application.* Let $A \equiv I_{[\rho,\infty)}$. Let Y be a bounded previsible process, and let f be a bounded Borel function on $\mathbb{R}$. Then the process

$$X \equiv f(B_t)\,Y$$

is previsible, so that

$$\mathbf{E}(X \bullet A)_\infty = \mathbf{E}(X \bullet A^p)_\infty.$$

Thus,

(23.3)
$$\mathbf{E}\,Y(\rho)f(B_\rho) = \mathbf{E}\int_0^\tau Y(t)f(B_t)\,dM(t)$$
$$= \mathbf{E}\int_0^1 Y(\tau_x)f(x)\,dx.$$

On taking $Y = 1$, we see that (as is otherwise trivial):

(23.4) *$B(\rho)$ is uniformly distributed on* (0, 1).

But, moreover, we see that:

(23.5) *a regular conditional distribution of the process $\{B(t \wedge \rho): t \geqslant 0\}$ given the value b (say) of $B(\rho)$ is the 'absolute' $\mathbf{P}^0$ law of the process $\{B(t \wedge \tau_b): t \geqslant 0\}$.*

Proof of (23.5). Let $t_1 < t_2 < \ldots < t_n \in (0, \infty)$, and let $f_1, f_2, \ldots, f_n$ be bounded Borel functions on $\mathbb{R}$. Let Y be the previsible process defined as follows:

$$Y(t) \equiv \prod_{i=1}^{n} f_i(B(t_i \wedge t)).$$

Now apply (23.3). □

Remark. Of course, result (23.5) is hardly a great surprise! But it does need proof, or else one commits the same type of error as in 'proving' the Cayley–Hamilton theorem by putting $\lambda = A$ in det $(\lambda I - A)$.

After our discussion of the exponential character of the distribution of A^p in the context of (22.7), it is amusing to discuss the distribution of A^p for the various raw increasing processes of this section.

It is clear that

(23.6) $$I^{\mathrm{p}}_{[\rho,\,\infty)}(\infty)=M_\tau=1.$$

Next,

(23.7) $$I^{\mathrm{p}}_{[\sigma,\,\infty)}(\infty)=L_\tau=L_\sigma.$$

We prove that

(23.8) *L_σ has the exponential distribution of rate* 1.

This fact is best seen in terms of excursion theory. However, here is a proof based on the ideas which we are now studying.

Proof of (23.8). Let $\lambda>0$, and let

$$X_t\equiv\lambda\exp(-\lambda L_t).$$

Then X is previsible, so that $X={}^{\mathrm{p}}X$. Hence, with $A\equiv I_{[\sigma,\,\infty)}$,

$$\mathbf{E}(X\bullet A)_\infty=\mathbf{E}(X\bullet A^{\mathrm{p}})_\infty,$$

so that

$$\lambda\mathbf{E}\exp(-\lambda L_\sigma)=\mathbf{E}\int_0^\sigma\lambda\exp(-\lambda L_s)\,dL_s=1-\mathbf{E}\exp(-\lambda L_\sigma).$$

Hence,

$$\mathbf{E}\exp(-\lambda L_\sigma)=\mathbf{E}\exp(-\lambda L_\tau)=(1+\lambda)^{-1}.$$

The distribution of $I^{\mathrm{p}}_{[\alpha,\,\infty)}(\infty)$ is obtained by taking $a=1$ and $\varphi(x)=-\log(x)$ in the following lemma.

(*23.9*) LEMMA. *If φ is a nonnegative function on* $(0,\infty)$, *then:*

$$\mathbf{E}\exp\left\{-\lambda\int_0^{\tau_a}\varphi(M_s)\,dL_s\right\}=\exp\left\{-\int_0^a\lambda\varphi(x)\,dx/[1+\lambda x\varphi(x)]\right\}.$$

Proof. Write

(23.10) $$\int_0^{\tau_a}\varphi(M_s)\,dL_s=\int_0^a\varphi(x)\,dV_x,$$

where

$$V_x\equiv L(\tau_x),\quad \tau_x\equiv\inf\{t\colon B_t=x\}.$$

The strong Markov theorem shows that the process $\{V_x\colon x\geqslant0\}$ has independent (but not stationary) increments. (This process has been studied in Problem 2.8.3 of Itô and McKean [1].) Moreover, for $y>x$,

$$\begin{aligned}\mathbf{E}^0\exp\{-\lambda c(V_y-V_x)\}&=\mathbf{E}^x\exp\{-\lambda c(V_y-V_x)\}\\&=\mathbf{P}^x[\tau_y<\tau_0]+\mathbf{P}^x[\tau_0<\tau_y]\,\mathbf{E}^0\exp\{-\lambda c V_y\}.\end{aligned}$$

By (7.13) and the trivial extension:

$$\mathbf{E}^0 \exp\{-\lambda c V_y\} = (1+\lambda c y)^{-1}$$

of (23.8), we have

$$\mathbf{E}^0 \exp\{-\lambda c(V_y - V_x)\} = (1+\lambda c x)/(1+\lambda c y). \tag{23.11}$$

But, because of (23.10), equation (23.11) is exactly what Lemma 23.9 says when φ has the form $cI_{(x,y]}$. The independent-increments property of V allows us to extend the result to the case when φ is an arbitrary nonnegative step function. A monotone-class argument finishes the proof. □

Remark. This proof was really a piece of excursion theory. Several closely related matters, including the Ray–Knight theorems, will be examined during the study of excursion theory in Part 8.

24. Decomposition of a continuous semimartingale. We present this further application before proving the Doléans theorems.

We can now settle a matter raised during §IV.31. Suppose that:

(i) L is a local martingale null at 0,
(ii) G is an FV process null at 0,
(iii) $X \equiv L + G$ is a continuous process, so that $\Delta L = -\Delta G$.

We wish to prove that we can write

$$X = M + A, \tag{24.1}$$

where M is a *continuous* local martingale null at 0, and A is a (necessarily continuous) FV_0 process.

For the moment, assume the following lemma.

(*24.2*) LEMMA. $G \in \mathrm{IV}_{0,\mathrm{loc}}$.

Then we can prove (24.1) as follows. We can write

$$X = L + (G - G^{\mathrm{p}}) + G^{\mathrm{p}},$$

and we know from Theorem 21.7 that $G - G^{\mathrm{p}}$ is a local martingale. Let

$$M \equiv L + (G - G^{\mathrm{p}}), \quad A \equiv G^{\mathrm{p}}.$$

Then A is certainly an FV_0 process. Moreover,

$$M = X - G^{\mathrm{p}},$$

and, since X is continuous adapted, M is a previsible local martingale. But we can now conclude from Theorem 19.4 that M is continuous. □

You have already been advised that it is not always too easy to decide whether a process is locally of integrable variation. So, watch the following steps required to prove Lemma 24.2. The argument has close links with that of

Theorem IV.12.1(ii) and with important work yet to come. We first need a most useful result.

(*24.3*) LEMMA (Yor). *Let N be a local martingale null at* 0. *Let*

$$N_t^* \equiv \sup\{|N_s|: s \leqslant t\}$$

as usual. Then N^* is locally integrable.

Proof of Lemma 24.3. First choose a sequence $(R(n))$ of stopping times such that $R(n)\uparrow\infty$ and

(24.4) each $N^{R(n)}$ is a uniformly integrable martingale.

Let

$$S(n) \equiv \inf\{t: N_t^* \geqslant n\},$$

and let $T_n \equiv R(n) \wedge S(n)$. Then $T_n \uparrow \infty$, and

$$N^*(T_n) \leqslant |N(T_n)| + n.$$

But, from (24.4) and the optional sampling theorem, $N(T_n) \in L^1$, and the result follows. □

Proof of Lemma 24.2. By Lemma 24.3, applied to $N = L$, we can choose a sequence T_n of stopping times such that $T_n \uparrow \infty$ and

$$(\Delta L)^*(T_n) \leqslant 2L^*(T_n) \in L^1.$$

Now, let V be the total-variation process of G, and set

$$U(n) \equiv \inf\{t: V(t) \geqslant n\}.$$

Then

$$V(U_n \wedge T_n) \leqslant n + (\Delta G)^*(T_n) = n + (\Delta L)^*(T_n),$$

and the lemma is proved. □

25. Proof of the basic (μ, A) correspondence. Suppose that A is a raw IV_0 process and that the signed measure μ_A on $((0,\infty)\times\Omega, \mathscr{B}(0,\infty)\times\mathscr{F})$ is induced by A in the usual way:

(25.1) $$\mu_A(X) \equiv \mathbf{E}(X \bullet A)_\infty = \mathbf{E}\int_0^\infty X(s)\,dA(s).$$

Let $\Lambda \in \mathscr{F}$, and let $t > 0$. By taking $X(s,\omega) \equiv I_{(0,t]}(s) I_\Lambda(\omega)$, we see that

$$\mu_A((0,t]\times\Lambda) = \mathbf{E}[A_t; \Lambda],$$

so that A_t is the Radon–Nikodym derivative

(25.2) $$A_t = d\nu_t/d\mathbf{P} \text{ on } \mathscr{F}, \text{ where } \nu_t(\Lambda) \equiv \mu_A((0,t]\times\Lambda).$$

Let us now recall Theorem 20.5.

(*25.3*) (Restatement of THEOREM 20.5). *A bounded signed measure μ on $((0,\infty)\times\Omega, \mathscr{B}(0,\infty)\times\mathscr{F})$ is of the form μ_A for some raw* IV_0 *process A if and only if μ is a signed* **P**-*measure in the sense that:*

(25.4) $$\mu(H)=0 \text{ for every evanescent set } H.$$

Moreover, A is then unique (modulo evanescence).

Proof of the 'only if' part. If $\mu=\mu_A$ for some raw IV_0 process A, and if $X=I_H$ for some evanescent set H, then $(X\bullet A)_\infty=0$ with probability 1, so that

$$\mu(H)=0.$$

Proof of the 'if' part. Suppose that μ is a bounded signed measure on $(0,\infty)\times\Omega$ such that (25.4) holds. We can decompose μ into its positive and negative parts: $\mu=\mu^+-\mu^-$, and since

$$\mu^+(H)=\sup\{\mu(J): J\in\mathscr{B}(0,\infty)\times\mathscr{F}, J\subseteq H\},$$

it is clear that μ^+ is a **P**-measure. Hence we can, *and do*, restrict attention to the case when μ is positive.

We proceed with (25.2) in mind. For each rational $q>0$, define a measure ν_q on $\mathscr{F}$ as follows:

$$\nu_q(\Lambda)\equiv\mu((0,q]\times\Lambda).$$

If Λ is **P**-null, then $(0,q]\times\Lambda$ is evanescent, so that $\nu_q(\Lambda)=0$. Hence ν_q is absolutely continuous with respect to **P**. Let A_q^* be (a version of) the Radon–Nikodym derivative

$$A_q^*=d\nu_q/d\mathbf{P}.$$

It is easy to prove from the fact that μ is positive that $q\mapsto A_q^*$ is almost surely non-decreasing. Now, for $t\geqslant 0$, define

$$A_t\equiv\limsup_{q\downarrow\downarrow t} A_q^*.$$

It is now routine to finish off the proof by monotone-class arguments, etc.. □

26. Proof of Doléans' 'optional' characterization result. Here and in the proof of the analogous 'previsible' result in the next section, we need the following handy lemma, which we seem determined that you shall remember by the end of the book!

(*26.1*) LEMMA. *Let $\beta\in L^1(\Omega,\mathscr{F},\mathbf{P})$, and let $\mathscr{G}$ be a sub-σ-algebra of $\mathscr{F}$ containing all* **P**-*null sets in $\mathscr{F}$. Suppose that, whenever $\xi\in L^\infty(\Omega,\mathscr{F},\mathbf{P})$ and $\eta\equiv\xi-\mathbf{E}[\xi|\mathscr{G}]$, then:*

$$\mathbf{E}(\beta\eta)=0.$$

Then β is $\mathscr{G}$ measurable.

Proof. We have for all ξ in $L^\infty(\Omega, \mathcal{F}, \mathbf{P})$,

$$\mathbf{E}[\beta\xi] = \mathbf{E}[\beta \mathbf{E}(\xi|\mathcal{G})]. \tag{26.2}$$

But, by the well-known property II.22(i) of conditional expectations,

$$\begin{aligned}\mathbf{E}[\beta \mathbf{E}(\xi|\mathcal{G})|\mathcal{G}] &= \mathbf{E}(\beta|\mathcal{G})\,\mathbf{E}(\xi|\mathcal{G}) \\ &= \mathbf{E}[\xi \mathbf{E}(\beta|\mathcal{G})|\mathcal{G}].\end{aligned} \tag{26.3}$$

From (26.2) and (26.3),

$$\mathbf{E}[\xi\{\beta - \mathbf{E}(\beta|\mathcal{G})\}] = 0, \quad \forall \xi \in L^\infty(\Omega, \mathcal{F}, \mathbf{P}),$$

whence $\beta = \mathbf{E}(\beta|\mathcal{G})$, a.s., and β is $\mathcal{G}$ measurable. □

Let us remind ourselves what we are trying to prove.

(*26.4*) (Restatement of THEOREM 20.6(i)). *Suppose that $\mu = \mu_A$ for some raw IV_0 process A. Then A is optional if and only if μ commutes with the optional projection in that*:

$$\mu(X) = \mu({}^\circ X) \quad \textit{for all bounded } X. \tag{26.5}$$

Proof of the 'if' part. Assume (26.5). Since A is an R-process, we need only show that A is adapted, that is, that for each $t > 0$, A_t is $\mathcal{F}_t$-measurable.

Let $t > 0$, and make the following definitions:

$$\begin{aligned}&\xi \in L^\infty(\Omega, \mathcal{F}, \mathbf{P}), \quad \eta \equiv \xi - \mathbf{E}[\xi|\mathcal{F}_t], \\ &X(s, \omega) \equiv I_{(0,\,t]}(s)\,\eta(\omega).\end{aligned} \tag{26.6}$$

Then, if T is a stopping time, $\{T \leqslant t\} \in \mathcal{F}_t$, and

$$\mathbf{E}[X_T; T < \infty] = \mathbf{E}[\eta; T \leqslant t] = 0,$$

because of the way in which η is defined. Thus,

$${}^\circ X = 0,$$

and

$$\mathbf{E}(X \bullet A)_\infty = \mathbf{E}[A_t \eta] = \mathbf{E}({}^\circ X \bullet A)_\infty = 0.$$

By Lemma 26.1, A_t is $\mathcal{F}_t$-measurable. □

Proof of the 'only if' part. This is achieved by reversing the above argument to establish (26.5) for X of the form (26.6), and then appealing to monotone-class theorems. □

27. Proof of the Doléans 'previsible' characterization result. This section, proving the central Doléans characterization result, shows the ideas developed so far as fitting together with a precision which is very satisfying.

(*27.1*) LEMMA. *Let A be an* FV_0 *process.* (*Recall that, by convention, A is an adapted R-process.*) *Then there exists a continuous* FV_0 *process α and a sequence* $(T(n))$ *of stopping times with disjoint graphs such that*:

$$A=\alpha+\sum_n(\Delta A)_{T(n)}I_{[T(n),\infty)},\tag{27.2}$$

the series converging absolutely for each t.

Proof. We use a familiar idea. Let $U(1,m)$, $U(2,m)$, ... be the first, second, ... times when

$$2^{-m}<|\Delta A|\leqslant 2^{-m-1}.$$

Now arrange the double sequence $(U(k,m))$ into a single sequence (T_n). The rest is elementary analysis. □

The following result contains in particular what we need now of a general characterization of previsible R-processes, Theorem 27.10 below.

(*27.3*) LEMMA. *Let A be an* FV_0 *process. Then A is previsible if and only if the following two conditions hold*:

(27.4(i)) *for every totally inaccessible stopping time T,*

$$\Delta A_T=0,\quad \text{a.s.};$$

(27.4(ii)) *for every previsible stopping time T, ΔA_T is $\mathscr{F}(T-)$-measurable.*

When this situation obtains, A has a representation of the form (27.2) *in which the stopping times $T(n)$ are all previsible.*

(*27.5*) COROLLARY. *If A is a previsible* FV_0 *process, then the total variation process V of A is previsible, and hence A is the difference*

$$A=\tfrac{1}{2}(V+A)-\tfrac{1}{2}(V-A)$$

of two previsible increasing processes null at 0.

Proof of the 'only if' part of Lemma 27.3. Suppose that A is previsible. Then it follows already from Theorem 12.3 that for a totally inaccessible stopping time T, $\Delta A_T=0$, a.s. Moreover, since $\Delta A=A-A_-$ is previsible, then for any random time T, ΔA_T is $\mathscr{F}(T-)$-measurable, by definition of $\mathscr{F}(T-)$. Thus, the two properties (27.4) hold.

Proof of the 'if' part of Lemma 27.3. If (27.4(i)) holds, then it is clear from the proof of Lemma 27.1 that (27.2) holds for a sequence $(T(n))$ of accessible stopping times with disjoint graphs. Now, by Lemma 14.2, any accessible stopping time is caught by a sequence of previsible stopping times. Hence,

$$\{(t,\omega)\colon \Delta A(t,\omega)\neq 0\}\subseteq\bigcup_n[S_n]$$

for some sequence (S_n) of previsible stopping times. Now redefine the (T_n) sequence as follows:

$$T_1 \equiv S_1, \quad [T_n] \equiv [S_n] \setminus \left(\bigcup_{m<n} [S_m] \right) \quad (n \geqslant 2).$$

Then (T_n) is a sequence of previsible stopping times with disjoint graphs such that (27.2) holds. Suppose also that (27.4(ii)) holds. It is clear from the representation (27.2) that to prove that A is previsible, we need only prove that

(27.6) *if T is previsible and β is $\mathscr{F}(T-)$-measurable, then*

$$Z \equiv \beta I_{[T, \infty)} \text{ is previsible.}$$

But, for a Borel subset B of $\mathbb{R}\setminus\{0\}$,

$$\{(t, \omega): Z(t, \omega) \in B\} = [T_\Lambda, \infty),$$

where $\Lambda \equiv \beta^{-1}(B) \in \mathscr{F}(T-)$; and we know from Lemma 18.10 that T_Λ is previsible. □

Proof of Corollary 27.5. We have:

$$V = V_A = V_\alpha + \sum |\Delta A_{T(n)}| I_{[T(n), \infty)}.$$

But V_α is continuous adapted (to see that it is adapted, think how it is defined) and therefore previsible. An application of (27.6) finishes the proof. □

(27.7) *Proof of the 'only if' part of Theorem 20.6(ii).* Suppose that A is a previsible IV_0 process. We must prove that for every bounded X,

$$\mathbf{E}(X \bullet A)_\infty = \mathbf{E}({}^{\mathrm{p}}X \bullet A)_\infty.$$

Because of Corollary 27.5, it is enough to prove this when A is increasing.

So, let A be a previsible increasing process null at 0. For $t > 0$, define

$$\tau_t \equiv \inf\{s: A_s \geqslant t\},$$

so that τ_t is the début of a previsible set which contains the graph of its début. By Lemma 12.2, τ_t is previsible. Hence, by definition of ${}^{\mathrm{p}}X$,

$$\mathbf{E}[{}^{\mathrm{p}}X(\tau_t); \tau_t < \infty] = \mathbf{E}[X(\tau_t); \tau_t < \infty].$$

Hence, using an elementary change-of-variables formula which you can easily prove (first for step functions, etc.),

$$\begin{aligned} \mathbf{E}\int_{(0,\infty)} X_s\, dA_s &= \int_0^\infty \mathbf{E}[X(\tau_t); \tau_t < \infty]\, dt \\ &= \int_0^\infty \mathbf{E}[{}^{\mathrm{p}}X(\tau_t); \tau_t < \infty]\, dt = \mathbf{E}\int_0^\infty {}^{\mathrm{p}}X_s\, dA_s. \end{aligned}$$

Thus, A commutes with the previsible projection.

(*27.8*) *Proof of the 'if' part of Theorem 20.6(ii).* Let A be a raw IV_0 process such that $\mu \equiv \mu_A$ commutes with the previsible projection:

$$\mu({}^{\mathrm{p}}X) = \mu(X).$$

We must prove that A is previsible. Since

$$\mu({}^{\mathrm{o}}X) = \mu({}^{\mathrm{p}}({}^{\mathrm{o}}X)) = \mu({}^{\mathrm{p}}X) = \mu(X),$$

μ commutes with the optional projection, so that A is optional. All that we need to do, therefore, is to prove that conditions (27.4(i)) and (27.4(ii)) are satisfied.

Let T be a totally inaccessible stopping time. Let $\Lambda \equiv \{\omega: \Delta A_T > 0\}$. Then $S \equiv T_\Lambda$ defines a totally inaccessible stopping time S. Let $X = I_{[S]}$. Then, obviously, ${}^{\mathrm{p}}X = 0$, and

$$0 = \mathbf{E}({}^{\mathrm{p}}X \bullet A)_\infty = \mathbf{E}(X \bullet A)_\infty = \mathbf{E}[\Delta A_S]$$

$$= \mathbf{E}[\Delta A_T; \Delta A_T > 0].$$

Hence, $\Delta A_T \leqslant 0$, a.s., and, likewise, $\Delta A_T \geqslant 0$, a.s. Thus condition (27.4(i)) holds.

Now let T be a previsible stopping time. We must prove that ΔA_T *is* $\mathscr{F}(T-)$-measurable. With Lemma 30.1 in mind, let

$$\xi \in L^\infty(\Omega, \mathscr{F}, \mathbf{P}) \quad \text{and} \quad \eta \equiv \xi - \mathbf{E}(\xi \mid \mathscr{F}(T-)).$$

We need only prove that $\mathbf{E}[\eta \Delta A_T] = 0$. Let $X = \eta I_{[T]}$. For any previsible time S,

$$\mathbf{E}[X_S: S < \infty] = \mathbf{E}[\eta I_{\{S=T\}}: S < \infty] = 0$$

because

$$I_{\{S=T<\infty\}} = I_{[S]}(T) \in m\mathscr{F}(T-).$$

Thus, ${}^{\mathrm{p}}X = 0$, and, since μ commutes with the previsible projection,

$$0 = \mathbf{E}({}^{\mathrm{p}}X \bullet A)_\infty = \mathbf{E}(X \bullet A)_\infty = \mathbf{E}[\eta \Delta A_T],$$

and the entire proof is finished. □

(*27.9*) *A general characterization of previsible R-processes.* You will find the following result as *Théorème C of Compléments au Chapitre* IV of Dellacherie and Meyer [1].

(*27.10*) THEOREM (Dellacherie and Meyer). *Let X be an adapted R-process. Then X is previsible if and only if the following two conditions are satisfied:*

(i) *for every totally inaccessible stopping time T, $X(T) = X(T-)$* a.s. on $\{T < \infty\}$;

(ii) *for every previsible stopping time T, $X(T)$ is $\mathscr{F}(T-)$-measurable.*

28. Lévy systems for Markov processes. The theory of Lévy systems was developed for a wide class of Markov processes by S. Watanabe—in association with Ikeda and, especially Motoo—back in the early 1960s. See the historic papers, S. Watanabe [1] and Motoo and S. Watanabe [1]. The approach via dual previsible projections originated in work of Walsh and Weil, and reached its definitive form in the paper, Benveniste and Jacod [1].

Theorem 28.1 is the main result for a general FD process X. Recall that the 'extended' state-space E_∂ of X is a compact metric space, and that $\mathscr{E}_\partial$ denotes $\mathscr{B}(E_\partial)$. By a kernel $N(\bullet, \bullet)$ on $(E_\partial, \mathscr{E}_\partial)$, we mean a function N from $E_\partial \times \mathscr{E}_\partial$ to $[0, \infty]$ such that (i) for each Γ in $\mathscr{E}_\partial$, the map $x \mapsto N(x, \Gamma)$ is $\mathscr{E}_\partial$ measurable, (ii) for each x in E, $N(x, \bullet)$ is a σ-finite measure on $\mathscr{E}_\partial$.

(*28.1*) THEOREM (S. Watanabe, . . .). *Let X be an FD process. Then there exist*

(i) *a kernel N on $(E_\partial, \mathscr{E}_\partial)$ with $N(x, \{x\})=0, \forall x \in E$;*

(ii) *a PCHAF H of X with $\mathbf{E}^\mu(H_t) < \infty, \forall t, \forall \mu$,*

such that the pair (N, H) is a Lévy system for X in the following sense.

Let f be any nonnegative $\mathscr{E}_\partial \times \mathscr{E}_\partial$ measurable function on $E_\partial \times E_\partial$ such that

$$f(x, x)=0, \quad \forall x \in E_\partial.$$

Put

$$A_t^f \equiv \sum_{s \leq t} f(X_{s-}, X_s),$$

$$\tilde{A}_t^f \equiv \int_{(0,t]} dH_s \int_{E_\partial} N(X_{s-}, dy) f(X_{s-}, y).$$

Then:

$$\mathbf{E}^\mu A_t^f = \mathbf{E}^\mu \tilde{A}_t^f \quad (\forall \mu, \forall t);$$

and if for some μ, $\mathbf{E}^\mu A_t^f < \infty$ $(\forall t)$, then $A^f - \tilde{A}^f$ is a $\mathbf{P}^\mu$ martingale.

We see, therefore, that $\tilde{A}^f$ *acts as a dual previsible projection under P^μ for A^f simultaneously for all f and all initial measures μ.*

For a proof, see Benveniste and Jacod [1].

(*28.2*) *Lévy systems for Markov chains with finite state–space.* Let X be a Markov chain with finite state–space and Q-matrix Q. Lemma IV.21.13 shows that a Lévy system for X consists of the pair (N, H), where

$$H(t)=t, \quad N(i, \{j\})=Q(i, j) \quad (i \neq j).$$

Observe that, therefore,

$$N(i, \{j\})=\lim_{t \downarrow 0} t^{-1} P(t, i, \{j\}). \tag{28.3}$$

(*28.4*) *Lévy systems for the symmetric Cauchy process.* Let B and W be independent Brownian motions started at 0. Let

$$\tau_t \equiv \inf\{s: W_s = t\}.$$

Then we saw in § II.59 that

$$(28.5) \qquad X_t \equiv B(\tau_t)$$

defines a symmetric Cauchy process X, that is, a Markov process X with transition function

$$(28.6) \qquad P(t, x, dy) = \pi^{-1} t[t^2 + (y-x)^2]^{-1}\, dy.$$

In particular, X has independent increments. In view of (28.3), it is plausible that a Lévy system of X is provided by the pair (N, H), where

$$(28.7) \qquad H_t = t, \quad N(x, dy) = \pi^{-1}(y-x)^{-2}\, dy,$$

and this is correct.

Let us see how this guess ties in with the representation (28.5) for X. We know from § II.37 that the τ process is the sum of jumps, the jumps of τ of size between b and $b+db$ ($b>0$) occurring as the points of a Poisson process of rate $(2\pi b^3)^{-1/2}\, db$. Now, from (28.5), such a jump of magnitude b in τ will result in a jump of X which is normally distributed with mean 0 and variance b. Hence the (signed) jumps of X of size between y and $y+dy$ ($y \in \mathbb{R}$) will occur as the points in a Poisson process of rate

$$\begin{aligned} dy \int_0^\infty & (2\pi b^3)^{-1/2}(2\pi b)^{-1/2} \exp(-y^2/2b)\, db \\ &= dy(\pi y^2)^{-1} \int_0^\infty \exp(-y^2/2b)(y^2/2b^2)\, db \\ &= dy(\pi y^2)^{-1} \int_0^\infty \exp(-u)\, du = dy(\pi y^2)^{-1}. \end{aligned}$$

We have proved the following lemma.

(*28.8*) LEMMA *Let X be a symmetric Cauchy process. Then the set of points*

$$\{(t, y): \Delta X_t \neq 0,\ y = \Delta X_t\} \subseteq (0, \infty) \times \mathbb{R}$$

are the atoms of a Poisson random measure on $(0, \infty) \times \mathbb{R}$ *with expectation measure* ν, *where*

$$\nu(dt, dy) = dt\, \mu(dy),$$
$$\mu(dy) = dy(\pi\, y^2)^{-1}.$$

The symmetric Cauchy process is an example of a Lévy process whose Lévy measure satisfies

$$\int_{\{|b| \leq 1\}} |b|^2 \mu(db) < \int_{\{|b| \leq 1\}} |b| \mu(db) = \infty,$$

so that, as we saw in § 2, the paths of the symmetric Cauchy process X are of unbounded variation on every interval, even though the continuous martingale part of X is zero.

6. THE MEYER DECOMPOSITION THEOREM

29. Introduction. The Meyer decomposition theorem is one of the central results of modern probability theory. It provided the foundation for the introduction of the general stochastic integral in Kunita and S. Watanabe [1]. Moreover—and this was its first motivation—it has deep connections with potential theory, and especially, as we shall see, with the Volkonskii–Šur–Meyer Theorem III.16.7 on additive functionals.

Let Z be a locally integrable increasing process (adapted!) null at 0. Then we know that we can write

$$Z = M + A,$$

where $A \equiv Z^p$ *is a previsible locally integrable increasing process null at* 0, and M *is a local martingale null at* 0; moreover, M and A are unique. Thus, *A compensates in a previsible way for the increase in Z.*

Now suppose that Z is a *submartingale* null at 0. Then *Z increases on average,* and it is reasonable to ask whether we can compensate for this tendency to increase by writing

$$Z = M + A, \tag{29.1}$$

where *A is a previsible locally integrable increasing process null at* 0, and *M is a local martingale null at* 0. The Meyer decomposition states that this can be done, and in a unique way.

(*29.2*) *The Doob decomposition in discrete time.* (See § II.54). Let $Z = (Z_n : n \geqslant 0)$ be a discrete-time submartingale relative to some set-up $(\Omega, \mathcal{F}, \{\mathcal{F}_n\}, \mathbf{P})$, with $Z_0 = 0$.

Suppose that we can write

$$Z = M + A, \tag{29.3}$$

where M is a martingale null at 0, and A is a process null at 0 and previsible in the sense that, for each n, A_n is $\mathcal{F}_{n-1}$-measurable. Then

$$\begin{aligned}\mathbf{E}[Z_n - Z_{n-1} | \mathcal{F}_{n-1}] &= \mathbf{E}[M_n - M_{n-1} | \mathcal{F}_{n-1}] + \mathbf{E}[A_n - A_{n-1} | \mathcal{F}_{n-1}] \\ &= 0 + (A_n - A_{n-1}) = A_n - A_{n-1}.\end{aligned}$$

Note that A is automatically increasing.

Conversely, if we set

$$A_n = \sum_{k=1}^{n} \mathbf{E}[Z_k - Z_{k-1} | \mathcal{F}_{k-1}], \tag{29.4}$$

then $M \equiv Z - A$ is a martingale.

We have proved the existence and uniqueness of the decomposition (29.3). In this discrete-time setting, this decomposition is called the *Doob decomposition of Z*.

Doob drew attention to the importance of obtaining a continuous-parameter analogue of his decomposition; and the result achieved by Meyer is sometimes called the Doob–Meyer decomposition, or (by Meyer) the Doob decomposition.

It must be appreciated that when Doob posed the problem, the concept of a 'previsible' process in continuous time was not known. A major part of Meyer's achievement was therefore that of finding an analogue of the discrete-parameter concept of 'previsible' which would allow formulation of the theorem. In fact, Meyer obtained his theorem with 'natural' instead of 'previsible'. But credit for the concepts of 'previsible process' and 'previsible stopping time' must also go to Meyer. It was, however, Cathérine Doléans who showed that

$$\text{Natural} = \text{Previsible}$$

for integrable increasing processes in continuous time. Doléans also give the definitive proof of the Meyer decomposition theorem, which we present below.

There is also a fine proof of the theorem by Rao [1], which proceeds by discrete approximation. The full connection between the discrete and continuous parameter cases is captured by Theorem 31.2 below, which is due to Doléans.

We first give the 'theorem with boundedness conditions', and later examine how it may be localized. We need the following definition.

(*29.5*) DEFINITION (class (D)). *A process X carried by our usual set-up* $(\Omega, \mathcal{F}, \{\mathcal{F}_t\}, \mathbf{P})$ *is said to be of* class (D) *if the family*

$$\{X_T\colon T \text{ a finite stopping time}\}$$

is uniformly integrable.

We have already seen the following important example of a class (D) process.

(*29.6*) LEMMA. *A uniformly integrable martingale M is of class* (D).

Proof. We have $M_T = \mathbf{E}[M_\infty | \mathcal{F}_T]$ by Theorem II.77.5. Now apply Lemma II.44. □

(*29.7*) THEOREM (Meyer decomposition theorem). *An adapted R-process Z is a submartingale of class (D) null at* 0 *if and only if Z may be written*

(29.8) $$Z = M + A,$$

where M is a uniformly integrable martingale null at 0, *and A is a previsible integrable increasing process null at* 0. *Moreover, the decomposition* (*29.8*) *is unique.*

Note that the theorem says that our class (D) submartingale Z is the optional projection of the raw increasing process Y, where

$$Y(t) = M(\infty) + A(t).$$

The 'if' part of the theorem is immediate because of Lemma (29.6) and the fact that $0 \leqslant A_T \leqslant A_\infty$ and $\mathbf{E}(A_\infty) < \infty$.

Note that if Z has the decomposition (29.8) and if μ is the measure $\mu = \mu_A$ associated with A, then for stopping times $S \leqslant T$,

(29.9) $$\mu((S, T]) = \mathbf{E}[A_T - A_S] = \mathbf{E}[Z_T - Z_S].$$

This is the clue to the Doléans proof of the 'if' part of the theorem.

(*29.10*) DEFINITION (compensator). *We shall call the process A in Theorem 29.7 the* compensator *of Z.*

This is compatible with our previous use of the term 'compensator'.

30. The Doléans proof of the Meyer decomposition. We are guided by Dellacherie [1].

Let Z be a submartingale of class (D) and null at 0. Then Z is uniformly integrable, and $Z_t \to Z_\infty$ a.s., and in L^1. Moreover, Fatou's lemma shows that the family

$$\{Z_T\colon T \text{ any stopping time, finite or infinite}\}$$

is uniformly integrable.

We proceed with (29.9) in mind.

Recall from Theorem II.77.5 that, if S and T are stopping times with $S \leqslant T$, then

(30.1) $$\mathbf{E}[Z_T - Z_S] \geqslant 0.$$

Let $\mathcal{J}$ now denote the class of previsible subsets H of $(0, \infty)\times\Omega$ of the form

$$H=(S_1, T_1]\cup(S_2, T_2]\cup\ldots\cup(S_n, T_n], \tag{30.2}$$

where $n\in\mathbb{N}$, $S_1\leqslant T_1\leqslant S_2\leqslant T_2\leqslant\ldots\leqslant S_n\leqslant T_n$, and $S_k<T_k$ on $\{S_k<\infty\}$, and $T_k<S_{k+1}$ on $\{T_k<\infty\}$. With (29.10) in mind, define a function μ on $\mathcal{J}$ by writing, for H as at (30.2),

$$\mu(H)=\sum_{k=1}^{n}\mathbf{E}[Z(T_k)-Z(S_k)].$$

Then $\mathcal{J}$ is an algebra generating $\mathcal{P}$, μ is finitely additive on $\mathcal{J}$, μ is nonnegative (by (30.1)), and

$$\mu((0,\infty)\times\Omega)=\mathbf{E}(Z_\infty)<\infty.$$

(*Comment.* It is rather messy to prove that $\mathcal{J}$ is an algebra—compare (IV.6.6)—and that μ is finitely additive on $\mathcal{J}$; but you can cope with these points as exercises if you so wish.)

We wish to prove that

$$\mu \textit{ is countably additive on } \mathcal{J}. \tag{30.3}$$

Then μ will extend to a measure on $\mathcal{P}$. To prove (30.3), we need the 'inner regularity' property of μ described in the following lemma. First observe the notation that for H as at (30.2), we write $\bar{H}$ for the closure of H:

$$\bar{H}\equiv[S_1, T_1]\cup[S_2, T_2]\cup\ldots\cup[S_n, T_n].$$

(*30.4*) LEMMA. *Let $H\in\mathcal{J}$, and let $\eta>0$. Then there exists a set $K\in\mathcal{J}$ such that $\bar{K}\subseteq H$ and $\mu(K)>\mu(H)-\eta$.*

Proof of lemma. It is clearly enough to be able to deal with the case when

$$H=(S, T], \text{ where } S\leqslant T \text{ and } S<T \text{ on } \{S<\infty\}. \tag{30.5}$$

So, let H be as at (30.5). For $n\in\mathbb{N}$,

if $S+1/n<T$, set $S_n\equiv S+1/n$, $T_n\equiv T$;
if $S+1/n\geqslant T$, set $S_n\equiv T_n\equiv\infty$.

Then $[S_n, T_n]\subseteq(S, T]$, and $S_n\downarrow S$, $T_n\downarrow T$. Since Z is right continuous and of class (D),

$$\mu(S_n, T_n]=\mathbf{E}[Z(T_n)-Z(S_n)]\to\mathbf{E}[Z(T)-Z(S)]=\mu(H),$$

so that we may take K to be $(S_n, T_n]$ for some large n.

Note. The class (D) assumption is not necessary for the validity of Lemma 30.4. We shall emphasize when it becomes crucial.

To prove that μ is countably additive on $\mathcal{J}$, we must show that if H_1, $H_2,\ldots\in\mathcal{J}$, and $H_1\supseteq H_2\supseteq\ldots$, and $\bigcap_n H_n=\varnothing$, then $\mu(H_n)\downarrow 0$. Let $\varepsilon>0$ be given.

Choose $K_1, K_2, \ldots$ in $\mathscr{J}$ such that $\bar{K}_n \subseteq H_n$ and $\mu(K_n) > \mu(H_n) - \varepsilon 2^{-n}$. Let

$$L_n \equiv K_1 \cap K_2 \cap \ldots \cap K_n \in \mathscr{J}.$$

Then

$$\mu(H_n) \leqslant \mu(L_n) + \varepsilon \quad (\forall n). \tag{30.6}$$

Since $\bar{L}_n \subseteq \bar{K}_n \subseteq H_n$, we have $\bar{L}_n \downarrow \varnothing$.

Let D_n be the debut of $\bar{L}_n$. Now, for each ω and each m, the section

$$\bar{L}_m^{(\omega)} \equiv \{t: (t, \omega) \in \bar{L}_m\}$$

is a *closed* subset of $[0, \infty)$. Moreover, $D_n(\omega) \in \bar{L}_m^{(\omega)}$ whenever $n \geqslant m$ and $D_n(\omega) < \infty$. Hence, if for some ω,

$$D_\infty(\omega) \equiv \uparrow\lim D_n(\omega) < \infty,$$

then $D_\infty(\omega) \in \bar{L}_m^{(\omega)}$ for every m, and we have a contradiction:

$$(D_\infty(\omega), \omega) \in \bigcap_m \bar{L}_m = \varnothing.$$

Hence, $D_n \uparrow \infty$.

Next, we have

$$\mu(L_n) \leqslant \mathbf{E}[Z(\infty) - Z(D_n)].$$

Since $D_n \uparrow \infty$, we have $Z(D_n) \to Z_\infty$, a.s.; and,

SINCE Z IS OF CLASS (D),

$Z(D_n) \to Z(\infty)$ in L^1. (An example in § 33 will clarify the fact that this use of the class (D) hypothesis is essential.) Hence, $\mu(L_n) \to 0$. From (30.6),

$$\limsup \mu(H_n) \leqslant \varepsilon,$$

and, since ε is arbitrary, $\mu(H_n) \downarrow 0$. Hence μ is countably additive on $\mathscr{J}$.

So far, we have proved the following theorem.

(*30.7*) THEOREM (Doléans). *Let Z be a submartingale of class* (D). *Then there exists a unique bounded measure* μ *on* $((0, \infty) \times \Omega, \mathscr{P})$ *such that*

$$\mu(S, T] = \mathbf{E}[Z(T) - Z(S)]$$

whenever S and T are stopping times with $S \leqslant T$.

(*30.8*) *Completion of proof of the Meyer theorem.* For a bounded (measurable) process X, define

$$\tilde{\mu}(X) \equiv \mu({}^{\mathrm{p}}X).$$

Then (compare our study of dual previsible projections) $\tilde{\mu}$ is a **P**-measure on

$((0, \infty)\times\Omega, \mathscr{B}(0, \infty)\times\mathscr{F})$. Since $\tilde{\mu}$ obviously commutes with the previsible projection,

$$\tilde{\mu}=\mu_A$$

for some integrable previsible increasing process A.

For any stopping time S, we have

$$\text{(30.9)}\qquad \mathbf{E}[A_\infty - A_S]=\tilde{\mu}((S, \infty))=\mathbf{E}[Z_\infty - Z_S].$$

Now, let $t\geqslant 0$, let $\Lambda\in\mathscr{F}_t$, and let $S\equiv t_\Lambda$. Then equation (30.9) reads

$$\mathbf{E}[A_\infty - A_t; \Lambda]=\mathbf{E}[Z_\infty - Z_t; \Lambda],$$

so that, if $M\equiv Z-A$, then

$$M_t=\mathbf{E}[M_\infty|\mathscr{F}_t],$$

and M is a uniformly integrable martingale.

Doléans' proof of the existence of the Meyer decomposition is complete.

(*30.10*) *Uniqueness of the decomposition.* This is now a familiar argument. If, with obvious notation,

$$Z=M+A=M'+A',$$

then the process Y with

$$Y\equiv M-M'=A'-A$$

is a previsible (and hence continuous) martingale which is of finite variation. Hence, Y is constant; and since Y is null at 0, Y is evanescent. □

31. Regular class (D) submartingales; approximation to compensators. Let Z be a submartingale of class (D) and null at 0. Let

$$Z=M+A$$

be the Meyer decomposition of Z.

(*31.1*) THEOREM. *The compensator A of Z is continuous if and only if Z is* regular *in the sense that for every finite previsible stopping time T,*

$$\mathbf{E}[Z(T)] = \mathbf{E}[Z(T-)].$$

Proof. By Corollary 17.10, $\mathbf{E}[\Delta M_T]=0$. By the previsible section theorem 19.1, if ΔA were not evanescent, there would be a previsible time on whose graph A jumped, contradicting the assumed regularity of Z. □

In the case of a regular submartingale of class (D), there is an especially nice tie-up, due to Doléans [1], between the discrete-parameter and continuous parameter versions of the (Doob–)Meyer decomposition theorem.

Let $(\Omega, \mathscr{F}, \{\mathscr{F}_t\}, \mathbf{P})$ satisfy the usual conditions. Let Z be a submartingale of class (D) with $Z_0=0$, and let

$$Z=M+A$$

be the Meyer decomposition of Z. Fix t, let S be a partition

$$S=\{t_0=0<t_1<\ldots t_n=t\}$$

of $[0, t]$ and (compare (29.4)) set

$$\begin{aligned}\Sigma_S &\equiv \sum_{i=1}^{n} \mathbf{E}[Z(t_i)-Z(t_{i-1})| \mathscr{F}(t_{i-1})] \\ &= \sum_{i=1}^{n} \mathbf{E}[A(t_i)-A(t_{i-1})| \mathscr{F}(t_{i-1})].\end{aligned}$$

(*31.2*) THEOREM (Doléans). *If Z is regular (equivalently: if A is continuous), then*

$$\lim \Sigma_S = A_t \quad \textit{in } L^1. \tag{31.3}$$

[*If Z is not assumed regular, then the best that can be said is that $\Sigma_S \to A_t$ in the weak (L^∞) topology of L^1.*]

Note. Let us clarify the statement in the 'regular' case, which is the only case in which we shall be interested. The theorem states that (if Z is regular) for given $\varepsilon>0$, $\exists\delta>0$ such that whenever $\max(t_i-t_{i-1})<\delta$, then $\mathbf{E}|\Sigma_S-A_t|<\varepsilon$.

Proof for the 'regular' case. Assume that Z is a regular submartingale of class (D) and null at 0, so that A is continuous.

Step 1. Suppose first that for some constant K, $A_\infty \leqslant K$. Let

$$W_i \equiv \mathbf{E}[A(t_i)-A(t_{i-1})| \mathscr{F}(t_{i-1})]-[A(t_i)-A(t_{i-1})].$$

Since $\mathbf{E}[W_i | \mathscr{F}(t_{i-1})]=0$, the variables $W_1, W_2, \ldots, W_n$ are orthogonal in L^2. Moreover, since $(a-b)^2 \leqslant 2a^2+2b^2$, we have

$$\begin{aligned}\mathbf{E}[W_i^2] &\leqslant 2\mathbf{E}\{\mathbf{E}[A(t_i)-A(t_{i-1})| \mathscr{F}(t_{i-1})]^2\}+2\mathbf{E}\{[A(t_i)-A(t_{i-1})]^2\} \\ &\leqslant 4\mathbf{E}\{[A(t_i)-A(t_{i-1})]^2\},\end{aligned}$$

using Jensen's inequality. Hence

$$\begin{aligned}\mathbf{E}[(\Sigma_S-A_t)^2] &= \mathbf{E}\left[\left(\sum_{i=1}^{n} W_i\right)^2\right]=\mathbf{E}\left(\sum_{i=1}^{n} W_i^2\right) \\ &\leqslant 4\mathbf{E}\sum_{i=1}^{n}[A(t_i)-A(t_{i-1})]^2 \\ &\leqslant 4K\mathbf{E}\{\max[A(t_i)-A(t_{i-1})]\}.\end{aligned}$$

But the paths of A are continuous (hence uniformly continuous) on $[0, t]$. Hence

$$\Sigma_S A_t \text{ in } L^2 \text{ and hence in } L^1.$$

Step 2. Return to the general case when A is integrable (but still continuous). Let

$$A^K(u) \equiv A(u) \wedge K, \quad B^K(u) \equiv A(u) - A^K(u).$$

Note that A^K and B^K are continuous and increasing. Put

$$\Sigma_{S,K} \equiv \sum_{i=1}^{n} \mathbf{E}[A^K(t_i) - A^K(t_{i-1}) | \mathcal{F}(t_{i-1})],$$

$$\Sigma'_{S,K} \equiv \sum_{i=1}^{n} \mathbf{E}[B^K(t_i) - B^K(t_{i-1}) | \mathcal{F}(t_{i-1})].$$

Then, by Step 1,

$$\lim_S \Sigma_{S,K} = A^K(t) \textit{ in } L^1,$$

while

$$\|\Sigma'_{S,K}\|_1 = \mathbf{E}[B^K(t)],$$

so that as $K \to \infty$, $\Sigma'_{S,K} \to 0$ in L^1 *uniformly* in S. Hence

$$\Sigma_S \to A_t \text{ in } L^1. \qquad \square$$

Note on Rao's work. Doléans' proof of Theorem 31.2 presupposes the Meyer decomposition theorem. Rao's paper (Rao [1]) establishes directly that Σ_S converges in the weak topology of L^1. An elaborate but clever argument enabled Rao to utilize the Dunford–Pettis weak compactness criterion.

(*31.4*) *Meyer's 'Laplaciens approchés'.* In his original proof of the decomposition theorem, Meyer used a different approximation, via '*laplaciens approchés*'. In some respects, Meyer's approximation is better than the 'discrete' approximation; and it is just as natural. We have already used *laplaciens approchés* for Markov processes in our proof of the Volkonskii–Šur–Meyer theorem.

See Dellacherie [1] for the general theory of these approximations.

32. The local form of the decomposition theorem. We need the following result.

(*32.1*) LEMMA. *A submartingale Z is locally of class (D).*

Proof. Let $U_n \equiv n \wedge \inf\{t: |Z_t| \geqslant n\}$. Then, for any stopping time S,

$$Z(S \wedge U_n) \leqslant n + |Z(U_n)|.$$

Hence, it will follow that $Z^{U(n)}$ is of class (D) once we show that $Z(U_n)$ is integrable. In other words, it is enough to show that if U is a bounded stopping time, then $Z(U)$ is integrable; and this was done in Theorem II.77.1. □

(*32.2*) DEFINITION (local submartingale). (i) *Let Z be a process null at* 0. *Then Z is called a* local submartingale null at 0 *if there exists a sequence* $(T(n))$ *of stopping times with* $T(n)$ *such that each* $Z^{T(n)}$ *is a submartingale. By Lemma 32.1, Z is then locally of class (D).*

(ii) *An adapted process Z is called a* local submartingale *if* $Z - Z_0$ *is a local submartingale null at* 0.

(*32.3*) THEOREM (Local form of the Meyer decomposition theorem). *Let Z be a local submartingale. Then Z may be written uniquely in the form*

$$Z = Z_0 + M + A,$$

where M is a local martingale null at 0, *and A is a previsible increasing process null at* 0.

Proof. We leave you to do the patching. □

(*32.4*) *Itô's formula for convex functions.* If $f\colon \mathbb{R} \to \mathbb{R}$ is convex and if M is a continuous local martingale, then $Z = f(M)$ is a local submartingale. Applying the generalized Itô formula IV.45.2, we see that

$$f(M_t) - f(M_0) = \int_0^t D_- f(M_s)\, dM_s + \tfrac{1}{2} \int_{\mathbb{R}} l_t^a \,\mu(da). \tag{32.5}$$

The first term on the right is a continuous local martingale, the second is a continuous increasing process, and hence (32.5) is the Meyer decomposition of the local submartingale $f(M)$.

33. An L^2 bounded local martingale which is not a martingale. We now simplify the Helms–Johnson example of § III.31.4. We construct a process Y which (like the Helms–Johnson example) shows that

(33.1(i)) *an L^2 bounded local martingale need not be a martingale,*

(33.1(ii)) *a UI supermartingale need not be of class (D).*

Let B be a Brownian motion in $\mathbb{R}^3$ started at the origin, and let

$$Z_t \equiv |B_t|^{-1}.$$

Let $\{\mathscr{B}_t\}$ be the augmented filtration generated by B.

We calculate:

$$\text{(33.2)} \qquad \mathbf{E}(Z_t^2) = \int_0^\infty r^{-2}.4\pi r^2.(2\pi t)^{-3/2}\exp(-r^2/2t)\,dr$$
$$= 4\pi(2\pi t)^{-1}\int_0^\infty (2\pi t)^{-1/2}\exp(-r^2/2t)dr$$
$$= 4\pi(2\pi t)^{-1}.\tfrac{1}{2} = t^{-1}.$$

Now, for $t \geqslant 0$, set

$$Y_t \equiv Z_{1+t}, \qquad \mathscr{F}_t \equiv \mathscr{B}_{1+t}.$$

Since the function $|x|^{-1}$ is harmonic away from the origin, and B never hits 0, Itô's formula shows that *Y is a local martingale* (relative to $\{\mathscr{F}_t\}$. Because, further, Y is positive and Y_0 integrable (from 33.2) *Y is a supermartingale by the 'Fatou lemma' IV.14.3.*

Calculation (33.2) shows that *Y is bounded in L^2*, whence *Y is certainly uniformly integrable.* We know that $Y_t \to 0$ (a.s.) as $t \to \infty$. Now if Y were a martingale, then, since Y is UI, we would have

$$Y_t = \mathbf{E}(Y_\infty | \mathscr{F}_t) = 0 \quad \text{(a.s.)},$$

which is absurd. Hence *Y is not a martingale.*

If the supermartingale Y were of class (D), then the Meyer decomposition would yield a 'global' decomposition

$$\text{(33.3)} \qquad Y = Y_0 + M - A$$

where M is a (UI) martingale and A is a previsible (integrable) increasing process. Since Y is not a martingale, A could not be zero; but since Y is a local martingale, the only Meyer decomposition of the local supermartingale Y is

$$Y = Y_0 + (Y - Y_0) + 0$$

in which A *is* zero. The only conclusion is that *Y is not of class (D).*

(33.4) **Exercise.** Show that if $T_n \equiv \inf\{t : Y_t \geqslant n\}$, then $T_n \uparrow \infty$. $Y(T_n) \to 0$, but $\mathbf{E}Y(T_n) = \mathbf{E}(Y_0)$.

34. The $\langle M \rangle$ process. Let M be a martingale null at 0 and bounded in L^2. Then M^2 is a submartingale, and by Doob's L^2 inequality,

$$\mathbf{E}\left[\sup_t M_t^2\right] \leqslant 4\mathbf{E}(M_\infty^2) < \infty.$$

Hence M^2 is a submartingale of class (D), and there exists a unique integrable previsible increasing process $\langle M \rangle$ null at 0 such that

(34.1) $M^2 - \langle M \rangle$ *is a uniformly integrable martingale.*

This fact was made the basis of modern stochastic integral theory by Kunita and S. Watanabe in their paper [1]. We know that the $[M]$ process of Theorem IV.26 is now used in preference.

For our martingale M bounded in L^2,

$$M^2 - [M] \text{ is a uniformly integrable martingale,}$$

so that

$$[M] - \langle M \rangle \text{ is a uniformly integrable martingale.}$$

Hence

(34.2) $\langle M \rangle$ *is the dual previsible projection of* $[M]$:

$$\langle M \rangle = [M]^p.$$

In particular, if M is continuous, then $[M]$ is continuous and hence previsible, so that $[M] = \langle M \rangle$.

(*34.3*) *Localization.* One of the reasons that the use of $[M]$ has superseded that of $\langle M \rangle$ is the following. We shall see in § 36 that $[M]$ may be defined for any local martingale M. However, the existence of $\langle M \rangle$ is more restricted, in a sense which the next result makes precise.

(*34.4*) THEOREM. *Let M be a local martingale null at* 0. *Then the following statements are equivalent*:

(i) *There exists a previsible increasing process $\langle M \rangle$ null at* 0 *such that $M^2 - \langle M \rangle$ is a local martingale.*

(ii) $M \in \mathscr{M}^2_{0,\text{loc}}$.

(iii) *The increasing process $[M]$ is locally integrable.*

When these conditions hold, $\langle M \rangle$ is the dual previsible projection of $[M]$.

We could prove this now, but each of these statements is equivalent to the statement that M^2 *is a special semimartingale*; once we have discussed special semimartingales in § 40, the proof of Theorem 34.4 will be immediate, so we defer the proof until then.

35. Last exits and equilibrium charge. The relationship between excessive functions and PCHAFs, which we started to explore in § III.16, is of central importance in probabilistic potential theory. Moreover, it provided much of the original motivation for the Meyer decomposition. In this section, we shall use dual previsible projections to explain the link between excessive functions, PCHAFs, and the integral representation of excessive functions in the case of

Brownian motion in $\mathbb{R}^d$; the extent to which we shall exploit special features of Brownian motion will convince you that the general story must be much more involved, as you will see if (as we hope) you choose to follow things up in Blumenthal and Getoor [1], Dynkin [2], or Getoor and Sharpe [1].

Fix some compact set $K \subseteq \mathbb{R}^d$, $d \geqslant 3$, and let B be BM($\mathbb{R}^d$). Define the last exit σ from K by

$$\sigma \equiv \sup\{t>0: B_t \in K\},$$

with the convention that $\sup \varnothing = 0$. If $H_K \equiv \inf\{t>0: B_t \in K\}$ is, as usual, the first hitting time of K, then we define

$$h(x) \equiv \mathbf{P}^x(\sigma > 0) = \mathbf{P}^x(H_K < \infty).$$

It is easy to see that $P_t h(x) = \mathbf{P}^x(\sigma > t) \uparrow h(x)$ as $t \downarrow 0$, so that *h is excessive, and $h(B_t)$ is a supermartingale.*

For the time being, fix the starting point x of Brownian motion, and notice that (see Remarks (35.13) for justification)

$$h(B_t) = \mathbf{P}^x[\sigma > t \mid \mathscr{F}_t] = \mathbf{E}^x[I_{[0,\sigma)}(t) \mid \mathscr{F}_t] = 1 - {}^{o}A_t,$$

where A is the raw increasing process

$$A_t \equiv I_{[\sigma,\infty)}(t).$$

From this, we can obtain an expression for the joint law of (σ, B_σ). Indeed, for $\varphi \in C_K^\infty(\mathbb{R}^d)$,

$$\begin{aligned} \text{(35.1)} \qquad \mathbf{E}^x[\mathrm{e}^{-\lambda\sigma}\varphi(B_\sigma); \sigma > 0] &= \mathbf{E}^x\left[\int_{(0,\infty)} \mathrm{e}^{-\lambda s}\varphi(B_s)\,dA_s\right] \\ &= \mathbf{E}^x\left[\int_{(0,\infty)} \mathrm{e}^{-\lambda s}\varphi(B_s)\,dA_s^{\mathrm{p}}\right], \end{aligned}$$

since $\mathrm{e}^{-\lambda t}\varphi(B_t)$ is a previsible process. Now by Theorem 21.4, A^{p} is the unique previsible increasing process such that

$$h(B_t) + A_t^{\mathrm{p}} \text{ is a martingale,}$$

and, *if h were* C^2, then by Itô's formula we immediately obtain

$$h(B_t) + \int_0^t (-\mathscr{G}h)(B_s)\,ds \text{ is a local martingale,}$$

where $\mathscr{G} \equiv \frac{1}{2}\Delta$ is the generator of B. Hence A^{p} is the PCHAF

$$A_t^{\mathrm{p}} = \int_0^t (-\mathscr{G}h)(B_s)\,ds,$$

related to the excessive function h by

(35.2) $$h(x)=\mathbf{E}^x[A^{\mathrm{p}}_\infty].$$

Further, from (35.1),

(35.3) $$\mathbf{E}^x[\mathrm{e}^{-\lambda\sigma}\varphi(B_\sigma);\ \sigma>0]=\mathbf{E}^x\left[\int_{(0,\,\infty)}\mathrm{e}^{-\lambda s}\varphi(B_s)(-\mathscr{G}h)(B_s)\,ds\right]$$
$$=\int r_\lambda(x,\ y)\varphi(y)(-\mathscr{G}h)(y)\,dy,$$

where $r_\lambda(\bullet,\bullet)$ is the resolvent density for $\mathrm{BM}(\mathbb{R}^d)$.

However, h will not be C^2 in general, nor even continuous (think of Lebesgue's thorn!). Nonetheless, if suitably interpreted, (35.3) remains true. We saw in §§ III.48 and I.22 that there exists a measure μ concentrated on ∂K such that

(35.4) $$h(x)=\int_{\partial K}\gamma(x,\ y)\,\mu(dy),$$

where $\gamma(x,\ y)=\int_0^\infty p_t(x,\ y)\,dt$ is the Green function for $\mathrm{BM}(\mathbb{R}^d)$; it is *twice* the Green function g appearing in §§ III.48. If μ had a continuous density ρ, then $-\mathscr{G}h=\rho$. Hence the appropriate interpretation of (35.3) when $h\notin C^2$ should be the following.

(*35.5*) THEOREM (Getoor and Sharpe [1]). *With the notation above,*

(35.6) $$\mathbf{P}^x(B_\sigma\in dy,\ \sigma\in dt)=p_t(x,\ y)\mu(dy)dt,\quad t>0,\quad y\in K.$$

Proof. The proof follows the lines sketched above, but applied to the approximations

(35.7) $$h_\varepsilon(x)\equiv P_\varepsilon h(x)=\mathbf{P}^x(\sigma>\varepsilon),$$

which are easily seen to be bounded C^2 (and even C^∞) functions. Now

$$h_\varepsilon(B_t)=\mathbf{P}[\sigma>t+\varepsilon\,|\,\mathscr{F}_t]=1-{}^{\circ}A(\varepsilon;\ t),$$

is a supermartingale, where $A(\varepsilon;\ \bullet)$ is the raw increasing process

$$A(\varepsilon;\ t)\equiv I_{[\sigma-\varepsilon,\,\infty)}(t)\,I_{\{\sigma>\varepsilon\}}.$$

Since h_ε *does* belong to C^2, we can deduce from Itô's formula that

$$A^{\mathrm{p}}(\varepsilon;\ t)=\int_0^t(-\mathscr{G}h_\varepsilon)(B_s)\,ds,$$

so that for any $\delta>0$, $\varphi\in C_K^\infty(\mathbb{R}^d)$,

$$\begin{aligned}\mathbf{E}^x[e^{-\lambda(\sigma-\varepsilon)}\varphi(B_{\sigma-\varepsilon});\ \sigma>\varepsilon+\delta]&=\mathbf{E}^x\left[\int_{(\delta,\infty)}e^{-\lambda s}\varphi(B_s)\,dA(\varepsilon;s)\right]\\&=\mathbf{E}^x\left[\int_{(\delta,\infty)}e^{-\lambda s}\varphi(B_s)(-\mathscr{G}h_\varepsilon)(B_s)\,ds\right]\\&=\int_{(\delta,\infty)}ds\,e^{-\lambda s}\int_{\mathbb{R}^d}p_s(x,y)\varphi(y)(-\mathscr{G}h_\varepsilon)(y)\,dy.\end{aligned}$$

However, from (35.4),

$$\begin{aligned}h_\varepsilon(x)&=\int_{\partial K}\mu(dy)\int_\varepsilon^\infty p_t(x,y)\,dt\\&=\int_{\mathbb{R}^d}\gamma(x,z)\,dz\int p_\varepsilon(z,y)\mu(dy)\end{aligned}$$

so that

$$(-\mathscr{G}h_\varepsilon)(x)=\int p_\varepsilon(x,y)\mu(dy).$$

Hence

$$\begin{aligned}&\mathbf{E}^x[e^{-\lambda(\sigma-\varepsilon)}\varphi(B_{\sigma-\varepsilon});\ \sigma>\varepsilon+\delta]\\&=\int_{\mathbb{R}^d}\int_{(\delta,\infty)}ds\,e^{-\lambda s}\left[\int_{\mathbb{R}^d}p_s(x,y)\varphi(y)p_\varepsilon(y,z)\,dy\right]\mu(dz).\end{aligned}$$

Letting $\varepsilon\downarrow 0$, we can use dominated convergence to conclude that

$$\mathbf{E}^x[e^{-\lambda\sigma}\varphi(B_\sigma);\ \sigma>\delta]=\int_{(\delta,\infty)}ds\,e^{-\lambda s}\int_{\mathbb{R}^d}p_s(x,z)\varphi(z)\mu(dz),$$

from which (35.6) follows. □

(35.8) *Hunt's Theorem I.22.7.* It should not surprise you that the approximations used to prove Theorem 35.5 can also be used to prove Hunt's Theorem, the conclusion of which we used at (35.4). We give such a proof now, to complement the time-reversal argument used in § III.48.

Since $h_\varepsilon(B_t)+\int_0^t(-\mathscr{G}h_\varepsilon)(B_s)\,ds$ is a UI martingale, for each $x\in\mathbb{R}^d$,

$$h_\varepsilon(x)=\mathbf{E}^x\int_0^\infty(-\mathscr{G}h_\varepsilon)(B_s)\,ds=\int\gamma(x,y)(-\mathscr{G}h_\varepsilon)(y)\,dy.$$

Hence for $\delta>0$ fixed and $\varepsilon>0$,

$$h_{\varepsilon+\delta}(x)=P_\delta h_\varepsilon(x)=\int dy\left[\int_\delta^\infty p_t(x,y)\,dt\right](-\mathscr{G}h_\varepsilon)(y).$$

Now let

(35.9) $$h^n_{\varepsilon+\delta}(x) \equiv \int dy\left[\int_\delta^n p_t(x,y)\,dt\right](-\mathscr{G}h_\varepsilon)(y)$$

and notice that for each $x\in\mathbb{R}^d$, $h^n_{\varepsilon+\delta}(x)$ is increasing in n and decreasing in ε (interchange the roles of δ and ε); thus

$$\begin{aligned} h_\delta(x) = \lim_{\varepsilon\downarrow 0} h_{\varepsilon+\delta}(x) &= \lim_{\varepsilon\downarrow 0}\lim_{n\to\infty} h^n_{\varepsilon+\delta}(x) \\ &= \lim_{n\to\infty}\lim_{\varepsilon\downarrow 0} h^n_{\varepsilon+\delta}(x). \end{aligned}$$

The only fact which we now need is that

(35.10) *for each $\eta>0$, the family of measures*

$$\{\exp(-\eta|y|^2)(-\mathscr{G}h_\varepsilon)(y)dy\colon 0<\varepsilon\leqslant 1\} \text{ is tight,}$$

so that by taking a subsequence ε_j, we have for some measure μ,

(35.11) $$\exp(-\eta|y|^2)(-\mathscr{G}h_{\varepsilon_j})(y)dy \Rightarrow \mu(dy)\exp(-\eta|y|^2)$$

for each $\eta=2^{-k}$, $k=0,1,\ldots$. Since

$$\int_\delta^n p_t(0,y)\,dt \leqslant n(2\pi\delta)^{-d/2}\exp(-|y|^2/2n),$$

we can use (35.11) in (35.9);

$$\lim_{\varepsilon\downarrow 0} h^n_{\varepsilon+\delta}(x) = \int\mu(dy)\left[\int_\delta^n p_t(x,y)\,dt\right],$$

and therefore

$$h_\delta(x) = \lim_{n\uparrow\infty}\lim_{\varepsilon\downarrow 0} h^n_{\varepsilon+\delta}(x) = \int\mu(dy)\left[\int_\delta^\infty p_t(x,y)\,dt\right],$$

whence

$$h(x) = \int\gamma(x,y)\mu(dy).$$

Thus h has an integral representation. By Theorem 35.5, the representing measure μ has the probabilistic interpretation (35.6), from which it is immediate that μ is unique, and is concentrated on K.

(*35.12*) **Exercise.** Prove (35.10). (*Hint*: multiply by a smooth bounded function which is zero in a large compact region, and integrate by parts.)

(*35.13*) *Remarks.* Return to the situation with which we began this section, where

$$\sigma \equiv \sup\{t>0: B_t \in K\}, \quad h(x) \equiv \mathbf{P}^x(\sigma>0), \quad A_t \equiv I_{[\sigma,\infty)}(t).$$

In general, we have to be rather careful about such 'obvious' facts as

$$h(B_\cdot) = 1 - {}^\circ A_\cdot. \tag{35.14}$$

Since the map $x \mapsto \mathbf{P}^x(\Lambda)$ can only be said to be universally measurable for $\Lambda \in \mathscr{F}$ (see §III.9), one has to check that h is Borel. In this case, this is easy, for two reasons: first, because (Exercise!) continuity of paths implies that $\{\sigma>0\} \in \mathscr{F}^\circ$, so that $x \mapsto \mathbf{P}^x(\sigma>0)$ is Borel; and second because the facts that $P_t h \uparrow h$ and $P_t h$ is continuous imply even that h is lower semicontinuous.

Since h is Borel and B is optional, $h(B)$ is optional. To check (35.14), we need to show that for any stopping time T,

$$h(B_T) = \mathbf{P}^x[\sigma > T \mid \mathscr{F}_T];$$

this is just the strong Markov theorem. (Strictly speaking, for each starting measure ν, we have to work with the set-up

$$(\Omega, \mathscr{F}^\nu, \{\mathscr{F}^\nu_t\}, \mathbf{P}^\nu)$$

to get the usual conditions; and so it goes on . . . .)

Finally, we note that since A is a right-continuous process, its optional projection ${}^\circ A$ is right-continuous by the 'Section Theorem' result (7.8). Hence, the supermartingale $h(B)$ is *already right-continuous*: it does not need to be modified.

For further comments on such matters and on the links with the *hypothèses droites*, see §46 below.

(*35.15*) *References.* Read Getoor and Sharpe [1]. This section is a key link with occupation times and quantum fields—see Dynkin [5, 6, 7].

7. STOCHASTIC INTEGRATION: THE GENERAL CASE

36. The quadratic variation process $[M]$. As you will remember from Chapter IV, the task of constructing the stochastic integral (Theorem IV.27.4) is largely completed once we have the existence (Theorem IV.26) of the quadratic variation process $[M]$ of the L^2-bounded martingale M. This section leads up to a proof of that existence result.

Firstly, we record some consequences of the definitions of previsible and dual previsible projections which will serve us well.

(*36.1*) PROPOSITION. (i) *The dual previsible projection of an IV_0 martingale is null.*

(ii) *If M is a UI martingale, then* ${}^{\mathrm{p}}(\Delta M) = 0$.

Proof. (i) This is a consequence of Theorem 21.4. (ii) The definition of the previsible projection together with Corollary 17.10 completes the proof of this assertion. □

(*36.2*) LEMMA. *Let $M \in \mathcal{M}^2 \cap \mathrm{IV}$, and define the increasing process*

$$A_t \equiv \sum_{0<s\leqslant t} \Delta M_s^2.$$

Then

$$N_t \equiv M_t^2 - M_0^2 - A_t$$

is a uniformly integrable martingale null at 0. In particular, $A_\infty \in L^1$.

Proof. We need only consider the case when M is null at 0. Let $T_n \equiv \inf\{t: |M_t| + A_t > n\}$. It is enough to prove the result for N^{T_n}, because then

$$0 = \mathbf{E}N(T_n) = \mathbf{E}\{M(T_n)^2 - M_0^2 - A(T_n)\} \leqslant \mathbf{E}M_\infty^2 - \mathbf{E}A(T_n)$$

and hence $\mathbf{E}A_\infty = \lim \mathbf{E}A(T_n) \leqslant \mathbf{E}M_\infty^2 < \infty$. Thus N is dominated by the integrable random variable $M_\infty^{*2} + |M_0|^2 + A_\infty$, and therefore

$$\mathbf{E}N(T) = \lim_n \mathbf{E}N(T \wedge T_n) = 0$$

for any stopping time T; Lemma II.77.6 implies that N is a uniformly integrable martingale.

To prove the result for N^{T_n}, let T be any stopping time such that $T \leqslant T_n$. Then

$$A(T) \leqslant A(T_n) \leqslant n + (2M_\infty^*)^2 \in L^1,$$

so N_T is integrable, and

$$\begin{aligned}
\mathbf{E}N_T &= \mathbf{E}\int_{(0,T]} 2M_{s-}\,dM_s \quad \text{by (IV.18.2);} \\
&= \mathbf{E}\int_{(0,T]} 2M_{s-}\,dM_s^{\mathrm{p}} \quad \text{since } M_- \text{ is previsible and bounded on } (0,T]; \\
&= 0,
\end{aligned}$$

by Proposition 36.1(i). □

Of course, another interpretation of Lemma 36.2 is that

$$\text{for } M \in \mathcal{M}^2 \cap \mathrm{IV}_0, \quad [M]_t = \sum_{0<s\leqslant t} \Delta M_s^2.$$

We have already constructed $[M]$ for $M \in c\mathcal{M}^2$; in view of the next result, this will prove to be all that is needed.

(*36.3*) THEOREM. *Let* $M \in \mathcal{M}^2$. *Then* $M \in c\mathcal{M}^2$ *if and only if* $MN \in \mathrm{UI}\,\mathcal{M}_0$ *for every* $N \in \mathcal{M}_0^2 \cap \mathrm{IV}$.

Proof. Suppose that $M \in c\mathcal{M}^2$, and let T be a stopping time. Then the process MN is dominated by $M_\infty^* N_\infty^* \in L^1$, so $M_T N_T$ is integrable and

$$\mathbf{E} M_T N_T = \mathbf{E} \int_{(0,T]} M_T \, dN_s$$
$$= \mathbf{E} \int_{(0,T]} M_s \, dN_s,$$

since N is an optional IV process, and the optional projection of $(0, T]M_T$ is $(0, T]M$. Thus,

$$\mathbf{E} M_T N_T = \mathbf{E} \int_{(0,T]} M_{s-} \, dN_s \quad \text{since } M \text{ is continuous;}$$
$$= \mathbf{E} \int_{(0,T]} M_{s-} \, dN_s^{\mathrm{p}} \quad \text{since } M_-(0, T] \text{ is previsible;}$$
$$= 0,$$

since $N \in \mathrm{IV}\mathcal{M}_0$. Thus (Lemma II.77.6) MN is a uniformly integrable martingale.

Suppose next that M is not continuous, and that $T = \inf\{u : \Delta M_u > \varepsilon\}$, where $\varepsilon > 0$ is so small that $\mathbf{P}(T < \infty) > 0$ (if this is impossible for M, apply the argument to $-M$!). Let $A \equiv [T, \infty)$, an IV_0 process, and let N be the martingale $A - A^{\mathrm{p}}$. Then as before,

$$\mathbf{E} M_T N_T = \mathbf{E} \int_{(0,T]} M_s \, dN_s$$
$$= \mathbf{E} \int_{(0,T]} \{\Delta M_s + M_{s-}\} \, d(A - A^{\mathrm{p}})_s$$
$$= \mathbf{E} \int_{(0,T]} \Delta M_s \, d(A - A^{\mathrm{p}})_s, \quad \text{since } M_-(0, T] \text{ is previsible;}$$
$$= \mathbf{E} \int_{(0,T]} \Delta M_s \, dA_s, \quad \text{since } {}^{\mathrm{p}}(\Delta M) = 0;$$
$$= \mathbf{E} \Delta M_T$$
$$\geqslant \varepsilon \mathbf{P}(T < \infty) > 0.$$

All that remains to prove is that $N \in \mathcal{M}^2$; but by (IV.18.2),

$$\begin{aligned}\mathbf{E}(A^{\mathrm{p}}_{\infty})^2 &= \mathbf{E}\left[\int_0^\infty A^{\mathrm{p}}_{s-}\,dA^{\mathrm{p}}_s + \int_0^\infty A^{\mathrm{p}}_s\,dA^{\mathrm{p}}_s\right] \\ &\leqslant \mathbf{E}\int_0^\infty 2A^{\mathrm{p}}_s\,dA^{\mathrm{p}}_s \\ &= \mathbf{E}\int 2A^{\mathrm{p}}_s\,dA_s \\ &= 2\mathbf{E}[A^{\mathrm{p}}_T;\, T<\infty] \\ &< \infty\end{aligned}$$

since $N = A - A^{\mathrm{p}}$ is an IV process. □

(*36.4*) *Remark.* We can restate Theorem 36.3 very succinctly; *the* IV_0 *martingales are dense in the orthogonal complement* $\mathrm{d}\mathcal{M}^2_0$ *of* $\mathrm{c}\mathcal{M}^2_0$. Of course, $MN \in \mathrm{UI}\mathcal{M}_0$ for $M \in \mathrm{c}\mathcal{M}^2_0$ and $N \in \mathrm{d}\mathcal{M}^2_0$ by definition. Combining Lemma 36.2 and Theorem 36.3 gives the following theorem.

(*36.5*) THEOREM. *Let* $M \in \mathrm{d}\mathcal{M}^2$, $N \in \mathcal{M}^2$. *Then*

$$M_t N_t - \sum_{0<s\leqslant t} \Delta M_s \Delta N_s$$

is a uniformly integrable martingale.

Proof. We may assume $M_0 = N_0 = 0$. Writing $N = N^{\mathrm{c}} + N^{\mathrm{d}}$ as the sum of its continuous and discontinuous parts, we see that it is enough to prove the theorem for M and N both in $\mathrm{d}\mathcal{M}^2$, since $N^{\mathrm{c}}M \in \mathrm{UI}\mathcal{M}_0$. If we can prove the result for $N = M \in \mathrm{d}\mathcal{M}^2$, polarization extends it to different N and M.

Because of (36.4), we can take $M^k \in \mathrm{IV}_0 \cap \mathcal{M}^2$ such that $\|(M-M^k)_\infty\|_2 \leqslant 2^{-k}$; then $M^k \to M$ uniformly almost surely and in L^2, and $\Delta M^k \to \Delta M$ uniformly almost surely and in L^2. If $A^k_t \equiv \sum_{0<s\leqslant t} (\Delta M^k_s)^2$, then, by Fatou's lemma,

$$\sum_{0<s\leqslant t} \Delta M^2_s \equiv A_t \leqslant \liminf A^k_t,$$

and so $\mathbf{E}A_t \leqslant \liminf \mathbf{E}A^k_t = \liminf \mathbf{E}(M^k_t)^2 = \mathbf{E}M^2_t$. In particular, A_t is finite a.s., and

$$\begin{aligned}\mathbf{E}\Big(\sup_t |A^k_t - A_t|\Big) &\leqslant \mathbf{E}\sum |\Delta(M^k - M)_s|.|\Delta(M^k+M)_s| \\ &\leqslant \mathbf{E}(\sum \Delta(M^k - M)^2_s)^{1/2}\,\mathbf{E}(\sum \Delta(M^k+M)^2_s)^{1/2}\end{aligned}$$

$$\leqslant \left[\liminf_{j\to\infty} \mathbf{E}\left(\sum \Delta(M^k - M^j)_s^2\right)\right]^{1/2}$$

$$\times \left[\liminf_{j\to\infty} \mathbf{E}\left(\sum \Delta(M^k + M^j)_s^2\right)\right]^{1/2}$$

$$= \left[\liminf_{j\to\infty} \mathbf{E}(M_\infty^k - M_\infty^j)^2\right]^{1/2}$$

$$\times \left[\liminf_{j\to\infty} \mathbf{E}(M_\infty^k + M_\infty^j)^2\right]^{1/2}$$

by Lemma 36.2, since each M^j is in IV_0;

$$\leqslant c2^{-k},$$

where $c = \|M_\infty\|_2 + \sup \|M_\infty^k\|_2$. Hence for any stopping time T,

$$\mathbf{E}(M_T^2 - A_T) = \lim_{k\to\infty} \mathbf{E}\{(M_T^k)^2 - A_T^k\} = 0,$$

using Lemma 36.2. Lemma II.77.6 implies that $M^2 - A$ is a uniformly integrable martingale. □

We are now in a position to prove Theorem IV.26, which we restate here.

(*36.6*) THEOREM. *Let $M \in \mathscr{M}_0^2$. Then there exists a unique increasing process $[M]$ null at 0 such that*:

(36.7(i)) $M^2 - [M]$ *is a uniformly integrable martingale*;

(36.7(ii)) $\Delta[M] = (\Delta M)^2$ *on* $(0, \infty)$.

Proof. We write $M = M^c + M^d$ in its unique decomposition, $M^c \in \mathrm{c}\mathscr{M}_0^2$, $M^d \in \mathrm{d}\mathscr{M}_0^2$. Then $\Delta M^d = \Delta M$ and by Theorem IV.30.1, there exists a unique continuous increasing process $[M^c]$ null at 0 such that $(M^c)^2 - [M^c]$ is a uniformly integrable martingale. Define

$$[M]_t \equiv [M^c]_t + \sum_{0<s\leqslant t} (\Delta M_s)^2 = [M^c]_t + \sum_{0<s\leqslant t} (\Delta M_s^d)^2. \tag{36.8}$$

Then

$$M_t^2 - [M]_t = (M_t^c)^2 + 2M_t^c M_t^d + (M_t^d)^2 - [M^c]_t - \sum_{0<s\leqslant t} (\Delta M_s^d)^2$$

$$= \{(M_t^c)^2 - [M^c]_t\} + 2M_t^c M_t^d + \left\{(M_t^d)^2 - \sum_{0<s\leqslant t} (\Delta M_s^d)^2\right\}$$

is the sum of three uniformly integrable martingales null at 0 (the second by Theorem IV.24.10, the third by Theorem 36.5), and so (36.8) defines an increasing process satisfying (36.7(i)) and (36.7(ii)). To establish uniqueness, suppose that A and B are increasing processes null at 0 satisfying the conditions of the Theorem. Then $\mathbf{E}A_\infty=\mathbf{E}B_\infty=\mathbf{E}M_\infty^2<\infty$, so $A-B$ is a process of integrable variation, which is continuous by (36.7(ii)). But

$$(A-B)=(M^2-B)-(M^2-A)$$

is also a martingale, and Theorem IV.30.4 implies that $A-B=0$. □

(*36.9*) *Remarks.* The stochastic integral $H\bullet M$ for $M\in\mathcal{M}^2$, $H\in L^2(M)$ is now defined; the proof of Theorem IV.26 completely justifies the construction and properties of §§IV.27–28.

Our task now is to extend the definition of $H\bullet X$ to the case of arbitrary semimartingales X and locally bounded previsible processes H. We take this up in the next section, but first we settle a point which ought to have been worrying you.

(*36.10*) LEMMA. *Let* $M\in\mathrm{IV}_0\cap\mathcal{M}^2$, *and* $H\in b\mathcal{P}$. *Then* $H\bullet M$ *is the pathwise Stieltjes integral* $\int H_s dM_s$.

Proof. To avoid confusion, we let $H*M$ stand for the pathwise Stieltjes integral. By Theorem IV.8.1, $H*M\in\mathrm{IV}\mathcal{M}_0$; we now show that $H*M\in\mathcal{M}^2$. Indeed, if $T_n\equiv\inf\{t: \int_{(0,t]}|dM_s|>n\}$, then the total variation of M, and of $H*M$, is bounded on $[0, T_n)$. Thus

$$\mathbf{E}(H*M)^2_{t\wedge T_n}=\mathbf{E}\left[\int_{(0,t\wedge T_n]}2(H*M)_{s-}H_s dM_s+\sum_{0<s\leqslant t\wedge T_n}H_s^2\Delta M_s^2\right]$$

from (IV.18.2);

$$=\mathbf{E}\left[\sum_{0<s\leqslant t\wedge T_n}H_s^2\Delta M_{s-}^2\right]$$

(since $(0, t\wedge T_n]H(H*M)_-$ is bounded and previsible, and the dual previsible projection of M is null by Proposition 36.1)

$$\leqslant c^2\mathbf{E}\left[\sum_{0<s\leqslant t\wedge T_n}\Delta M_s^2\right]\leqslant c^2\mathbf{E}M_\infty^2$$

where $c=\sup_{s,\omega}|H(s,\omega)|$. Now let $n\to\infty$; by Fatou's lemma, $H*M$ is bounded in L^2. Finally, we use the Kunita–Watanabe characterization of $H\bullet M$ (Theorem IV.28.1). For $N\in\mathcal{M}^2$, $[M,N]_t=\sum_{0<s\leqslant t}\Delta M_s\Delta N_s$ by Theorem 36.5, since $M\in\mathrm{IV}_0\cap\mathcal{M}^2\subseteq \mathrm{d}\mathcal{M}_0^2$. Thus

$$\mathbf{E}(H*M)_\infty N_\infty = \mathbf{E}\int_{(0,\infty)} N_\infty H_s\,dM_s$$

$= \mathbf{E}\int_{(0,\infty)} N_s H_s\,dM_s$, since M is optional, and the optional projection of N_∞ is N;

$= \mathbf{E}\int_{(0,\infty)} \Delta N_s H_s\,dM_s$, since $N_- H$ is previsible, and the dual previsible projection of M is 0;

$$= \mathbf{E}\sum H_s \Delta M_s \Delta N_s$$

$$= \mathbf{E}\int_{(0,\infty)} H_s\,d[M,N]_s. \qquad \square$$

37. Stochastic integrals with respect to local martingales. There are local martingales which are not locally in $\mathscr{M}^2$, and this is why we cannot define $\langle M\rangle$ for all local martingales. However, every local martingale is locally the sum of an IV martingale and a bounded martingale; because we know how to perform stochastic integrals with respect to both classes, we can make up a stochastic integral $H\bullet M$ for $H \in \mathrm{lb}\,\mathscr{P}$, $M \in \mathscr{M}_{0,\mathrm{loc}}$.

(*37.1*) DEFINITION (strong reduction). *Let T be a stopping time, $M \in \mathscr{M}_{0,\mathrm{loc}}$. We say T* reduces M strongly *if there exist a bounded martingale U and an* IV_0 *martingale V such that*

$$M^T = U + V.$$

This concept of strong reduction is stronger than that of Dellacherie and Meyer [1].

(*37.2*) PROPOSITION. *Let $M \in \mathscr{M}_{0,\mathrm{loc}}$, and suppose that T reduces M strongly. If S is a stopping time, then $S \wedge T$ reduces M strongly. If S reduces M strongly, then so does $S \vee T$.*

Proof. The first statement is trivial; for the second, notice that

$$M^{S\vee T} = M^S + (M^T - M^{S\wedge T}). \qquad \square$$

The next result explains why we introduced strong reduction.

(*37.3*) THEOREM. *Let $M \in \mathscr{M}_{0,\mathrm{loc}}$. Then there exist stopping times $T_n \uparrow \infty$ each of which reduces M strongly.*

Proof. We can find stopping times $S_n \uparrow \infty$ such that M^{S_n} is a uniformly integrable martingale, and $|M_t|[0, S_n)$ is bounded. Let $A^n \equiv \Delta M_{S_n}[S_n, \infty)$, an IV_0 process.

Then we may write M^{S_n} as the sum of $U^n \in \mathrm{UI}\mathscr{M}_0$ and $V^n \in \mathrm{IV}\mathscr{M}_0$, where

$$U^n = M^{S_n} - A^n + (A^n)^{\mathrm{p}}, \quad V^n = A^n - (A^n)^{\mathrm{p}}.$$

Notice that $M^{S_n} - A^n$ is bounded. Let $S_{nk} \equiv \inf\{u: (A^n)^{\mathrm{p}}_u \geqslant k\}$, a previsible stopping time by Lemma 12.2; and for each n, k, let $(R'_{nkl})_{l \in \mathbb{N}}$ be a sequence announcing S_{nk}. Then $(A^n)^{\mathrm{p}}$ is bounded on $[0, R'_{nkl}]$, and so $R_{nkl} \equiv R'_{nkl} \wedge S_n$ reduces M strongly. Now let $T_m \equiv \max\{R_{nkl}: n, k, l \leqslant m\}$. □

We deduce immediately an important consequence.

(*37.4*) COROLLARY. $\mathscr{M}_{0,\,\mathrm{loc}} = (\mathrm{bc}\mathscr{M}_0 + \mathrm{bd}\mathscr{M}_0 + \mathrm{IV}\mathscr{M}_0)_{\mathrm{loc}}$.

Proof. If T reduces M strongly, $M^T = U + V$, with U a bounded martingale, then U is in $\mathscr{M}_0^2$, so has a unique decomposition $U = U^{\mathrm{c}} + U^{\mathrm{d}}$. We can now stop so as to make U^{c} bounded. □

If $M = U^{\mathrm{c}} + U^{\mathrm{d}} + V \in \mathrm{b}\mathscr{M}_0 + \mathrm{IV}\mathscr{M}_0$, then, although U^{d} and V are not uniquely defined, U^{c} is; indeed, if $M = \tilde{U}^{\mathrm{c}} + \tilde{U}^{\mathrm{d}} + \tilde{V}$ is another decomposition, then $U^{\mathrm{c}} - \tilde{U}^{\mathrm{c}} = \tilde{U}^{\mathrm{d}} - U^{\mathrm{d}} + \tilde{V} - V$ is a bounded martingale which is both continuous and purely discontinuous, therefore null. By extension, we call U^{c} *the continuous martingale part of* M, and denote it by M^{c}. The proof of the next result is an easy exercise.

(*37.5*) PROPOSITION. *Let* $M \in \mathscr{M}_{0,\,\mathrm{loc}}$. *Then there exists a unique* $M^{\mathrm{c}} \in \mathrm{c}\mathscr{M}_{0,\,\mathrm{loc}}$ *such that, for any* T *which reduces* M *strongly,*

$$(M^{\mathrm{c}})^T = (M^T)^{\mathrm{c}}.$$

Likewise, there exists a unique continuous increasing process $[M^{\mathrm{c}}]$ *such that, for any* T *reducing* M *strongly,*

$$[M^{\mathrm{c}}]^T = [(M^T)^{\mathrm{c}}].$$

With (36.8) in mind, we make the following definition, which obviously agrees with the earlier one when $M \in \mathscr{M}_0^2$.

(*37.6*) DEFINITION (quadratic variation process $[M]$). *Let* $M \in \mathscr{M}_{0,\,\mathrm{loc}}$. *We define the* quadratic variation process $[M]$ of M *by*

$$[M]_t = [M^{\mathrm{c}}]_t + \sum_{0 < s \leqslant t} \Delta M_s^2. \tag{37.7}$$

We can obviously define the quadratic covariation process $[M, N]$ of two local martingales by polarization as before. Here now is the full form of pre-Theorem IV.26.4.

(*37.8*) THEOREM. *Let* $M, N \in \mathscr{M}_{0,\,\mathrm{loc}}$. *Then* $[M, N]$ *is the unique* FV_0 *process* A *with the properties*

(i) *$MN-A$ is a local martingale;*
(ii) $\Delta A=\Delta M\Delta N$.

Proof. Uniqueness follows from Theorem IV.30.4. To prove that $[M, N]$ has the stated properties, we can suppose that $M=N$. Stopping at a strongly reducing time, we can suppose that $M=U^c+U^d+V$ as in Corollary 37.4. Then we express

$$M^2-[M]=\{(U^c)^2-[M^c]\}+2U^c(U^d+V)+\{(U^d+V)^2-\sum\Delta M^2\}$$

as the sum of three terms, the first of which is a uniformly integrable martingale by the definition of $\langle M^c\rangle$, and the second of which is also a uniformly integrable martingale by (IV.24.12) and Theorem 36.3. As for the third, we know that $(U^d)^2-\sum(\Delta U^d)^2$ is a uniformly integrable martingale (Theorem 36.5), and $U^dV-\sum\Delta U^d\Delta V$ is a uniformly integrable martingale, since for any stopping time T, just as in the proof of Lemma 36.10,

$$\begin{aligned}\mathbf{E}U^d_T V_T&=\mathbf{E}\int_{(0,T]} U^d_T dV_s\\ &=\mathbf{E}\int_{(0,T]} U^d_s dV_s\\ &=\mathbf{E}\int_{(0,T]} \Delta U^d_s dV_s\\ &=\mathbf{E}\sum_{0<s\leq T}\Delta U_s\Delta V_s.\end{aligned}$$

Finally, stopping V at $S_n\equiv\inf\{u:\int_{(0,u]}|dV_s|>n\}$, we know that

$$V(t\wedge S_n)^2-\sum_{0<s\leq t\wedge S_n}\Delta V_s^2=2\int_{(0,t\wedge S_n]}V_{s-}dV_s,$$

which is in IV$\mathscr{M}_0$, by Theorem IV.8.1. Thus $V^2-\sum\Delta V^2$ is a local martingale. □

Suppose that $H\in b\mathscr{P}$ and $M\in bc\mathscr{M}_0+bd\mathscr{M}_0+IV\mathscr{M}_0$, $M=U^c+U^d+V$. Then we may define the stochastic integral $H\bullet M$ as

$$H\bullet M=H\bullet U^c+H\bullet U^d+H\bullet V.$$

That this is independent of the representation of M follows from Lemma 36.10 and the fact that U^c is uniquely defined. That

$$(H\bullet M)^T=H(0,T]\bullet M^T$$

for each stopping time T follows from Theorem IV.27.6(i) and properties of the

Stieltjes integral. Thus the localization Lemma IV.13.5 allows us to extend the definition of $H{\cdot}M$ to $H \in \mathrm{lb}\mathscr{P}$ and $M \in (\mathrm{bc}\mathscr{M}_0 + \mathrm{bd}\mathscr{M}_0 + \mathrm{IV}\mathscr{M}_0)_{\mathrm{loc}} = \mathscr{M}_{0,\,\mathrm{loc}}$. The stochastic integral has the following properties.

(*37.9*) THEOREM. *Let* $H, K \in \mathrm{lb}\mathscr{P}$, $M, N \in \mathscr{M}_{0,\,\mathrm{loc}}$. Then:

(i) $(H{\cdot}M)^T = H(0, T]{\cdot}M^T = H(0, T]{\cdot}M = H{\cdot}M^T$ *for any stopping time* T;
(ii) $(H{\cdot}M)^2 - \int H^2 d[M]$ *is a local martingale*;
(iii) $H{\cdot}(K{\cdot}M) = (HK){\cdot}M$;
(iv) $\Delta(H{\cdot}M) = H\Delta M$;
(v) $(H{\cdot}M)^{\mathrm{c}} = H{\cdot}M^{\mathrm{c}}$;
(vi) $(H{\cdot}M)N - \int H\,d[M, N]$ *is a local martingale*;
(vii) $[H{\cdot}M, N] = \int H\,d[M, N]$;
(viii) if $M \in \mathrm{FV}_0$, *then* $H{\cdot}M$ *is the pathwise Stieltjes integral.*

Proof. By localization, these reduce to known properties of the L^2 stochastic integral (Theorem IV.27.6) and the Stieltjes integral, except for (viii). We leave the details to you as an exercise. For (viii), use the fact that $\mathrm{FV}\mathscr{M}_{0,\,\mathrm{loc}} = \mathrm{IV}\mathscr{M}_{0,\,\mathrm{loc}}$ and that $H{\cdot}M$ is independent of the representation of $M \in \mathrm{b}\mathscr{M}_0 + \mathrm{IV}\mathscr{M}_0$. □

38. Stochastic integrals with respect to semimartingales. The final extension of the stochastic integral to $H \in \mathrm{lb}\mathscr{P}$ and $X \in \mathscr{S}_0$ is something of an anticlimax.

(*38.1*) DEFINITION. *Suppose* $H \in \mathrm{lb}\mathscr{P}$, $X = M + A \in \mathscr{S}_0$, *where* $M \in \mathscr{M}_{0,\,\mathrm{loc}}$ *and* $A \in \mathrm{FV}_0$. *We define the* stochastic integral $H{\cdot}X$ *by*

$$H{\cdot}X = H{\cdot}M + H{\cdot}A.$$

Theorem 37.9(viii) shows that $H{\cdot}X$ is the same whatever decomposition $X = M + A$ is used. In fact, the only 'uniqueness' in the decomposition of X is that M^{c} is unique; it is called the *continuous-martingale part of* X and is written X^{cm}. The proof of this is a trivial exercise. We also write $[X^{\mathrm{cm}}]$ for $[M^{\mathrm{c}}]$, and define

$$[X]_t \equiv [X^{\mathrm{cm}}]_t + \sum_{0 < s \leqslant t} (\Delta X_s)^2,$$

extending to $[X, Y]$ by polarization. Properties of the Stieltjes integral and Theorem 37.9 give us some properties of $H{\cdot}X$.

(*38.2*) THEOREM. *Let* $H, K \in \mathrm{lb}\mathscr{P}$, $X, Y \in \mathscr{S}_0$. *Then*:

(i) $(H{\cdot}X)^T = H(0, T]{\cdot}X = H(0, T]{\cdot}X^T = H{\cdot}X^T$ *for any stopping time* T;
(ii) $H{\cdot}(K{\cdot}X) = (HK){\cdot}X$;
(iii) $\Delta(H{\cdot}X) = H\Delta X$;
(iv) $(H{\cdot}X)^{\mathrm{cm}} = H{\cdot}X^{\mathrm{cm}}$;
(v) $[H{\cdot}X, Y] = \int H\,d[X, Y]$.

Proof. **Exercise.** □

As in the earlier developments of stochastic integration, the integration-by-parts formula which we now prove is the essence of the stochastic calculus.

(*38.3*) THEOREM (Integration-by-parts formula). *Let X and Y be semimartingales. Then*:

$$X_tY_t-X_0Y_0=\int_{(0,t]}X_{s-}\,dY_s+\int_{(0,t]}Y_{s-}\,dX_s+[X,Y]_t. \tag{38.4}$$

Proof. Firstly note that the stochastic integrals are well-defined, since $X_-\in\mathrm{lb}\mathscr{P}$. It is enough to establish (38.4) for X and Y in $\mathscr{S}_0$, and since both sides are bilinear in X and Y, it is enough to prove three cases of the integration-by-parts formula:

(i) when $X, Y\in\mathrm{FV}_0$;

(ii) when $X\in\mathscr{M}_{0,\,\mathrm{loc}}$, $Y\in\mathrm{FV}_0$;

(iii) when $X, Y\in\mathscr{M}_{0,\,\mathrm{loc}}$.

(i) The first case is already done (Theorem IV.18.4).

(ii) By stopping, we may reduce the second to $X=U^{\mathrm{c}}+U^{\mathrm{d}}+V$, where $U^{\mathrm{c}}\in\mathrm{bc}\mathscr{M}_0$, $U^{\mathrm{d}}\in\mathrm{bd}\mathscr{M}_0$, $V\in\mathrm{IV}\mathscr{M}_0$, and we may even suppose that Y is of bounded variation. This is because if $T_n\equiv\inf\{t: \int_{(0,t]}|dY_s|>n\}$, then the jumps of the two sides of (38.4) agree at time T_n; the integration-by-parts formula is clearly satisfied for X and $Y'\equiv Y(T_n)\,[T_n,\infty)$, so it is enough to prove it for X and the bounded variation process $Y^{T_n}-Y'$. We prove it separately for U^{c} and Y, and for U^{d} and Y, assuming Y is of variation bounded by $c<\infty$.

Firstly, take $X=U^{\mathrm{c}}$, let $t_i^n\equiv(i2^{-n})\wedge t$, and let $\Delta_k^nX\equiv X(t_k^n)-X(t_{k-1}^n)$, $\Delta_k^nY\equiv Y(t_k^n)-Y(t_{k-1}^n)$. Then:

$$X_tY_t=\sum_{k\geq1}X(t_{k-1}^n)\Delta_k^nY+\sum_{k\geq1}Y(t_{k-1}^n)\Delta_k^nX+\sum_{k\geq1}\Delta_k^nX\,\Delta_k^nY \tag{38.5}$$

$$\to\int_{(0,t]}X_{s-}\,dY_s+\int_{(0,t]}Y_{s-}\,dX_s$$

as $n\to\infty$, dominated convergence dealing with the first integral, an obvious L^2 estimate of the difference between the second sum and the second integral (both L^2-bounded martingales) coping with that term, and the third term going to zero because

$$|\Sigma\,\Delta_k^nX\,\Delta_k^nY|\leq\sup\{|X_u-X_s|: u, s\leq t, |u-s|\leq2^{-n}\}\cdot V_Y(t)$$

and $X=U^{\mathrm{c}}$ is continuous.

Next, if $X=U^{\mathrm{d}}\in\mathrm{d}\mathscr{M}_0$, take $X^n\in\mathscr{M}^2\cap\mathrm{IV}_0$, $X^n\to X$. Then (38.4) holds for X^n and Y, and

$$\mathbf{E}\left|\int X_{s-}^n\,dY_s-\int X_{s-}\,dY\right|\leq\mathbf{E}(X^n-X)_\infty^*\,V_Y(\infty)$$

$$\leq c\cdot\mathbf{E}(X^n-X)_\infty^*\to0,$$

$$\mathbf{E}\left(\left(\int Y_{s-}\,d(X^n-X)_s\right)^2\right)=\mathbf{E}\left(\int Y_{s-}^2\,d[X^n-X]_s\right)$$
$$\leqslant c^2\,\mathbf{E}[X^n-X]_\infty$$
$$=c^2\,\mathbf{E}(X^n-X)_\infty^2\to 0,$$

and

$$\mathbf{E}\left|\sum\Delta(X^n-X)_s\Delta Y_s\right|\leqslant\mathbf{E}\{(\sum\Delta(X^n-X)_s^2)^{1/2}(\sum\Delta Y_s^2)^{1/2}\}$$
$$\leqslant c\mathbf{E}(\sum\Delta(X^n-X)_s^2)^{1/2},$$
$$\leqslant c\mathbf{E}(X^n-X)_\infty^2\to 0,$$

establishing (38.4) for X and Y.

(iii) By polarization, it is enough to take $X=Y$; by stopping, we may assume that $X\in\mathrm{bc}\mathcal{M}_0+\mathrm{bd}\mathcal{M}_0+\mathrm{IV}\mathcal{M}_0$. Because of (i) and (ii), we have only to deal with the case where $X\in\mathrm{bc}\,\mathcal{M}_0+\mathrm{bd}\,\mathcal{M}_0$. We shall be finished if we can deal with the three cases:

(a) $X=Y\in\mathrm{bc}\mathcal{M}_0$;
(b) $X\in\mathrm{bc}\mathcal{M}_0$, $Y\in\mathrm{bd}\mathcal{M}_0$;
(c) $X=Y\in\mathrm{bd}\mathcal{M}_0$.

Case (*a*) was done in Theorem IV.32.4. For case (*b*), take $Y^n\in\mathcal{M}_0^2\cap\mathrm{IV}$, $Y^n\to Y$. By stopping, we may assume that $X_\infty^*+[X]_\infty\leqslant c<\infty$ for all ω. Now (38.4) holds for X, Y^n, and

$$\mathbf{E}\left[\int X_{s-}\,d(Y-Y^n)_s\right]^2\leqslant c^2\mathbf{E}[Y-Y^n]_\infty=c^2\mathbf{E}(Y_\infty-Y_\infty^n)^2$$
$$\to 0,$$

$$\mathbf{E}\left[\int(Y-Y^n)_{s-}\,dX_s\right]^2=\mathbf{E}\int(Y-Y^n)_{s-}^2\,d[X]_s$$
$$\leqslant\mathbf{E}(Y-Y^n)_\infty^{*2}[X]_\infty$$
$$\leqslant c\mathbf{E}(Y-Y^n)_\infty^{*2}\to 0,$$

dealing with case (*b*).

Finally for case (*c*), we use the summation-by-parts formula (38.5) again. If $A_t^n\equiv\sum_{k\geqslant 1}(\Delta_k^n X)^2$, then

$$\sup_t\left|\sum_{k\geqslant 1}X(t_{k-1}^n)\Delta_k^n X-\int_{(0,t]}X_{s-}\,dX_s\right|\to 0\text{ in }L^2$$

and so $A_t^n\to A_t$ uniformly in L^2, and A is an increasing process (since A^n is increasing on the subsequence $i2^{-n}$, $i\in\mathbb{Z}^+$). Moreover, it is clear that A must jump whenever X jumps, and at such times, $\Delta A_t = \Delta X_t^2$. Thus $A - [X]$ is increasing, and we have

$$X^2 - A = 2\int X_{s-}\,dX_s, \text{ a martingale in } \mathscr{M}_0^2.$$

But Theorem 36.5 implies that $X^2 - [X]$ is a martingale in $\mathscr{M}_0^2$. Hence the increasing process $A - [X]$ is a martingale in $\mathscr{M}_0^2$, so $A = [X]$. □

39. Itô's formula for semimartingales. As before, the integration-by-parts formula extends to the more general change-of-variables formula, or Itô's formula.

(*39.1*) THEOREM. *Let $f: \mathbb{R}^n \to \mathbb{R}$ be C^2, and suppose $X = (X^1, \ldots, X^n)$ is a semimartingale in $\mathbb{R}^n$. Then*

$$f(X_t) - f(X_0) = \int_{(0,t]} D_i f(X_{s-})\,dX_s + \tfrac{1}{2}\int_{(0,t]} D_{ij} f(X_{s-})\,d[(X^i)^{cm}, (X^j)^{cm}]_s + \sum_{0<s\leq t} \{f(X_s) - f(X_{s-}) - D_i f(X_{s-})\,\Delta X_s^i\}. \quad (39.2)$$

Remarks. We have used $(X^i)^{cm}$ to denote the continuous-martingale part of the semimartingale X^i.

Proof. The argument runs exactly as before (Theorem IV.18.4, Theorem IV.32.8): the change-of-variables formula holds for polynomial f, and an arbitrary C^2 function f can be uniformly approximated (together with its derivatives of order up to 2) on compacts by polynomials. The only point requiring any alteration is the localization; we may by stopping (at time T, say) assume that X is the sum of a bounded martingale and a finite-variation process, but this does not ensure that X is bounded, which is needed to make the approximation argument work. However, it is easy to see that the jumps of both sides of (39.2) agree, so by applying the argument to $X - X_T[T, \infty)$, a bounded process, the result follows. (We used the same device to prove case (ii) of Theorem 38.3.) □

40. Special semimartingales. Amongst the class $\mathscr{S}$ of all semimartingales, there is a particularly nice subclass for which the finite-variation process is of locally integrable variation.

(*40.1*) DEFINITION (special semimartingale). *The semimartingale X is said to be a* special semimartingale *if there is a decomposition $X = X_0 + M + A$, where $M \in \mathscr{M}_{0,\text{loc}}$ and $A \in \text{IV}_{0,\text{loc}}$.*

The main facts about special semimartingales are summarized in the following theorem.

(*40.2*) THEOREM. *Let X be a semimartingale. Then the following are equivalent:*
(i) *X is special.*
(ii) *There is a decomposition of X in which the FV process is previsible.*
(iii) *There is a decomposition of X in which the FV process is of locally integrable variation.*
(iv) *The process $(\sum \Delta X_s^2)^{1/2}$ is locally integrable.*

If X is special, then in every decomposition

$$X = X_0 + M + A$$

the process A is in $\mathrm{IV}_{0,\,\mathrm{loc}}$. There is a unique decomposition (called the canonical *decomposition) in which A is previsible.*

Proof. (i) ⇒ (ii). If $X = X_0 + M + A$, $A \in \mathrm{IV}_{0,\,\mathrm{loc}}$, then by Theorem 21.7 there exists a unique previsible increasing process A^{p} such that $A - A^{\mathrm{p}}$ is a local martingale. Thus

$$X = X_0 + (M + A - A^{\mathrm{p}}) + A^{\mathrm{p}}$$

is a decomposition of X in which the finite-variation process is previsible.
(ii) ⇒ (iii). If $X = X_0 + M + A$, and A is previsible, the stopping times

$$T_n \equiv \inf\left\{t: \int_{(0,\,t]} |dA_s| \geqslant n\right\}$$

are previsible by (27.5) and (12.2), so each T_n is announced by some sequence $(S_{nk})_{k \geqslant 0}$. If we let $R_n = \max\{S_{mk}: m, k \leqslant n\}$, then the stopping times R_n increase to ∞, and A is of bounded variation on $(0, R_n]$.
(iii) ⇒ (i). This is a restatement of the definition of a special semimartingale.
The equivalence of (i) and (iv) follows from a result of interest in its own right.

(*40.3*) LEMMA. *If $M \in \mathscr{M}_{\mathrm{loc}}$, then $[M]^{1/2} \in \mathrm{IV}_{0,\,\mathrm{loc}}$.*

Proof. We may suppose without loss of generality that $M_0 = 0$. By stopping, we can reduce to the case where $M = U + V$, U a bounded martingale null at 0, V an IV martingale null at 0 (Theorem 37.3). Then

$$[M] = [U + V] \leqslant 2[U] + 2[V],$$

so that

$$[M]^{1/2} \leqslant \sqrt{2}\{[U]^{1/2} + [V]^{1/2}\}.$$

Since U is bounded, U is bounded in L^2 and $[U]_\infty \in L^1$, therefore $[U]_\infty^{1/2} \in L^2 \subseteq L^1$. As for V, since it is an IV martingale,

$$[V]_\infty = \sum(\Delta V_s)^2 \leqslant \left(\sum|\Delta V_s|\right)^2$$

by (36.8). Hence $[V]_\infty^{1/2}$ is bounded by the variation of V, which is integrable. Thus $[M]_\infty^{1/2}$ is integrable. □

(i) ⇒ (iv).
$$\begin{aligned}(\sum \Delta X_s^2)^{1/2} &\leqslant (\sum \Delta M_s^2)^{1/2} + (\sum \Delta A_s^2)^{1/2}\\ &\leqslant [M]_\infty^{1/2} + \sum |\Delta A_s|\\ &\leqslant [M]_\infty^{1/2} + \int_0^\infty |dA_s|\end{aligned}$$

and so $(\sum \Delta X_s^2)^{1/2}$ is locally integrable if X is special, by Lemma 40.3.
(iv) ⇒ (i). Suppose that $X = X_0 + M + A$, $M \in \mathscr{M}_{0,\text{loc}}$, $A \in \text{FV}_0$. The hypothesis and Lemma 40.3 ensure that by stopping we may ensure that $Y \equiv [M]_\infty^{1/2} + (\sum \Delta X_s^2)^{1/2}$ is integrable. But then

$$(\sum \Delta A_s^2)^{1/2} \leqslant (\sum \Delta M_s^2)^{1/2} + (\sum \Delta X_s^2)^{1/2} \leqslant Y$$

is integrable, and in particular, $\sup |\Delta A_t| \leqslant Y$. Thus if $T_n \equiv \inf\left\{t: \int_0^t |dA_s| > n\right\}$, $\int_0^{T_n} |dA_s| \leqslant n + Y$ is integrable, and so A is of locally integrable variation, and X is special.

The argument just used to prove that (iv) ⇒ (i) also proves that the FV_0 process in any decomposition of X is in $\text{IV}_{0,\text{loc}}$. The uniqueness of the canonical decomposition is a (by now) routine application of Theorems 19.4 and IV.30.4. □

(*40.4*) *Examples.* (i) By Theorem 40.2, X is special if and only if $(\sum \Delta X_s^2)^{1/2}$ is locally integrable, so any continuous semimartingale is special. In the canonical decomposition $X = X_0 + M + A$, it is evident that $M = X - X_0 - A$ is previsible, and therefore (Theorem 19.4) continuous. Hence A is also continuous.

(ii) *Proof of Theorem 34.4.* By Theorem 40.2, each of the statements (i) and (iii) of Theorem 34.4 is equivalent to the statement that M^2 is a special semimartingale. To prove that (ii) and (iii) are equivalent, let N denote the local martingale $M^2 - [M]$. For any stopping time T,

$$N^T = (M^T)^2 - [M]^T,$$

so if $[M]$ is locally integrable, take a sequence (T_n) of stopping times, $T_n \uparrow \infty$, which reduce N and for which $[M](T_n)$ is integrable; thus $M \in \mathscr{M}^2_{0,\text{loc}}$. Conversely, if $M \in \mathscr{M}^2_{0,\text{loc}}$, take $T_n \uparrow \infty$ such that T_n reduce N and such that $M^{T_n} \in \mathscr{M}_0^2$. Then $[M](T_n) \in L^1$ and $[M]$ is locally integrable. □

41. Quasimartingales. Given an adapted R-process X, it is obviously of importance to be able to decide whether X is a semimartingale. If this is not obvious from inspection, then the easiest way in practice is to check whether X is a quasimartingale.

(*41.1*) DEFINITION. *An adapted R-process is a* quasimartingale *if*

$$\mathrm{Var}(X) \equiv \sup \mathbf{E}(A_\tau) < \infty,$$

where the supremum is taken over all dissections $\tau \equiv (t_0, t_1, \ldots, t_n)$, $0 \leqslant t_0 < t_1 < \ldots < t_n$, *and where* A_τ *is defined as*

(41.2) $$A_\tau \equiv \sum_{i=0}^{n-1} |\mathbf{E}[X(t_{i+1}) - X(t_i) | \mathscr{F}(t_i)]| + |X(t_n)|.$$

Notice that the space $\mathcal{Q}$ of quasimartingales is a vector space, and that Var (•) is a norm on $\mathcal{Q}$. Every L^1-bounded martingale is a quasimartingale, as is every IV process, so that each semimartingale of the form $X = M + A$, where M is an L^1-bounded martingale and $A \in \mathrm{IV}_0$, is also a quasimartingale. Notice also that if X is a nonnegative supermartingale, then X is a quasimartingale, and

$$\mathrm{Var}(X) = \mathbf{E}(X_0).$$

Thus the difference of two nonnegative supermartingales is a quasimartingale; for us, the most important fact about quasimartingales is the converse, that *every quasimartingale is the difference of two nonnegative supermartingales*, and hence is a semimartingale. This result is due in this generality to K. Murali Rao [2]; we prove it now, following Dellacherie-Meyer [1].

(*41.3*) THEOREM (Rao, Stricker). *If X is a quasimartingale, then there exist nonnegative supermartingales Y and Z such that*

(41.4) $$X = Y - Z,$$

and

(41.5) $$\mathrm{Var}(X) = \mathrm{Var}(Y) + \mathrm{Var}(Z).$$

Moreover, if $\bar{X} = \bar{Y} - \bar{Z}$ *is another representation of X as the difference of two nonnegative supermartingales, then* $\bar{Y} - Y = \bar{Z} - Z$ *is a nonnegative supermartingale.*

Proof. Fix $s > 0$, let $\tau \equiv (t_0, t_1, \ldots, t_n)$ be a dissection, $s = t_0 < t_1 < \ldots < t_n$, and define

$$A_\tau^+ \equiv \sum_{i=0}^{n-1} \mathbf{E}(X(t_i) - X(t_{i+1}) | \mathscr{F}(t_i))^+ + X(t_n)^+.$$

Define A_τ^- analogously, and define $Y_s^\tau \equiv \mathbf{E}(A_\tau^+ | \mathscr{F}_s)$, $Z_s^\tau \equiv \mathbf{E}(A_\tau^- | \mathscr{F}_s)$. Evidently, $Y_s^\tau - Z_s^\tau = X_s$. Next, if σ is a refinement of τ, then $Y_s^\tau \leqslant Y_s^\sigma$. To prove this, it is enough to take the case where σ is obtained from τ by the addition of one point $t \in (t_i, t_{i+1})$; but in this case, if $\xi \equiv \mathbf{E}[X(t_i) - X(t) | \mathscr{F}(t_i)]$, $\eta \equiv \mathbf{E}[X(t) - X(t_{i+1}) | \mathscr{F}(t)]$, and $\zeta \equiv \mathbf{E}[X(t_i) - X(t_{i+1}) | \mathscr{F}(t_i)]$, then $\zeta = \xi + \mathbf{E}[\eta | \mathscr{F}(t_i)]$, so that $\zeta^+ \leqslant \xi^+ + \mathbf{E}[\eta | \mathscr{F}(t_i)]^+ \leqslant \xi^+ + \mathbf{E}[\eta^+ | \mathscr{F}(t_i)]$, and the inequality $Y_s^\tau \leqslant Y_s^\sigma$ follows immediately. Hence $Y_s' \equiv \lim Y_s^\tau$ exists in L^1, the limit being taken as the mesh of the dissection tends to zero.

The process Y' is a supermartingale; indeed, if $s<t$ are fixed and if τ is any dissection of (s,∞) containing the point t, then $\mathbf{E}[Y_t^\tau|\mathcal{F}_s]\leqslant Y_s^\tau$, and on letting the mesh of the dissection tend to zero, we obtain $\mathbf{E}[Y_t'|\mathcal{F}_s]\leqslant Y_s'$. Defining Z' analogously, it is clear that for each s, $X_s=Y_s'-Z_s'$ a.s., since $X_s=Y_s^\tau-Z_s^\tau$. Now Y' may not have R-paths, but the process Y defined by

$$Y_t\equiv \liminf_{q\downarrow\downarrow t,\ q\in\mathbb{Q}} Y_q'$$

is a supermartingale, and almost all of its paths are R-paths. Defining the supermartingale Z analogously, we can represent X as $Y-Z$, the difference of two nonnegative supermartingales.

To establish (41.5), notice that $\operatorname{Var}(Y)=\mathbf{E}\,Y_0\leqslant\liminf\mathbf{E}\,Y_q'\leqslant\mathbf{E}\,Y_0'$, and evidently $\mathbf{E}\,Y_0'+\mathbf{E}Z_0'\leqslant\operatorname{Var}(X)$. The inequality $\operatorname{Var}(X)\leqslant\operatorname{Var}(Y)+\operatorname{Var}(Z)$ is immediate from the definition of Var $(\cdot)$.

Finally, if $X=\bar{Y}-\bar{Z}$ is another representation of X as the difference of two nonnegative supermartingales, then for $s<t$,

$$\begin{aligned}\mathbf{E}[\bar{Y}_s-\bar{Y}_t|\mathcal{F}_s]&=\mathbf{E}[X_s-X_t|\mathcal{F}_s]+\mathbf{E}[\bar{Z}_s-\bar{Z}_t|\mathcal{F}_s]\\&\geqslant\mathbf{E}[X_s-X_t|\mathcal{F}_s].\end{aligned}$$

Hence, since Y is also a supermartingale,

$$\mathbf{E}[\bar{Y}_s-\bar{Y}_t|\mathcal{F}_s]\geqslant\mathbf{E}[X_s-X_t|\mathcal{F}_s]^+.$$

Thus for any dissection τ containing s and t,

$$\mathbf{E}[\bar{Y}_s-\bar{Y}_t|\mathcal{F}_s]\geqslant\mathbf{E}[Y_s^\tau-Y_t^\tau|\mathcal{F}_s],$$

and so

$$\mathbf{E}[\bar{Y}_s-\bar{Y}_t|\mathcal{F}_s]\geqslant\mathbf{E}[Y_s'-Y_t'|\mathcal{F}_s].$$

Thus $\bar{Y}-Y'$ is a nonnegative supermartingale, and since

$$(\bar{Y}-Y)_t=\liminf_{q\downarrow\downarrow t,\ q\in\mathbb{Q}}(\bar{Y}-Y')_q\quad\text{a.s.},$$

the process $\bar{Y}-Y$ is also a nonnegative supermartingale. □

(*41.6*) *Remarks.* If the quasimartingale X is an L^1-bounded martingale, then the Rao decomposition (41.4) of X is the same as the Krickeberg decomposition.

8. ITÔ EXCURSION THEORY

42. Introduction. Let X be a Markov process (with R-paths), and let K be a compact subset of the state-space of X. Then

(42.1) $$M\equiv\{t: X_t\in K \text{ or } X_{t-}\in K\}$$

defines a random closed set, the complement of which may be written as a countable disjoint union of open intervals, called *excursion intervals.* The central theme of excursion theory is the analysis of the random closed set M and the behaviour of X on the excursion intervals.

To cover applications to chains, etc., one should ideally take X to be a Ray process with compact metric state–space F. Though we shall formulate and prove results in that context, we would not wish to force you to study Ray processes (fun though they are—see Part 5 of Chapter III). If you wish, you can pretend that X is a nice FD process; every point of the state–space of X is then 'extremal' in Ray terms—there are no branch-points.

(*42.2*) *Some history, references, advice, etc.* The first explicit appearance of formal excursion theory was in Itô [6], where K was taken to be a singleton. Itô's discoveries that the excursions of X over different excursion intervals are independent and identically distributed (in a sense soon to be made precise), together with his Poisson-point-process picture of excursions, remain the most powerful computational techniques.

It must be emphasized, however, that the Poisson-point-process picture of excursions was very clear to Lévy at an intuitive level, and runs throughout much of his work (Lévy [3, 4, 5]) on chains. Just how fully understood the picture was in chain theory can be seen from the very penetrating analysis made in 1954 by Kendall and Reuter [1] of some famous 'pathological' chain examples due to Kolmogorov. As explained in § III.57, Neveu's work [2, 3, 4, 5] essentially completed our probabilistic understanding of excursion theory for chains. Kingman's work [3] on *regenerative phenomena* showed that the idea of concentrating on the random closed set M, where K is a singleton or a finite set, could eventually lead to a more-or-less complete description of analytic properties of transition functions. Excursion ideas, especially applied via the amazing branching procedures of D. G. Kendall [4], helped provide a proof (Williams [9]) of Theorem III.55.1 on Q-matrices.

The 'Poisson' approach seems to have come rather late to Brownian motion. There has, however, been very extensive study of 'excursions straddling fixed times'. Of course, it was Lévy who led the way—see Lévy [2] and § 2.9 of Itô and McKean [1]. For very valuable later work, see Chung [3] and Vervaat [1]. The description of the Itô excursion law for Brownian motion in the original Volume 1 will be proved at § 55 below.

An exceptionally attractive approach to excursions from a single point may be found in Greenwood and Pitman [1], and one of the most important applications may be found in their paper [2].

Excursion theory is very closely related to *last-exit decompositions* and to *Lévy systems.* Again, it was in chain theory that last-exit decompositions were first developed—see Chung [1], Neveu [2, 3, 4, 5] and Kingman [3]. For more general processes, Getoor and Sharpe [1, 2, 3, 4] are the key references, and

Sharpe [1] is sure to be an authoritative account. Lévy systems were mentioned in § 28, and references were given there. See also references at the end of § 55.

A theoretical synthesis of various approaches and of the theory of *incursions* was achieved by Maisonneuve [1] and Maisonneuve and Meyer [1].

We are here concerned only with getting you started via discussion of the case when K is a singleton: $K=\{a\}$.

You should then study what happens for general K. The abstract theory is covered by the work of Maisonneuve and Meyer mentioned above. With extra structure, a very rich harvest of results becomes available: see, for example, Motoo's astonishing paper [1], Section V.6 of Ikeda and Watanabe [1], and Hsu [1]. Also, you must read Silverstein's profound analysis (Silverstein [1, 2]) for symmetrizable processes via the theory of Dirichlet forms.

43. Excursion theory for a finite Markov chain. To motivate and illustrate excursion theory, we begin by seeing what it has to say for a finite Markov chain.

Let X be an irreducible Markov chain on the finite state–space I, with Q-matrix Q. For any $a \in I$, we define the (local) time at a by

$$L^a(t) \equiv \int_0^t I_{\{X_s=a\}}\,ds,$$

with the corresponding inverse process

$$\gamma_t^a \equiv \inf\{u>0\colon L_u^a>t\}.$$

If $H_b \equiv \inf\{t>0\colon X_t=b\}$, what can we say about the $\mathbf{P}^a$-law of $L^a(H_b)$, the time spent in a before the chain reaches b, or of $L^b(\gamma_t^a)$, the time spent in b before time t has been spent in a? The excursion standpoint allows us to say already that, for some positive α, β, for all $\lambda>0$,

$$\mathbf{E}^a \exp(-\lambda L^a(H_b)) = \alpha(\alpha+\lambda)^{-1}, \tag{43.1}$$

$$\mathbf{E}^a \exp(-\lambda L^b(\gamma_t^a)) = \exp[-\alpha t \lambda(\beta+\lambda)^{-1}]. \tag{43.2}$$

We begin the explanation of (43.1)–(43.2) with a simple result, announced in § III.57.

(*43.3*) PROPOSITION. *Under* $\mathbf{P}^a$, γ^a *is a subordinator.*

Proof. For each $t \geqslant 0$, γ_t^a is a finite stopping time and $X(\gamma_t^a)=a$. Since L^a is a continuous additive functional, $L^a(\gamma_t^a)=t$ for all t, and for all $s, t \geqslant 0$,

$$\gamma_{t+s}^a(\omega) - \gamma_t^a(\omega) = \gamma_s^a(\theta_{\gamma_t^a}\omega),$$

where θ_t is the usual shift operator. Hence by the strong Markov property, for each $\lambda \in \mathbb{R}$,

$$\mathbf{E}^a[\exp\{i\lambda(\gamma_{t+s}^a - \gamma_t^a)\} \mid \mathcal{F}(\gamma_t^a)] = \mathbf{E}^a \exp\{i\lambda\gamma_s^a\}, \quad \text{a.s.}. \qquad \square$$

(*43.4*) **Exercise.** Let

(43.5) $$R(\lambda)=\{r_{ij}(\lambda)\}=(\lambda-Q)^{-1},$$

as usual. Show that $f(X_t)\exp(-\lambda t+\beta_\lambda L_t^a)$ is a martingale, where $\beta_\lambda=r_{aa}(\lambda)^{-1}$, and $f.\equiv r_{\cdot a}(\lambda)$. Hence deduce that γ^a is a subordinator under $\mathbf{P}^a$, and that β_λ is the *Laplace exponent* of γ^a in the sense that

$$\mathbf{E}^a\exp(-\lambda\gamma_t^a)=\exp(-t\beta_\lambda).$$

This provides another proof of (III.57.4). □

Now let $T_0=0$, and define inductively the stopping times T_n, S_n by

$$S_n\equiv\inf\{t>T_{n-1}: X_t\neq a\},\quad T_n\equiv\inf\{t>S_n: X_t=a\},\quad n\geqslant 1.$$

The open intervals (S_n, T_n) are the excursion intervals, and the processes

$$\{\xi_t^n: t\geqslant 0\}\equiv\{X((t+S_n)\wedge T_n): t\geqslant 0\}$$

are independent and identically distributed, by the strong Markov property; indeed, the processes $\{\eta_t^n: t\geqslant 0\}\equiv\{X((t+T_{n-1})\wedge T_n): t\geqslant 0\}$ are i.i.d., and ξ^n is a function of η^n. We call ξ^n the *n*th *excursion*, and we think of it as a *point in the excursion space*:

$$U\equiv\{\text{R-paths } f: \mathbb{R}^+\to I \text{ such that } f(t)=a\Rightarrow f(s)=a \text{ for } s>t\}.$$

We can be quite explicit about the law of ξ^n; by §IV.21.4,

(43.6) $$\mathbf{P}^a(\xi_0^1=b)=q_{ab}/q_a\quad(b\neq a),$$

and, since S_n is a stopping time, the law of ξ^n is simply the law of the chain started with the distribution (43.6) and stopped when it first reaches a. Moreover, the ξ^n are independent of the holding times S_n-T_{n-1}.

A picture illustrates these points (Figure VI.3). If we think of I as $\{0, 1, \ldots, N\}$ for some N, with $a=0$, the lower graph is the path of $X+L^a$ turned on its side, and the upper graph represents the (excursion) point process $(L^a(S_n), \xi^n)$.

Let μ denote the law of ξ^1. Almost every useful application of excursion theory is based on the following result, or its generalizations.

(*43.7*) THEOREM. *Under* $\mathbf{P}^a$, *the point process* $\{(L^a(S_n), \xi^n): n\geqslant 1\}$ *is a Poisson point process in* $\mathbb{R}^+\times U$ *with expectation measure* $q_a dt\times\mu=dt\times n$, *where* $n\equiv q_a\mu$.

Proof. Abbreviate $L^a(S_n)$ to τ_n, and let N denote the random measure on $\mathbb{R}^+\times U$ derived from $\{(\tau_n, \xi^n): n\geqslant 1\}$:

$$N(A)=\sum_{n\geqslant 1} I_A(\tau_n, \xi^n).$$

According to the construction of Poisson random measures in §II.37, it is sufficient to prove that, for each fixed $K>0$:

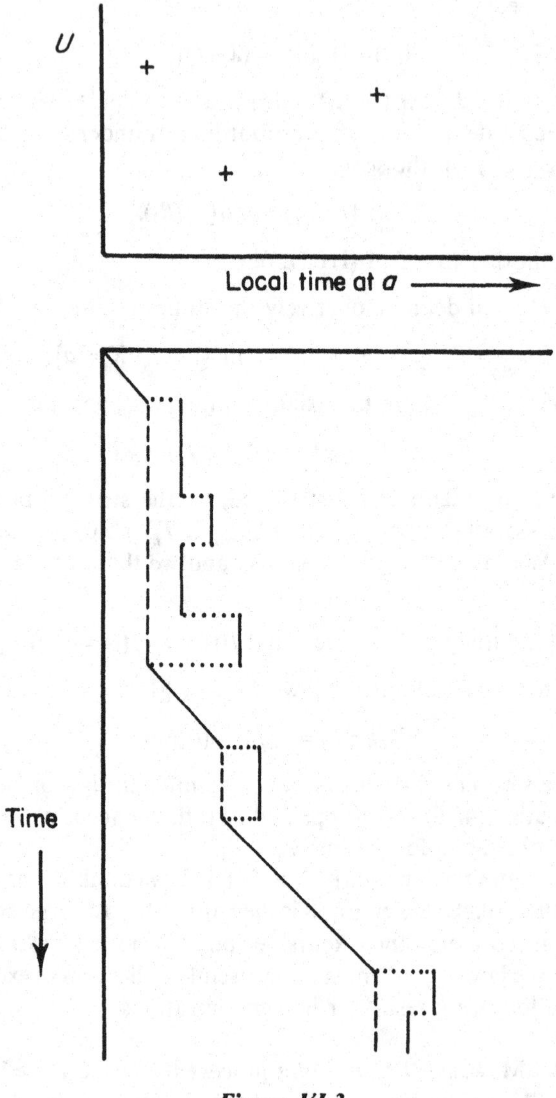

Figure VI.3

(i) $N([0, K] \times U)$ has a Poisson distribution of mean $q_a K$.
(ii) Given that $N([0, K] \times U) = k$, the random set

$$\{(\tau_n, \xi^n): n = 1, \ldots, k\}$$

of k points has the same law as that of a set of k points independently distributed in $[0, K] \times U$ with common law

$$K^{-1} dt \times \mu.$$

The first of these follows immediately from the fact that $\tau_n - \tau_{n-1}$ are independent exponentials of mean q_a^{-1}. As for the second, it is an elementary fact that, given $N([0, K] \times U) = k$, the law of $\tau_1, \ldots, \tau_k$ is the same as the law of the order statistics of a sample of size k with the uniform distribution over $[0, K]$. Since the ξ^n are independent of the τ_n, the second condition also follows. □

(*43.8*) *Remarks.* Look back at Neveu's Theorem III.57.8. We have now proved for a finite chain (writing a here in place of b in §,III.57) that the excursion measure n has the Markovian form stated in Theorem III.57.8, with

(43.9)
$$g_{aj}(t) \equiv \sum_{i \neq a} q_{ai}\, {}_a p_{ij}(t) \quad (j \neq a).$$

Statements III.57.9–12 follow directly from this, and to complete the proof of Neveu's Theorem for finite chains, we have just got to verify (III.56.1), the relation by which $g_{aj}(\bullet)$ was defined in §§ 56–57.

(*43.10*) **Exercise.** Confirm that $g_{aj}(\bullet)$ defined by (43.9) satisfies the *last-exit decomposition* formula (III.56.1):

(43.11)
$$p_{aj}(t) = \int_0^t p_{aa}(s) g_{aj}(t-s)\, ds \quad (j \neq a).$$

Later, in § 50, we shall give an excursion proof of this. (*Hint*: Use Laplace transforms and the formula (43.5).) □

Now we show how to deduce results (43.1–2) from Theorem 43.7 using only simple properties of Poisson processes. At the end of the section, we give a much quicker proof, still using excursion ideas.

Here is the proof of (43.1). If $B \subset U$, and if N denotes the Poisson process $\{(\tau_n, \xi^n): n \geqslant 1\}$ as before, then:

$$\mathbf{P}^a(\inf\{t: N([0, t] \times B) > 0\} > s) = \mathbf{P}^a(N([0, s] \times B) = 0)$$
$$= \exp\{-q_a s \mu(B)\}$$

since the number of points in $[0, s] \times B$ is a Poisson random variable with mean $(q_a dt \times \mu)([0, s] \times B) = q_a s \mu(B)$. But if we take

(43.12)
$$B \equiv \{f \in U : f(t) = b \text{ for some } t\},$$

then:

$$L^a(H_b) = \inf\{t: N([0, t] \times B) > 0\}.$$

(Look at Figure VI.3 and see what the two random variables are!) Hence $L^a(H_b)$ has an exponential distribution with parameter $\alpha = q_a \mu(B)$, which is what (43.1) says.

To prove (43.2), notice that

$$L^b(\gamma_t^a) = \sum_{n \geq 1} \{L^b(T_n) - L^b(S_n)\} I_{\{L_a(S_n) < t\}}. \tag{43.13}$$

Now the random variables

$$L^b(T_n) - L^b(S_n) \equiv \int_0^\infty I_{\{b\}}(\xi_s^n) ds$$

are independent identically distributed, since the ξ^n are. Moreover, they are independent of the events

$$\{L^a(S_n) < t\} = \left\{ \sum_{j=1}^n (S_j - T_{j-1}) < t \right\}$$

which are determined by the holding times $S_j - T_{j-1}$ in a. Since

$$L^a(S_n) < t \text{ if and only if } N([0,t] \times U) \geq n, \quad \text{a.s.,}$$

$L^b(\gamma_t^a)$ is the sum of $N([0,t] \times U)$ independent random variables each with the law of $L^b(T_1) - L^b(S_1)$. However, we can obtain the law of $\int_0^\infty I_{\{b\}}(\xi_s^n) ds$ explicitly:

$$\begin{aligned} \mathbf{E}\exp\left[-\lambda \int_0^\infty I_{\{b\}}(\xi_s^n) ds \right] &= \mathbf{E}\left[I_{\{\xi \notin B\}} + I_{\{\xi \in B\}} \exp\left(-\lambda \int_0^\infty I_{\{b\}}(\xi_s^n) ds \right) \right] \\ &= \mu(B^c) + \mu(B) \mathbf{E}^b \exp(-\lambda L^b(H_a)) \end{aligned}$$

where $\mathbf{E}$ denotes integration over U with the measure μ. The last equality follows from the fact that ξ behaves like the chain X stopped when it reaches a; we use the strong Markov property at H_b. But (43.1) says

$$\mathbf{E}^a \exp(-\lambda L^a(H_b)) = \alpha/(\alpha + \lambda);$$

reversing the rôles of a and b, we obtain for some $\beta > 0$,

$$\mathbf{E}^b \exp(-\lambda L^b(H_a)) = \beta/(\beta + \lambda).$$

Hence, since $\alpha = q_a \mu(B)$,

$$\mathbf{E}\exp\left(-\lambda \int_0^\infty I_{\{b\}}(\xi_s^n)\, ds \right) = 1 - \frac{\alpha}{q_a} \cdot \frac{\lambda}{\beta + \lambda}.$$

Since $L^b(\gamma_t^a)$ is the sum of a Poisson number (with mean $q_a t$) of random variables with this law,

$$\begin{aligned} \mathbf{E}^a \exp(-\lambda L^b(\gamma_t^a)) &= \sum_{n \geq 0} \frac{(q_a t)^n}{n!} e^{-q_a t} \left[1 - \frac{\alpha}{q_a} \cdot \frac{\lambda}{\beta + \lambda} \right]^n \\ &= \exp(-\alpha t \lambda (\beta + \lambda)^{-1}), \end{aligned}$$

as stated.

Though these arguments may take a little time to write out in complete detail, the essence of excursion ideas is contained in Figure VI.3; we calculate the laws of functionals of the process X by translating them into equivalent functionals of the Poisson process of excursions.

(*43.14*) *Evaluation of the constants* α *and* β. Let $J \equiv I \setminus \{a, b\}$, and partition Q as follows:

$$Q = \begin{pmatrix} -q_a & u & \rho \\ x & D & y \\ \sigma & v & -q_b \end{pmatrix} \quad \begin{matrix} \{a\} \\ J \\ \{b\} \end{matrix}$$

Then $\alpha^{-1} = \mathbf{E}^a L^a(H_b)$. If we stop the chain when it hits b, we modify Q to

$$\tilde{Q} = \begin{pmatrix} -q_a & u & \rho \\ x & D & y \\ 0 & 0 & 0 \end{pmatrix}$$

and we have the well-known Green's function formula:

$$\begin{aligned} \alpha^{-1} = \mathbf{E}^a L^a(H_b) &= \lim_{\lambda \downarrow 0} \tilde{R}_\lambda I_{\{a\}}(a) \\ &= \lim_{\lambda \downarrow 0} (\lambda - \tilde{Q})^{-1}(a, a) \\ &= (q_a + uD^{-1}x)^{-1}. \end{aligned}$$

Thus $\alpha = q_a + uD^{-1}x$ and similarly $\beta = q_b + vD^{-1}y$. □

(*43.15*) *Quick proof of* (*43.1*) *and* (*43.2*). All we need to say is that the chain on the set $\{a, b\}$ obtained by observing X in the clock $L_a + L_b$ has Q-matrix:

$$\begin{pmatrix} -\alpha & \alpha \\ \beta & -\beta \end{pmatrix}.$$

Thus, by the Lévy–Hincin formula (II.64.1), the subordinator $\{L^b(\gamma_t^a) \colon t \geqslant 0\}$, which has jumps of exponential size (of mean β^{-1}) occurring at rate α, satisfies

$$\begin{aligned} \mathbf{E}^a \exp\{-\lambda L^b(\gamma_t^a)\} &= \exp\left\{-t \int (1 - e^{-\lambda c}) \alpha \beta e^{-\beta c} dc\right\} \\ &= \exp\{-\alpha t \lambda / (\beta + \lambda)\}. \end{aligned}$$ □

44. Taking stock. In developing excursion theory from state a for a finite Markov chain, we found a PCHAF L^a whose set of growth points is M; we showed that γ^a, the inverse to L^a, is a subordinator; and we showed that the point process of excursions, with values in $\mathbb{R}^+ \times U$, is a Poisson process. The aim now is to develop the same ideas in a more general context. Parts of the programme (notably the existence and properties of L^a) are technically more difficult, but the lines of the argument are exactly the same.

45. Local time L at a regular extremal point a. Let X be an honest Ray process (see remarks at the beginning of §42 on how to feel at ease with the 'Ray' hypothesis) with compact metric state–space K with metric d. Let a denote a point of K on which we fix our attention. Let M denote the random closed set

$$M \equiv \{t \geqslant 0: X_t = a \text{ or } X_{t-} = a\},$$

and let

$$H \equiv \inf\{t > 0: t \in M\}.$$

Notice that if $t_n \downarrow\downarrow t$, $t_n \in M$, then $X_t = a$, by right continuity of the paths of X. Here are two properties of H, the non-trivial second of which uses the 'Ray' analogue (III.41.3) of the Blumenthal quasi-left-continuity property (III.9).

(*45.1*) THEOREM. (i) *H is a stopping time.*

(ii) *Suppose that a is extreme, so that:*

$$\mathbf{P}_0(a, \{a\}) = 1.$$

Then $X(H) = a$ on $\{H < \infty\}$, a.s.

Proof. (i) Rather than appeal to general theorems, we give a bare-hands proof for later use. For $\varepsilon > 0$, let

$$H^\varepsilon \equiv \inf\{t \geqslant \varepsilon: X_t = a \text{ or } X_{t-} = a\},$$

and let

$$H_n^\varepsilon = \inf\{t \geqslant \varepsilon: d(X_t, a) < n^{-1}\}.$$

Then, from Lemma II.74.4, each H_n^ε is a stopping time, and if we can prove that each H^ε is a stopping time, then it will follow that $H = \downarrow \lim_{\varepsilon \downarrow 0} H^\varepsilon$ is another. We shall prove that $H_n^\varepsilon \uparrow H^\varepsilon$. Clearly the H_n^ε increase to some limit $S(\leqslant H^\varepsilon)$ and for each real t,

$$\begin{aligned} S < t < H^\varepsilon &\Rightarrow \varepsilon \leqslant H_n^\varepsilon < t < H \text{ for all } n \\ &\Rightarrow \text{there exists } t_n \in [\varepsilon, t) \text{ such that } d(X(t_n), a) < n^{-1} \\ &\Rightarrow \text{there exists } s \in [\varepsilon, t] \text{ such that either } X_s = a \text{ or } X_{s-} = a \\ &\Rightarrow H^\varepsilon \leqslant t, \end{aligned}$$

a contradiction.

(ii) Certainly, $X(H) = a$ on $\{H = 0\}$. Now, for $\varepsilon > 0$, $H = H^\varepsilon$ on $B = \{\varepsilon < H < \infty\}$, and either $X(H^\varepsilon) = a$, when there is nothing more to be done, or else $X(H^\varepsilon -) = a$. Now it is not hard to prove that on B, $X(H^\varepsilon -) = a$ if and only if $H_n^\varepsilon \uparrow\uparrow H^\varepsilon$. So, defining

$$A = \{\varepsilon < H < \infty, X(H^\varepsilon -) = a\} = \{\varepsilon < H < \infty, H_n^\varepsilon < H^\varepsilon \text{ for all } n\},$$

and letting $A_n \equiv \{\varepsilon \leqslant H_n^\varepsilon < H\}$, we find that $T_n' \equiv (H_n^\varepsilon)_{A_n}$ are stopping times increasing to $(H^\varepsilon)_A$. Thus, if we fix $N \in \mathbb{N}$, $N > \varepsilon$, and define $T_n = T_n' \wedge (N - n^{-1})$, then we obtain $T_n \uparrow\uparrow T \equiv (H^\varepsilon)_A \wedge N$. Theorem III.41.3 now shows that, for each $x \in F$,

$$\mathbf{P}^x(X_T = a \mid \mathscr{F}_{T-}) = \mathbf{P}_0(X_{T-}, \{a\}).$$

The event $A \cap \{H^\varepsilon < N\} = \{T < N\}$ is an element of $\mathscr{F}_{T-}$, and on this event, $X(T-) = a$. Since a is extreme, $X_T = a$ a.s. on $A \cap \{H^\varepsilon < N\}$. Now let $N \to \infty$, then let $\varepsilon \downarrow 0$. □

Blumenthal's 01 law (Theorem III.9) extends to extreme points of Ray processes, so, if the point a is extreme, then:

either $\mathbf{P}^a(H=0)=1$ (in which case we say that a is *regular*) *or* $\mathbf{P}^a(H=0)=0$.

Until further notice, we make the following assumption.

(*45.2*) ASSUMPTION. *The point a is a regular extreme point.*

Almost all examples of interest (for excursion theory for a singleton!) satisfy this assumption. In § 50, we shall discuss the (much simpler) situation where it fails.

(*45.3*) *The* 1-*potential* ψ *of* $\{a\}$. We define

$$\psi(x) \equiv \mathbf{E}^x[e^{-H}] \quad (x \in F)$$

and assume that ψ is not identically zero.

From the Markov property at time $t > 0$, it follows that for $x \in F$,

$$\begin{aligned} e^{-t} P_t \psi(x) &= \mathbf{E}^x[\exp(-\inf\{u > t \colon X_u = a \text{ or } X_{u-} = a\})] \\ &\uparrow \psi(x) \quad \text{as } t \downarrow 0. \end{aligned}$$

Thus ψ is 1-excessive. But more than this is true; following (V.3.13) of Blumenthal and Getoor [1], we show below that ψ is *uniformly* 1-excessive (see § III.16). The significance of this is, of course, that we can use Theorem III.16.7 to deduce that ψ *is the* 1-*potential of some PCHAF*, which turns out to be the local time at a.

(*45.4*) THEOREM. *Suppose that a is a regular extreme point. Then the* 1-*excessive function ψ is uniformly* 1-*excessive.*

Proof. For any stopping time T, define P_T^1 by

$$P_T^1 f(x) \equiv \mathbf{E}^x[e^{-T} f(X_T)]$$

for bounded Borel f. Since a is regular, $\psi(a) = 1$, and by Theorem 45.1, we know that $X(H) = a$ on $\{H < \infty\}$, a.s. Thus

$$P_H^1 \psi(x) = \mathbf{E}^x[e^{-H} \psi(X(H))] = \psi(x).$$

Given $\varepsilon > 0$, there exists $\delta > 0$ such that, for $t < \delta$,

$$\psi(a) - \varepsilon \leqslant e^{-t} P_t \psi(a) \leqslant \psi(a)$$

since ψ is 1-excessive. Now, the strong Markov property shows that $P^1_{H+t} = P^1_H P^1_t$ (see § III.10.6), and, since $P^1_H(x, \cdot)$ is concentrated on a, we find that, for $t \leqslant \delta$,

$$\psi(x) = P^1_H \psi(x) \leqslant P^1_H(e^{-t} P_t \psi + \varepsilon)(x) \leqslant P^1_{H+t} \psi(x) + \varepsilon.$$

But, by the strong Markov property applied at time $T \equiv H + t$,

$$P^1_T \psi(x) = \mathbf{E}^x[\exp(-\inf\{u > T: X_u = a \text{ or } X_{u-} = a\})] \leqslant P^1_t \psi(x),$$

and hence ψ is uniformly 1-excessive. □

(*45.5*) *Existence of L.* From Theorems 45.4 and § III.16.7, there exists a unique PCHAF L whose 1-potential is ψ:

$$\psi(x) = \mathbf{E}^x \int_0^\infty e^{-s} dL_s \text{ for every } x.$$

We call L the *local time of X at a in the Blumenthal–Getoor normalization.* We shall show that, almost surely, *the set of points of increase of L is exactly the closed random set M*, justifying the name 'local time at a'. We shall also show that a PCHAF which grows only on M must be a multiple of L, and (abusing the terminology slightly) we shall frequently also call a multiple of L a (or the!) local time of X at a; the precise normalization is rarely important, as you will have gathered. In particular, if $A_t \equiv \int_0^t I_{\{a\}}(X_s)ds$ is not identically zero, then A is a PCHAF growing only on M and therefore is a multiple of L; if the process spends a positive time in a, then the local time is simply the true time spent in a. The example of Brownian motion on $\mathbb{R}$ shows that this does not always happen.

(*45.6*) *Important technical note.* A technical difficulty mentioned in § III.33 haunts us here, and in connection with (45.9) below. It is discussed in § 46.

(*45.7*) *The growth set of L.* Let

$$J_+ \equiv \{t \geqslant 0: \text{for all } h > 0,\ L_{t+h} > L_t\},$$

$$J_- \equiv \{t > 0: \text{for all } 0 < h < t,\ L_{t-h} < L_t\}$$

denote the set of points of right and left increase (respectively) of L, and let $J = J_+ \cup J_-$ denote the set of points of increase. The set J is closed. Similarly, we let $M_+ \equiv \{t \geqslant 0: X_t = a\}$, $M_- \equiv \{t > 0: X_{t-} = a\}$.

(*45.8*) THEOREM. *For each* $\mu \in \Pr(F)$,

$$\mathbf{P}^\mu[J = M] = \mathbf{P}^\mu[J_+ \subseteq M_+] = 1.$$

Proof. The random variable $C_\infty \equiv \int_0^\infty e^{-s} dL_s$ is in $L^2(\mathbf{P}^\mu)$ (see the proof of Theorem III.16.7) so

(45.9) $$C_t \equiv \mathbf{E}^\mu[C_\infty | \mathscr{F}_t] = \int_0^t e^{-s} dL_s + e^{-t}\psi(X_t)$$

is an L^2-bounded martingale. (Re-read Note 45.6.) The result follows from four inclusions.

(i) $J_- \subseteq \bar{J}_+$, which is an easy piece of analysis.

(ii) $M_- \subseteq \bar{M}_+$, a.s.. For this,

$$\{M_- \not\subseteq \bar{M}_+\} \subseteq \bigcup_{p<q} \{X_t \neq a \text{ for } p \leqslant t < q,\ X_{s-} = a \text{ for some } p < s < q\}$$

$$\subseteq \bigcup_{p<q} \{H \circ \theta_p < q-p,\ (\theta_p X)_t \neq a \text{ for all } 0 \leqslant t < q-p\},$$

where p and q run through $\mathbb{Q}$. Theorem 45.1 implies that each of this countable collection of sets is null.

(iii) $J_+ \subseteq M_+$, a.s.. Notice that each point of J_+ is a limit point from the right of points of J_+. Therefore

$$\{J_+ \not\subseteq M_+\} \subseteq \bigcup_{p<q} \{J_+ \cap [p,q] \neq \varnothing,\quad M_+ \cap [p,q] = \varnothing\}$$

$$\subseteq \bigcup_{p<q} \{L_q > L_p,\ H \circ \theta_p \geqslant q-p\},$$

where p and q run through $\mathbb{Q}$. Each of the sets of this last countable collection is null. Indeed, for any $x \in F$,

$$C_0 = \psi(x) = \mathbf{E}^x\left[\int_0^H e^{-s} dL_s + e^{-H}\right],$$

since $\psi(X_H) = \psi(a) = 1$ a.s. on $\{H < \infty\}$, by Theorem 45.1 and regularity of a; moreover, $\psi(x) = \mathbf{E}^x(e^{-H})$, so $\mathbf{P}^x(L_H = 0) = 1$.

(iv) $M_+ \subseteq \bar{J}_+$, a.s.. Let $T \equiv \inf\{t > 0\colon L_t > 0\}$. Then

$$1 = \mathbf{E}^a C_T = \mathbf{E}^a[e^{-T}\psi(X_T)],$$

so $\mathbf{P}^a(T=0) = 1$. Hence

$$\{M_+ \not\subseteq \bar{J}_+\} \subseteq \bigcup_{p<q} \{L_p = L_q,\ X_t = a \text{ for some } p < t < q\},$$

a countable union of null sets. □

(*45.10*) PROPOSITION. *If A is a PCHAF which grows only on M, then for some* $\alpha > 0$, $\mathbf{P}^\mu(A_t = \alpha L_t \text{ for all } t) = 1$ *for each* $\mu \in \Pr(F)$.

Proof. Since A is constant on $[t, H\circ\theta_t]$, the martingale C' defined by $C'_t \equiv \mathbf{E}^\mu\left[\int_0^\infty e^{-s}dA_s \,|\, \mathscr{F}_t\right]$ has the form

$$C'_t = \int_0^t e^{-s}dA_s + \alpha e^{-t}\psi(X_t),$$

where $\alpha \equiv \mathbf{E}^a\left[\int_0^\infty e^{-s}dA_s\right]$. Hence $C' - \alpha C$ is a continuous finite-variation local martingale null at zero, therefore null. □

In particular, $\psi_\lambda(x) \equiv \mathbf{E}^x[e^{-\lambda H}]$ is uniformly λ-excessive and so is the λ-potential of some PCHAF L_λ which grows exactly on M. Proposition 45.10 says that $L_\lambda = c_\lambda L$, for some c.

(*45.11*) **Exercise.** Prove that $c_\lambda = \left[\mathbf{E}^a\left(\int_0^\infty e^{-\lambda s}dL_s\right)\right]^{-1}$.

(*45.12*) *Remark.* This section obviously has very close connections with § 35, in which, you may remember, we also encountered a technical difficulty 'to be discussed later'.

46. Some technical points; hypothèses droites, etc. The points which we are now going to discuss were the original *raison d'être* for the general theory, and are fundamental for a rigorous theory of Markov processes. Let us see what the general theory has to say about the last section, thereby treating in some detail a point which has been rather glossed over at earlier stages (including in the proof of the Volkonskii–Šur–Meyer theorem in § III.17).

Look at the definition (45.3):

$$\psi(x) \equiv \mathbf{E}^x[e^{-H}].$$

What can we say about the function ψ? Well, we know from (III.9.8) that ψ is universally measurable on F, and so such expressions as $P_t\psi$ are at least well-defined. Now there are special reasons why we can say that in the particular case which we are now studying, ψ is Borel (see below). However, in many important applications in Markov process theory, we cannot guarantee the Borel property, so that we cannot then say without further deep analysis that $\psi(X_t)$ is even optional.

But let us bypass that problem by explaining why our ψ *is* Borel. The point is that since we have defined H as

$$H \equiv \inf\{t > 0 : X_{t-} = a \text{ or } X_t = a\},$$

then, as we saw in the proof of part (i) of Theorem 45.1,

$$H = \downarrow\lim_{\varepsilon\downarrow 0} H^{\varepsilon}, \quad H^{\varepsilon} = \uparrow\lim H^{\varepsilon}_{n},$$

$$\{H^{\varepsilon}_{n} < t\} = \bigcup_{\varepsilon \leqslant q < t} \{d(a, X_q) < n^{-1}\} \quad (q \text{ denoting a rational}),$$

so that H is measurable on the *uncompleted* σ-algebra $\mathscr{F}^{\circ}$. Hence, by a standard monotone-class argument (see § III.9 for the type of thing),

(46.1) *ψ is a Borel function on F.*

Now consider the process V, where

(46.2) $$V_t = \mathrm{e}^{-t}\psi(X_t).$$

Because ψ is 1-excessive, V is a supermartingale. In the proof of part (iv) of Theorem 45.8, we use via the optional-sampling theorem the fact that V is *already right-continuous*—it is *not* enough to say that V has a right-continuous modification. It is, in fact, possible to rephrase the proof of part (iv) of Theorem 45.8 so as to avoid using right-continuity of V. However, the very existence of the local time L rests on arguments in § III.17 which used the same right-continuity property in an essential way. The treatment we now give for V can also be applied in § III.17, clinching the existence of L.

(*46.3*) *Proof that the process V is right-continuous.* Work with a fixed μ in $\Pr(F)$ and the corresponding set-up

(46.4) $$(\Omega, \mathscr{F}^{\mu}, \mathscr{F}^{\mu}_{t}, \mathbf{P}^{\mu})$$

which satisfies the usual conditions. Since ψ is Borel, the function $R_{\lambda+1}\psi$ is Borel, so that the process

(46.5) $t \mapsto \mathrm{e}^{-(\lambda+1)t} R_{\lambda+1}\psi(X_t)$ *is optional.*

We are going to use Theorem 6.3 to prove that the process at (46.5) is right-continuous.

By the strong Markov theorem, for any stopping time T, we have the Dynkin formula:

(46.6) $$\mathbf{E}\left[\int_0^{\infty} \mathrm{e}^{-(\lambda+1)s}\psi(X_s)\,ds \,\middle|\, \mathscr{F}_T\right] = \int_0^T \mathrm{e}^{-(\lambda+1)s}\psi(X_s)\,ds + \mathrm{e}^{-(\lambda+1)T} R_{\lambda+1}\psi(X_T), \quad \text{a.s.}$$

Let (T_n) be a sequence of stopping times with $T_n \downarrow T$. On writing T_n instead of T in (46.6) and letting $n \uparrow \infty$, we find (on using the convergence Theorem

II.51.1 for martingales filtering to the left for the left-hand side) that

$$e^{-(\lambda+1)T(n)}R_{\lambda+1}\psi(X_{T(n)}) \to e^{-(\lambda+1)T}R_{\lambda+1}\psi(X_T), \quad \text{a.s.}$$

Hence, by Theorem 6.3, the process at (46.5) is right-continuous.

Next, since ψ is uniformly 1-excessive,

$$\lambda R_{\lambda+1}\psi \uparrow \psi \textit{ uniformly on } F.$$

It now follows that $\psi(X_t)$ is right-continuous in t. □

(46.7) *Use of the Meyer decomposition theorem.* At several stages, we have commented on the close relationship between the Volkonskii–Šur–Meyer Theorem III.16.7 and the Meyer decomposition Theorem 29.7. Let us now examine the relationship in the present context.

The process V is now known to be a (bounded) right-continuous supermartingale (with limiting value 0 at $t = \infty$). The Meyer decomposition theorem allows us to conclude that there is a unique previsible increasing process A^μ and a UI martingale M such that

$$V = M - A^\mu,$$

and since $M_\infty = A^\mu_\infty$,

$$e^{-t}\psi(X_t) = \mathbf{E}^\mu[A^\mu_\infty | \mathscr{F}_t] - A^\mu_t.$$

This is exactly (45.9), so that

$$A^\mu_t = \int_0^t e^{-s} dL_s, \quad L_t = \int_0^t e^s dA^\mu_s.$$

One difficulty with the general theory approach is created by the need to show that A^μ can be chosen independently of μ. We do not discuss the resolution of this difficulty here. Sharpe [1] should be the definitive place to study it.

We know that L is continuous, and therefore so is A^μ. By Theorem 31.1, this corresponds exactly to the regularity property:

(46.8) *for any previsible stopping time* T, $\mathbf{E}V(T) = \mathbf{E}V(T-)$,

$$(\mathbf{E} \equiv \mathbf{E}^\mu).$$

We can establish this directly via the quasi-left-continuity Theorem III. 65.3, as a consequence of which we have

$$\mathbf{E}[V(T) | \mathscr{F}(T-)] = e^{-T}\mathbf{E}^{X(T-)}\psi(X_0).$$

The definition of ψ shows that even if $X(T-)$ is a branch-point,

$$\mathbf{E}^{X(T-)}\psi(X_0) = \psi(X_{T-}).$$

Property (46.8) now follows. □

We leave this topic with that indication of the fundamental nature in Markov process theory of the section theorems and of the the Meyer decomposition. For the theory of Meyer's *hypothèses droites* of right process, we refer to Getoor [1] and Sharpe [1], mentioning only that the property

$$t \mapsto f(X_t) \textit{ is right-continuous for any } \alpha\textit{-excessive } f$$

is the fundamental *hypothèse*.

47. The Poisson point process of excursions. We continue to mimic the development for finite chains.

(*47.1*) PROPOSITION. *The process* $\gamma_t \equiv \inf\{u>0: L_u>t\}$ *is a subordinator.*

Remark. We have to interpret 'subordinator' in a generalized sense: the Lévy measure may put positive mass on $\{\infty\}$, allowing a jump to ∞. This arises because L_∞ may be finite.

Proof. For each t, $\gamma_t \in J_+$, so by Theorem 45.8, $X(\gamma_t)=a$. Now the proof is exactly the same as in Proposition 43.3. □

(*47.2*) **Exercise.** For each $\lambda>0$, prove that if c_λ is defined as at (45.11), then $\psi_\lambda(X_t)\exp(-\lambda t+c_\lambda L_t)$ is a local martingale, and deduce from this that γ is a subordinator with Laplace exponent $\lambda \mapsto c_\lambda$.

(*47.3*) DEFINITION (excursion, excursion space U). *An* excursion *is an R-function* f: $\mathbb{R}^+ \to F$ *satisfying the coffin condition*:

$$f(t)=f(H)=a \quad \textit{for all } t \geqslant H,$$

where

$$H \equiv H(f) = \inf\{t>0: f(t)=a \text{ or } f(t-)=a\}.$$

The lifetime $H=H(f)$ *of the excursion* f *must be positive. We denote the set of all excursions by* U.

The Borel σ-algebra $\mathscr{B}(U)$ on U is the smallest σ-algebra which makes each evaluation map $f \mapsto f(t)$ measurable. [Technical note—see also §46. This σ-algebra is indeed the Borel σ-algebra of the topological space U if we take the Skorokhod metric (see Billingsley [2]) on U. It follows from the proof of part (i) of Theorem 45.1 that H is a Borel function on the Skorokhod space $D(\mathbb{R}^+, F)$ of all R-functions from $\mathbb{R}^+$ to F if one uses the Skorokhod metric on D. Hence,

$$U = \bigcap_{p \in \mathbb{Q}} (\{H(f) \geqslant p\} \cup \{H(f)<p, f(p)=a\})$$

is a Borel subset of D.]

(47.4) DEFINITION (point process of excursions). *The* point process of excursions from a *is defined to be*

(47.5) $$\Pi \equiv \{(t, e_t): \gamma_t \neq \gamma_{t-}\},$$

where $e_t \in U$ *is defined as follows:*

$$\begin{aligned} e_t(s) = &\quad X(\gamma_{t-}+s) \quad \text{for } 0 \leqslant s < \gamma_t - \gamma_{t-};\\ = &\quad a \qquad\qquad\quad \text{for } s \geqslant \gamma_t - \gamma_{t-}. \end{aligned}$$

For each t *such that* $\gamma_t \neq \gamma_{t-}$, *we call* e_t the excursion at local time t.

Notice that each e_t is in U: $L(\gamma_t) = L(\gamma_{t-})$, so if $\gamma_t > \gamma_{t-}$, then $J \cap (\gamma_{t-}, \gamma_t) = M \cap (\gamma_{t-}, \gamma_t) = \varnothing$, and e_t satisfies the coffin condition.

One can think of a point process as a $\mathbb{Z}^+$-valued random measure, and this is often notationally easier. We therefore define, for Borel $A \subseteq \mathbb{R}^{++} \times U$,

$$N(A) \equiv |A \cap \Pi|.$$

Let $U_\infty \equiv \{f \in U: H(f) = \infty\}$, $U_0 \equiv U \setminus U_\infty$. The following result is the heart of Itô excursion theory.

(47.6) THEOREM (Itô). *There exists a* σ*-finite measure* n *on* U *such that* $\delta \equiv n(U_\infty) < \infty$ *with the following property: if* N' *is a Poisson random measure on* $\mathbb{R}^{++} \times U$ *with expectation measure* $\mu = dt \times n$, *and if*

$$\zeta' = \inf\{t > 0: N'((0, t] \times U_\infty) > 0\},$$

then under $\mathbf{P}^a$,

$$N \stackrel{\mathcal{D}}{=} N'|_{(0, \zeta'] \times U}.$$

The measure n *is called the* characteristic measure *or* Itô excursion law *of the excursion process.*

Remark. Itô [6] assumed that $\delta = 0$; in this case, $\zeta' = \infty$ a.s., and the theorem says that the excursion point process N is a Poisson point process, exactly analogous to Theorem 43.7.

Proof. We shall assume that

$$\zeta \equiv \inf\{t: N((0, t] \times U_\infty) > 0\} = \sup_t L_t$$

is finite with positive probability (the case when ζ is almost surely infinite follows similar lines but is much easier). For $t > 0$, define the shifted point process $\theta_t N$ by

$$\theta_t N((s, u] \times B) = N((t+s, t+u] \times B)$$

for all $0 \leqslant s \leqslant u$, and Borel $B \subseteq U$. The essence of the proof is the remark that

(47.7) *under* $\mathbf{P}^a$, *conditional on* $\{\zeta > t\}$, *the process* $\theta_t N$ *is independent of* $N|_{(0, t] \times U}$ *and has law equal to the* $\mathbf{P}^a$*-distribution of* N.

This is simply the strong Markov property at γ_t; since $X(\gamma_t)=a$ a.s. on $\{\gamma_t<\infty\}=\{\zeta>t\}$, $\theta_{\gamma_t}X$ is independent of $\mathscr{F}(\gamma_t)$, with law $\mathbf{P}^a$ on $\{\gamma_t<\infty\}$. Hence

$$\mathbf{P}^a(\zeta>t+s)=\mathbf{P}^a(\zeta>t)\mathbf{P}^a(\zeta>s)$$

for all $t, s\geqslant 0$, and so $\mathbf{P}^a(\zeta>t)=e^{-\delta t}$ for some $\delta\geqslant 0$. Since ζ is not a.s. infinite, $\delta>0$.

Now we can define the measure n; for Borel $B\subseteq U_\infty$, we set

$$n(B)\equiv\delta\mathbf{P}^a(N(\mathbb{R}^+\times B)=1),$$

and for Borel $B\subseteq U_0$, we define

$$n(B)\equiv\delta(1-e^{-\delta})^{-1}\mathbf{E}^a[N((0,1]\times B)].$$

It will be sufficient to prove that for $0=t_0<t_1<\ldots<t_n=t$, for disjoint Borel $B_j\subseteq U_0$, for $B\subseteq U_\infty$ and for $k_{ij}\in\mathbb{Z}^+$,

(47.8) $$\mathbf{P}^a(\{N(A_{ij})=k_{ij}:\ i,j=1,\ldots,n\}\cap\{\zeta>t, N(\mathbb{R}^+\times B)=1\})$$
$$=\prod_{i,j=1}^{n}(\mu(A_{ij})^k e^{-\mu(A_{ij})}/k_{ij}!)e^{-\delta t}n(B)/n(U_\infty),$$

where $A_{ij}=(t_{i-1},t_i]\times B_j$. In fact, we may restrict the B_j to be subsets of $\{f\in U_0: H(f)>\varepsilon\}$; the advantage in doing so is that there can only be finitely many excursions of lifetime greater than ε in any finite (local time) interval, and so $N(A_{ij})$ is a.s. finite for all i, j.

For any $B\subseteq U_\infty$ and $t>0$,

$$\mathbf{P}^a\{\zeta>t, N(\mathbb{R}^+\times B)=1\}=e^{-\delta t}\mathbf{P}^a\{N(\mathbb{R}^+\times B)=1\}$$
$$=e^{-\delta t}n(B)/n(U_\infty).$$

Thus repeated application of the strong Markov property in the form (47.7) to the left side of (47.8) reduces it to

$$\prod_{i=1}^{n}\mathbf{P}^a(\{N((0,s_i]\times B_j)=k_{ij}: j=1,\ldots,n\}\cap\{\zeta>s_i\})n(B)/n(U_\infty)$$

where $s_i=t_i-t_{i-1}$.

But it is easy to deduce from (47.7) that, conditional on $\zeta>t$,

$$\{Z_s: 0\leqslant s\leqslant t\}\equiv\left\{\sum_{j=1}^{n}jN((0,s]\times B_j): 0\leqslant s\leqslant t\right\}$$

is a subordinator. Hence, by the Lévy–Itô decomposition in §II.37.4, the processes $N((0,s]\times B_j)$, $j=1,\ldots,n$, are independent Poisson processes of appropriate rate λ_j, which establishes (47.8) apart from identifying the means $\mu(A_{ij})=s_i\lambda_j$ of the Poisson distributions. But $\{N((0,s]\times B_j): s\geqslant 0\}$ is a Poisson counting process of rate λ_j killed at an independent exponential time ζ of rate δ. Therefore

$$\mathbf{E}^aN((0,1]\times B_j)=\delta^{-1}(1-e^{-\delta})\lambda_j,$$

and $\lambda_j=n(B_j)$. You can check that $\mathbf{E}(\zeta\wedge 1)=\delta^{-1}(1-e^{-\delta})$.

The σ-finiteness of n is now immediate from the fact that, for each $\varepsilon>0$, and with the notation $B_\varepsilon \equiv \{f \in U_0: H(f)>\varepsilon\}$, the *Poisson* variable $N((0, 1] \times B_\varepsilon)$ is finite. □

Note. We have already done enough for interesting applications. For example, you can now read § 51 on the Skorokhod embedding theorem. However, it seems best for order of presentation to complete the general theory now.

(*47.9*) *References.* See Itô [6] and Greenwood and Pitman [1].

48. Markovian character of n. Continue with the notation of §§ 45–47. Let ${}_a\mathbf{Q}^x$ denote the law of the stopped process $X^H \equiv X(t \wedge H)$. The following theorem makes precise the intuitive idea that excursions are Markovian.

We are not going to enter into the matter of which 'completions' of $\mathscr{F}_t^\circ$, etc., might be used. We use $\mathscr{F}_t$ for an 'appropriate' completion.

(*48.1*) THEOREM. *Let G be an open neighbourhood of a, and define*

$$\tau(f) \equiv \inf\{t>0: f(t) \notin G\} \quad (f \in U_0).$$

Then:

(i) $n(H>t)<\infty$ *for each* $t>0$.

(ii) $n(\tau<\infty)<\infty$.

(iii) *For any* $A \in \mathscr{F}_\tau$, $B \in \mathscr{F}$,

$$n(\{f: f \in A, \tau<\infty, \theta_\tau f \in B\}) = \int_F n(f \in A, \tau<\infty, f_\tau \in dx)\, {}_a\mathbf{Q}^x(B).$$

(iv) *For any* $t>0$, $A \in \mathscr{F}_t$, $B \in \mathscr{F}$,

$$n(\{f: f \in A, H>t, \theta_t f \in B\}) = \int_F n(f \in A, H>t, f_t \in dx)\, {}_a\mathbf{Q}^x(B).$$

Proof. (i)–(ii). We know that $n(U_\infty)<\infty$. Now, if $A \in \mathscr{B}(U_0)$, and $\zeta \equiv \inf\{t: N((0, t] \times U_\infty)>0\}$, then, because of the Poisson property, $n(A)<\infty$ if and only if $\mathbf{P}^a(N((0, t] \times A)<\infty | \zeta>t)=1$ for all $t>0$. But if $A=\{f: \tau(f)<\infty\}$, then given $\zeta>t$, $N((0, t] \times A)$ is the number of crossings from a to G^c before the finite time γ_t. Since X has R-paths, this number must be finite. Assertion (i) is proved similarly.

(iii)–(iv). If we define $\sigma_u \equiv \sup\{s \leqslant u: X_s=a \text{ or } X_{s-}=a\}$, then σ is an adapted R-process. Now fix $t>0$. If $T \equiv \inf\{s: s-\sigma_s>t\}$, then T is a stopping time, which occurs in the first excursion of duration greater than t, just when time t of that excursion has elapsed. By the Poisson-process picture, given that $T<\infty$, the law of the first excursion of duration greater than t is $n(\bullet \cap \{H>t\})/n(H>t)$, and so

$$\begin{aligned} &n(\{f: f \in A, \theta_t f \in B\} \cap \{H>t\})/n(H>t) \\ &= \mathbf{P}^a(C; (\theta_T X)^H \in B | T<\infty), \end{aligned}$$

where C is an obvious element of $\mathscr{F}_T$. Then result (iv) is exactly equivalent to the strong Markov property at T:

$$\mathbf{P}^a(C;\ (\theta_T X)^H \in B;\ T<\infty)$$
$$= \int_F \mathbf{P}^a(C;\ T<\infty;\ X_T \in dx)\ {}_a\mathbf{Q}^x(B).$$

The proof of (iii) is similar. □

Note. More applications now become available. In particular, you can now read the proof of the Ray–Knight theorem in § 52.

The Markov property (iv) of Theorem 48.1 can be stated as follows:

(48.2) *on* $\{f\colon H(f)>t\}$, ${}_a\mathbf{Q}^{f(t)}$ *is a regular conditional distribution of* $n\circ\theta_t^{-1}$ *given* $\mathscr{F}_t$.

In particular, for $t>0$, $x\in F\backslash\{a\}$,

(48.3) $$n(\{f\colon f_t\in dx;\ \theta_t f\in B;\ t<H\}) = n_t(dx)\,.\,{}_a\mathbf{Q}^x(B),$$

where

(48.4) $$n_t(\Gamma)\equiv n(\{f\colon f_t\in\Gamma;\ t<H\}),\quad \Gamma\in\mathscr{B}(F\backslash\{a\}).$$

A very familiar argument now shows that

(48.5) *the characteristic measure n is completely determined when the family of measures* $\{n_t\colon t>0\}$ *on* $\mathscr{B}(F\backslash\{a\})$ *is known.*

This family of measures clearly constitutes an entrance law:

(48.6) $$n_t\cdot {}_aP_s = n_{t+s}\quad\text{on}\quad F\backslash\{a\},$$

where $\{{}_aP_u\colon u\geqslant 0\}$ is the transition semigroup on $F\backslash\{a\}$ for the *killed* or *taboo* semigroup associated with

$$\{X_t\colon t<H\}.$$

Obviously, we wish to be able to calculate the entrance law $\{n_t\}$, which amounts exactly to establishing a *last-exit decomposition*:

(48.7) $$P_t(a,\Gamma) = \mathbf{E}^a\int_0^t n_{t-s}(\Gamma)\,dL_s,\quad \Gamma\in\mathscr{B}(F\backslash\{a\}).$$

See § 50. Once we have done this, we have a description of n itself. The best way to understand (48.7) is via the technique of 'marking excursions'.

Advice. Though we present that technique next so as to keep the theoretical treatment (and its notation) in a block, we strongly advise you to consider reading § 51 on the Skorokhod embedding and § 52 on the Ray–Knight theorem to familiarize yourself with the use of ordinary 'unmarked' excursions before you study the extra structure brought in by the marking procedure.

49. Marking the excursions. Let T be an exponential random variable independent of X. Then the idea of looking at X at time T is very familiar; indeed, the $\mathbf{P}^x$ law of X_T is simply $\lambda R_\lambda(x, \cdot)$, where $\lambda = (\mathbf{E}T)^{-1}$. However, it is not hard to see that *the* $\mathbf{P}^a$*-distribution of* $L^a(T)$ *is again exponential* (though with a different mean in general), and so T is a 'good' time for the local time scale as well. This remark, which allows us to link the real and local time scales, is very powerful, both conceptually and computationally, as examples will show.

Let $\mathbf{P}^a$ denote the law on $\Omega \equiv \{\text{R-paths } f\colon \mathbb{R}^+ \to F\}$ of our Ray process with compact metric state–space F started at the regular extremal point a, the excursions from which we are considering. Let

J denote the set of R-paths $g\colon \mathbb{R}^+ \to \mathbb{Z}^+$ such that $g(0)=0$ and $\Delta g_t = 0$ or 1 for all t.

Fix $\lambda > 0$, and let m be the law on J under which the canonical process is a Poisson process of rate λ. Now we consider the process $\{(X_t, \nu_t)\colon t \geqslant 0\}$ with values in $F \times \mathbb{Z}^+$ defined on $\tilde{\Omega} \equiv \Omega \times J$ by

$$(X_t(f, g), \nu_t(f, g)) = (f(t), g(t)),$$

with the product σ-algebras, and the product law $\tilde{\mathbf{P}}^a \equiv \mathbf{P}^a \times m$. Thus ν is a Poisson process of rate λ independent of X, and X has a local time L^a at a with inverse γ, as before.

(49.1) Some definitions. We defined the *excursion e_t of the path f at local time t* after (47.4):

$$e_t(\cdot) \equiv f((\gamma_{t-} + \cdot) \wedge \gamma_t).$$

We define analogously the *increment function η_t of ν over the excursion at local time t:*

$$\eta_t(\cdot) \equiv \nu((\gamma_{t-} + \cdot) \wedge \gamma_t) - \nu(\gamma_{t-}).$$

We call (e_t, η_t) the *marked excursion at local time t*. Next, we define the *marked-excursion point process* $\tilde{\Pi}$ by

$$\tilde{\Pi} \equiv \{(t, (e_t, \eta_t))\colon \gamma_t \neq \gamma_{t-}\}.$$

We let $\tilde{N}$ denote the associated random measure; for Borel $\tilde{B} \subseteq \mathbb{R}^+ \times \tilde{\Omega}$,

$$\tilde{N}(\tilde{B}) = |\tilde{B} \cap \tilde{\Pi}|.$$

Hence for Borel $A \subseteq \Omega$, $N(A) = \tilde{N}(A \times J)$. Now since the numbers of points of a Poisson process (on $\mathbb{R}^+$) in disjoint sets are independent random variables, it follows that, conditional on X, the increments of ν over different excursion intervals are independent Poisson.

(*49.2*) THEOREM. *Under the law* $\mathbf{P}^a \times m$, $\tilde{N}$ *is a Poisson process with expectation measure* $dt \times \tilde{n}$, *where*

$$\tilde{n}(A \times B) = \int_A n(df)m(\{g \in J: g(\cdot \wedge H(f)) \in B\})$$

for Borel $A \subseteq \Omega$, $B \subseteq J$.

This is just a rather pedantic way of stating the intuitively obvious idea that *if we firstly mark the time axis* $\mathbb{R}^+$ *of* X *with the points of an independent Poisson process of rate* λ, *and then break the path of* X *into its marked excursions* (e_t, η_t), *the result is the same as if we were firstly to break the path of* X *into its unmarked excursions* e_t, *and then mark each excursion independently with the points of an independent Poisson process of rate* λ.

Let $\tilde{U} \equiv \{(f, v) \in U \times J: v(t) = v(H(f))$ for all $t \geqslant H(f)\}$, the natural excursion space for marked excursions (e_t, η_t). In analogy with what we did before, we let $\tilde{U}_\infty \equiv \{(f, v) \in \tilde{U}: H(f) = \infty\}$, $\tilde{U}_0 \equiv \tilde{U} \setminus \tilde{U}_\infty$, and we define

$$\tilde{U}^* \equiv \{(f, v) \in \tilde{U};\ v(\infty) > 0\},$$

the set of 'starred' excursions, *where a starred excursion is one which contains at least one mark before its lifetime* H.

(*49.3*) *The dead zone* $\tilde{\partial}$. You may have wondered about the case when (under $\mathbf{P}^a$), X spends positive real time at a, so that X might well be at state a at the time of a Poisson mark. To deal with this possibility, *we create a point* $\tilde{\partial}$ *and extend the point process* $\tilde{\Pi}$ *to have values in* $\mathbb{R}^{++} \times (\tilde{U} \cup \tilde{\partial})$ *by adding to the marked excursions already considered points (null excursions)*

$$(L(S), \tilde{\partial}),$$

where S *is the time of a Poisson mark* $(\Delta v(S) = 1)$ *of the* v*-process such that* $X(S) = a$. Then:

$$\tilde{n}(\tilde{\partial}) = \begin{cases} \lambda \text{ in the case when } L_t = \int_0^t I_a(X_s)\,ds, \\ 0 \text{ when } X \text{ spends zero real time at } a. \end{cases}$$

(*49.4*) THEOREM. *The following results hold:*

(i) $\tilde{n}(\tilde{U}^*) < \infty$.

(ii) $\tilde{n}(\tilde{U}^*) = \int_U n(df)[1 - \exp(-\lambda H(f))]$.

(iii) $\tilde{n}(\tilde{U}^* \cup \tilde{\partial}) = \left\{\mathbf{E}^a \int_0^\infty e^{-\lambda t}\,dL(t)\right\}^{-1}$.

Proof. (i) If $\tilde{n}(\tilde{U}^*) = \infty$, then for each $t > 0$,

$$\tilde{N}((0, t] \times \tilde{U}^*) = \text{number of starred excursions before real time } \gamma_t$$
$$= \infty, \quad \text{a.s.}$$

But, if $\gamma_t < \infty$, an event of positive probability, then the number of points in a Poisson process of rate λ in the interval $(0, \gamma_t]$ must be finite and we have a contradiction.

(ii) The equality follows from Theorem 49.2 on taking $A = U$, $B = \{g \in J: g(\infty) > 0\}$.

(iii) Let T be the time of the first Poisson mark of the v process on the real axis, so that T is exponentially distributed with rate λ. Then $L(T)$ is the local time at a corresponding to the first excursion which is either starred or null. Thus,

(49.5) $L(T)$ *is exponentially distributed with rate* $\tilde{n}(\tilde{U}^* \cup \tilde{\partial})$.

Hence,

$$\tilde{n}(\tilde{U}^* \cup \tilde{\partial})^{-1} = \mathbf{E}^a L(T) = \mathbf{E}^a \int_0^\infty \lambda e^{-\lambda t} L(t)\, dt \tag{49.6}$$

$$= \mathbf{E}^a \int_0^\infty e^{-\lambda t}\, dL(t).$$

□

50. Last-exit decomposition; calculation of the excursion law n. Continue with the notation of the last few sections. Let T be the time of the first mark of the v process, so that T is an exponentially distributed random variable of rate λ, independent of X.

The law of X_T is a functional of the first point of $\tilde{N}$ in $\tilde{U}^* \cup \tilde{\partial}$, the law of which is

$$\tilde{n}(\cdot \cap (\tilde{U}^* \cup \tilde{\partial}))/\tilde{n}(\tilde{U}^* \cup \tilde{\partial}),$$

and hence, for $\Gamma \in \mathscr{B}(F \setminus \{a\})$, Theorem 49.2 yields:

$$\mathbf{P}^a\{X_T \in \Gamma\} = \int_0^\infty \lambda e^{-\lambda t} n_t(\Gamma)\, dt / \tilde{n}(\tilde{U}^* \cup \tilde{\partial}). \tag{50.1}$$

We have

$$\mathbf{P}^a\{X_T = a\} = \tilde{n}(\tilde{\partial})/\tilde{n}(\tilde{U}^* \cup \tilde{\partial}).$$

The *last-exit decomposition* (48.7):

$$\mathbf{P}^a\{X_t \in \Gamma\} = \mathbf{E}^a \int_0^t n_{t-s}(\Gamma)\, dL_s, \quad \Gamma \in \mathscr{B}(F \setminus \{a\}) \tag{50.2}$$

follows immediately on substituting (49.6) into (50.1) and inverting the Laplace transforms. And since, for $\Gamma \in \mathscr{B}(F \setminus \{a\})$,

$$\int_0^\infty e^{-\lambda t} n_t(\Gamma)\, dt = R_\lambda(a, \Gamma) \Big/ \left(\mathbf{E}^a \int_0^\infty e^{-\lambda s} dL_s \right), \tag{50.3}$$

the entrance law $\{n_t: t > 0\}$ is determined, and hence, as was explained at the end of § 48, the excursion law n is determined.

(*50.4*) **Exercise.** Prove that under $\mathbf{P}^a$, X_T and L_T are independent!

We now look at two important examples.

(*50.5*) *Excursion law n from a point a in the minimal state–space of a Markov chain.* If (in this case) we take $j \neq a$, then equation (50.2) identifies $n_t(\{j\})$ with $g_{aj}(t)$, because it reads:

$$p_{aj}(t) = \int_0^t g_{aj}(t-s)p_{aa}(s)\,ds. \tag{50.6}$$

Now re-read §III.57 for the full story for chains.

(*50.7*) *Excursion law n from* 0 *for a* BM($\mathbb{R}$) *process B. We shall work with the local time l for B at* 0 *in its semimartingale normalization, so that* $|B|-l$ *is a martingale.* Then:

$$\mathbf{E}^0 \int_0^\infty e^{-\lambda t} dl(t) = \int_0^\infty e^{-\lambda t} p_t(0,0)\,dt = (2\lambda)^{-1/2},$$

p denoting the Brownian transition density function. Hence, from (49.6), since $\tilde{n}(\tilde{\partial})=0$ for Brownian motion, we have the important result

$$\tilde{n}(\tilde{U}^*) = \tilde{n}(\tilde{U}^* \cup \tilde{\partial}) = \gamma \equiv (2\lambda)^{1/2}. \tag{50.8}$$

It now follows from (50.3) that $n_t(dx) = n_t(x)dx$, where

$$\int_0^\infty e^{-\lambda t} n_t(x)\,dt = \gamma r_\lambda(0,x) = e^{-\gamma|x|}, \quad \gamma = (2\lambda)^{1/2},$$

r denoting the Brownian resolvent density. Thus (see (I.9.2))

$$n_t(x) = |x|(2\pi t^3)^{-1/2} \exp(-x^2/2t). \tag{50.9}$$

There is a very good reason (time reversibility!) why the result from §I.9 should be relevant; but we do not go into details about that now.

So, in exact analogy with the results for chains, *the restriction of n to the upward excursions (in the obvious sense) may be described as follows: for* $0<t_1<t_2<\ldots<t_m$ *and* $x_1, x_2, \ldots, x_m > 0$,

$$n\{f(t_k) \in dx_k : 1 \leqslant k \leqslant m\} \tag{50.10}$$

$$= n_{t_1}(x_1)dx_1 \prod_{k=2}^m p^-(t_k - t_{k-1}, x_{k-1}, x_k)dx_k,$$

where $n_t(x)$ is as at (50.9), and p^- is the transition density function for Brownian motion killed at 0:

$$p^-(t,x,y) = (2\pi t)^{-1/2}[\exp\{-(y-x)^2/2t\} - \exp\{-(y+x)^2/2t\}]$$

(see §I.13.7).

Though equation (50.10) provides a complete description of n, it is not as useful for calculations as the description of n in § 55 below.

(50.11) Reconstructing $\{P_t\}$ *from* $\{{}_aP_t\}$ *and* $\{n_t: t>0\}$. Return to the general situation with which we began this section. We explain how to reconstruct the resolvent R_λ of $\{P_t\}$ on F from the resolvent $\{{}_aR_\lambda\}$ of $\{{}_aP_t\}$ on $F\backslash\{a\}$ and the entrance law $\{n_t: t>0\}$ on $F\backslash\{a\}$. In this connection, we note for future use that the entrance law property (48.6):

(50.12)
$$n_t \cdot {}_aP_t = n_{t+s}$$

transforms (on multiplying by $e^{-\lambda t-\mu s}$ and integrating over $(s,t)\in(\mathbb{R}^{++})^2$) to the form:

(50.13)
$$(\lambda-\mu)n_\lambda \cdot {}_aR_\mu = n_\mu - n_\lambda, \quad \lambda,\mu>0,$$

where n_λ $(\lambda>0)$, which we recognize as being a Laplace transform because it has a Greek parameter, is defined by

$$n_\lambda(\Gamma) \equiv \int_0^\infty e^{-\lambda t} n_t(\Gamma)\,dt, \quad \Gamma\in\mathscr{B}(F\backslash\{a\}).$$

It is convenient to extend the measures $n_\lambda(\cdot)$ and ${}_aR_\lambda(x,\cdot)$ by setting

$$n_\lambda(\{a\})\equiv 0, \quad {}_aR_\lambda(x,\{a\})\equiv 0 \ (x\in F), \quad {}_aR_\lambda(a,F)\equiv 0.$$

Because of the strong Markov property at time H_a, we then have, for all (bounded measurable) f on F,

(50.14)
$$R_\lambda f(x) = {}_aR_\lambda f(x) + \psi_\lambda(x) R_\lambda f(a), \quad \psi_\lambda(x)\equiv \mathbf{E}^x(e^{-\lambda H_a}).$$

Hence, we need only construct $R_\lambda f(a)$.

From (50.3),

(50.15)
$$n_\lambda(\Gamma) = R_\lambda(a,\Gamma) \Big/ \left(\mathbf{E}^a \int_0^\infty e^{-\lambda s} dL_s\right), \quad \Gamma\in\mathscr{B}(F\backslash\{a\}).$$

We consider the two possible cases, writing

$$R_\lambda(a,a) \text{ for } R_\lambda(a,\{a\}), \text{ and } F\backslash a \text{ for } F\backslash\{a\}$$

for convenience.

Case 1: $R_\lambda(a,a)=0$ *for some (then all)* $\lambda>0$. In this case, since X is honest, we have

$$\lambda R_\lambda(a, F\backslash a) = \lambda R_\lambda(a,F) = 1,$$

so that (50.15) yields:

$$R_\lambda(a,\Gamma) = n_\lambda(\Gamma)/\lambda n_\lambda(F\backslash a), \quad \Gamma\in\mathscr{B}(F\backslash a).$$

In functional notation:

$$n_\lambda f \equiv n_\lambda(f) = \int_F f(x) n_\lambda(dx),$$

we have

$$R_\lambda f(a) = n_\lambda f/(\lambda n_\lambda 1). \tag{50.16}$$

Case 2: $R_\lambda(a, a) > 0$ *for some (then all)* $\lambda > 0$. In this case, we can (and do) take L to be real time in a. Then (50.3) reads:

$$n_\lambda(\Gamma) = R_\lambda(a, \Gamma)/R_\lambda(a, a), \quad \Gamma \in \mathscr{B}(F\backslash a)$$

and since X is assumed honest, we have

$$\lambda R_\lambda(a, a) n_\lambda(F\backslash a) = \lambda R_\lambda(a, F\backslash a) = 1 - \lambda R_\lambda(a, a),$$

so that

$$R_\lambda(a, a) = (\lambda + \lambda n_\lambda 1)^{-1}$$

and

$$R_\lambda f(a) = \frac{f(a) + n_\lambda f}{\lambda + \lambda n_\lambda 1}. \tag{50.17}$$

An interesting question now arises. Suppose that we know nothing of R_λ, but are given only a resolvent $\{{}_aR_\lambda\}$ on $F\backslash a$ and some solution $\{n_\lambda\}$ of (50.12). Do the prescriptions (50.14, 16) and (50.14, 17) produce resolvents R_λ? Showing that indeed they do is just a rather complicated exercise in algebra (see Rogers [4]). It certainly need not be the case that the point a is a regular extreme point: a might be a branch point or an irregular extreme point.

In the remainder of this section, which you might well choose to skip on a first reading, we therefore comment briefly on the theory of excursions from a state a (of an honest Ray process X) which is not regular extremal.

(*50.18*) *Excursions from a point 'a' which is not regular and extremal.* Suppose that a is a state of our Ray process X which is not regular extremal. Broadly speaking, the set M of visits or approaches to a is discrete, and this breaks the time axis into a sequence of excursion intervals: the excursions of X over the different excursion intervals are independent, and identically distributed. In more detail, we must consider two cases.

Case (i): *a is a branch point.* Exploiting the fact that, for all x, $\mathbf{P}^x(X_t \neq a \text{ for all } t) = 1$, by (III.41.1), it is easy to show that $H = \inf\{t > 0: X_{t-} = a\}$ is positive a.s., and $X(H-) = a$ on $\{H < \infty\}$. If $H_k \equiv \inf\{t > 0: d(X_t, a) < k^{-1}\}$, then $H_k \uparrow\uparrow H$ on $\{H < \infty\}$, so that H is previsible, and by Theorem III.41.3, for any $\eta \in b\mathscr{F}$, for any x,

$$\mathbf{E}^x(\theta_H \eta \mid \mathscr{F}_{H-}) = \mathbf{E}^a(\eta) \text{ on } \{H < \infty\}. \tag{50.19(i)}$$

Case (ii): *a is an irregular extreme point*. By Theorem (45.1(ii)), $X(H)=a$ a.s. on $\{H<\infty\}$, so that for all $x\in F$, $\eta\in b\mathscr{F}$,

(50.19(ii)) $$\mathbf{E}^x(\theta_H\eta|\mathscr{F}_H)=\mathbf{E}^a(\eta) \text{ on } \{H<\infty\}.$$

In each of the two cases, if $T_0\equiv 0$, $T_{k+1}\equiv\inf\{u>T_k\colon u\in M\}$ $(k\geqslant 0)$, then by hypothesis $T_k<T_{k+1}$ on $\{T_k<\infty\}$, and $\rho\equiv\mathbf{E}^a[e^{-H}]<1$. By the strong Markov property, $\mathbf{E}^a[e^{-T_k}]=\rho^k\to 0$ as $k\to\infty$, so that $\sup T_k=\infty$, a.s. Define

$$\nu\equiv\sup\{k\geqslant 1\colon T_{k-1}<\infty\},$$

and define the excursions $\xi^k\in U$ of X by

$$\begin{aligned}\xi^k(t)&\equiv X(T_{k-1}+t) \quad \text{if } 0\leqslant t<T_k-T_{k-1} \text{ and } T_{k-1}<\infty,\\ &\equiv a \qquad\qquad\qquad \text{otherwise.}\end{aligned}$$

Let n be the $\mathbf{P}^a$-distribution of ξ^1, a probability measure on U. Then the analogue of Theorem 47.6 is the following.

(*50.20*) THEOREM: *If a is not a regular extreme point, then*

(i) $\mathbf{P}^a(\nu>k)=n(U_0)^k$, $k\in\mathbb{Z}^+$.
(ii) *Conditional on* $\{\nu>k\}$, $\xi^1,\xi^2,\dots,\xi^k$ *are independent and identically distributed with law* $n(U_0)^{-1}n(\cdot\cap U_0)$, *independent of* $\{\xi^r\colon r>k\}$;
(iii) *Conditional on* $\{\nu=k\}$, ξ^k *has law* $n(U_\infty)^{-1}n(\cdot\cap U_\infty)$.

Proof. The proof proceeds by repeated application of results (50.19); we leave the details as an exercise.

We define the 'local time at a' before t to be the number of visits to M during the time interval $(0, t]$. All of the algebra of equations (50.14–16) then holds, as you can check (but see Remark 50.21).

Exercise III.26.4 is an illustration.

(*50.21*) *Remark.* In order for equation (50.14) to make sense when $x=a$, we need to make the convention

$$\psi_\lambda(a)=1.$$

This requires us to think of $\psi_\lambda(a)$ as $\mathbf{E}^a[\exp(-\lambda T_a)]$, where

$$T_a\equiv\inf\{t\geqslant 0\colon X_t=a \text{ or } X_{t-}=a\},$$

X being considered as having parameter set $\{0-\}\cup[0,\infty)$, with

$$\mathbf{P}^x[\cdot]\equiv\mathbf{P}[\cdot|X(0-)=x].$$

This is the correct set-up for Ray processes.

(*50.22*) **Exercise.** Rethink Exercise III.26.4, showing that the μ_k can be constants summing to infinity but satisfying certain constraints (what?), in which case ∞

becomes a regular extreme point. Sketch the sample paths of the process for this case.

51. The Skorokhod embedding theorem. Before the main event of this section, some preliminaries . . . .

We use the semimartingale normalization of local time l at 0 for Brownian motion B on $\mathbb{R}$. Thus, by Tanaka's formula,

$$|B|-l \quad \text{and} \quad B^+-\tfrac{1}{2}l$$

are martingales. We define the excursion law n for the BM($\mathbb{R}$) process B in terms of the normalization l (exactly as in § 50).

(*51.1*) **Exercise.** Show that $\psi(x)\equiv \mathbf{E}^x \exp(-H_0)=\exp(-2^{1/2}|x|)$. By applying Itô's formula to $\psi(B)$, show that the Blumenthal–Getoor normalization of local time at 0 is $2^{1/2}l$. □

Now for a very useful little result. (On comparing results, always be careful about one-sided and two-sided situations and about associated problems of normalization of local time.)

(*51.2*) PROPOSITION. *For* $x>0$,

(i) $n(\{f\in U: \sup_t f(t)>x\})=(2x)^{-1}$.

(ii) $n(\{f\in U: \sup_t |f(t)|>x\})=x^{-1}$.

Proof. Let $G_x\equiv\{f\in U: \sup f(t)>x\}$. Then:

$$T\equiv\inf\{t: N((0,t]\times G_x)>0\}$$

is an exponential random variable of rate $n(G_x)$, where we have constructed N using the right-continuous inverse γ to l. Indeed,

$$\begin{aligned}\mathbf{P}(T>t)&=\mathbf{P}\{N((0,t]\times G_x)=0\}\\ &=\exp[-tn(G_x)]\end{aligned}$$

since the number of points in $(0,t]\times G_x$ is a Poisson random variable of mean $\mu((0,t]\times G_x)$. Thus $\mathbf{E}(T)=n(G_x)^{-1}$. However, if $\tau\equiv\inf\{t: B_t=x\}$, then $T=l_\tau$, and since $B^+-\frac{1}{2}l$ is a martingale, the optional sampling theorem yields $x=\frac{1}{2}\mathbf{E}T$. This establishes result (i). You prove result (ii), both in direct fashion and also by deducing it from result (i). □

(*51.3*) **Exercise.** Let B be Brownian motion on $\mathbb{R}$, $B_0=1$, $T\equiv\inf\{t>0: B_t=0\}$, and let $X_t\equiv B(\sigma_t\wedge T)$, where σ is inverse to the strictly increasing PCHAF $A_t\equiv\int m(da)l_t^a$ of B. Here, m is a measure on $(0,\infty)$ which is finite and positive on any interval (a, b), where $0<a<b<\infty$; m is the speed measure of the diffusion X. Prove that

$$\text{(51.4)} \qquad \mathbf{E}\int_0^{A(T)} I_{(0,1]}(X_s)\,ds = \mathbf{E}\int_0^1 m(da)l_T^a = 2\int_0^1 am(da).$$

This result is used in § V.51 to characterize the boundary point 0 of the diffusion X.

(*51.5*) *The Skorokhod embedding theorem.* The simple result of Proposition 51.2 is all we shall need to prove the celebrated theorem of Skorokhod about the embedding of a given law in Brownian motion. The key idea in this proof is due to Azéma and Yor [1], whose proof uses martingale calculus. A nice feature is that there is no need for any of the untidy approximation employed in early proofs of the result; the stopping time is defined by a global recipe. An excursion proof of the Azéma–Yor results was given in Rogers [1]. We prove the result here only under the simplifying assumptions that the distribution function F is continuous and strictly increasing; the general case needs a bit more care, as you will see if you consult Azéma and Yor [1] or Rogers [1].

(*51.6*) THEOREM (Skorokhod; Azéma and Yor). *Let F be a distribution function on $\mathbb{R}$ such that*

$$\int x^2 F(dx) < \infty, \quad \int xF(dx) = 0.$$

Then there exists a Brownian stopping time T (non-randomized in the sense that it is a stopping time relative to the natural augmented filtration of B) such that B_T has distribution F and T is minimal *in the sense that*

$$\mathbf{E}T = \int x^2 F(dx).$$

(*51.7*) *Remark* (Doob). As Doob has remarked, the problem of embedding a given law in Brownian motion is trivial if no minimality criterion is imposed; just take $T=\inf\{t>1: B_t = h(B_1)\}$, where h is chosen so that $h(B_1)$ has the given law!

Proof. Define

$$\psi(x) \equiv \int_x^\infty tF(dt)(1-F(x))^{-1}.$$

Under the assumption that F is continuous and strictly increasing, the function ψ is well-defined, continuous and strictly increasing, with continuous strictly increasing inverse φ. Moreover, $\psi(x) > x$ for all x, and $\lim_{x\to-\infty} \psi(x) = 0$. Now define

$$T \equiv \inf\{t: \psi(B_t) < S_t\}$$

where $S_t \equiv \sup\{B_s: s \leq t\}$. Then $\psi(B_T) = S_T$ and $B_T = \varphi(S_T)$ on $\{T < \infty\}$; and to calculate the law of B_T, it is enough to find the law of S_T. But notice that

$$T = \inf\{t: S_t - B_t > S_t - \varphi(S_t)\}$$

and recall the result of Lévy (V.6.5) that $(S, S-B)$ has the same law as $(l, |B|)$; the law of S_T is then the same as the law of $l(T')$, where

$$T' \equiv \inf\{t: |B_t| > l_t - \varphi(l_t)\}.$$

We obtain the law of $l(T')$ by excursion theory. If N is the Poisson process of excursions of $|B|$ from zero, let K be the Poisson process on $\mathbb{R}^{++} \times \mathbb{R}^+$ defined by

$$K((0, t] \times A) \equiv N((0, t] \times m^{-1}(A)) \quad (A \text{ Borel}, t > 0)$$

where $m: U_0 \to \mathbb{R}^+$ is the measurable map $m(f) \equiv \sup_t |f(t)|$. By Proposition 51.2, K has expectation measure $dt \cdot x^{-2} dx$. If $C \equiv \{(l, y) \in \mathbb{R}^{+2}: y > l - \varphi(l)\}$, then

$$l(T') \equiv \inf\{t: K\{((0, t] \times \mathbb{R}^+) \cap C\} > 0\}.$$

Figure VI.4 shows the point process K in $\mathbb{R}^{++} \times \mathbb{R}^+$, and the time $l(T')$. The curve is the boundary of the region C, the graph of the function $l \mapsto l - \varphi(l)$.

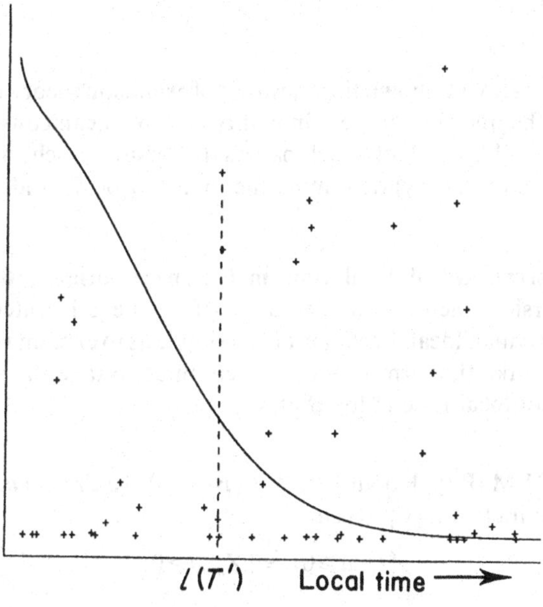

Figure VI.4

Hence

$$\begin{aligned}\mathbf{P}(B_T > x) &= \mathbf{P}(S_T > \psi(x)) \\ &= \mathbf{P}(l(T') > \psi(x)) \\ &= \mathbf{P}(K(((0, \psi(x)] \times \mathbb{R}^+) \cap C) = 0).\end{aligned}$$

The number of points of K in $((0, \psi(x)] \times \mathbb{R}^+) \cap C$ is a Poisson random variable with mean

$$\begin{aligned}\alpha &\equiv \int_{((0,\psi(x)] \times \mathbb{R}^+) \cap C} dl\, y^{-2} dy \\ &= \int_0^{\psi(x)} dl \int_{l-\varphi(l)}^{\infty} y^{-2} dy \\ &= \int_0^{\psi(x)} (l - \varphi(l))^{-1} dl \\ &= \int_{-\infty}^{x} (\psi(u) - u)^{-1} \psi(du) \\ &= -\log(1 - F(x))\end{aligned}$$

by elementary calculus. Hence

$$\mathbf{P}(B_T > x) = e^{-\alpha} = 1 - F(x)$$

as required.

Minimality of T. Having shown the relevance of excursion theory to the problem, we leave aside the question of the minimality of T. You can consult Azema and Yor [1] or Rogers [1], both of which papers (the first by stochastic calculus, the second by excursion theory) determine the joint law of B_T and T. □

52. Diffusion properties of local time in the space variable; the Ray–Knight theorem. Excursion theory provides a proof of the celebrated Ray–Knight theorem on Brownian local times, a proof so simple as to explain very clearly why the result must take the form it does. We continue to take the semimartingale normalization of local time l^x for B at x.

(*52.1*) THEOREM (Ray, Knight). (i) *Let* $\{B_t : t \geq 0\}$ *be Brownian motion started at* 1, *and let* $\tau \equiv \inf\{t : B_t = 0\}$. *Then:*

$$\{l^a_\tau : a \geq 0\} \overset{\mathcal{D}}{=} \{Z_a : a \geq 0\}$$

($\overset{\mathcal{D}}{=}$ *signifying identity in law*), *where Z is the solution of SDE*

(52.2)
$$Z_a = 2\int_0^a (Z_b^+)^{1/2}\, dW_b + 2(a \wedge 1).$$

(ii) *Let* $\{B_t : t \geqslant 0\}$ *be Brownian motion started at* 0, *and let* $T \equiv \inf\{t : l_t^0 > 1\}$. *Then:*

$$\{l_T^a : a \geqslant 0\} \stackrel{\mathscr{D}}{=} \{Y_a : a \geqslant 0\},$$

where Y *is the solution of the SDE*

(52.3)
$$Y_a = 1 + 2\int_0^a (Y_b^+)^{1/2}\, dW_b.$$

(*52.4*) *Remarks.* Pathwise uniqueness holds for both SDEs (see Theorem V.40.1), so the laws of Z and Y are uniquely defined. *The solution* Y *has the law of a squared zero-dimensional Bessel process,* and Z *has the law of a squared two-dimensional Bessel process up to time* 1, *after which it evolves like a squared zero-dimensional Bessel process.* Essentially, the reason that $\{l_\tau^a : a \geqslant 0\}$ and $\{l_T^a : a \geqslant 0\}$ are so nice (diffusions, in fact) is that the stopping times τ and T are nice times in the local time scale. In contrast, constant stopping times are not particularly nice in the local time scale, and $\{l_t^a : a \geqslant 0\}$ is not a Markov process, though it is a semimartingale (Perkins [2], Jeulin [2]). Stopping times S can be constructed such that $\{l_S^a : a \geqslant 0\}$ is not even a semimartingale; Barlow [4] has an example.

Proof (. . . , Walsh [1], . . .) (i) Fix x in (0, 1), and consider the Poisson process of excursions from x. This takes values in the space

$$U \equiv \{f \in C(\mathbb{R}^+, \mathbb{R}) : f(0) = x, \quad 0 < H_x < \infty, \ f(t) = x \text{ for } t \geqslant H_x\},$$

where $H_x(f) \equiv \inf\{t > 0 : f(t) = x\}$, and U splits into the disjoint union $U^+ \cup U^- \cup U^\partial$, where

$$U^+ = \{f \in U : f(t) \geqslant x \text{ for all } t\},$$
$$U^- = \{f \in U : 0 < f(t) \leqslant x \text{ for all } t\},$$
$$U^\partial = \{f \in U : f(t) \leqslant 0 \text{ for some } t\}.$$

We are implicitly setting aside the piece $\{B_t : 0 \leqslant t < H_x\}$ of the path before H_x and are decomposing into the point process of excursions the process $\{B(t + H_x) : t \geqslant 0\}$. The pre-$H_x$ piece of the path is independent of this, so we can easily deal with it separately later.

The following paragraph is a striking illustration of the power of excursion theory.

Now since $\mathbb{R}^+ \times U^+$, $\mathbb{R}^+ \times U^-$ *and* $\mathbb{R}^+ \times U^\partial$ *are disjoint, the restrictions of* N *to these sets will be independent; in particular,*

$$l_\tau^x = \inf\{t > 0 : N((0, t] \times U^\partial) > 0\}$$

is independent of the restrictions of N *to* $\mathbb{R}^+\times U^+$ *and* $\mathbb{R}^+\times U^-$. Next, $\{l^a_\tau: 0\leqslant a\leqslant x\}$ is a functional of N restricted to $(0, l^x_\tau]\times U^-$, and $\{l^a(\tau)-l^a(H_x): a\geqslant x\}$ is a functional of N restricted to $(0, l^x_\tau]\times U^+$. Thus *conditional on* l^x_τ, *the processes* $\{l^a_\tau: 0\leqslant a\leqslant x\}$ *and* $\{l^a(\tau)-l^a(H_x): a\geqslant x\}$ *are independent.* Since $\{l^a(H_x): a\geqslant x\}$ is a functional of $\{B_t: 0\leqslant t\leqslant H_x\}$ which is independent of N, it follows that $\{l^a_\tau: 0\leqslant a\leqslant x\}$ *and* $\{l^a_\tau: a\geqslant x\}$ *are conditionally independent given* l^x_τ, which is a statement of the fact that $\{l^a_\tau: 0\leqslant a\leqslant 1\}$ is a Markov process. A similar argument for $a\geqslant 1$ allows us to conclude that $\{l^a_\tau: a\geqslant 0\}$ is Markovian.

It remains to identify the transition mechanism of $\{l^a_\tau: 0\leqslant a\leqslant 1\}$ and $\{l^a_\tau: a\geqslant 1\}$. We consider the first of these. Suppose that $0<x<y<1$. Given that $l^x_\tau=l$, Proposition 51.2 shows that the number R of excursions before local time l for which $H_y<\infty$ is a Poisson random variable with mean $ln(H_y<\infty)=l/2\delta$, where $\delta=y-x$. Only on such excursions can l^y increase. These excursions are i.i.d. (independent and identically distributed) with law $n(\bullet\cap\{H_y<\infty\})/n(H_y<\infty)$. According to Theorem 48.1(iii), an excursion for which $H_y<\infty$ behaves after H_y like Brownian motion started at y run until it reaches x; thus the increment of l^y during such an excursion has the law of $l^y(H_x)$ for a Brownian motion started at y – and that is exponential with mean 2δ, by Proposition 51.2. Thus $l^y(\tau)-l^y(H_x)$ is made up of the sum of R independent exponential random variables of mean 2δ which are independent of R. Hence, by an elementary argument used twice in § 43, we have, for $\lambda>0$,

$$\mathbf{E}^1[\exp\{-\lambda(l^y(\tau)-l^y(H_x))\}\,|\,l^x_\tau=l]=\exp\{-l\lambda/(1+2\delta\lambda)\}.$$

Lastly, $l^y(H_x)$ is exponential with mean 2δ and is independent of $\{l^a(\tau)-l^a(H_x): a\geqslant 0\}$, so that, for $0<x<y<1$,

$$\text{(52.5)}\qquad \mathbf{E}^1[\exp(-\lambda l^y_\tau)|l^x_\tau=l]=(1+2\delta\lambda)^{-1}\exp\{-l\lambda/(1+2\delta\lambda)\}.$$

It is a simple exercise in Itô calculus to verify that if for fixed y in (0, 1), we define

$$M_x\equiv(1+2(y-x)\lambda)^{-1}\exp[-Z_x\lambda/(1+2(y-x)\lambda)]\quad(0\leqslant x\leqslant y)$$

where Z solves (52.2), then $\{M_x: 0\leqslant x\leqslant y\}$ is a bounded martingale, and hence the diffusion Z has the same transition mechanism in [0, 1] as l^x_τ.

The argument for $a\geqslant 1$ is identical except that now for $1\leqslant x<y$ there is no contribution to l^y_τ from the piece $\{B_t: t<H_x\}$ of the path before H_x. Thus

$$\mathbf{E}^1[\exp(-\lambda l^y_\tau)|l^x_\tau=l]=\exp\{-l\lambda/(1+2\delta\lambda)\},$$

which we check is the transition mechanism of the squared zero-dimensional Bessel process in the same manner as before.

(ii) The techniques of proof are the same as for (i); we leave the details to you. □

(*52.6*) *References.* There is a huge literature on the Ray–Knight theorem. See, for example, Ray [2], Knight [3], § 2.8 of Itô and McKean [1], Williams [4, 6],

McKean [3], Walsh [1, 3] and other fine papers in Azema and Yor [2], McGill [2], and, for an interesting connection with quantum-field theory using results of Dynkin, Sheppard [1]. (As mentioned in the preface, local time and quantum fields are closely interrelated.) For applications of the Ray–Knight theorems to 'self-avoiding Brownian motions', see Norris, Rogers and Williams [2].

For 'nice times in the local time scale'—technically, *identifiable* times—see Walsh [2].

53. Arcsine law for Brownian motion. Let B denote a Brownian motion on $\mathbb{R}$ with $B_0=0$, and let

$$A_t=\int_0^t I_{[0,\infty)}(B_s)\,ds.$$

We shall calculate the law of A_t:

$$\mathbf{P}^0(A_t\leqslant s)=(2/\pi)\arcsin[(s/t)^{1/2}]\quad (s\leqslant t), \tag{53.1}$$

by taking an exponential random variable T of rate λ independent of B, and calculating (for $\alpha>0$)

$$\mathbf{E}^0\exp(-\alpha A_T)=\int_0^\infty \lambda e^{-\lambda t}\mathbf{E}^0\exp(-\alpha A_t)\,dt.$$

If (53.1) is correct, then a few lines of calculation show that

$$\mathbf{E}^0\exp(-\alpha A_T)=\gamma/\beta,\quad \gamma\equiv(2\lambda)^{1/2},\quad \beta\equiv(2\lambda+2\alpha)^{1/2};$$

it is this which we prove.

We mark the time axis with red points according to a Poisson process of rate λ independent of B, and we let T denote the time of the first red mark. We also mark the time-axis with blue points, independently of the red points, in a Poisson process whose rate varies with time as $\alpha I_{[0,\infty)}(B_t)$. (Thus the number of blue points in $[0,t]$ is, conditional on B, Poisson with mean αA_t.) Then

$$\mathbf{E}^0\exp(-\alpha A_T)=\mathbf{P}^0(\text{the first red point is before the first blue point}).$$

We utilize the fact that the coloured marks can be created in the following alternative way. Mark the time axis with Red points at rate $\lambda I_{(-\infty,0)}(B_t)$ and independently with Green points at rate $(\lambda+\alpha)I_{[0,\infty)}(B_t)$. Then recolour each Green point red with probability $\lambda(\lambda+\alpha)^{-1}$, independently from one point to the next; finally, recolour all remaining Green points blue, and class Red points as being red. Then:

$$\begin{aligned}&\mathbf{P}^0(\text{the first red point is before the first blue point})\\&\quad=\mathbf{P}^0(\text{the first Red point is before the first Green point})\\&\qquad+\mathbf{P}^0(\text{the first Green point is before the first Red point, and is recoloured red})\end{aligned} \tag{53.2}$$

$= \mathbf{P}^0$(the first Red point is before the first Green point)

$+ \mathbf{P}^0$ (the first Green point is before the first Red point). $\lambda(\lambda+\alpha)^{-1}$.

Now consider the n-measure of the set of those excursions which initially contain a Green point: it must be $\frac{1}{2}\beta$. Indeed, if we inserted Green points at constant rate $\alpha+\lambda$, the rate would be β (by (50.8)); but we are only inserting marks into excursions into $\mathbb{R}^+$, so by symmetry the mass must be halved. Likewise, the measure of those excursions which initially contain a Red point is $\frac{1}{2}\gamma$. No excursion can contain initially both a Red point and a Green point, so the local times at which the first Green-marked excursion and the first Red-marked excursion occur will be independent exponential random variables of rates $\frac{1}{2}\beta$ and $\frac{1}{2}\gamma$ respectively. Hence:

$$\mathbf{P}^0(\text{the first Red point is before the first Green point})$$
$$= \gamma/(\gamma+\beta),$$

whence, by (53.2),

$$\mathbf{P}^0(\text{the first red point is before the first blue point})$$
$$= \frac{\gamma}{\gamma+\beta} + \frac{\beta}{\gamma+\beta} \cdot \frac{\lambda}{\lambda+\alpha} = \frac{\gamma}{\beta}$$

as asserted. □

This proof may seem at first reading to be more tortuous and less rigorous than, for example, the (Feynman–Kac) proof in § 2.6 of Itô and McKean [1]. Read our proof again and convince yourself of the fundamental *simplicity* of the ideas used. To attempt to give more rigorous detail would kill this simplicity; to confine ourselves to stochastic-calculus arithmetic will only allow us to prove things which excursion theory also *explains*. Moreover, given familiarity with excursion thinking, you will be able to see the whole proof of the arcsine law and many other results 'in a moment, in the twinkling of an eye'. Excursion theory is sorcery too powerful to be dismissed—so work your apprenticeship.

The proof just given is really another way of formulating the proof mentioned at § III.24.4 which will be discussed again in § 59 below.

The huge literature on the arcsine law is well surveyed in Bingham and Doney [1], which also poses interesting questions on possible extensions to higher dimensions.

54. Resolvent density of a 1-dimensional diffusion. Let m be the speed measure of a diffusion X in natural scale on an interval I. We shall use the notation of § V.50. To complete the proof of Theorem V.50.7, we need the following result.

(*54.1*) LEMMA. *For* $x \in \operatorname{int}(I)$,

$$r_\lambda(x,x) = c_\lambda \psi_\lambda^+(x)\psi_\lambda^-(x),$$

where

$$c_\lambda^{-1} = \tfrac{1}{2}\{\psi_\lambda^-(x)D\psi_\lambda^+(x+) - \psi_\lambda^+(x)D\psi_\lambda^-(x+)\}.$$

Proof. From (V.50.10),

$$\begin{aligned} r_\lambda(x,x) &= \mathbf{E}^x \int_0^\infty e^{-\lambda t} L^x(dt) \\ &= \mathbf{E}^x \int_0^\infty \lambda e^{-\lambda u} L^x(u)\,du \\ &= \mathbf{E}^x L^x(T) \end{aligned}$$

where T is an exponential random variable of rate λ independent of X. Thus if we mark the time axis with the points of a Poisson process of rate λ, the rate in local time at x of starred excursions is $(r_\lambda(x,x))^{-1}$. But by looking at those excursions which get at least ε from x, we see that (if $H_x \equiv \inf\{t > 0 \colon X_t = x\}$)

$$\begin{aligned} &\text{rate in local time of starred excursions} \\ &= \lim (2\varepsilon)^{-1}[(1 - \mathbf{E}^{x+\varepsilon}(e^{-\lambda H_x})) + (1 - \mathbf{E}^{x-\varepsilon}(e^{-\lambda H_x}))] \\ &= \lim \frac{1}{2\varepsilon}\left[1 - \frac{\psi_\lambda^-(x+\varepsilon)}{\psi_\lambda^-(x)} + 1 - \frac{\psi_\lambda^+(x-\varepsilon)}{\psi_\lambda^+(x)}\right] \end{aligned}$$

from which the result follows. □

Hint. In case this seems a little hasty, just recall that X is a time-transformation of Brownian motion B, the local time of X being inherited from that of B via the time-transformation as described in § V.49. Thus, Proposition 51.2 is applicable, and, for small ε, the contribution to the lifetime of an excursion from x made by the escape time from $(x-\varepsilon, x+\varepsilon)$ can be ignored. The probability which underlies the appearance (in the analytical treatment) of the Wronskian is therefore clear.

55. Path decomposition of Brownian motions and of excursions. In this section, we prove some path-decomposition results for Brownian motions, and recall and prove the description of the Itô excursion law for reflecting Brownian motion given in the first edition of Volume 1.

Take Ω to be the canonical path space $C(\mathbb{R}^+, \mathbb{R})$ with the canonical process $\{X_t \colon t \geqslant 0\}$, and carrying the law $\mathbf{W}_c$ of Brownian motion with drift c started at 0. (Thus, $X_t - ct$ is a Brownian motion under $\mathbf{W}_c$.) Define $L_t \equiv \sup\{-X_s \colon s \leqslant t\}$, and let $Y_t \equiv X_t + L_t$.

(55.1) LEMMA. *The process Y is a diffusion on $\mathbb{R}^+$, and L is the local time at 0 for Y. The taboo transition density function of 'Y killed at 0' is:*

$$ {}_0p_t^c(x, y) = (2\pi t)^{-1/2} e^{c(y-x) - \frac{1}{2}c^2 t} [e^{-(y-x)^2/2t} - e^{-(y+x)^2/2t}], \tag{55.2} $$

and the entrance law $\{n_t^c\}$ associated with excursions of Y from 0 for the normalization L is given by $n_t^c(dy) = n_t^c(y)dy$, where:

$$ n_t^c(y) = 2y(2\pi t^3)^{-1/2} \exp[-(y-ct)^2/2t]. \tag{55.3} $$

These facts specify the excursion law n^c from 0 for Y.

Proof. If $c=0$, then Y is a reflecting Brownian motion byLévy's result proved at § V.6.5, and we already know from subsection (50.7) that the lemma is true in this case.

By the Cameron–Martin–Girsanov formula (§ IV.38),

$$ d\mathbf{W}_c/d\mathbf{W}_0 = h(t, X_t) \quad \text{on } \mathscr{F}^\circ_{t+}, $$

where

$$ h(t, x) = \exp(cx - \tfrac{1}{2}c^2 t). $$

Under each $\mathbf{W}_c$, the bivariate process (Y_t, L_t) is a Markov process; because of the translation invariance of X, it is easy to see that Y is also a Markov process, using the criterion of Dynkin [2], Theorem 10.13, to decide when a function of a Markov process is again Markov. It is also easy to see that the semigroup of Y has to be FD, so that Y is an FD process (and therefore strong Markov). Hence Y is a diffusion. The formula (55.2) follows from the case $c=0$ by Cameron–Martin–Girsanov.

Next, since $\mathbf{W}_c$ is equivalent to $\mathbf{W}_0$ on each $\mathscr{F}^\circ_{t+}$, it follows that, under $\mathbf{W}_c$, L is a PCHAF of Y which grows only when $Y=0$; thus L is (a multiple of) the local time at zero for Y. By the reflection principle,

$$ \mathbf{W}_0(Y_t \in dy, L_t \in dl)/dydl = 2g(t, y+l), $$

where

$$ g(t, a) \equiv (2\pi t^3)^{-1/2} |a| \exp(-a^2/2t) $$

is the Brownian first-passage density. Hence

$$ \mathbf{W}_c(Y_t \in dy, L_t \in dl)/dydl = 2g(t, y+l)h(t, y-l), $$

and, if T is an exponential random variable of mean λ^{-1} independent of X, we have

$$ \mathbf{W}_c(Y_T \in dy, L_T \in dl)/dydl = 2\lambda \exp\{-(\alpha - c)y - (\alpha + c)l\}, $$

where $\alpha \equiv (c^2 + 2\lambda)^{1/2}$. Hence Y_T and L_T are independent exponential random variables,

$$ \mathbf{E}_c L_T = \mathbf{E}_c \int_0^\infty e^{-\lambda s} dL_s = (\alpha + c)^{-1}, $$

and, by the last-exit decomposition (50.3),

$$\int_0^\infty e^{-\lambda t} n_t(\Gamma)\, dt = R_\lambda^c(0, \Gamma) \Big/ \left(\mathbf{E}_c \int_0^\infty e^{-\lambda s} dL_s \right)$$

$$= \int_\Gamma \lambda^{-1} (\alpha - c) e^{-(\alpha - c)y} dy / (\alpha + c)^{-1},$$

which, upon inversion of the Laplace transform, yields (55.3). □

If $c \leqslant 0$, then $L_\infty = \infty$ a.s., and there are no infinite excursions.

Suppose now that $c > 0$. Then $L_\infty < \infty$ a.s., and the excursion measure n^c puts positive mass on $U_\infty = \{f \in U: H_0 = \infty\}$; indeed,

(55.4)
$$n^c(U_\infty) = n^c(\{f: H_0 > t, \text{ and } H_0 = \infty\})$$
$$= \int_0^\infty n_t^c(dy) \mathbf{W}_c(H_{-y} = \infty)$$
$$= \int_0^\infty n_t^c(dy)(1 - e^{-2cy})$$
$$= 2c.$$

Thus,

(55.5)
$$L_\infty = \inf\{t: N((0, t] \times U_\infty) > 0\}$$

is exponentially distributed with rate $2c$.

The law $n^c(\bullet \cap U_\infty)/n^c(U_\infty)$ is determined by the probability entrance law (look again at (55.4))

$$(2c)^{-1} n_t^c(dy)(1 - e^{-2cy})$$

and the transition density function

$$(1 - e^{-2cx})^{-1} {}_0p_t^c(x, y)(1 - e^{-2cy}),$$

(corresponding to conditioning Y to hit infinity before 0).

But this shows that:

(55.6) $(2c)^{-1} n^c$ *restricted to* U_∞ *is just the law of the nonnegative diffusion* R *determined by the SDE:*

(55.7)
$$dR_t = dB_t + c.\coth(cR_t)dt, \quad R_0 = 0.$$

Similarly, on U_0 the law of the excursions is determined by the entrance law

$$n_t^c(dy) e^{-2cy} = n_t^{-c}(dy)$$

and the transition density

$$e^{2cx} {}_0p_t^c(x, y) e^{-2cy} = {}_0p_t^{-c}(x, y).$$

Thus

(55.8) $\qquad$ *n^c restricted to U_0 is n^{-c}.*

In view of the Poisson nature of the excursion point process (Theorem 47.6), on combining results (54.4–8), we deduce the following result of Williams [1].

(*55.9*) THEOREM *Let $c>0$. On some probability space, take three independent random elements:*

$\{X_t: t\geqslant 0\}$, a Brownian motion with constant drift $-c$;
$\{R_t: t\geqslant 0\}$, a solution of (55.7),
and γ an exponential random variable with mean $(2c)^{-1}$.

Define $\tau\equiv\inf\{t: X_t=-\gamma\}$, and

$$\tilde{X}_t\equiv\begin{cases}X_t & (0\leqslant t\leqslant\tau)\\ R_{t-\tau}-\gamma & (t\geqslant\tau).\end{cases}$$

Then $\{\tilde{X}_t: t\geqslant 0\}$ is a Brownian motion with constant drift c.

By the techniques of scale and speed (see §§ V.7.46–47), we can deduce the corresponding decomposition at the minimum (respectively, maximum) of the path of any transient 1-dimensional diffusion whose minimum (maximum) is finite almost surely; see Williams [1] for more details. In particular, if we take Brownian motion started at $\varepsilon>0$ and stopped at 0, the path decomposition has the following form.

(*55.10*) COROLLARY. *Take a* $\mathrm{BES}^0(3)$ *process $\{\tilde{R}_t: t\geqslant 0\}$ independent of the* $\mathrm{BES}^\varepsilon(3)$ *process $\{R_t: t\geqslant 0\}$, both independent of the random variable S, which has the distribution*

$$\mathbf{P}(S>x)=\varepsilon/x \quad (x\geqslant\varepsilon).$$

Set $\tau\equiv\inf\{t: R_t=S\}$, $\tilde{\tau}\equiv\inf\{t: \tilde{R}_t=S\}$. Then the process Y, where

$$Y_t\equiv\begin{cases}R_t & (0\leqslant t<\tau),\\ S-\tilde{R}_{t-\tau} & (\tau\leqslant t<\tau+\tilde{\tau}),\\ 0 & (t\geqslant\tau+\tilde{\tau}),\end{cases}$$

is a Brownian motion started at $\varepsilon>0$ and stopped at 0.

We use this to prove the following characterization of the excursion law for reflecting Brownian motion.

(*55.11*) THEOREM. *Let X be a reflecting Brownian motion on $\mathbb{R}^+$, and let L denote its local time at 0 normalized so that $X-L$ is a (Brownian) martingale. Let U*

be the excursion space of continuous functions f from $\mathbb{R}^+$ to $\mathbb{R}^+$ such that $f(0)=0$ and for some $\zeta>0$,

$$f(t)>0, \quad 0<t<\zeta \quad \text{and} \quad f(t)=0, \quad t\geqslant\zeta.$$

For $f\in U$, let $S\equiv\sup_t f(t)$.

Then the excursion law n for X from 0 for the normalization L may be described as follows:

(55.12(i)) $$n(S>x)=x^{-1} \quad (x>0).$$

(55.12(ii)) *The regular conditional probability distribution of n given S is the law of the process Y described as follows: take two independent* $\mathrm{BES}^0(3)$ *processes R and $\tilde{R}$; set:*

$$\tau\equiv\inf\{t: R_t=S\}, \quad \tilde{\tau}\equiv\inf\{t: \tilde{R}_t=S\}$$

and define

$$Y_t\equiv\begin{cases} R_t & (0\leqslant t<\tau), \\ S-\tilde{R}_{t-\tau} & (\tau\leqslant t<\tau+\tilde{\tau}), \\ 0 & (t\geqslant\tau+\tilde{\tau}). \end{cases}$$

Proof. Let $\tau_\varepsilon(f)\equiv\inf\{t: f(t)=\varepsilon\}$ $(f\in U)$, and let $U^\varepsilon\equiv\{f\in U: \tau_\varepsilon(f)<\infty\}$. Define the shift operator θ_ε: $U^\varepsilon\to C(\mathbb{R}^+,\mathbb{R}^+)$ by $(\theta_\varepsilon f)(t)\equiv f(\tau_\varepsilon(f)+t)$, and let $\mathscr{F}^\varepsilon$ denote $\theta_\varepsilon^{-1}(\mathscr{F})$, where $\mathscr{F}$ is the Borel σ-algebra on $C(\mathbb{R}^+,\mathbb{R}^+)$. Then for $0<\delta<\varepsilon$, we have $U^\varepsilon\subseteq U^\delta$, and $\mathscr{F}^\varepsilon\subseteq\mathscr{F}^\delta$. If $\mathscr{U}$ denotes the Borel σ-algebra of U, then $\mathscr{U}=\sigma(\mathscr{G})$, where $\mathscr{G}$ is the algebra $\bigcup_{\varepsilon>0}\mathscr{F}^\varepsilon$; indeed,

$$\{a<f_t<b\}=\bigcup_m\bigcup_k\bigcap_{j\geqslant k}\{a+m^{-1}\leqslant(\theta_{1/j}f)(t)\leqslant b-m^{-1}\},$$

for example.

Now let n be the σ-finite measure on $(U,\mathscr{U})$ defined in (55.12), and let n' be the (possibly different) characteristic measure of the excursions for X. We shall prove that n and n' agree on $\mathscr{F}^\varepsilon$ for each $\varepsilon>0$, whence they agree on $\mathscr{U}$.

Firstly, $n(U^\varepsilon)=\varepsilon^{-1}$ from (55.11(i)), and $n'(U^\varepsilon)=\varepsilon^{-1}$ by Proposition 51.2, so n and n' both give the same mass to U^ε. Next, for $B\in\mathscr{F}$,

$$\begin{aligned} n'(\theta_\varepsilon^{-1}B,\tau_\varepsilon<\infty)&=n'(\tau_\varepsilon<\infty).{}_0\mathbf{P}^\varepsilon(B) \\ &=n'(U^\varepsilon).{}_0\mathbf{P}^\varepsilon(B) \end{aligned}$$

by Theorem 48.1(iii), where ${}_0\mathbf{P}^\varepsilon$ is the law of Brownian motion started at ε and stopped at zero. But Corollary 55.10 reveals that Theorem 55.11 is 'cooked', so that

$${}_0\mathbf{P}^\varepsilon(B)=n((\theta_\varepsilon^{-1}B)\cap U^\varepsilon)/n(U^\varepsilon),$$

and the truth of the theorem is now self-evident. □

(*55.13*) *References.* There are an enormous number of papers on path decomposition. Here is a small sample.

For papers on the concrete diffusion case, see Williams [7, 4], Pitman [1], Walsh [2], Rogers [1], Cutland and Kendall [1]. This last is a nice use of non-standard analysis. The grossissement approach is given in Jeulin [1] and Jeulin and Yor [2].

Many attempts have been made to give the general result for Markov processes. See Jacobsen [1], Millar [1, 2] and Getoor [3]. A good survey of the literature is given in Pitman [3].

Greenwood and Pitman [2] is a striking application to Wiener–Hopf theory.

56. An illustrative calculation. Here is a problem typical of those where excursion theory takes in its stride what stochastic calculus cannot do except by *grossissement.*

Let B be a $\mathrm{BM}^0(\mathbb{R})$ process. Define

$$T \equiv \inf\{t: |B_t| = 1\}, \quad \sigma \equiv \sup\{s < T: B_s = 0\},$$

$$A_t^+ \equiv \int_0^t I_{[0,\infty)}(B_s)\,ds, \quad A_t^- \equiv t - A_t^+.$$

What is the joint law of $(A_\sigma^+, A_\sigma^-, T)$?

For fixed α, β, $\lambda > 0$, we shall calculate the expression

$$\xi \equiv \mathrm{E}\exp\{-\alpha A_\sigma^+ - \beta A_\sigma^- - \lambda(T - \sigma)\}$$

by interpreting it in excursion terms. It is perhaps worth noting now that we shall use the following notation:

$$\gamma \equiv (2\alpha)^{1/2}, \quad \delta \equiv (2\beta)^{1/2}, \quad \theta \equiv (2\lambda)^{1/2}.$$

We split U into the disjoint union $U \equiv U^+ \cup U^- \cup U^1$, where

$$U^+ \equiv \{f \in U: 0 \leqslant f(t) < 1 \text{ for all } t\},$$
$$U^- \equiv \{f \in U: 0 \geqslant f(t) > -1 \text{ for all } t\},$$
$$U^1 \equiv U \backslash (U^+ \cup U^-).$$

Splitting the path of B into excursions, we mark at rate α those excursions which lie in U^+, we mark at rate β those which lie in U^-, and we mark at rate λ those in U^1. We introduce the following 'starred' sets:

$$U^{\pm *} \equiv \{f \in U^\pm: f \text{ contains a mark}\},$$
$$U^{1*} \equiv \{f \in U^1: f \text{ contains a mark before it reaches } 1\}.$$

Lastly, we define $U_0^1 \equiv U^1 \setminus U^{1*}$. Then – and this needs thought –

$$\xi = \mathbf{P}\,(\text{the first excursion in } U^{+*} \cup U^{-*} \cup U^1 \text{ lies in } U_0^1)$$
$$= n(U_0^1)\{n(U^{+*}) + n(U^{-*}) + n(U^1)\}^{-1}$$

and it is now just a question of calculating the excursion measures, which we do with the aid of the characterization (55.11) of the Brownian excursion law. Clearly $n(U^1) = 1$, from (55.11(i)). Next,

$$n(U_0^1) = n(U^1)\mathbf{P}\,(\text{excursion which escapes from } [-1, 1] \text{ has no } \lambda\text{-mark before it leaves } [-1, 1])$$
$$= n(U^1)\mathbf{E}\exp\{-\lambda H^3(1)\},$$

where $H^3(x) \equiv \inf\{t: R_t = x\}$, and R is a $\mathrm{BES}^0(3)$ process. This follows from (55.12(ii)). But (Exercise !) $e^{-\lambda t} R_t^{-1} \sinh \theta R_t$ is a local martingale ($\theta \equiv (2\lambda)^{1/2}$), so we deduce that

$$n(U_0^1) = \theta \operatorname{cosech} \theta. \tag{56.1}$$

To calculate $n(U^{+*})$, we firstly condition the excursion to have maximum $x \in (0, 1)$; given that the maximum is x, the excursion splits into two pieces, the piece before the maximum is reached, and the piece after the maximum is reached. The lengths of these two intervals are independent each with distribution equal to the law of the first passage time to x for a $\mathrm{BES}_0(3)$ process, by (55.12(ii)). Thus

$$n(U^{+*}) = \int_0^1 \tfrac{1}{2} x^{-2} [1 - (\gamma x \operatorname{cosech} \gamma x)^2]\, dx \tag{56.2}$$
$$= \tfrac{1}{2}[\gamma \coth \gamma - 1],$$

where $\gamma \equiv (2\alpha)^{1/2}$, since $\alpha \mapsto \gamma x \operatorname{cosech} \gamma x$ is the Laplace transform of the first passage time to x for a $\mathrm{BES}^0(3)$. We calculate $n(U^{-*})$ similarly to obtain

$$\mathbf{E}\exp\{-\alpha A_\sigma^+ - \beta A_\sigma^- - \lambda(T - \sigma)\} = \frac{2\theta \operatorname{cosech} \theta}{\gamma \coth \gamma + \delta \coth \delta}$$

where $\gamma = (2\alpha)^{1/2}$, $\delta = (2\beta)^{1/2}$, $\theta = (2\lambda)^{1/2}$.

57. Feller Brownian motions. At the end of § 50, we mentioned the idea of constructing resolvents from excursion laws. Both in the Markov-chain and diffusion settings, it was Feller [3, 4] who made the first major breakthrough.

Let F be the compact metric space

$$F \equiv [0, \infty].$$

One of Feller's achievements was to solve completely the problem of describing the most general honest strong Markov process X on F which behaves as BM($\mathbb{R}$) within $(0, \infty]$, the point ∞ being absorbing. Our treatment exactly parallels the work of Reuter [1] and Neveu [2, 3, 4] on chains.

We know from the strong Markov theorem that the resolvent $\{R_\lambda\}$ of such a process satisfies

(57.1) $$R_\lambda f(x) = R_\lambda^- f(x) + \psi_\lambda(x) R_\lambda f(0),$$

where $\{R_\lambda^-\}$ is the resolvent of BM($\mathbb{R}$) killed at 0: if $\theta \equiv (2\lambda)^{1/2}$, then

$$r_\lambda^-(x, y) = \theta^{-1}[e^{-\theta|y-x|} - e^{-\theta|y+x|}], \quad x, y \in (0, \infty),$$

$$\lambda R_\lambda^-(\infty, \{\infty\}) = 1, \quad \forall \lambda,$$

$$R_\lambda^-(0, \cdot) = 0, \quad R_\lambda^-(x, \{0\}) = 0,$$

and

$$\psi_\lambda(x) = \mathbf{E}^x \exp(-\lambda H_0) = e^{-\theta x}.$$

We also know (from § 50) that if n_λ is the Laplace transform of the entrance law on $[0, \infty]$ derived from the excursion law of X from 0, then

(57.2) $$(\lambda - \mu) n_\lambda R_\mu^- = n_\mu - n_\lambda, \quad n_\lambda(\{0\}) = 0.$$

Note that

$$n_\lambda 1 = \int_{[0, \infty]} n_\lambda(dx) = \int_{(0, \infty)} n_\lambda(dx) + n_\lambda(\{\infty\})$$

may have a non-zero contribution from ∞.

We know from § 50 that there are two cases: one described by (50.16) in which

(57.3(i)) $$R_\lambda f(0) = (n_\lambda f)/(\lambda n_\lambda 1),$$

and the other by (50.17) in which

(57.3(ii)) $$R_\lambda f(0) = \{f(0) + n_\lambda f\}/\{\lambda + \lambda n_\lambda 1\}.$$

Following the treatment in Rogers [2], we now describe the most general family of (finite) measures $\{n_\lambda\}$ on F which satisfy (57.2). Let

$$k_\lambda(dx) = \lambda(1 - \psi_1(x)) n_\lambda(dx).$$

Then we may rewrite (57.2) as follows:

(57.4) $$k_\lambda\{(1 - \psi_1)^{-1} R_\mu^- f\} = \lambda(\lambda - \mu)^{-1}(n_\mu f - n_\lambda f), \quad f \in C(F).$$

As $\lambda \to \infty$, the right-hand side tends to $n_\mu f$. As for the left-hand side, we notice first that

$$k_\lambda 1 = \lambda n_\lambda(1 - \psi_1) = \lambda n_\lambda R_1^- 1 = \lambda(n_1 1 - n_\lambda 1)/(\lambda - 1),$$

which remains bounded as $\lambda \to \infty$, so that we can find a subsequential weak limit k of the measures k_λ. Now, if $\beta \equiv (2\mu)^{1/2}$, then

$$\lim_{x\downarrow 0}\{(1-\psi_1)^{-1}R_\mu^- f\}(x)=2^{1/2}\int_0^\infty e^{-\beta y}f(y)\,dy,$$

$$\lim_{x\uparrow\infty}\{(1-\psi_1)^{-1}R_\mu^- f\}(x)=\mu^{-1}f(\infty),$$

so that $(1-\psi_1)^{-1}R_\mu^- f$ extends to a continuous function on F. By letting $\lambda\to\infty$ (through an appropriate subsequence) in (57.4), we obtain

$$\begin{aligned} n_\mu f &= k\{(1-\psi_1)^{-1}R_\mu^- f\} \\ &= k(\{0\})2^{1/2}\int_0^\infty e^{-\beta y}f(y)\,dy+\int_{(0,\infty]}\{(1-\psi_1)^{-1}R_\mu^- f\}(x)k(dx). \end{aligned}$$

Writing $p_2=2^{-1/2}k(\{0\})$ and $p_4(dx)\equiv\{1-\psi_1(x)\}^{-1}k(dx)$ on $(0,\infty]$, we see that

(57.5) $$\int(1-e^{-\mu x})p_4(dx)<\infty \text{ for some (then all) } \mu>0,$$

and

(57.6) $$n_\mu f=2p_2\int_0^\infty e^{-\beta y}f(y)\,dy+\int_{(0,\infty]}p_4(dx)R_\mu^- f(x).$$

Conversely, it is immediately verified that if a measure p_4 on $(0,\infty]$ satisfies (57.5) and $p_2\geqslant 0$, then $\{n_\lambda\}$ defined via (57.6) satisfies the entrance law property (57.2); and, as mentioned in § 50, a rather complicated piece of algebra (see Rogers [2]) then shows that $\{R_\lambda\}$, as defined by (57.1) and either (57.3(i)) or (57.3(ii)), is a resolvent. That the constructed resolvent $\{R_\lambda\}$ is a Ray resolvent may be verified directly.

The point 0 is a branch point for the Ray process X with resolvent $\{R_\lambda\}$ if and only if the following three conditions are satisfied:

(i) prescription (57.3(i)) is used;
(ii) $p_2=0$;
(iii) the total mass $p_4((0,\infty])$ of p_4 is finite;

in this case, on approaching 0, the process X branches to a point of $(0,\infty]$ according to the branching measure $p_4(\cdot)/p_4((0,\infty])$.

In every other case, 0 is extremal, and X is an FD process. If prescription (57.3(ii)) is used, then the process X spends positive real time in 0 (the boundary point 0 is then said to be sticky), and $\{n_\lambda\}$ is the Laplace transform of the entrance law corresponding to taking local time at 0 to be true time at 0. If (57.3(i)) is used, then, for some normalization of local time at 0, $\{n_\lambda\}$ will have the correct significance.

In every case in which 0 is regular and extremal, we have for the appropriate normalization of local time:

(Killing) $p_4(\{\infty\})=n(\{f\in U: f(t)=\infty,\ t>0\})$;

(Continuous exit) $n(\{f\in U\colon f(0+)=0\})=0$ or ∞ according as $p_2=0$ or $p_2>0$;
(Lévy measure) $n(\{f\in U\colon f(0+)\in dx\})=p_4(dx)\quad(0<x<\infty)$.

Thus we have a full understanding of the excursion behaviour.

Feller achieved a complete understanding of the analytic side of things, describing the generator $\mathscr{G}$ of X. Since the generator $\mathscr{G}$ acts as $\mathscr{G}f=\frac{1}{2}f''$ on $(0,\infty)$, the crucial thing is to describe the domain $R_\lambda C(F)$ of the generator in terms of boundary and lateral conditions. Feller showed, and you can confirm, that $R_\lambda C(F)$ consists exactly of those functions f such that f is C^2 on $(0,\infty)$, and

$$-p_2f'(0)+\tfrac{1}{2}p_3f''(0+)=\int_{(0,\infty]}p_4(dx)[f(x)-f(0)],$$

where $p_3=0$ if (57.3(i)) is used, and $p_3=1$ if (57.3(ii)) is used. The domain $R_\lambda C(F)$ is dense in $C(F)$ if and only if 0 is not a branchpoint.

(*57.7*) *Example.* An amusing way in which Feller Brownian motions arise in the following. Let B be a BM($\mathbb{R}$), let m be a measure on $(-\infty,0]$. Define the *fluctuating* PCHAF φ of B via

$$\varphi_t\equiv\int_0^t I_{(0,\infty)}(B_s)\,ds-\int_{-\infty}^0 l_t^a m(da),$$

and the corresponding time-substitution:

$$\tau_t\equiv\inf\{u\colon \varphi_u>t\}.$$

Then $X_t\equiv B(\tau_t)$ defines a Feller Brownian motion X, and via this construction we obtain precisely those Feller Brownian motions for which $p_3=0$ and the measure p_4 is totally monotonic. See London, McKean, Rogers and Williams [1].

58. Example: censoring and reweighting of excursion laws. Let U be the space of excursions from 0 of Brownian motion, and let n be the Brownian excursion law on U. Let $K>0$, and define

$$C\equiv\{f\in U\colon 0\leqslant f(t)<1,\ \forall t\},$$

$$G\equiv\{f\in U\colon f(t)\leqslant 0,\ \forall t\},$$

and define, for $\Gamma\in\mathscr{B}(U)$,

$$\tilde{n}(\Gamma)=Kn(\Gamma\cap C)+n(\Gamma\cap G).$$

Construct a Poisson process of excursions from 0 using the excursion measure $\tilde{n}$.

What is the process which we get by 'putting all these excursions together'? (This is an example of a general type of transformation studied by Knight and Pittenger [1], where one deletes certain excursions of a given Markov process.) It is intuitively obvious that the resulting process will be a diffusion in $(-\infty, 1)$

behaving like 1 − BES(3) in (0, 1) and like BM($\mathbb{R}$) in $(-\infty, 0)$; but the problem is to prove this, and to deal with the 'skewness' at 0.

The behaviour of the excursion process in $U^- \equiv \{f\colon f_t \leqslant 0 \text{ for all } t\}$ is already well understood, so we concentrate on the behaviour on C. Obviously for $x_i \in (0, 1)$,

$$n(\{f \in C\colon f(t_i) \in dx_i,\ i = 1, \ldots, k;\ H > t_n\})$$
$$= n'_{t_1}(dx_1)\left\{\prod_{i=2}^{k} q(s_i, x_{i-1}, x_i)\,dx_i\right\}(1 - x_k),$$

where $0 < t_1 < t_2 < \ldots < t_n$, $s_i \equiv t_i - t_{i-1}$, q is the transition density of Brownian motion killed at 0 and 1, and n'_t is the appropriate entrance law:

$$q(t, x, y) \equiv \mathbf{P}^x\left(B_t \in dy,\ \sup_{s \leqslant t} B_s < 1,\ \inf_{s \leqslant t} B_s > 0\right),$$

$$n'_t(dx) = n\left(\left\{f \in U\colon f_t \in dx,\ \sup_{s \leqslant t} f_s \leqslant 1,\ H > t\right\}\right).$$

Thus under n, the restriction to C of the excursion evolves in a Markovian fashion with transition density $(1-x)^{-1} q(t, x, y)(1-y)$ and entrance law $n'_t(dx)(1-x) = \nu_t(dx)$ (say). The resolvent density of Brownian motion killed at 0 and 1 is

$$\theta^{-1}\left[e^{-\theta|x-y|} - \left(\frac{\sinh \theta x}{\sinh \theta}\right) e^{-\theta(1-y)} - \left(\frac{\sinh \theta(1-x)}{\sinh \theta}\right) e^{-\theta y}\right], \quad \theta \equiv (2\lambda)^{1/2},$$

which is, for $0 < x < y < 1$,

$$\frac{\sinh \theta x \sinh \theta(1-y)}{\theta \sinh \theta}.$$

Thus the resolvent density ρ_λ of the evolution of the excursion is

$$\rho_\lambda(x, y) \equiv \int_0^\infty e^{-\lambda t}(1-x)^{-1} q(t, x, y)(1-y)\,dt \tag{58.1}$$
$$= \left(\frac{1}{1-x}\right)\left[\frac{\sinh \theta x \sinh \theta(1-y)}{\theta \sinh \theta}\right](1-y) \quad (0 < x < y < 1).$$

To find the Laplace transform of the entrance law, introduce an independent exponential random variable T of rate λ; then

$$\int_0^\infty \lambda e^{-\lambda t} n'_t(dx)\,dt = \lim_{\varepsilon \downarrow 0} n\left(\left\{f \in U\colon f_{T+\varepsilon} \in dx,\ \sup_{\varepsilon \leqslant s \leqslant T+\varepsilon} f_s \leqslant 1,\ H > T + \varepsilon\right\}\right)$$
$$= \lim_{\varepsilon \downarrow 0} \int_0^\infty \frac{y e^{-y^2/2\varepsilon}}{(2\pi\varepsilon^3)^{1/2}} \lambda \left[\frac{\sinh \theta y \sinh \theta(1-x)}{\theta \sinh \theta}\right] dy$$
$$= \tfrac{1}{2}\lambda \frac{\sinh \theta(1-x)}{\sinh \theta}.$$

To summarize, the excursions in (0, 1) evolve according to the resolvent ρ_λ given by (58.1) with entrance law whose Laplace transform is

$$\nu_\lambda(dx)=\tfrac{1}{2}\left[\frac{\sinh\theta(1-x)}{\sinh\theta}\right](1-x)\,dx\quad(0<x<1).$$

Excursions in $(-\infty, 0)$ evolve as ordinary Brownian excursions.

So, we have all the information necessary to construct the resolvent of the desired process via prescription (50.14, 16). The process will clearly be an FD diffusion. Let us compute its scale function, s. If $H_a\equiv\inf\{u>0\colon X_u=a\}$, then for $0<a<x<b<1$,

$$\frac{s(x)-s(a)}{s(b)-s(a)}=\mathbf{P}^x(H_b<H_a|H_0<H_1)$$

$$=\frac{x-a}{b-a}\cdot\frac{1-b}{1-x}$$

whence $s(x)=(1-x)^{-1}$ for $x\in(0, 1)$. For $x<0$, the scale function is the Brownian scale function, so

$$s(x)=\begin{cases}(1-x)^{-1} & \text{for } (x\in(0,1))\\ cx+1 & \text{for } (x<0)\end{cases}$$

for some $c>0$. To find c, we use the facts (which you can check) that, for $0<a<1$ and $b>0$,

$$n(\{f\in C\colon \sup f_t>a\})=(1-a)/2a$$

$$n(\{f\in G\colon \inf f_t<-b\})=1/2b,$$

so

$$\mathbf{P}(H_a<H_{-b})=\frac{K(1-a)/2a}{K(1-a)/2a+1/2b}$$

$$=\frac{s(0)-s(-b)}{s(a)-s(-b)}$$

whence $c=K$. Hence the generator $\mathscr{G}$ of the diffusion is

$$\mathscr{G}=\tfrac{1}{2}\frac{d^2}{dx^2}-\left(\frac{1}{1-x}\right)I_{(0,1)}(x)\frac{d}{dx};$$

with the following boundary condition at 0:

(58.2) $$f'(0-)=Kf'(0+).$$

(*58.3*) **Exercise.** Check the boundary condition by using Dynkin's characteristic operator (§III.12):

$$\mathscr{G}f(0)=\lim_{\varepsilon\downarrow 0}\varepsilon^{-2}\left\{\frac{f(\varepsilon)[s(0)-s(-\varepsilon)]+f(-\varepsilon)[s(\varepsilon)-s(0)]}{s(\varepsilon)-s(-\varepsilon)}-f(0)\right\}$$

$$=\lim_{\varepsilon\downarrow 0}\varepsilon^{-2}\left\{\frac{(1-\varepsilon)Kf(\varepsilon)+f(-\varepsilon)}{(1-\varepsilon)K+1}-f(0)\right\}.$$

Note the way in which the different forms of $\mathscr{G}$ on the two sides of 0 feature when you evaluate the limit.

59. Excursion theory by stochastic calculus: McGill's lemma. In the case of 1-dimensional diffusions, stochastic calculus provides an alternative method of computing the laws of certain functionals of the excursion process. The heart of the matter is a lemma of McGill [1] which allows a systematic approach to such calculations.

Let B be a $\mathrm{BM}^0(\mathbb{R})$, and let $\{\mathscr{B}_t\}$ denote its natural (augmented) filtration. Define

$$\varphi_t=\int_0^t I_{(-\infty,0)}(B_s)\,ds,\quad \tau_t\equiv\inf\{u:\ \varphi_u>t\}.$$

Define the σ-algebra $\mathscr{E}$ generated by the excursions below 0 by

$$\mathscr{E}\equiv\bigvee_t\tilde{\mathscr{B}}_t,\quad \tilde{\mathscr{B}}_t\equiv\sigma(\{B(\tau_s):\ s\leqslant t\}).$$

Then by Tanaka's formula,

$$\tilde{B}_t\equiv B(\tau_t)=B(\tau_t)\wedge 0=\beta_t-\tfrac{1}{2}\tilde{l}_t,$$

where $\tilde{l}_t\equiv l^0(\tau_t)$, and β_t is the process

$$\beta_t=\int_0^{\tau_t} I_{(-\infty,0]}(B_s)\,dB_s.$$

We shall show that β is a $\{\tilde{\mathscr{B}}_t\}$-Brownian motion. It then follows from § V.6 that β and $\tilde{B}$ generate the same (Brownian) filtration.

To see that β is a $\{\tilde{\mathscr{B}}_t\}$-Brownian motion, let

$$N_t\equiv\int_0^t I_{(-\infty,0]}(B_s)\,dB_s=(B_t\wedge 0)+\tfrac{1}{2}l_t^0.$$

Then if $T_n\equiv\inf\{t:\ B_t<-n \text{ or } l_t^0>n\}$, and $S_n\equiv\varphi(T_n)$, the conditions of Proposition IV.30.10 are satisfied, and so β is a $\{\tilde{\mathscr{B}}_t\}$-local martingale. It is continuous, because it could only jump when τ jumps; and a jump of τ is an interval in which B makes an excursion into $(0,\infty)$, throughout which both $B_\cdot\wedge 0$ and $l_\cdot^0$ are constant, and therefore N is also constant. Finally, $N_t^2-\varphi_t$ is a continuous $\{\mathscr{B}_t\}$-local martingale, from which we deduce similarly that $N(\tau_t)^2-\varphi(\tau_t)=\beta_t^2-t$ is a continuous $\{\tilde{\mathscr{B}}_t\}$-local martingale, and β is a $\{\tilde{\mathscr{B}}_t\}$-Brownian motion.

(*59.1*) LEMMA (McGill). *Suppose that M is an L^2-bounded $\{\mathscr{B}_t\}$-martingale null at zero, (strongly) orthogonal to the martingale $B^- - \frac{1}{2}l^0$. Then:*

$$\mathbf{E}[M_\infty | \mathscr{G}] = 0.$$

Proof. By Theorem IV.36.1, we can express M as

$$M_t = \int_0^t H_s \, dB_s$$

for some $H \in L^2(B)$, and the fact that M is orthogonal to $B^- - \frac{1}{2}l^0$ is the same as saying that

(59.2) $$\mathbf{P}(\{H_s I_{(-\infty, 0]}(B_s) = 0 \text{ for a.e. } s\}) = 1.$$

Since $\mathscr{G}$ is the σ-algebra generated by the Brownian motion β, Theorem IV.36.1 shows that any element Y of $L^2(\mathscr{G})$ is expressible as

$$Y = a + \int_0^\infty K_s \, d\beta_s$$

for some $\tilde{\mathscr{B}}$-previsible K such that $\mathbf{E}\int K^2 ds < \infty$. By Proposition IV.30.10,

$$Y = a + \int_0^\infty K(\varphi_s) I_{(-\infty, 0]}(B_s) \, dB_s,$$

and

$$\begin{aligned} \mathbf{E}M_\infty Y &= \mathbf{E}\langle M, Y \rangle_\infty \\ &= \mathbf{E}\int_0^\infty H_s K(\varphi_s) I_{(-\infty, 0]}(B_s) \, ds \\ &= 0, \end{aligned}$$

by (59.2). Here, $\{Y_t : t \geqslant 0\}$ is, of course, the martingale $\mathbf{E}(Y | \mathscr{B}_t)$. □

(*59.3*) *Example.* The easiest application of McGill's lemma is to proving formula (III.24.2):

(59.4) $$\mathbf{E}(e^{-\lambda \tau_t} | \mathscr{G}) = \exp(-\lambda t - \tfrac{1}{2}\theta \tilde{l}_t)$$

where $\theta \equiv (2\lambda)^{1/2}$. Recall from § III.24 that the arcsine law follows immediately from this result. Notice that

$$N_t \equiv \exp(-\theta B_t^+ + \tfrac{1}{2}\theta l_t^0 - \lambda(t - \varphi_t))$$

satisfies

$$dN_t = -\theta N_t (dB_t^+ - \tfrac{1}{2} dl_t^0) = -\theta N_t I_{(0, \infty)}(B_t) dB_t,$$

and is therefore orthogonal to $B^- - \frac{1}{2}l^0$. Stop N at the time $T_n = \tau_t \wedge \inf\{u : l_u^0 > n\}$ to make it bounded, and use McGill's lemma to obtain

$$\mathbf{E}[N(T_n) | \mathscr{G}] = 1.$$

Since T_n is $\mathscr{G}$-measurable, so is $l^0(T_n)$, and since $B^+(T_n)=0$, we therefore have

$$\mathbf{E}[\exp\{-\lambda(T_n-\varphi(T_n))\}|\mathscr{G}]=\exp\{-\tfrac{1}{2}\theta l^0(T_n)\}.$$

Letting $n\to\infty$ gives the result. □

(*59.5*) *Example.* We can deduce that for any bounded measurable f,

$$(59.6)\quad \mathbf{E}\left[\int_0^\infty \lambda e^{-\lambda t}f(B_t)\,dt\,\middle|\,\mathscr{G}\right]=\int_0^\infty \lambda\exp(-\lambda t-\tfrac{1}{2}\theta\tilde{l}_t)f(\tilde{B}_t)\,dt$$
$$+\left(\int_0^\infty \theta e^{-\theta x}f(x)\,dx\right)\int_0^\infty \tfrac{1}{2}\theta\exp(-\lambda t-\tfrac{1}{2}\theta\tilde{l}_t)\,d\tilde{l}_t.$$

An excursion argument runs as follows. If f is supported in $(-\infty, 0]$, then

$$\int_0^\infty \lambda e^{-\lambda t}f(B_t)\,dt=\int_0^\infty \lambda e^{-\lambda t}f(B_t)\,d\varphi_t$$
$$=\int_0^\infty \lambda e^{-\lambda\tau_t}f(\tilde{B}_t)\,dt$$

and the result follows from (59.4). If f is supported in $(0,\infty)$, introduce a random variable T independent of B with exponential distribution of rate λ; in terms of this, we want $\mathbf{E}(f(B_T)|\mathscr{G})$. Now from (59.4), $\mathbf{P}(T>\tau_t|\mathscr{G})=\exp(-\lambda t-\tfrac{1}{2}\theta\tilde{l}_t)$, so

$$\mathbf{P}(B_T<0|\mathscr{G})=\int_0^\infty \lambda\exp(-\lambda t-\tfrac{1}{2}\theta\tilde{l}_t)\,dt$$

and

$$\mathbf{P}(B_T>0|\mathscr{G})=\int_0^\infty \tfrac{1}{2}\theta\exp(-\lambda t-\tfrac{1}{2}\theta\tilde{l}_t)\,d\tilde{l}_t.$$

But the process of excursions into $\mathbb{R}^+$ is independent of the process of excursions into $\mathbb{R}^-$, and has the law of the excursion process of reflecting Brownian motion. Thus, given $B_T>0$, the law of B_T is the law of reflecting Brownian motion at an independent exponential time of rate λ, that is, the law of B_T is exponential rate θ. □

(*59.7*) **Exercise.** Prove (59.6) using McGill's lemma.

(*59.8*) *Excursion filtration; Walsh's conjecture.* One reason for interest in Examples 59.6 and 59.8 is the following.

Let X be a regular diffusion on $\mathbb{R}$. Define

$$\varphi_t^x\equiv\text{meas}\{s\leqslant t\colon X_s\leqslant x\},\quad \tau_t^x\equiv\inf\{u\colon \varphi_u^x>t\}.$$

Now define, for $x\in\mathbb{R}$,

$$\mathscr{G}^x\equiv\sigma\{X(\tau_s^x)\colon s\geqslant 0\},$$

the σ-algebra generated by excursions below x. Walsh ([1]) conjectured that any martingale relative to the 'excursion filtration' $\{\mathscr{E}^x: x \in \mathbb{R}\}$ is continuous except perhaps at the 'stopping place' ρ relative to $\{\mathscr{E}^x: x \in \mathbb{R}\}$ defined by

$$\rho \equiv \inf_t X(t).$$

Williams [14] explained that one could in principle prove this in analogy with the proof of Meyer's previsibility theorem given in § 15. However, as described by Williams, this method would require doing many calculations beginning with Examples (59.3) and (59.5) and becoming horribly complicated. Some better way of adapting Meyer's method was needed, and this has been found by Rogers [4].

See Walsh [2] and McGill [3] for fascinating papers on the local-time sheet.

60. Our best wishes for your own efforts at stochastics.

From the Book of Ezekiel, Chapter 37:

> "*The hand of the Lord was upon me, and carried me out in the spirit of the Lord, and set me down in the midst of the valley which was full of bones. And caused me to pass by them round about: and, behold, there were very many in the open valley; and lo, they were very dry. And he said unto me, Son of man, can these bones live? And I answered, O Lord God, thou knowest.*"

References

ABRAHAMS, R. and ROBBIN, J.
[1] *Transversal Mappings and Flows*, Benjamin, New York, Amsterdam, 1967.

AIZENMANN, M. and SIMON, B.
[1] Brownian motion and the Harnack inequality for Schrödinger operators, *Comm. Pure and Appl. Math.*, **35**, 209–273 (1982).

ALBEVERIO, S., BLANCHARD, PH. and HØEGH-KROHN, R.
[1] Newtonian diffusions and planets, with a remark on non-standard Dirichlet forms and polymers, *Stochastic Analysis and Applications: Lecture Notes in Mathematics 1095*, Springer, Berlin, 1984, pp. 1–24.

ALBEVERIO, S., FENSTAD, I. E., HØEGH-KROHN, R. and LINDSTRÖM, T.
[1] *Non-standard Methods in Probability and Mathematical Physics.*

ALDOUS, D.
[1] Stopping times and tightness, *Ann. Prob.*, **6**, 335–40 (1978).

ANCONA, A.
[1] Negatively curved manifolds, elliptic operators and Martin boundary (to appear).

ARNOLD, L. and WIHSTUTZ, V. (editors)
[1] *Lyapunov Exponents (Proceedings): Lecture Notes in Mathematics 1186*, Springer, Berlin, 1986.

AZEMA, J. and YOR, M.
[1] Une solution simple au problème de Skorokhod, *Séminaire de Probabilités XIII: Lecture Notes in Mathematics 721*, Springer, Berlin, 1979, pp. 90–115, 625–633.
[2] (Editors) Temps locaux, *Astérisque* **52–53**, Société Mathématique de France, 1978.

AZENCOTT,
[1] Grandes déviations et applications, *Ecole d'Été de Probabilités de Saint-Flour VIII: Lecture Notes in Mathematics 774*, Springer, Berlin, 1980.

BARLOW, M. T.
[1] Study of a filtration expanded to include an honest time, *Z. Wahrscheinlichkeitstheorie*, **44**, 307–323 (1978).
[2] Decomposition of a Markov process at an honest time (unpublished).
[3] One dimensional stochastic differential equation with no strong solution, *J. London Math. Soc.*, **26**, 335–347 (1982).
[4] On Brownian local time, *Séminaire de Probabilités XV: Lecture Notes in Mathematics 850*, Springer, Berlin, 1981, pp. 189–190.

BARLOW, M. T., JACKA, S. and YOR, M.
[1] Inequalities for a pair of processes stopped at a random time, *Proc. London Math. Soc.*, **52**, 142–172 (1986).
[2] Inégalities pour un couple de processus arrêtés à un temps quelconque, *C. R. Acad. Sci.*, **299**, 351–354 (1984).

BARLOW, M. T. and PERKINS, E.
[1] One-dimensional stochastic differential equations involving a singular increasing process, *Stochastics*, **12**, 229–249 (1984).
[2] Strong existence, uniqueness and non-uniqueness in an equation involving local time, *Séminaire de Probabilités XVII: Lecture Notes in Mathematics 986*, Springer, Berlin, 1983, pp. 32–66.

BARLOW, M. T. and YOR, M.
[1] (Semi-) martingale inequalities and local times, *Z. Wahrscheinlichkeitstheorie*, **55**, 237–254 (1981).
[2] Semi-martingale inequalities via the Garsia–Rodemich–Rumsey lemma and applications to local times, *J. Funct. Anal.*, **49**, 198–229 (1982).

BASS, R. and CRANSTON, M.
[1] The Malliavin calculus for pure jump processes and applications to local time, *Ann. Prob.*, **14**, 490–532 (1986).

BAXENDALE, P.
[1] Asymptotic behaviour of stochastic flows of diffeomorphisms; two case studies, *Probab. Th. Rel. Fields*, **73**, 51–85 (1986).
[2] Moment stability and large deviations for linear stochastic differential equations,
[3] The Lyapunov spectrum of a stochastic flow of diffeomorphisms.
[4] Brownian motions on the diffeomorphism group, I, *Compos. Math.*, **53**, 19–50 (1984).

BAXENDALE, P. and STROOCK, D. W.
[1] Paper on Lyapunov exponents (to appear).

BENSOUSSAN, A.
[1] Lectures on stochastic control, *Nonlinear Filtering and Stochastic Control: Lecture Notes in Mathematics 972*, Springer, Berlin, 1982, pp. 1–62.

BENEŠ, V. E., SHEPP, L. A. and WITSENHAUSEN, H. S.
[1] Some solvable stochastic control problems, *Stochastics*, **4**, 39–83 (1980).

BENVENISTE, A. and JACOD, J.
[1] Systèmes de Lévy des processus de Markov, *Invent. Math.*, **21**, 183–198 (1973).

BERMAN, S. M.
[1] Local times and sample function properties of stationary Gaussian processes, *Trans. Amer. Math. Soc.*, **137**, 277–300 (1969).
[2] Harmonic analysis of local times and sample functions of Gaussian processes, *Trans. Amer. Math. Soc.*, **143**, 269–281 (1969).
[3] Gaussian processes with stationary increments: local times and sample function properties, *Ann. Math. Statist.*, **41**, 1260–1272 (1970).

BICHTELER, K.
[1] Stochastic integration and L^p-theory of semi-martingales, *Ann. Prob.*, **9**, 49–89 (1981).

BICHTELER, K. and FONKEN, D.
[1] A simple version of the Malliavin calculus in dimension one, *Martingale Theory in Harmonic Analysis and Banach Spaces: Lecture Notes in Mathematics 939*, Springer, Berlin, 1982, pp. 6–12.

BICHTELER, K. and JACOD, J.
[1] Calcul de Malliavin pour les diffusions avec sauts: Existence d'une densité dans le cas unidimensionnel, *Séminaire de Probabilités XVII: Lecture Notes in Mathematics 986*, Springer, Berlin, 1983, pp. 132–157.

BILLINGSLEY, P.
[1] *Ergodic Theory and Information*, Wiley, New York, 1965.
[2] *Convergence of Probability Measures*, Wiley, New York, 1968.
[3] Conditional distributions and tightness, *Ann. Prob.*, **2**, 480–485 (1974).

BINGHAM, N. H. and DONEY, R. A.
[1] On fluctuation theory in higher dimensions.

BISHOP, R. and CRITTENDEN, R. J.
[1] *Geometry of Manifolds*, Academic Press, New York, 1964.

BISMUT, J.-M.
[1] *Méchanique Aléatoire: Lecture Notes in Mathematics 866*, Springer, Berlin, 1981.
[2] Martingales, the Malliavin calculus and hypoellipticity under general Hörmander's conditions, *Z. Wahrscheinlichkeitstheorie*, **56**, 469–505 (1981).
[3] Calcul de variations stochastiques et processus de sauts, *Z. Wahrscheinlichkeitstheorie*, **56**, 469–505 (1983).
[4] Large deviations and the Malliavin calculus, *Progress in Math.*, Birkhauser, Boston, 1984.
[5] The Atiyah–Singer theorems; a probabilistic approach: I, The index theorem, *J. Funct. Anal.*, **57**, 56–98 (1984); II, The Lefschetz fixed-point formulas, *ibid.*, 329–348.

BISMUT, J.-M. and MICHEL, D.
[1] Diffusions conditionnelles, I, II, *J. Funct. Anal.*, **44**, 174–211 (1981); **45**, 274–292 (1981).

BLACKWELL, D. and KENDALL, D. G.
[1] The Martin boundary for Polya's urn scheme and an application to stochastic population growth, *J. Appl. Prob.*, **1**, 284–296 (1964).

BLUMENTHAL, R. M. and GETOOR, R. K.
[1] *Markov Processes and Potential Theory*, Academic Press, New York, 1968.

BREIMAN, L.
[1] *Probability*, Addison-Wesley, Reading, Mass., 1968.

BREMAUD, P.
[1] *Point Processes and Queues: Martingale Dynamics*, Springer, New York, 1981.

BOURBAKI, N.
[1] Topologie générale, in *Eléments de Mathématique*, Hermann, Paris, 1958, Chap. IX, 2nd edition.

BOUGEROL, P. and LACROIX, J.
[1] *Products of Random Matrices with Applications to Schrödinger Operators*, Birkhäuser, Boston, 1985.

BURKHOLDER, D.
[1] Distribution function inequalities for martingales, *Ann. Prob.*, **1**, 19–42 (1973).

CARLEN, E. A.
[1] Conservative diffusions, *Comm. Math. Phys.*, **94**, 293–315 (1984).
[2] Potential scattering in quantum mechanics, *Ann. Inst. H. Poincaré*, **42**, 407–428 (1985).

CARVERHILL, A. P.
[1] Flows of stochastic dynamical systems: ergodic theory, *Stochastics*, **14**, 273–318 (1985).
[2] A formula for the Lyapunov exponents of a stochastic flow. Application to a perturbation theorem, *Stochastics*, **14**, 209–226 (1985).

[3] A 'Markovian' approach to the multiplicative ergodic (Oseledeč) theorem for nonlinear stochastic dynamical systems.

CARVERHILL, A. P., CHAPPELL, M. J. and ELWORTHY, K. D.
[1] Characteristic exponents for stochastic flows, *Proceedings, BIBOS I: Stochastic Processes.*

CARVERHILL, A. P. and ELWORTHY, K. D.
[1] Flows of stochastic dynamical systems; the functional analytic approach, *Z. Wahrscheinlichkeitstheorie*, **65**, 245–268 (1983).

CHALEYAT-MAUREL, MIREILLE
[1] La condition d'hypoellipticité d'Hörmander, *Astérisque*, **84–85**, 189–202 (1981).

CHALEYAT-MAUREL, MIREILLE and EL KAROUI, NICOLE
[1] Un problème de réflexion et ses applications au temps local et aux équations différentielles stochastiques sur $\mathbb{R}$, cas continu. In Azema and Yor [2], pp. 117–144.

CHEEGER, J. and EBIN, D. G.
[1] *Comparison Theorems in Riemannian Geometry*, North-Holland, Amsterdam, Oxford, New York, 1975.

CHUNG, K. L.
[1] *Markov Chains with Stationary Transition Probabilities*, 2nd edition, Springer, Berlin, 1967.

[2] Probabilistic approach in potential theory to the equilibrium problem, *Ann. Inst. Fourier, Grenoble*, **23**, 313–322 (1973).
[3] Excursions in Brownian motion, *Ark. Mat.*, **14**, 155–177 (1976).

CHUNG, K. L. and GETOOR, R. K.
[1] The condenser problem, *Ann. Prob.*, **5**, 82–86 (1977).

CHUNG, K. L. and WALSH, J. B.
[1] To reverse a Markov process, *Acta Math.*, **123**, 225–251 (1969).
[2] Meyer's theorem on previsibility, *Z. Wahrscheinlichkeitstheorie*, **29**, 253–256 (1974).

CHUNG, K. L. and WILLIAMS, R. J.
[1] *Introduction to Stochastic Integration*, Birkhaüser, Boston, 1983.

CIESIELSKI, Z. and TAYLOR, S. J.
[1] First passage times and sojourn times for Brownian motion in space and the exact Hausdorff measure of the sample path, *Trans. Am. Math. Soc.*, **103**, 434–450 (1962).

ÇINLAR, E., JACOD, J., PROTTER, P. and SHARPE, M. J.
[1] Semimartingales and Markov processes, *Z. Wahrscheinlichkeitstheorie*, **54**, 161–220 (1980).

ÇINLAR, E., CHUNG, K. L. and GETOOR, R. K. (editors)
[1] *Seminars on Stochastic Processes* 1981, 1982, 1983, 1984 (four volumes), Birkhäuser, Boston.

CLARK, J. M. C.
[1] The representation of functionals of Brownian motion by stochastic integrals, *Ann. Math. Stat.*, **41**, 1282–1295 (1970); **42**, 1778 (1971).
[2] An introduction to stochastic differential equations on manifolds, in *Geometric Methods in Systems Theory* (eds. D. Q. Mayne and R. W. Brockett), Reidel, Dordrecht, 1973.

[3] The design of robust approximations to the stochastic differential equations of non-linear filtering, in *Communication Systems and Random Process Theory* (ed. J. Skwirzynski), Sijthoff and Noordhoff, Alphen an den Rijn, 1978.

CLARKSON, B. (editor)
[1] *Stochastic problems in dynamics*, Pitman, London, 1977.

COCOZZA, C. and YOR, M.
[1] Démonstration simplifiée d'un thèoreme de Knight, *Séminaire de Probabilités XIV: Lecture Notes in Mathematics 721*, Springer, Berlin, 1980, pp. 496–499.

CRANSTON, M.
[1] Means of approach of two-dimensional Brownian motion (to appear in *Ann. Probab.*).

CUTLAND, N.
[1] Non-standard measure theory and its applications, *Bull. London Math. Soc.*, **15**, 529–589 (1983).

CUTLAND, N. and KENDALL, W. S.
[1] A non-standard proof of one of David Williams' splitting-time theorems, in D. G. Kendall [5], pp. 37–48.

DARLING, R. W. R.
[1] Martingales in manifolds—definition, examples, and behaviour under maps, *Séminaire de Probabilités XVI Supplement: Lecture Notes in Mathematics 921*, Springer, Berlin, 1982, pp. 217–236.

DAVIES, E. B. and SIMON, B.
[1] Ultracontractivity and the heat kernel for Schrödinger operators and Dirichlet Laplacians, *J. Funct. Anal.* **59**, 335–395 (1984).

DAVIES, B.
[1] Picard's theorem and Brownian motion, *Trans. Amer. Math. Soc.*, **213**, 353–362 (1975).

DAVIS, M. H. A.
[1] On a multiplicative functional transformation arising in non-linear filtering theory, *Z. Wahrscheinlichkeitstheorie*, **54**, 125–139 (1980).
[2] Pathwise non-linear filtering, in *Stochastic Systems: the Mathematics of Filtering and Identification and Applications* (eds. M. Hazewinkel and J. C. Willems), Reidel, Dordrecht, 1981.
[3] Some current issues in stochastic control theory, *Stochastics*.

DAVIS, M. H. A. and VARAIYA, P.
[1] Dynamic programming conditions for partially observed stochastic systems, *SIAM J. Control*, **11**, 226–261 (1973).

DAWSON, D. A. and GÄRTNER, J.
[1] Large deviations from the McKean–Vlasov limit for weakly-interacting diffusions, *Stochastics*, **20**, 247–308 (1987).

DELLACHERIE, C.
[1] *Capacités et Processus Stochastiques*, Springer, Berlin, 1972.
[2] Quelques exemples familiers en probabilités d'ensembles analytiques non-Boréliens, *Séminaire de Probabilités XII: Lecture Notes in Mathematics*, Springer, Berlin, 1978, pp. 742–745.
[3] Un survoi de la theorie de l'intégrale stochastique, *Stoch. Proc. Appl.*, **10**, 115–144 (1980).

DELLACHERIE, C., DOLEANS(-DADE), CATHERINE, LETTA, G. and MEYER, P.-A.
[1] Diffusions à coefficients continus d'après D. W. Stroock et S. R. S. Varadhan, *Séminaire de probabilités IV: Lecture Notes in Mathematics 124*, Springer, Berlin, 1970, pp. 241–282.

DELLACHERIE, C. and MEYER, P. A.
[1] *Probabilités et Potentiel*, Chaps. I–VI, Hermann, Paris, 1975; Chaps. V–VIII, Hermann, Paris, 1980; Chaps. IX–XI, Hermann, Paris, 1983; Chapters XII–XVI (1987).

DE WITT-MORETTE, CECILE and ELWORTHY, K. D. (editors)
[1] New stochastic methods in physics, *Physics Reports*, **77**, (3), 121–382 (1981).

DOLÉANS(-DADE), CATHERINE
[1] Existence du processus croissant naturel associé à un potentiel de la classe (D), *Z. Wahrscheinlichkeitstheorie*, **9**, 309–314 (1968).
[2] Quelques applications de la formule de changement de variables pour les semimartingales, *Z. Wahrscheinlichkeitsth.*, **16**, 181–194 (1970).

DOLÉANS-DADE, C. and MEYER, P. A.
[1] Equations différentielles stochastiques, *Sém. de Probabilités XI: Lecture Notes in Mathematics 581*, Springer, Berlin, 1977, pp. 376–382.

DOOB, J. L.
[1] *Stochastic Processes*, Wiley, New York, 1953.
[2] State-spaces for Markov chains, *Trans. Am. Math. Soc.*, **149**, 279–305 (1970).
[3] *Classical Potential Theory and its Probabilistic Counterpart*, Springer, New York, 1981.

DOSS, H.
[1] Liens entre équations différentielles stochastiques et ordinaires, *Ann. Inst. Henri Poincaré B*, **13**, 99–126 (1977).

DUBINS, L. and SCHWARZ, G.
[1] On continuous martingales, *Proc. Nat. Acad. Sci. USA*, **53**, 913–916 (1965).

DUNFORD, N. and SCHWARTZ, J. T.
[1] *Linear operators: Part I, General Theory*, Interscience, New York, 1958.

DURRETT, R.
[1] *Brownian Motion and Martingales in Analysis*, Wadsworth, Belmont, Ca., 1984.
[2] (Editor) *Particle systems, random media, large deviations, Contemporary Maths.* **41**, Amer. Math. Soc., Providence, RI, 1985.

DYNKIN, E. B.
[1] *Theory of Markov Processes*, English translation, Pergamon Press, Oxford, 1960.
[2] *Markov Processes*, English translation in two volumes, Springer, Berlin, 1965.
[3] Non-negative eigenfunctions of the Laplace–Beltrami operator and Brownian motion in certain symmetric spaces (in Russian), *Dokl. Akad. Naud SSSR*, **141**, 288–291 (1961).
[4] Diffusion of tensors, *Dokl. Acad. Nauk. SSSR*, **179**, 1264–1267 (1968).
[5] Local times and quantum fields, in Çinlar, Chung and Getoor [1, 1983].
[6] Gaussian and non-Gaussian random fields associated with Markov processes, *J. Funct. Anal.*, **55**, 344–376 (1984).
[7] Self-intersection local times, occupation fields and stochastic integrals, (to appear in *Advances in Appl. Math.*).

ELLIOTT, R. J.
[1] *Stochastic Calculus and Applications*, Springer, Berlin, 1982.

ELLIOTT, R. J. and ANDERSON, B. D. O.
[1] Reverse time diffusions, *Stochastic Processes and their Applications*, **19**, 327–339 (1985).

ELWORTHY, K. D.
[1] *Stochastic Differential Equations on Manifolds*, London Mathematical Society Lecture Note Series 20, Cambridge University Press, Cambridge, 1982.
[2] (Editor) *From local times to global geometry, control and physics*, Proceedings, Warwick Symposium 1984/85, Longman, Harlow and Wiley, New York, 1986.

ELWORTHY, K. D. and STROOCK, D. W.
[1] Large deviation theory for mean exponents of stochastic flows, Appendix to Carverhill, Chappell and Elworthy [1].

ELWORTHY, K. D. and TRUMAN, A.
[1] Classical mechanics, the diffusion (heat) equation and the Schrödinger equation on a Riemannian manifold, *J. Math. Phys.*, **22**, (10), 2144–2166 (1981).
[2] The diffusion equation and classical mechanics: an elementary formula, in *Stochastic processes in quantum theory and statistical physics* (ed. S. Albeverio et al.), *Lecture Notes in Physics*, **173**, Springer, Berlin, 1982, pp. 136–146.

EMERY, M.
[1] Annoncabilité des temps prévisibles: deux contre-exemples, *Séminaire de Probabilités IV: Lecture Notes in Mathematics 784*, Springer, Berlin, 1980, pp. 318–323.

ETHIER, S. N. and KURTZ, T. G.
[1] *Markov Processes: Characterization and Convergence*, Wiley, New York, 1986.

FELLER, W.
[1] *Introduction to Probability Theory and its Applications, Vol. 1, 2nd edn.*, Wiley, New York, 1957; *Vol. 2*, Wiley, New York, 1966.
[2] Boundaries induced by non-negative matrices, *Trans. Am. Math. Soc.*, **83**, 19–54 (1956).
[3] On boundaries and lateral conditions for the Kolmogorov equations, *Ann. Math.*, Ser. II, **65**, 527–570 (1957).
[4] Generalized second-order differential operators and their lateral conditions, *Illinois J. Math.*, **1**, 459–504 (1957).

FLEMING, W. H. and RISHEL, R. W.
[1] *Deterministic and Stochastic Optimal Control*, Springer, Berlin, 1975.

FÖLLMER, H.
[1] Calcul d'Itô sans probabilités, *Sém. de Probabilités XV: Lecture Notes in Mathematics 850*, Springer, Berlin, 1981, pp. 143–150.

FREEDMAN, D.
[1] *Brownian Motion and Diffusion*, Holden-Day, San Francisco, 1971.
[2] *Approximating Countable Markov Chains*, Holden-Day, San Francisco, 1972.

FRIEDMAN, A.
[1] *Stochastic Differential Equations and Applications*, in two volumes, Academic Press, New York, 1975.

FUJISAKI, M., KALLIANPUR, G. and KUNITA, H.
[1] Stochastic differential equations for the non-linear filtering problem, *Osaka J. Math.*, **9**, 19–40 (1972).

FUKUSHIMA, M.
[1] *Dirichlet Forms and Markov Processes*, Kodansha, Tokyo, 1980.
[2] Basic properties of Brownian motion and a capacity on the Wiener space, *J. Math. Soc. Japan*, **36**, 161–176 (1984).

GARCIA ALVAREZ, M. A. and MEYER, P. A.
[1] Une théorie de la dualité à un ensemble polaire près: I, *Ann. Prob.*, **1**, 207–222 (1973).

GARSIA, A.
[1] *Martingale Inequalities: Seminar Notes on Recent Progress*, Benjamin, Reading, Ma., 1973.

GEMAN, D. and HOROWITZ, J.
[1] Occupation densities, *Ann. Prob.*, **8**, 1–67 (1980).

GEMAN, D., HOROWITZ, J. and ROSEN, J.
[1] A local time analysis of intersections of Brownian paths in the plane, *Ann. Prob.*, **12**, 86–107 (1984).

GETOOR, R. K.
[1] *Markov processes: Ray Processes and Right Processes: Lecture Notes in Mathematics 440*, Springer, Berlin, 1975.
[2] Excursions of a Markov process, *Ann. Prob.*, **8**, 244–266 (1979).
[3] Splitting times and shift functionals, *Z. Wahrscheinlichkeitstheorie*, **47**, 69–81 (1979).

GETOOR, R. K. and SHARPE, M. J.
[1] Last exit times and additive functionals, *Ann. Prob.*, **1**, 550–569 (1973).
[2] Excursions of Brownian motion and Bessel process, *Z. Wahrscheinlichkeitstheorie*, **47**, 83–106 (1979).
[3] Last exit decompositions and distributions, *Indiana Univ. Math. J.*, **23**, 377–404 (1973).
[4] Excursions of dual processes, *Advances in Math.*, **45**, 259–309 (1982).

GIKHMAN, I. I. and SKOROKHOD, A. V.
[1] *The Theory of Stochastic Processes* (three volumes), Springer, Berlin, 1979.

GRAY, A., KARP, L. and PINSKY, M. A.
[1] The mean exit time from a ball in a Riemannian manifold.

GRAY, A. and PINSKY, M. A.
[1] The mean exit time from a small geodesic ball in a Riemannian manifold, *Bull. Sc. Math.*, **107**, 345–370 (1983).

GREENWOOD, P. and PITMAN, J. W.
[1] Construction of local time and Poisson point processes from nested arrays, *J. London Math. Soc.* (2), **22**, 182–192 (1980).
[2] Fluctuation identities for Lévy processes and splitting at the maximum, *Adv. Appl. Prob.*, **12**, 893–902 (1980).

GRENANDER, U.
[1] *Probabilities on Algebraic Structures*, Wiley, New York, 1963.

GRIFFEATH, D.
[1] Coupling methods for Markov processes, in *Advances in Mathematics Supplementary Studies: Studies in Probability and Ergodic Theory*, Vol. 2, Academic Press, New York, 1978, pp. 1–43.

GROMOV, M. and ROHLIN, V. A.
[1] *Russian Math. Surveys*, **25**, 1–57 (1970).

HALMOS, P.
[1] *Measure Theory*, Van Nostrand, Princeton, NJ, 1959.

HAUSSMANN, U.
[1] On the integral representation of Itô processes, *Stochastics*, **3**, 17–27 (1979).
[2] *A stochastic maximum principle for optimal control of diffusions*, Longman, Harlow, 1986.

HAWKES, J.
[1] Multiple points for symmetric Lévy processes, *Math. Proc. Camb. Phil.*, **83**, 83–90 (1978).
[2] The measure of the range of a subordinator, *Bull. London Math. Soc.*, **5**, 21–28 (1973).

HAZEWINKEL, M. and WILLEMS, J. C. (editors)
[1] *Stochastic Systems: the Mathematics of Filtering and Identification and Applications*, Reidel, Dordrecht, 1981.

HELGASON, S.
[1] *Differential Geometry and Symmetric Spaces*, Academic Press, New York, 1962.

HILLE, E. and PHILLIPS, R. S.
[1] *Functional Analysis and Semigroups*, American Mathematical Society Colloquium Publications, Providence, RI, 1957.

HOLLEY, R., STROOCK, D. W. and WILLIAMS, D.
[1] Applications of dual processes to diffusion theory, *Proc. AMS Prob. Symp.*, Urbana, 1976, pp. 23–36.

HÖRMANDER, L.
[1] Hypoelliptic second-order differential equations, *Acta Math.*, **117**, 147–171 (1967).

HSU, P.
[1] On excursions of reflecting Brownian motion, *Trans. Amer. Math. Soc.*, **296**, 239–264 (1986).
[2] Brownian motion and the index theorem (to appear).

HUNT, G. A.
[1] Markoff processes and potentials: I, II, III, *Illinois J. Math.*, **1**, 44–93; 316–369 (1957); **2**, 151–213 (1958).

IKEDA, N. and WATANABE, S.
[1] *Stochastic Differential Equations and Diffusion Processes*, North Holland–Kodansha, Amsterdam and Tokyo, 1981.
[2] Malliavin calculus of Wiener functionals and its applications, in Elworthy [2], pp. 132–178.

ITÔ, K.
[1] Stochastic integral, *Proc. Imp. Acad. Tokyo*, **20**, 519–524 (1944).
[2] On a stochastic integral equation, *Proc. Imp. Acad. Tokyo*, **22**, 32–35 (1946).
[3] Stochastic differential equations in a differential manifold, *Nagoya Math. J.*, **1**, 35–47 (1950).
[4] The Brownian motion and tensor fields on a Riemannian manifold, *Proc. Intern. Congr. Math., Stockholm*, 1963, pp. 536–539.

[5] Stochastic parallel displacement, in *Probabilistic Methods in Differential Equations: Lecture Notes in Mathematics 451*, Springer, Berlin, 1975, pp. 1–7.
[6] Poisson point processes attached to Markov processes, *Proc. 6th Berkeley Symp. Math. Statist. Prob.*, Vol. 3, University of California Press, 1971, pp. 225–240.
[7] (editor) *Proceedings of the 1982 Taniguchi Intern. Symp. on Stochastic Analysis*, Kinokuniya–Wiley, 1984.

ITÔ, K. and MCKEAN, H. P.
[1] *Diffusion Processes and their Sample Paths*, Springer, Berlin, 1965.

JACKA, S.
[1] A finite fuel stochastic control problem, *Stochastics*, **10**, 103–113 (1983).
[2] A local time inequality for martingales, *Sém. de Probabilités XVII: Lecture Notes in Mathematics 986*, Springer, Berlin, 1983.

JACOBSEN, M.
[1] Splitting times for Markov processes and a generalised Markov property for diffusions, *Z. Wahrscheinlichkeitstheorie*, **30**, 27–43 (1974).
[2] *Statistical Analysis of Counting Processes: Lecture Notes in Statistics 12*, Springer, New York, 1982.

JACOD, J.
[1] A general theorem of representation for martingales, *Proc. AMS Prob. Symp., Urbana*, 1976, pp. 37–53.
[2] *Calcul Stochastique et Problèmes de Martingales: Lecture Notes in Mathematics 714*, Springer, Berlin, 1979.

JACOD, J. and YOR, M.
[1] Etude des solutions extrémales et représentation intégrale des solutions pour certains problèmes de martingales, *Z. Wahrscheinlichkeitsth.*, **38**, 83–125 (1977).

JEULIN, T.
[1] *Semimartingales et Grossissement d'une Filtration: Lecture Notes in Mathematics 833*, Springer, Berlin, 1980.

JEULIN, T. and YOR, M.
[1] Grossissement d'une filtration et semi-martingales: formules explicites, *Séminaire de Probabilités XII: Lecture Notes in Mathematics 649*, Springer, Berlin, 1978, pp. 78–97.
[2] (editors) *Grossissements de Filtrations: Examples et Applications: Lecture Notes in Mathematics 1118*, Springer, Berlin, 1985.

JOHNSON, G. and HELMS, L. L.
[1] Class (D) supermartingales, *Bull. Amer. Math. Soc.*, **69**, 59–62 (1963).

KAILATH, T.
[1] An innovations approach to least squares estimation, Part I: Linear filtering with additive white noise, *IEEE Trans. Automatic Control*, **13**, 646–655 (1968).

KALLIANPUR, G.
[1] *Stochastic Filtering Theory*, Springer, Berlin, 1980.

KENDALL, D. G.
[1] Pole-seeking Brownian motion and bird navigation (with discussion), *J. Roy. Statist. Soc.* B, **36**, 365–417 (1974).
[2] The diffusion of shape, *Adv. Appl. Prob.*, **9**, 428–430 (1979).
[3] Shape manifolds, Procrustean metrics, and complex projective spaces, *Bull. London Math. Soc.*, **16**, 81–121 (1984).

[4] A totally unstable Markov process, *Quarterly J. Math. Oxford*, **9**, (34), 149–160 (1958).
[5] (Editor) *Analytic and geometric stochastics* (special supplement to Adv. Appl. Prob. to honour G. E. H. Reuter), Applied Prob. Trust, 1986.

KENDALL, D. G. and REUTER, G. E. H.
[1] Some pathological Markov processes with a denumerable infinity of states and the associated contraction semigroups of operators on *l*, *Proc. Intern. Congress Math.* 1954 (Amsterdam), **3**, 377–415 (1956).

KENDALL, W. S.
[1] Knotting of Brownian motion in 3-space, *J. London Math. Soc.* (2), **19**, 378–384 (1979).
[2] Brownian motion, negative curvature, and harmonic maps. *Stochastic Integrals: Lecture Notes in Mathematics 851*, Springer, Berlin, 1981, pp. 479–491.
[3] Brownian motion on a surface of negative curvature, *Séminaire de probabilités XVIII: Lecture Notes in Mathematics 1059*, Springer, Berlin, 1984, pp. 70–76.
[4] Survey article on stochastic differential geometry (to appear).

KENT, J.
[1] Some probabilistic properties of Bessel functions, *Ann. Prob.*, **6**, 760–770 (1978).
[2] The infinite divisibility of the von Mises-Fisher distribution for all values of the parameter in all dimensions. *Proc. London Math. Soc.*, **3**, (35), 359–384 (1977).

KHASMINSKII, R. Z.
[1] Ergodic properties of recurrent diffusion processes and stabilization of the solution of the Cauchy problem for parabolic equations, *Th. Prob. and Appl.*, **5**, 179–196 (1960).
[2] *Stochastic stability of differential equations*, Sijthoff and Noordhoff, Alphen aan den Rijn, 1980.

KIFER, Y.
[1] Brownian motion and positive harmonic functions on complete manifolds of non-positive curvature, in Elworthy [2], pp. 187–232.

KINGMAN, J. F. C.
[1] Subadditive ergodic theory, *Ann. Prob.*, **1**, 883–909 (1973).
[2] Completely random measures, *Pacific J. Math.*, **21**, 59–78 (1967).
[3] *Regenerative Phenomena*, Wiley, New York, 1972.

KNIGHT, F. B.
[1] Note on regularisation of Markov processes, *Illinois J. Math.*, **9**, 548–552 (1965).
[2] A reduction of continuous square-integrable martingales to Brownian motion, in *Martingales: a Report on a Meeting at Oberwolfach (ed. H. Dinges): Lecture Notes in Mathematics 190*, Springer, Berlin, 1971, pp. 19–31.
[3] Random walks and the sojourn density process of Brownian motion, *Trans. Amer. Math. Soc.*, **107**, 56–86 (1963).

KNIGHT, F. B. and PITTENGER, A. O.
[1] Excision of a strong Markov process, *Z. Wahrscheinlichkeitsth.*, **23**, 114–120 (1972).

KOBAYASHI, S. and NOMIZU, K.
[1] *Foundations of Differential Geometry*, Wiley–Interscience, New York, 1969.

KOZIN, F. and PRODROMOU, S.
[1] Necessary and sufficient conditions for almost sure sample stability of linear Itô equations, *SIAM J. Appl. Math.*, **21**, 413–425 (1971).

KRYLOV, N. V.
[1] *Controlled Diffusion Processes*, Springer, New York, 1980.

KUELBS, J.
[1] The law of the iterated logarithm for Banach space valued random variables, in *Probability in Banach Spaces: Lecture Notes in Mathematics 526*, Springer, Berlin, 1976, pp. 131–142.

KUNITA, H.
[1] On the decomposition of the solutions of stochastic differential equations, in *Stochastic Integrals: Lecture Notes in Mathematics 851*, Springer, Berlin, 1981, pp. 213–255.
[2] On backward stochastic differential equations, *Stochastics*, **6**, 293–313 (1982).
[3] Stochastic differential equations and stochastic flows of homeomorphisms.
[4] Stochastic partial differential equations connected with nonlinear filtering, in Mitter and Moro [1].

KUNITA, H. and WATANABE, S.
[1] On square integrable martingales, *Nagoya Math. J.*, **30**, 209–245 (1967).

KUNITA, H. and WATANABE, T.
[1] Some theorems concerning resolvents over locally compact spaces, in *Proc. 5th Berkeley Symp. Math. Statist. Prob.*, Vol. 2, Part 2, University of California Press, 1967, pp. 131–164.
[2] Markov processes and Martin boundaries, I, *Illinois J. Math.*, **9**, 485–526 (1965).
[3] On certain reversed processes and their application to potential theory and boundary theory, *J. Math. Mech.*, **15**, 393–434 (1966).

KUSUOKA, S. and STROOCK, D.
[1] Applications of the Malliavin calculus, Part I, *Proceedings of the 1982 Taniguchi Intern. Symp. on Stochastic Analysis* (ed. K. Itô), Kinokuniya–Wiley, 1984, pp. 271–306.
[2] Applications of the Malliavin calculus, Part II, *J. Fac. Sci. U. of Tokyo* (IA), **32**, 1–76 (1985).

LE GALL, J.-F.
[1] Applications du temps local aux equations différentielles stochastiques unidimensionelles, *Séminaire de Probabilités XVII: Lecture Notes in Mathematics 986*, Springer, Berlin, 1983, pp. 15–31.
[2] Sur la saucisse de Wiener et les points multiples du mouvement Brownien, *Ann. Prob.*, **14**, 1219–1244 (1986).
[3] Sur les temps local d'intersection du mouvement Brownien plan et la méthode de renormalization de Varadhan, *Séminaire de Probabilités XIX: Lecture Notes in Mathematics 1123*, Springer, Berlin, 1985, pp. 314–331.

LENGLART, E., LEPINGLE, D. and PRATELLI, M.
[1] Présentation unifiée de certaines inégalités de la théorie des martingales, *Séminaire de Probabilités XIV: Lecture Notes in Mathematics 784*, Springer, Berlin, 1980.

LEVY, P.
[1] *Théorie de l'Addition des Variables Aléatoires*, Gauthier Villars, Paris, 1954.
[2] *Processus Stochastiques et Mouvement Brownien*, Gauthier Villars, Paris, 1965.
[3] Systèmes markoviens et stationnaires. Cas dénombrable, *Ann. Ecole Norm. Sup.* (3), **68**, 327–381 (1951); **69**, 203–212 (1952).
[4] Processus markoviens et stationnaires due cinquième type (infinité dénombrable des états possibles, parametre continu). *C. R. Acad. Sci. Paris*, **236**, 1630–1632 (1953).
[5] Processus markoviens et stationnaires. Cas dénombrable. *Ann. Inst H. Poincaré*, **16**, 7–25 (1958).

LEWIS, J. T.
[1] Brownian motion on a submanifold of Euclidean space, *Bull. London Math. Soc.*, **18**, 616–20 (1986).

LIGGETT, T.
[1] *Interacting Particle Systems*, Springer, New York, 1985.

LINDVALL, T.
[1] On coupling of diffusion processes, *J. Appl. Probab.*, **20**, 82–93 (1983).

LIPSTER, R. S. and SHIRYAYEV, A. N.
[1] *Statistics of Random Processes*, I (English translation), Springer, Berlin, 1977.

LONDON, R. R., MCKEAN, H. P., ROGERS, L. C. G. and WILLIAMS, D.
[1] A martingale approach to some Wiener–Hopf problems, I, *Séminaire de Probabilités XVI: Lecture Notes in Mathematics 920*, Springer, Berlin, 1982, pp. 41–67.

LYONS, T. J.
[1] Finely holomorphic functions, *J. Funct. Anal.*, **37**, 1–18 (1980).
[2] Instability of the Liouville property for quasi-isometric Riemannian manifolds and reversible Markov chains, (to appear).
[3] The critical dimension at which quasi-every path is self-avoiding, in D. G. Kendall [5], pp. 87–100.

LYONS, T. J. and MCKEAN, H. P.
[1] Windings of the plane Brownian motion. *Adv. Math.*, **51**, 212–225 (1984).

MAISONNEUVE, B.
[1] Systèmes régéneratifs, *Asterique*, **15**, Société Mathematique de France, 1974.

MAISONNEUVE, B. and MEYER, P.-A.
[1] Ensembles aléatoires markoviens homogènes, in *Séminaire de probabilités VIII: Lecture Notes in Mathematics* 381, Springer, Berlin, 1974, pp. 172–261.

MALLIAVIN, P.
[1] Stochastic calculus of variation and hypo-elliptic operators, *Proc. Intern. Symp. Stoch. Diff. Equations*, Kyoto, 1976 (ed. K. Itô), Kinokuniya–Wiley, 1978, pp. 195–263.
[2] C^k-hypoellipticity with degeneracy, in *Stochastic Analysis* (ed. A. Friedman and M. Pinsky), Academic Press, New York, 1978, pp. 199–214.
[3] Formule de la moyenne, calcul de perturbations et théorèmes d'annulation pour les formes harmoniques, *J. Funct. Anal.*, **17**, 274–291 (1974).

MALLIAVIN, M. P. and MALLIAVIN, P.
[1] Factorisations et lois limites de la diffusion horizontale au dessus d'un espace riemmanien symmetrique, *Lecture Notes in Mathematics 404*, Springer, Berlin, 1974, pp. 166–217.

MARTIN, R. S.
[1] Minimal positive harmonic functions, *Trans. Am. Math. Soc.*, **49**, 137–164 (1941).

MCGILL, P.
[1] Calculation of some conditional excursion formulae, *Z. Wahrscheinlichkeitstheorie*, **61**, 255–260 (1982).
[2] Markov properties of diffusion local time: a martingale approach, *Adv. Appl. Prob.*, **14**, 789–810 (1980).
[3] Integral representation of martingales in the Brownian excursion filtration, *Séminaire de Probabilités XX: Lecture Notes in Mathematics 1204*, Springer, Berlin, 1986, pp. 465–502.

McKean, H. P.
[1] *Stochastic Integrals*, Academic Press, New York, 1969.
[2] Excursions of a non-singular diffusion, *Z. Wahrscheinlichkeitstheorie*, **1**, 230–239 (1963).
[3] Brownian local times, *Adv. Math.*, **16**, 91–111 (1975).

McNamara, J. M.
[1] A regularity condition on the transition probability measure of a diffusion process. *Stochastics*, **15**, 161–182 (1985).

Mandl, P.
[1] *Analytic Treatment of One-Dimensional Markov Processes*, Springer, Berlin, 1968.

Meléard, S.
[1] Application du calcul stochastique à l'étude de processus de Markov réguliers sur [0, 1], *Stochastics*, **19**, 41–82 (1986).

Metivier, M. and Pellaumail, J.
[1] *Stochastic Integration*, Academic Press, New York, 1979.

Meyer, P. A.
[1] Un cours sur les intégrales stochastiques, *Séminaire de Probabilités X: Lecture Notes in Mathematics 511*, Springer, Berlin, 1976, pp. 245–400.
[2] *Probability and Potential* (English translation), Blaisdell, Waltham, Mass., 1966.
[3] *Processus de Markov: Lecture Notes in Mathematics 26*, Springer, Berlin, 1967.
[4] *Processus de Markov: la Frontière de Martin: Lecture Notes in Mathematics 77*, Springer, Berlin, 1970.
[5] Démonstration simplifiée d'un théorème de Knight, *Sém. de Probabilités V: Lecture Notes in Mathematics 191*, Springer, Berlin, 1971, pp. 191–195.
[6] Démonstration probabiliste de certaines inégalités de Littlewood–Paley, *Sém. de Probabilités X: Lecture Notes in Mathematics 511*, Springer, Berlin, 1976, pp. 125–183.
[7] Flot d'un equation différentielle stochastique, *Sém. de Probabilitès XV: Lecture Notes in Mathematics 850*, Springer, Berlin, 1981, pp. 103–117.
[8] Sur la démonstration de prévisibilité de Chung and Walsh, *Sém. de Probabilités IX: Lecture Notes in Mathematics 465*, Springer, Berlin, 1975, pp. 530–533.
[9] Géometrie stochastique sans larmes, *Séminaire de Probabilités XV: Lecture Notes in Mathematics 850*, Springer, Berlin, 1981, pp. 44–102.
[10] Géometrie stochastique sans larmes (bis), *Séminaire de Probabilités XVI: Supplément, Lecture Notes in Mathematics* 921, Springer, Berlin, 1982, pp. 165–207.
[11] Eléments de probabilités quantiques, *Séminaire de Probabilités XX: Lecture Notes in Mathematics 1204*, Springer, Berlin, 1986, pp. 186–312.

Mihlstein, G. N.
[1] Approximate integration of stochastic differential equations, *Th. Prob. Appl.*, **19**, 557–562 (1974).

Millar, P. W.
[1] Random times and decomposition theorems, in *Probability: Proc. Symp. Pure Math. XXXI*, American Mathematical Society, Providence, RI, 1977, pp. 91–103.
[2] A path decomposition for Markov processes, *Ann. Prob.*, **6**, 345–348 (1978).

Mitter, S. K.
[1] Lectures on non-linear filtering and stochastic control, in Mitter and Moro [1], pp. 170–207.

MITTER, S. K. and MORO, A. (editors)
[1] Non-linear filtering and stochastic control, *Lecture Notes in Mathematics 972*, Springer, Berlin, 1982.

MOTOO, M.
[1] Application of additive functionals to the boundary problem of Markov processes (Lèvy's system of U-processes), *Proc. Fifth Berkeley Symposium Math. Statist. Prob.* II(2), Univ. Calif. Press, Berkeley, 1967, pp. 75–110.

MOTOO, M. and WATANABE, S.
[1] On a class of additive functionals of Markov processes, *J. Math. Kyoto Univ.*, **4**, 429–469 (1965).

NAKAO, S.
[1] On the pathwise uniqueness of solutions of one-dimensional stochastic differential equations, *Osaka J. Math.*, **9**, 513–518 (1972).

NASH, J. F.
[1] The imbedding problem for Riemannian manifolds, *Ann. of Math.*, **63**, 20–63 (1956).

NELSON, E.
[1] *Dynamical Theories of Brownian Motion*, Princeton Univ. Press, 1967.
[2] *Quantum Fluctuations*, Princeton Univ. Press, 1984.

NEVEU, J.
[1] *Bases Mathématiques du Calcul des Probabilités*, Masson, Paris, 1964.
[2] Sur les états d'entrée et les états fictifs d'un processus de Markov, *Ann. Inst. Henri Poincaré*, **17**, 323–337 (1962).
[3] Lattice methods and submarkovian processes, *Proc. 4th Berkeley Symp. Math. Statist. Prob.*, Vol. 2, University of California Press, 1960, pp. 347–391.
[4] Une généralisation des processus à accroissements positifs indépendants, *Abh. Math. Sem. Univ. Hamburg*, **25**, 36–61 (1961).
[5] Entrance, exit and fictitious states for Markov chains, *Proc. Aarhus Colloq. Combin. Prob.*, 1962, pp. 64–68.

NORRIS, J. R.
[1] Simplified Malliavin calculus, *Séminaire de Probabilités XX: Lecture Notes in Mathematics 1204*, Springer, Berlin, 1986, pp. 101–130.

NORRIS, J. R., ROGERS, L. C. G. and WILLIAMS, D.
[1] Brownian motion of ellipsoids, *Trans. Amer. Math. Soc.*, **294**, 757–765 (1986).
[2] Self-avoiding random walk: a Brownian motion model with local time drift, *Prob. Thy. and Rel. Fields*, **74**, 271–287 (1987).

OCONE, D.
[1] Malliavin's calculus and stochastic integral: representation of functionals of diffusion processes, *Stochastics*, **12**, 161–185 (1984).

ORIHARA, A.
[1] On random ellipsoid, *J. Fac. Sci. Univ. Tokyo, Sect. 1A Math.*, **17**, 73–85 (1970).

PARDOUX, E.
[1] Stochastic differential equations and filtering of diffusion processes, *Stochastics*, **3**, 127–167 (1979).
[2] Grossissement d'une filtration et retournement du temps d'une diffusion, *Sém. de Probabilités XX: Lecture Notes in Mathematics 1204*, Springer, Berlin, 1986, pp. 48–55.

[3] Equations of non-linear filtering, and applications to stochastic control with partial observations, in Mitter and Moro [1], pp. 208–248.

PARDOUX, E. and TALAY, D.
[1] Discretization and simulation of stochastic differential equations, to appear in *Acta Appl. Math.*.

PARTHASARATHY, K. R.
[1] *Probability Measures on Metric Spaces*, Academic Press, New York, 1967.

PAUWELS, E. and ROGERS, L. C. G.
[1] Paper on Brownian motions on homogeneous spaces (to appear).

PERKINS, E.
[1] Local time and pathwise uniqueness for stochastic differential equations, *Sém. de Probabilités XVI: Lecture Notes in Mathematics 920*, Springer, Berlin, 1982, pp. 201–208.
[2] Local time is a semimartingale, *Z. Wahrscheinlichtkeitsth.*, **60**, 79–117 (1982).

PHELPS, R. R.
[1] *Lectures on Choquet's Theorem*, Van Nostrand, Princeton, NJ, 1966.

PINSKY, M. A.
[1] Homogenization and stochastic parallel displacement, in Williams [13], pp. 271–284.
[2] Stochastic Riemannian geometry, in *Probabilistic Analysis and Related Topics*, 1 (ed. A. T. Bharucha-Reid), Academic Press, New York, 1978.

PITMAN, J. W.
[1] One-dimensional Brownian motion and the three-dimensional Bessel process, *J. Appl. Prob.*, **7**, 511–526 (1975).
[2] Path decomposition for conditional Brownian motion, *Inst. Math. Statist. Univ. Copenhagen*, Preprint No. 11 (1974).
[3] Lévy systems and path decompositions, in Çinlar, Chung and Getoor [1, 1981].

PITMAN, J. W. and YOR, M.
[1] Bessel processes and infinitely divisible laws, in *Stochastic Integrals* (*ed. D. Williams*), *Lecture Notes in Mathematics 851*, Springer, Berlin, 1981.
[2] A decomposition of Bessel bridges. *Z. Wahrscheinlichkeitsth.*, **59**, 425–457 (1982).
[3] The asymptotic joint distribution of windings of planar Brownian motion, *Bull. Amer. Math. Soc.*, **10**, 109–111 (1984).
[4] Asymptotic laws of planar Brownian motion, *Ann. Probab.*, **14**, 733–779 (1986).

PITTENGER, A. O. and SHIH, C. T.
[1] Coterminal families and the strong Markov property, *Trans. Amer. Math. Soc.*, **182**, 1–42 (1973).

POOR, W. A.
[1] *Differential Geometric Structures*, McGraw-Hill, New York, 1981.

PORT, S. C. and STONE, C. J.
[1] Classical potential theory and Brownian motion, *Proc. 6th Berkeley Symp. Math. Statist. Prob.*, Vol. 3, University of California Press, 1972, pp. 143–176.
[2] Logarithmic potentials and planar Brownian motion, *Proc. 6th Berkeley Symp. Math. Statist. Prob.*, Vol. 3, University of California Press, 1972, pp. 177–192.
[3] *Brownian Motion and Classical Potential Theory*, Academic Press, New York, 1978.

PRICE, G. C. and WILLIAMS, D.
[1] Rolling with 'slipping': I, *Sém. de Probabilités XVII: Lecture Notes in Mathematics 986*, Springer, Berlin, 1983, pp. 194–197.

PROHOROV, YU. V.
[1] Convergence of random processes and limit theorems in probability, *Theor. Prob. Applic.*, **1**, 157–214 (1956).

PROTTER, P.
[1] On the existence, uniqueness, convergence and explosions of solutions of stochastic differential equations, *Ann. Probab.*, **5**, 243–261 (1977).

RAO, K. M.
[1] On decomposition theorems of Meyer, *Math. Scand.*, **24**, 66–78 (1969).
[2] Quasimartingales, *Math. Scand.*, **24**, 79–92 (1969).

RAY, D. B.
[1] Resolvents, transition functions and strongly Markovian processes, *Ann. Math.*, **70**, 43–72 (1959).
[2] Sojourn times of a diffusion process, *Illinois J. Math.*, **7**, 615–630 (1963).

REUTER, G. E. H.
[1] Denumerable Markov processes, II, *J. London Math. Soc.*, **34**, 81–91 (1959).

REVUZ, D.
[1] The Martin boundary of a recurrent random walk has one or two points, in *Probability: Proc. Symp. Pure Math. XXXI*, American Mathematical Society, Providence, RI, 1977, pp. 125–130.

ROGERS, L. C. G.
[1] Williams' characterization of the Brownian excursion law: proof and applications, *Séminaire de Probabilités XV: Lecture Notes in Mathematics 850*, Springer, Berlin, 1981, pp. 227–250.
[2] Itô excursion theory via resolvents, *Z. Wahrscheinlichkeitstheorie*, **63**, 237–255 (1983).
[3] Smooth transition densities for one-dimensional diffusions, *Bull. London Math. Soc.*, **17**, 157–161 (1985).
[4] Continuity of martingales in the Brownian excursion filtration, to appear.

ROSEN, J.
[1] A local time approach to self-intersections of Brownian paths in space, *Comm. Math. Phys.*, **88**, 327–338 (1983).

SCHWARTZ, L.
[1] Géometrie différentielle du 2ième ordre, semimartingales et équations différentielles stochastiques sur une variété différentielle, *Sém. de Probabilités XVI: Supplément, Lecture Notes in Mathematics 921*, Springer, Berlin, 1982, pp. 1–148.

SHARPE, M. J.
[1] Forthcoming book on Markov processes.

SHEPPARD, P.
[1] On the Ray–Knight property of local times, *J. London Math. Soc.*, **31**, 377–384 (1985).

SHIGA, T. and WATANABE, S.
[1] Bessel diffusions as a one-parameter family of diffusion processes, *Z. Wahrscheinlichkeitstheorie*, **27**, 37–46 (1973).

SHIGEKAWA, I.
[1] Derivatives of Wiener functionals and absolute continuity of induced measure, *J. Math. Kyoto Univ.*, **20**, 263–289 (1980).

SILVERSTEIN, M. L.
[1] *Symmetric Markov Processes: Lecture Notes in Mathematics 426*, Springer, Berlin, 1974.
[2] *Boundary Theory for Symmetric Markov Processes: Lecture Notes in Mathematics 516*, Springer, Berlin, 1976.

SIMON, B.
[1] *Functional Integration and Quantum Physics*, Academic Press, New York, 1979.
[2] Paper on tunnelling (to appear, *Ann. Inst. H. Poincare*?)

SKOROKHOD, A. V.
[1] Limit theorems for stochastic processes, *Theor. Prob. Applic.*, **1**, 261–290 (1956).
[2] Limit theorems for Markov processes, *Theor. Prob. Applic.*, **3**, 202–246 (1958).

SPITZER, F.
[1] *Principles of Random Walk*, Van Nostrand, Princeton, NJ, 1964.

STRASSEN, V.
[1] An invariance principle for the law of the iterated logarithm, *Z. Wahrscheinlichkeitstheorie*, **3**, 211–226 (1964).
[2] Almost sure behaviour of sums of independent random variables and martingales, *Proc. 5th Berkeley Symp. Math. Statist. Prob.*, Vol. 2, Part 1, University of California Press, 1966, pp. 315–343.

STROOCK, D. W.
[1] The Malliavin calculus and its applications to second-order parabolic differential operators I, II, *Math. System Theory*, **14**, 25–65, 141–171 (1981).
[2] The Malliavin calculus; a functional analytical approach, *J. Funct. Anal.*, **44**, 217–257 (1981).
[3] Diffusion processes associated with Lévy generators, *Z. Wahrscheinlichkeitstheorie*, **32**, 209–244 (1975).
[4] *An Introduction to the Theory of Large Deviations*, Springer, Berlin, New York, 1984.

STROOCK, D. W. and VARADHAN, S. R. S.
[1] *Multidimensional Diffusion Processes*, Springer, New York, 1979.
[2] On the support of diffusion processes with applications to the strong maximum principle, *Proc. 6th Berkeley Symp. Math. Statist. Prob.*, III, Univ. Calif. Press, Berkeley, 1972, pp. 333–359.
[3] Diffusion processes with boundary conditions, *Comm. Pure Appl. Math.*, **24**, 147–225 (1971).

STROOCK, D. W. and YOR, M.
[1] Some remarkable martingales, *Sem. de Probabilités XV: Lecture Notes in Mathematics 850*, Springer, Berlin, 1981, pp. 590–603.

SUSSMANN, H. J.
[1] On the gap between deterministic and stochastic ordinary differential equations, *Ann. Prob.*, **6**, 19–41 (1978).

TAYLOR, H. M.
[1] A stopped Brownian motion formula, *Ann. Prob.*, **3**, 234–246 (1975).

TAYLOR, S. J.
[1] Sample path properties of processes with stationary independent increments, in *Stochastic Analysis*, eds D. G. Kendall and E. F. Harding, Wiley, New York, 1973, pp. 387–414.

TSIREL'SON, B. S.
[1] An example of the stochastic equation having no strong solution, *Teoria Verojatn. i Primenen.*, **20**, (2), 427–430 (1975).

VAN DEN BERG, M. and LEWIS, J. T.
[1] Brownian motion on a hypersurface, *Bull. London Math. Soc.*, **17**, 144–150 (1985).

VARADHAN, S. R. S.
[1] Large deviations and applications, *SIAM*, Philadelphia, 1984.

WALSH, J. B.
[1] Excursions and local time, in Azema and Yor [2], pp. 159–192.
[2] Stochastic integration with respect to local time, in Çinlar, Chung and Getoor [1; 1983]

WARNER, F. W.
[1] *Foundations of Differentiable Manifolds and Lie Groups*, Springer, Berlin, New York, 1983.

WATANABE, S.
[1] On discontinuous additive functionals and Lévy measures of a Markov process, *Jap. J. Math.*, **34**, 53–79 (1964).

WHITNEY, H.
[1] *Geometric Integration Theory*, Princeton University Press, Princeton, N. J., 1957.

WHITTLE, P.
[1] *Optimization Over Time* (two volumes), Wiley, Chichester, 1982, 1983.

WILLIAMS, D.
[1] Brownian motions and diffusions as Markov processes, *Bull. London Math. Soc.*, **6**, 257–303 (1974).
[2] Some basic theorems on harnesses, in *Stochastic Analysis*, eds. D. G. Kendall and E. F. Harding, Wiley, New York, 1973, pp. 349–366.
[3] On Lévy's downcrossing theorem, *Z. Wahrscheinlichkeitstheorie*, **40**, 157–158 (1977).
[4] Path decomposition and continuity of local time for one-dimensional diffusions, I, *Proc. London Math. Soc.*, Ser. 3, **28**, 738–768 (1974).
[5] On a stopped Brownian motion formula of H. M. Taylor, *Séminaire de Probabilités X: Lecture Notes in Mathematics 511*, Springer, Berlin, 1976, pp. 235–239.
[6] Markov properties of Brownian local time, *Bull. Am. Math. Soc.*, **75**, 1035–1036 (1969).
[7] Decomposing the Brownian path, *Bull. Am. Math. Soc.*, **76**, 871–873 (1970).
[8] The *Q*-matrix problem for Markov chains, *Bull. Am. Math. Soc.*, **81**, 1115–1118 (1975).
[9] The *Q*-matrix problem, *Séminaire de Probabilités X: Notes in Mathematics 511*, Springer, Berlin, 1976, pp. 216–234.
[10] A note on the *Q*-matrices of Markov chains, *Z. Wahrscheinlichkeitstheorie*, **7**, 116–121 (1967).
[11] Some *Q*-matrix problems, in *Probability: Proc. Symp. Pure Math. XXXI*, American Mathematical Society, Providence, RI, 1977, pp. 165–169.

[12] *Diffusions, Markov Processes, and Matingales, Volume 1: Foundations*, Wiley, Chichester, New York, 1979.
[13] (editor) Stochastic integrals: *Proceedings, LMS Durham Symposium, Lecture Notes in Mathematics 851*, Springer, Berlin 1981.
[14] Conditional excursion theory, *Sém. de Probabilités XIII: Lecture Notes in Mathematics 721*, Springer, Berlin, 1979, 490–494.

YAMADA, T.
[1] On a comparison theorem for solutions of stochastic differential equations and its applications, *J. Math. Kyoto Univ.*, **13**, 497–512 (1973).

YAMADA, T. and OGURA, Y.
[1] On the strong comparison theorems for solutions of stochastic differential equations, *Z. Wahrscheinlichkeitstheorie*, **56**, 3–19 (1981).

YAMADA, T. and WATANABE, S.
[1] On the uniqueness of solutions of stochastic differential equations, *J. Math. Kyoto Univ.*, **11**, 155–167 (1971).

YOR, M.
[1] Sur certains commutateurs d'une filtration, *Sém. de Probabilités XV: Lecture Notes in Mathematics 850*, Springer, Berlin, 1981, pp. 526–528.
[2] Sur la continuité des temps locaux associés à certaines semimartingales, in Azema and Yor [2], pp. 23–35.
[3] Rappel et préliminaires généraux, in Azema and Yor [2], pp. 17–22.
[4] Précisions sur l'existence et la continuité des temps locaux d'intersection du mouvement Brownien dans $\mathbb{R}^2$, *Sém. de Probabilités XX: Lecture Notes in Mathematics 1204*, Springer, Berlin, 1986, pp. 532–542.
[5] Sur la répresentation comme intégrales stochastiques des temps d'occupation du mouvement Brownien dans $\mathbb{R}^d$, *ibid.*, pp. 543–552.

YOSIDA, K.
[1] *Functional Analysis*, Springer, Berlin, 1965.
[2] Brownian motion in homogeneous Riemannian space, *Pacific J. Math.*, **2**, 263–296 (1952).

ZAKAI, M.
[1] The Malliavin calculus, *Acta Appl. Math.*, **3**, 175–207 (1985).

ZHENG, W. A. and MEYER, P.-A.
[1] Quelques résultats de 'méchanique stochastique', *Sém. de Probabilités XVIII: Lecture Notes in Mathematics 1059*, Springer, Berlin, 1984, pp. 223–244.

ZVONKIN, A. K.
[1] A transformation of the phase space of a diffusion process that removes the drift, *Math. USSR Sbornik*, **22**, 129–149 (1974).

Index to Volumes 1 and 2

Printed in the United States
By Bookmasters